Materials for Infrared Windows and Domes

Properties and Performance

About the cover: The micrograph shows damage on the (100) face of a diamond crystal impacted at 130 m/s by a 1-mm-diameter sapphire sphere. The width of the undamaged central region is 200 μm. The effects of particle impact on infrared window materials are discussed in Chapter 7, along with coatings and claddings designed to protect windows. Chapter 9 describes the properties and uses of diamond as an infrared window. [Photograph kindly provided by Robert H. Telling, Cavendish Laboratory, University of Cambridge. Thanks, Rob!]

Materials for Infrared Windows and Domes

Properties and Performance

DANIEL C. HARRIS

SPIE Optical Engineering Press
A Publication of SPIE—The International Society for Optical Engineering
Bellingham, Washington USA

Library of Congress Cataloging-in-Publication Data

Harris, Daniel C., 1948–
Materials for infrared windows and domes: properties and performance / Daniel C. Harris
p. cm.
Includes bibliographical references and index.
ISBN 0-8194-3482-5
1. Guided missiles—Optical equipment. 2. Infrared detectors. 3. Noses (Aircraft) I. Title.
UG1310.H37 1999
629.1'2—dc21 99-36434
CIP

Published by

SPIE—The International Society for Optical Engineering
P.O. Box 10
Bellingham, Washington 98227-0010
Phone: 360/676-3290
Fax: 360/647-1445
Email: spie@spie.org
WWW: http://www.spie.org/

Printed in the United States of America.

CONTENTS

As-grown sapphire boule

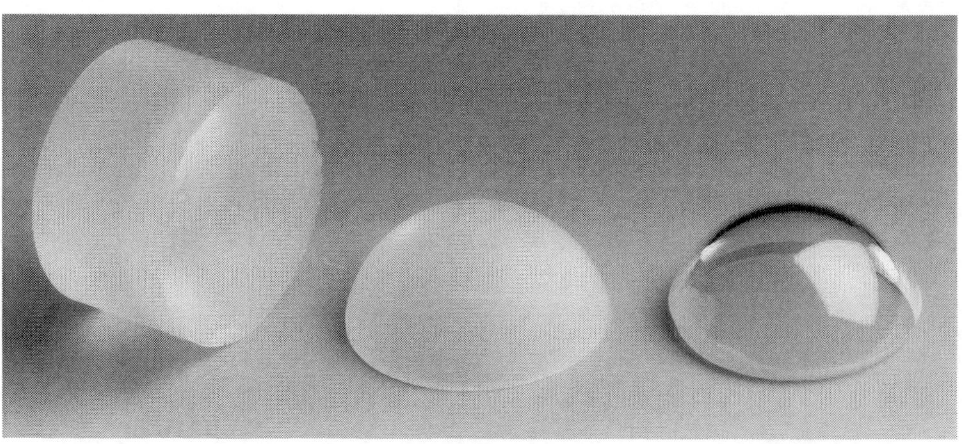

Rod removed from boule by core drill | Dome blank scooped with diamond saw | Polished dome

Genesis of a sapphire dome. [Photos courtesy Crystal Systems, Salem, Massachusetts.]

PREFACE

This text is intended to provide a comprehensive introduction to infrared-transparent materials for windows and domes that must withstand harsh environmental conditions, such as high-speed flight or high-temperature process monitoring. Each section contains sufficient introductory explanation so that the book should be readable by anyone with a background in science or engineering. The current volume builds on its predecessor, *Infrared Window and Dome Materials*, published in 1992 as part of the SPIE Tutorial Text series. The book you are holding incorporates seven years of new developments in the field of infrared windows and includes additional reference information and some more theory to make it more useful.

My wife, Sally, prepared many of the illustrations and contributed in many ways to the production of this volume. The manuscript benefited from critical reviews by Mike Thomas of the Applied Physics Lab and Lee Goldman of Raytheon and many helpful discussions with Claude Klein. Mel Nadler, Mike Seltzer and Andy Wright recorded some of the spectra that appear in this book. I also wish to express my appreciation to the management at China Lake, which values creative and scholarly activities, and to the Office of Naval Research for ongoing support of window material research and development.

I welcome your comments, corrections and suggestions. I can be reached at dan_harris@alum.mit.edu.

Dan Harris
China Lake, California

June 1999

After finishing work on windows, the author is seen beginning the floors.

Materials for Infrared Windows and Domes

Properties and Performance

Chapter 0

THE HEAT OF THE NIGHT AND THE DUST OF THE BATTLEFIELD

All objects above absolute zero temperature emit infrared radiation. This radiation can be used to measure the temperature of an object in a laboratory or factory, or can be used to observe military targets on a battlefield. "Heat-seeking" missiles use infrared radiation from the hot exhaust of a target aircraft to guide themselves to their prey.

The nose of the infrared-guided missile in Fig. 0.1 is a hemispheric, infrared-transparent dome made of magnesium fluoride. The ceramic dome protects a delicate, hermetically sealed infrared seeker from the harsh environment of high speed missile flight. The dome must withstand rapid aerothermal heating when the missile is launched and must resist long term erosion from raindrops and dust encountered during captive carry of the missile beneath the wing of an aircraft.

This book discusses optical, mechanical and thermal properties of infrared window and dome materials. It describes fabrication techniques and coatings required to enhance optical transmission and mechanical durability and to reject undesired radiation. We will emphasize the few rugged materials meant for external use in demanding environments, but will also mention important materials that can only be used in benign situations.

Fig. 0.1. Nose of a Sidewinder missile showing dome that protects the infrared seeker. (Photograph courtesy Naval Air Warfare Center, China Lake, California.)

0.1 Electromagnetic spectrum and atmospheric transmission

Electromagnetic radiation is characterized by a *frequency*, ν (number of oscillations of the electric field per second) and *wavelength*, λ (distance between crests of the electric field). The speed of light, c, is the product of wavelength and frequency:

$$c = \lambda \nu . \tag{0-1}$$

The speed of light in vacuum is exactly 2.997 924 58 × 10^8 m/s (which defines the length of a meter). The energy, E, of a photon (a "particle" of light) is proportional to its frequency, and therefore inversely proportional to its wavelength:

$$E = h \nu = \frac{h c}{\lambda} . \tag{0-2}$$

The constant of proportionality, *Planck's constant*, has the value $h = 6.626 \times 10^{-34}$ J·s. Physical constants and conversion factors are listed in Appendix A at the end of the book.

Table 0.1 shows the names, wavelengths and frequencies for various parts of the electromagnetic spectrum. Infrared wavelengths are often expressed in micrometers, μm (10^{-6} m), while visible wavelengths are typically given in nanometers, nm (10^{-9} m). We see that infrared radiation spans the wavelength region between 0.780 and 1000 μm. Visible radiation has higher energy than infrared, while microwave radiation has lower energy.

Table 0.1. Electromagnetic spectrum

Type of radiation	Wavelength	Frequency (s^{-1} ≡ hertz = Hz)
Cosmic rays	$<10^{-12}$ m	$>3 \times 10^{20}$
Gamma rays	10^{-12} - 10^{-11} m	3×10^{20} - 3×10^{19}
X-rays	10^{-11} - 10^{-8} m	3×10^{19} - 3×10^{16}
Ultraviolet	10^{-8} - 3.80×10^{-7} m	3×10^{16} - 7.89×10^{14}
Visible	3.80×10^{-7} - 7.80×10^{-7} m	7.89×10^{14} - 3.84×10^{14}
violet	380 - 430 nm	
blue	430 - 480 nm	
green	480 - 530 nm	
yellow	530 - 580 nm	
orange	580 - 620 nm	
red	620 - 780 nm	
Infrared	7.80×10^{-7} m - 10^{-3} m (0.78 μm - 1000 μm)	3.84×10^{14} - 3×10^{11}
midwave window	3 - 5 μm	
long wave window	8 - 14 μm	
Microwave	10^{-3} - 10^{-1} m	3×10^{11} - 3×10^{9}
Radio	$>10^{-1}$ m	$<3 \times 10^{9}$

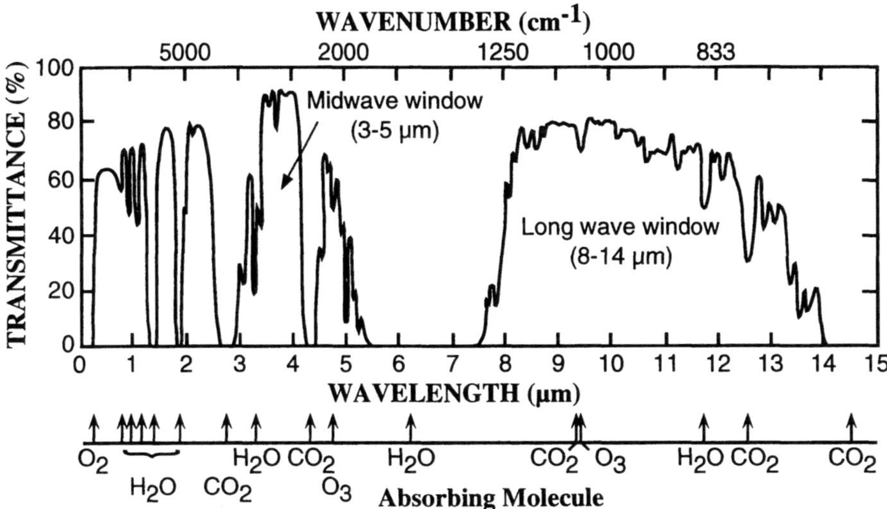

Fig. 0.2. Infrared transmission spectrum of the atmosphere for a 1.8 km horizontal path at sea level with 40% relative humidity.[1,2] Good references to read about atmospheric transmission are *Infrared System Engineering*,[2] *The Infrared Handbook*,[3] and Volume 2 of *The Infrared & Electro-Optical Systems Handbook*.[4]

Figure 0.2 is a transmission spectrum of the atmosphere, with absorption features by various atmospheric constituents noted. The transmittance of a perfectly transparent medium is 100%, while an opaque medium has a transmittance of 0%. In Fig. 0.2 we see a transmission window in the visible region (below 1 μm) and several windows in the infrared. Two important regions for infrared sensing are the *long wave infrared window* between 8 and 14 μm and the *midwave infrared window* from 3 to 5 μm. The midwave "window" is interrupted by a strong absorption band from carbon dioxide near 4.3 μm. Atmospheric transmission varies with pathlength (how far you are looking through the atmosphere), altitude and humidity.

At the top of Fig. 0.2 is a *wavenumber* scale. Wavenumber is defined as the reciprocal of wavelength, with cm^{-1} being the most common units:

$$\text{Wavenumber} \equiv 1 / \lambda \, . \tag{0-3}$$

The unit cm^{-1} is usually read as "reciprocal centimeters" or "wavenumbers." Since energy is inversely proportional to wavelength [Eq. (0-2)], energy is directly proportional to wavenumber. A 3000 cm^{-1} photon has three times the energy of a 1000 cm^{-1} photon.

0.2 Blackbody radiation

A *blackbody* absorbs all radiation striking its surface. If it is at constant temperature, but thermally isolated from its surroundings, it must emit the same amount of energy that it absorbs. Hence, a blackbody is a perfect emitter as well as a perfect absorber. A real object that you might observe with an infrared sensor is not a blackbody, but its

Fig. 0.3. Night landing of space shuttle at Edwards Air Force Base, as seen by 8-12 μm Magnavox IR-18 forward looking infrared camera. Notice that the hot underside of the aircraft emits the greatest intensity. (Photograph courtesy Naval Air Warfare Center.)

behavior is often qualitatively similar to that of a blackbody. For example, a sheet of white bond paper has about 93% of the emission of a blackbody. Human skin has about 98% of the emission of a blackbody. Figure 0.3 shows 8-12 μm radiation emitted from a space shuttle as seen by an infrared camera during a night landing. We will discuss the radiant emission of real materials further in the next chapter, but for now we explore the properties of ideal blackbodies.

The power per unit area, designated *exitance* (or emittance in the older literature), radiating from a blackbody is proportional to the fourth power of absolute temperature:

$$\text{Exitance} \equiv M = \sigma T^4 \qquad \sigma = \frac{2\pi^5 k^4}{15 c^2 h^3} = 5.670 \times 10^{-8} \frac{\text{W}}{\text{m}^2\ \text{K}^4} \tag{0-4}$$

where k is Boltzmann's constant (1.3807×10^{-23} J/K), c is the speed of light and h is Planck's constant. The temperature, T, is given in kelvins (K).

The power radiating from a blackbody is a function of wavelength, λ, as well as temperature. The wavelength dependence is expressed in the *Planck distribution:*

$$M_\lambda = \frac{2\pi hc^2}{\lambda^5}\left(\frac{1}{e^{hc/\lambda kT} - 1}\right) = \frac{3.74177487 \times 10^8}{\lambda^5}\left(\frac{1}{e^{1.43876866 \times 10^4/\lambda T} - 1}\right) \tag{0-5}$$

in which M_λ, the power per unit area per unit wavelength emitted by the blackbody, is called the *spectral exitance* or spectral emittance. The numbers on the right side of Eq. (0-5) give M_λ, in units of W/m^2/μm when λ is expressed in micrometers. Figure 0.4 shows that:

1. The exitance at any wavelength increases as temperature increases.
2. The peak exitance shifts to shorter wavelength as the temperature increases.

For temperatures above 100 K, the wavelength of maximum emission, λ_{max}, is given by the *Wein displacement law:*

$$\lambda_{max}\, T \approx 2.878 \times 10^{-3}\ \text{m·K}\ . \tag{0-6}$$

Equation (0-6) tells us that the wavelength of maximum emission at 300 K is $\lambda_{max} \approx (2.878 \times 10^{-3}\ \text{m·K}) / 300\ \text{K} = 9.6 \times 10^{-6}\ \text{m} = 9.6\ \mu\text{m}$, while the wavelength of

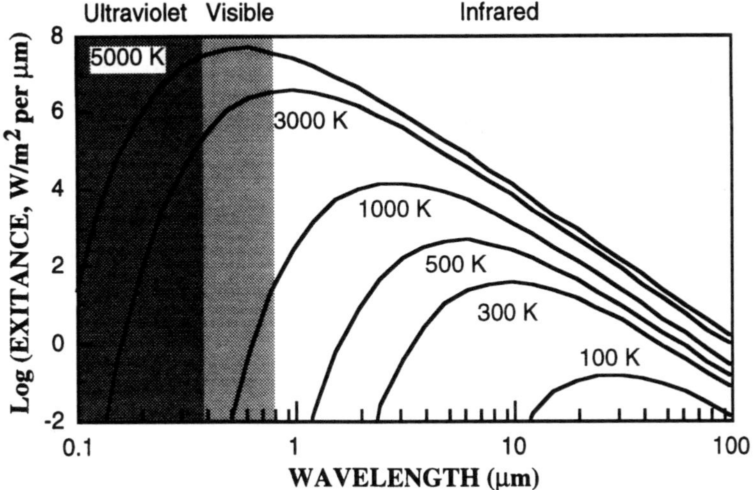

Fig. 0.4. Planck distribution [Eq. (0-5)] of light emitted from a blackbody. The ordinate gives the watts per square meter per micrometer of wavelength emitted from the surface of the blackbody. Note that both axes are logarithmic.

maximum emission at 800 K is (2.878×10^{-3} m·K) / 800 K = 3.6 µm. Thus, objects at room temperature have maximum emission in the long wave infrared region, while the 500°C exhaust of a jet engine has maximum emission in the midwave infrared region.

Example: Total power emitted by a blackbody. How much radiant energy is emitted by one square meter of an object at 77 K? At 300 K? At 2000 K? This is easy to answer with Eq. (0-4):

$$\text{At 77 K: Exitance} = \sigma T^4 = [5.670 \times 10^{-8} \frac{\text{W}}{\text{m}^2\ \text{K}^4}]\ (77\ \text{K})^4 = 1.99 \frac{\text{W}}{\text{m}^2}\ .$$

$$\text{At 300 K: Exitance} = \sigma\,(300)^4 = 459 \frac{\text{W}}{\text{m}^2}\ .$$

$$\text{At 2000 K: Exitance} = \sigma\,(2000)^4 = 9.07 \times 10^5 \frac{\text{W}}{\text{m}^2}\ .$$

The point of this example is that radiant emission from an object increases very rapidly with increasing temperature. A "red hot" object at 2000 K emits $(2000/300)^4$ = 2000 times as much energy as the same object at 300 K.

In Fig. 0.4 the area under each curve gives the total exitance. The area of a given wavelength interval gives the energy emitted in that wavelength interval:

$$\text{Total exitance} = \int_0^\infty M_\lambda \, d\lambda \qquad \text{Exitance } (\lambda_1 \text{ to } \lambda_2) = \int_{\lambda_1}^{\lambda_2} M_\lambda \, d\lambda\ . \qquad (0\text{-}7)$$

Table 0.2. Blackbody emission

	Exitance (W/m^2)		
Temperature (K)	3-5 μm	8-10 μm	8-14 μm
300	5.9	61.1	172.6
600	1719	957	1937
2000	1.60×10^5	1.07×10^4	1.86×10^4

The integral on the right of Eq. (0-7) allows us to evaluate the radiant energy emitted by a blackbody in different spectral regions at different temperatures.

Example: Blackbody emission in midwave (3-5 μm) and long wave (8-14 μm) infrared windows. Let's find the radiant energy per square meter emitted by a blackbody in the regions 3-5 μm, 8-10 μm and 8-14 μm at 300 K, 600 K and 2000 K. We find the 3-5 μm emission at 300 K from the integral

$$\text{Exitance (3-5 μm)} = \int_3^5 M_\lambda \, d\lambda = 5.86 \, \frac{\text{W}}{\text{m}^2}$$

where M_λ has the numerical constants on the right side of Eq. (0-5) and $T = 300$ K. In a similar manner, you could fill in Table 0.2 These calculations show that near room temperature there is much more emission from a blackbody in the long wave infrared region than in the midwave region. Near 600 K the two regions have about equal radiant energy, whereas at 2000 K the midwave exitance is much greater than the long wave exitance.

Some long wave infrared systems are limited to the 8-10 μm range by their detector or their window. Comparison of the energy available in the 8-10 μm and 8-14 μm regions in Table 0.2 shows that the 8-10 μm portion of the window contains only one-third to one-half of the total long wave infrared energy available from a blackbody in the range 300 to 2000 K.

0.3 Transmission through rain, snow, fog and dust

The effective range of infrared (and other optical) sensing systems is limited by absorption and scatter of electromagnetic radiation by the atmosphere. To discuss this subject, we first need to define transmittance with the help of Fig. 0.5. Consider light of radiant power P_o (watts per square meter) incident on a cylinder of air or some optical material with a pathlength b. If the transmitted light has a radiant power P (which is $\leq P_o$), then the transmittance is defined as

$$\text{Transmittance} = t = \frac{P}{P_o} . \tag{0-8}$$

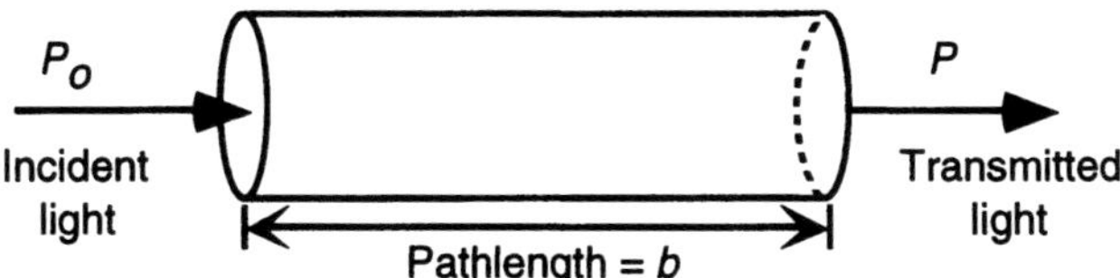

Fig. 0.5. Light of radiant power P_o (W/m^2) strikes a cylinder of air or other material of pathlength b. The transmitted radiant power is P.

Transmittance ranges from 0 to 1. A transmittance of 1 means that all light is transmitted and a transmittance of 0 means that no light is transmitted.

Suppose that the cylinder in Fig. 0.5 represents a section of the atmosphere through which infrared light passes from an object to a detector. The transmitted power P is less than the initial power P_o because of absorption by the atmosphere or scatter by rain, snow, fog, or dust. The decrease in transmission is commonly measured by an *attenuation coefficient*, γ, expressed in units of *decibels per kilometer* (dB/km). An attenuation of 10 dB corresponds to one power-of-ten decrease in transmittance. The relationship between attenuation coefficient and transmittance is

$$t = 10^{-\gamma b/10} \tag{0-9}$$

where b is the pathlength expressed in km. Equation (0-9) applies when the loss of transmission is due to *absorption* of electromagnetic radiation by components of the atmosphere such as water vapor and carbon dioxide. If the transmission loss is predominantly due to optical *scatter* (from dust in the air, for example), then Eq. (0-9) is an approximation that is valid for a transmittance above ~80% (or $\gamma b/10 \le 0.1$), which corresponds to ~1 dB of loss.*

Table 0.3 gives formulas for computing approximate atmospheric attenuation at a wavelength of 10.6 μm for different weather conditions. This is a commonly used infrared wavelength emitted by a carbon dioxide laser. The formulas in Table 0.3 are only approximate because atmospheric effects vary widely and because attenuation generally has contributions from both absorption and scatter. Eq. (0-9) is valid for absorption but only applies to scatter losses not exceeding 1 dB.

Example: Atmospheric attenuation. A CO_2 laser beam is directed at a mirror located 1.0 km away and the reflected signal is measured by a detector placed next to the laser. Estimate the atmospheric transmittance over the 2.0-km pathlength if the atmospheric condition is (a) raining at a rate of 2.0 mm/h or (b) dust with a reported visibility of 1.2 km.

(a) For rain, the attenuation coefficient computed from Table 0.3 is

*Eq. (0-9) is valid for absorption and single scattering events. If the concentration of particles in the atmosphere is high, it is likely that a single ray of light will be scattered multiple times and Eq. (0-9) breaks down.

Table 0.3. Atmospheric attenuation coefficients at 10.6 μm wavelength[5]

Condition	Attenuation coefficient (dB/km)*	
Clear air	$\gamma = (1.084 \times 10^{-5}) p(P+193p)\left(\frac{296}{T}\right)^{5.25} + 625\left(\frac{296}{T}\right)^{1.5}(10^{-970/T}) + \frac{1.4}{V}$	
Rain	$\gamma = 1.9\, R^{0.63}$	(Observed attenuation from rain does not vary significantly between 0.63 and 10.6 μm)
Snow	$\gamma = 2\, S^{0.75}$	
Fog	$\gamma = \frac{1.7}{V^{1.5}}$	
Dust	$\gamma = \frac{5}{V}$	

Symbols: p = partial pressure of water (mbar); P = atmospheric pressure (mbar); T = temperature (K); V = visual visibility (km); R = rainfall rate (mm/h); S = snowfall rate (mm/h)

*Attenuation coefficients are approximate. Ranges of attenuation coefficients for a given weather type are shown on pages 24-25 of Reference 5.

$$\gamma = 1.9\, R^{0.63} = 1.9\,(2.0)^{0.63} = 2.94 \text{ dB/km}$$

and the transmittance is calculated with Eq. (0-9):

$$t = 10^{-\gamma b/10} = 10^{-(2.94 \text{ dB/km})(2.0 \text{ km})/10} = 0.26 = 26\% \ .$$

Only 26% of the laser light reaches the detector if 100% of the light is returned by the mirror.

(*b*) The attenuation coefficient for dust (and fog) is based on the visibility reported by a human observer using visible light. If the reported visibility is 8 km, the attenuation coefficient computed with the formula in Table 0.3 is

$$\gamma = 5/V = 5/8 = 0.62 \text{ dB/km} \ .$$

The transmittance is:

$$t \approx 10^{-\gamma b/10} = 10^{-(0.62 \text{ dB/km})(2.0 \text{ km})/10} = 0.75 = 75\% \ .$$

In this calculation for dust, we have slightly exceeded the valid range of Eq. (0-9), which is good for transmittance ≥80%, or a total loss of ≤ 1 dB.

Figures 0.6 - 0.10 show atmospheric transmittance for different weather conditions calculated with Eq. (0-9) and attenuation coefficients from Table 0.3. In Figure 0.6 we see that humidity has a strong effect on infrared transmittance of clear air. The greater the humidity, the lower the transmittance because water absorbs infrared radiation. Figure 0.7 indicates that infrared transmittance is significantly degraded by any rainfall rates ≥1 mm/h. Similarly, snow, fog and dust substantially decrease atmospheric transmittance.

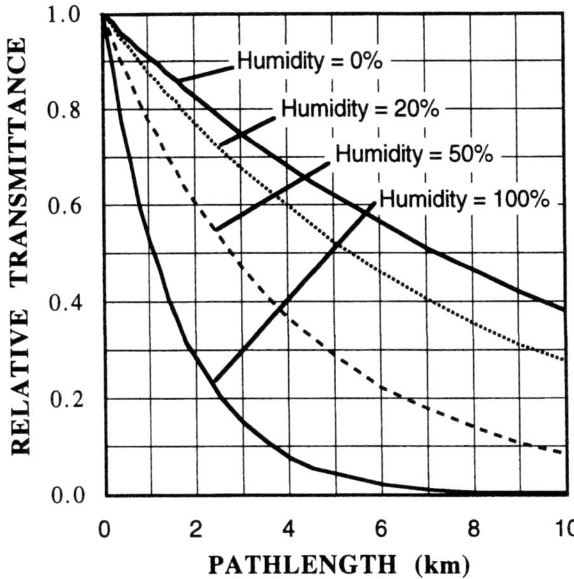

Fig. 0.6. Transmittance of 10.6 μm infrared radiation in clear air as a function of humidity. Curves were computed from the first equation in Table 0.3 with the following parameters: P = atmospheric pressure = 1000 mbar, T = temperature = 298.15 K, V = visibility = 20 km. The partial pressure of water vapor, p, was computed from the relationship p = humidity × p_o, where p_o is the vapor pressure of air saturated with water vapor at the existing temperature. For T = 298.15 K, p_o = 31.67 bar.

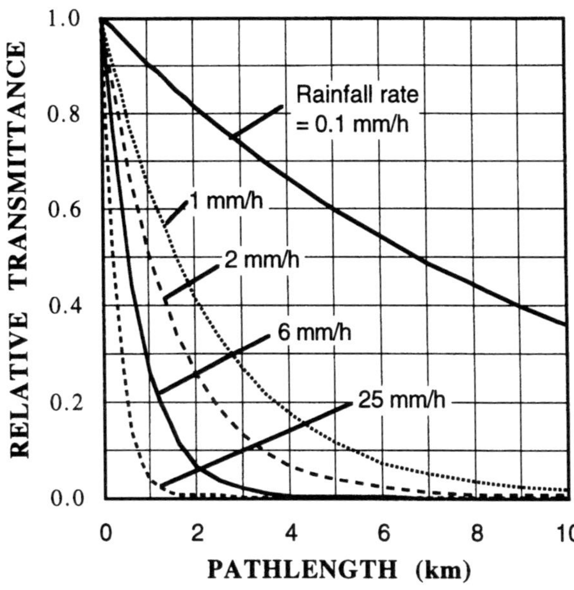

Fig. 0.7. Transmittance of 10.6 μm infrared radiation in rain. Curves were computed with the second equation in Table 0.3 with different rainfall rates. Attenuation from rain does not vary significantly with wavelength between 0.63 and 10.6 μm.[6]

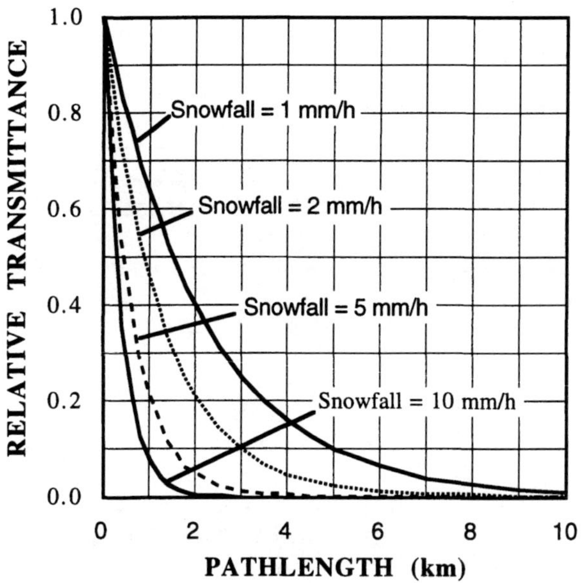

Fig. 0.8. Transmittance of 10.6 μm infrared radiation in snow. Curves were calculated from the third equation in Table 0.3 with different snowfall rates.

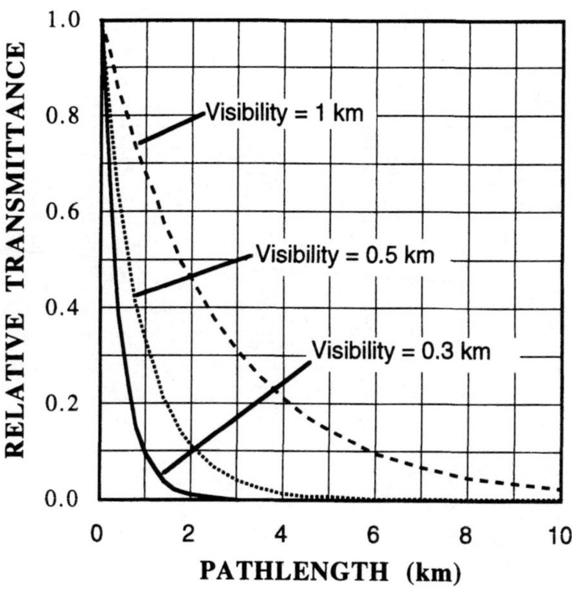

Fig. 0.9. Transmittance of 10.6 μm infrared radiation in fog with different visibilities reported by a person based on visual sighting of distant objects. Curves were computed from the fourth equation in Table 0.3.

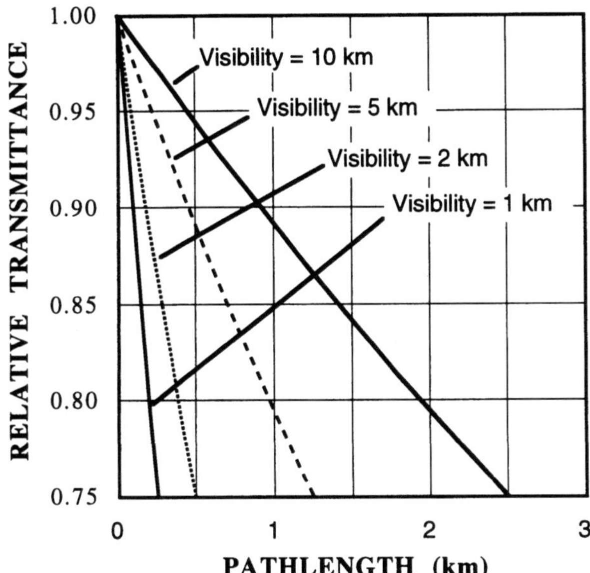

Fig. 0.10. Transmittance of 10.6 μm infrared radiation in dust with different visibilities reported by a person based on visual sighting of distant objects. Curves were computed from the fifth equation in Table 0.3.

References

1. H. A. Gebbie, W. R. Harding, C. Hilsum, A. W. Pryce and V. Roberts, "Atmospheric Transmission in the 1 to 14 μ Region," *Proc. Roy. Soc.*, **A206**, 87-107 (1951).
2. R. H. Hudson, Jr., *Infrared System Engineering*, Wiley, New York (1969).
3. W. L. Wolfe and G. J. Zissis, eds., *The Infrared Handbook*, Environmental Research Institute of Michigan, Ann Arbor, Michigan (1985).
4. M. E. Thomas and D. D. Duncan, "Atmospheric Transmission," in J. S. Accetta and D. L. Shumaker, eds., *The Infrared & Electro-Optical Systems Handbook* (Vol. 2: *Atmospheric Propagation of Radiation*, F. G. Smith, ed.), Environmental Research Institute of Michigan, Ann Arbor, Michigan, and SPIE Press, Bellingham, Washington (1993).
5. G. W. Kamerman, "Laser Radar," in J. S. Accetta and D. L. Shumaker, eds., *The Infrared & Electro-Optical Systems Handbook* (Vol. 6: *Active Electro-Optical Systems*, C. S. Fox, ed.), p. 26, Environmental Research Institute of Michigan, Ann Arbor, Michigan, and SPIE Press, Bellingham, Washington (1993).
6. D. G. Rensch and R. K. Long, "Comparative Studies of Extinction and Backscattering by Aerosols, Fog, and Rain at 10.6 μm and 0.63 μm," *Appl. Opt.*, **9**, 1563-1573 (1970).

Chapter 1

OPTICAL PROPERTIES OF INFRARED WINDOWS

In addition to the obvious need for transparency, a critical requirement for a window in a hot environment is that it does not emit light that would obscure the scene being viewed. In this chapter we will explore transmission, emission and the related properties of reflection, refraction and scatter.

1.1 A day in the life of a photon

Consider what happens when light passes through an optical window with thickness b in Fig. 1.1. Suppose for the moment that the material can transmit, absorb or reflect the light, but it cannot scatter light away from the incident direction. Radiant power P_o (W/m^2) strikes the first surface, where radiant power R_1 is reflected and radiant power P_1 enters the sample. Some of P_1 is absorbed, so power P_2 ($<P_1$) arrives at the second surface. Some power is transmitted through the second surface, and R_2 is reflected. This process of partial reflection and partial transmission continues *ad infinitum* as the light ray bounces back and forth inside the window and eventually dies to near zero intensity. The net transmitted power, P, is the sum of all the partially transmitted light at the second surface.

The attenuation of light as it passes through an absorber is exponential:

$$\text{Internal transmittance} = \frac{P_2}{P_1} = e^{-\alpha b} \qquad (1\text{-}1)$$

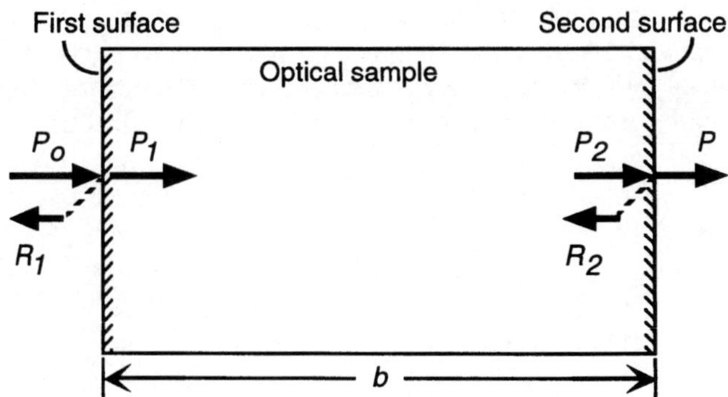

Fig. 1.1. Light passing through a slab of optical material.

where b is the thickness of sample (customarily expressed in cm) and α is called the *absorption coefficient*, with units of cm^{-1}. The fraction P_2/P_1 is called *internal transmittance*, while the fraction P/P_o is referred to as *external transmittance*, *in-line transmittance*, or most commonly just *transmittance*. For a window with a thickness of b = 1.0 cm, an absorption coefficient of $\alpha = 1.0\ cm^{-1}$ gives an internal transmittance of $e^{-(1.0\ cm)\cdot(1.0\ cm^{-1})} = 0.37$, or 37%. If the absorption coefficient is 0.10 cm^{-1}, the internal transmittance is increased to $e^{-(0.10\ cm)\cdot(1.0\ cm^{-1})} = 90\%$.

The external transmittance, P/P_o, is the fraction of incident light that is transmitted through the window. The *absorptance* is the fraction of incident light that is absorbed:

$$\text{Absorptance} = \frac{P_1 - P_2}{P_o} . \tag{1-2}$$

An absorptance of 0.10 means that 10% of the incident light is absorbed by the window, and usually converted to heat. One of the most sensitive methods for measuring low levels of absorptance is to measure the rise in temperature of the sample.

Let's recap the confusing distinction between absorptance and absorption coefficient: *Absorptance* is a number between 0 and 1 that tells us what fraction of the incident light is absorbed by the window. An *absorption coefficient* of $\alpha\ cm^{-1}$ means that $e^{-\alpha b}$ is the fraction of light passing through the first surface that is transmitted through a b-cm-thick window to the second surface.

By analogy to the definition of absorptance, *reflectance* is defined as the fraction of incident light reflected back toward the source. At the first surface in Fig. 1.1, the radiant power R_1 is reflected. At the second surface, R_2 is reflected. When R_2 arrives back at the first surface, some is transmitted back toward the original source, and some is reflected again. There are an infinite number of reflections at each surface, but their magnitude decreases rapidly, so only the first few are important. The quantity that we call reflectance is the total power from multiple internal reflections that is eventually transmitted back toward the source.

Light is absorbed at the surface of a material, as well as inside.* For highly transparent materials, surface absorption can be significant in comparison to bulk absorption. Surface absorption depends on surface polishing and cleaning. To quantify surface absorption, we measure absorptance as a function of sample thickness for a series of samples with the same quality and surface finish. The y-intercept in a graph of absorptance vs. thickness is the thickness-independent surface absorptance (Fig. 1.2). In a study of 25-mm-thick zinc selenide at a wavelength of 8 μm, the two surfaces absorbed a total of 0.4% of the incident light, while the bulk absorbed 0.5% of the light.[1] A 14-cm-thick specimen of potassium chloride at a wavelength of 3.8 μm absorbed a total of 0.0050% of incident light at the two surfaces and 0.0084% in the bulk.[2] The unpolished cylindrical surface of commercial Nd:YAG (neodymium doped yttrium aluminum garnet) laser rods absorbs several percent of 1.06 μm radiation incident on the side of the rod.[3]

---

*A surface absorption coefficient, $\alpha_{surface}$, is defined through the equation $\alpha = \alpha_{bulk} + \alpha_{surface}/2b$, where α is the measured total absorption coefficient [defined in Eq. (1-1)], α_{bulk} is the bulk absorption coefficient, and b is the thickness of the sample.

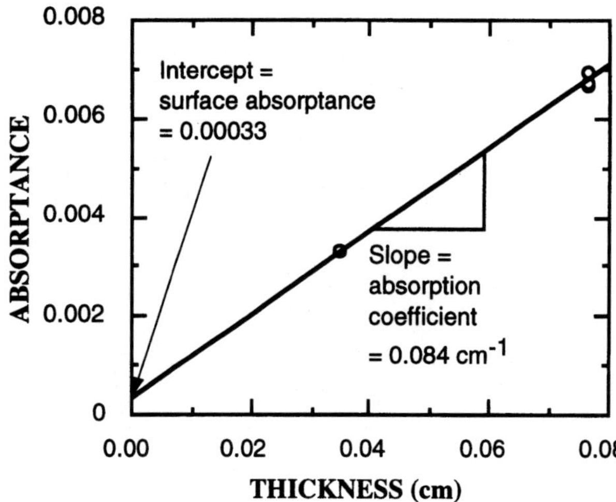

Fig. 1.2. Absorptance of high quality chemical vapor deposited diamond measured by laser calorimetry as a function of sample thickness. There are three replicate data points at each of the two thicknesses. The nonzero intercept is attributed to a small amount of surface absorption. [Raytheon data.]

In addition to being transmitted, absorbed or reflected, light can be scattered away from the incident direction. This happens when light passes into inhomogeneous regions of the sample, such as grain boundaries, impurities and voids. The fraction of incident light diverted from the incident direction is called *scatter*. Scatter can occur on both surfaces of a window, and inside the bulk. In practice, it may be difficult to distinguish bulk scatter from absorption, because both processes decrease the radiant power that is transmitted through the sample. The effective absorption coefficient for a sample has one component from true absorption and another from bulk scatter:

$$\text{Effective absorption coefficient} = \alpha = \alpha_{absorption} + \alpha_{scatter} \qquad (1\text{-}3)$$

where $\alpha_{absorption}$ is the true absorption coefficient and $\alpha_{scatter}$ is the contribution from bulk scatter.*

Since transmission, reflection, absorption and scatter represent everything that can happen to light as it passes through a material, the sum of transmittance (t), reflectance (r), absorptance (a) and scatter (s) must be unity:

$$t + r + a + s = 1 \; . \qquad (1\text{-}4)$$

If a sample has negligible absorption and scatter, then the sum of transmittance and reflectance must be unity:

$$t + r = 1 \qquad \text{(if absorption and scatter are negligible).} \qquad (1\text{-}5)$$

A high quality optical window with 20% reflectance will have 80% transmittance, since absorption and scatter are negligible for such a window.

---

*As stated beneath Eq. (0-9), use of a scatter coefficient in the exponent of Eq. (1-1) is only valid for a scattering loss of ~20% in transmission. If the scatter contribution to optical loss is >20%, then Eq. (1-1) is not a good approximation.

1.2 Refraction and refractive index

Refraction is the bending of a light ray when it passes from one medium to another of different *refractive index*. In Fig. 1.3 we see a light ray pass from a medium (such as air) of refractive index n_1 into another medium (such as an optical window) with refractive index n_2. The angle of reflection (θ_1) is equal to the angle of incidence. The angle of refraction (θ_2) is not, in general, equal to the angle of incidence.

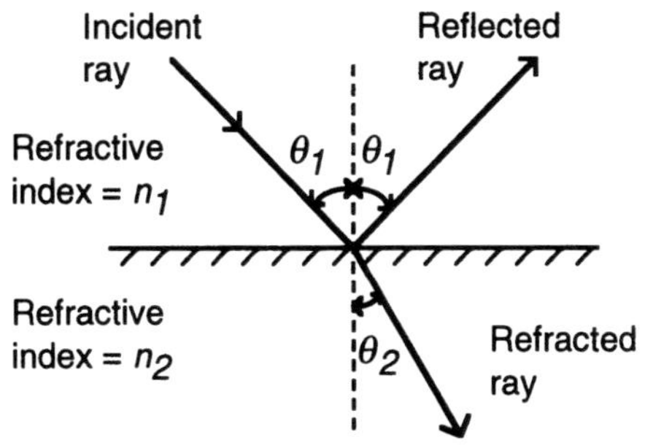

Fig. 1.3. Refraction (bending) of light at the interface between two different media.

The relation between the angle of incidence, θ_1, and the angle of refraction, θ_2, in Fig. 1.3 is given by *Snell's law*:

$$n_1 \sin \theta_1 = n_2 \sin \theta_2 \ . \qquad (1\text{-}6)$$

The greater the ratio n_2/n_1, the more the light ray is bent from its incident direction.

Refractive index also relates the speed of light (c) in a medium of refractive index, n, to the speed of light (c_o) in vacuum (refractive index = 1):

$$c = c_o \,/\, n \ . \qquad (1\text{-}7)$$

That is, when light travels through a quartz window with refractive index 1.5, its speed is reduced to 1/1.5 = 67% of its speed in vacuum. The frequency of light (ν) does not change, but the wavelength (λ) decreases to 67% of its vacuum value, since $\lambda \, \nu = c$.

Example: Angle of refraction. Suppose that a light ray passes from air ($n \approx 1.0$) into zinc sulfide ($n = 2.2$), with an angle of incidence of 45°. What will be the angle of refraction? To answer this question, we use Snell's law, Eq. (1-6):

$$n_1 \sin \theta_1 = n_2 \sin \theta_2 \;\Rightarrow\; 1.0 \sin 45° = 2.2 \sin \theta_2 \;\Rightarrow\; \theta_2 = 18.7° \ .$$

What would be the angle of refraction if the window were made from zinc selenide (n = 2.4) instead of zinc sulfide?

$$1.0 \sin 45° = 2.4 \sin \theta_2 \;\Rightarrow\; \theta_2 = 17.1° \ .$$

As the index of refraction increases, the angle of refraction decreases. What if the incident ray is normal to the interface between the two media?

$$1.0 \sin 0° = n_2 \sin \theta_2 \Rightarrow \theta_2 = 0° .$$

For any value of n_2, there is no refraction of a perpendicular ray.

Whenever $n_2 < n_1$ in Fig. 1.4, there is a *critical angle of incidence*, θ_c, beyond which all light is reflected, and none is refracted. This explains how the *optical fiber* in Fig. 1.4 works.[4] The index of refraction of the cladding is less than that of the core. When light inside the core strikes the cladding at the angle θ_i in Fig. 1.4, it can be reflected back into the core or refracted at the angle θ_r. No light enters the cladding when θ_r is 90°. To find the critical angle of incidence for the case $n_2 = 1.0$ and $n_1 = 2.0$, we insert $\theta_2 = 90°$ into Snell's law:

$$n_1 \sin \theta_c = n_2 \sin 90° \Rightarrow 2.0 \sin \theta_c = 1.0 \cdot 1 \Rightarrow \theta_c = 30° . \quad (1\text{-}8)$$

That is, if the angle of incidence, θ_i, is ≥30°, all light will be reflected back into the core. A ray entering one end of the fiber within the cone of acceptance will emerge from the other end of the fiber with little loss, providing the core is transparent. Optical fibers are flexible and can be bent (within reason) to transmit light from one point to another.

Figure 1.5 shows the refractive index of numerous infrared window materials as a function of wavelength. Many materials, especially oxides and fluorides (containing the elements oxygen or fluorine), have refractive indexes less than 2. Heavier atoms with more polarizable electron clouds (more easily deformed by an oscillating electric field) give materials with higher refractive index. Sulfides and selenides (containing the elements sulfur or selenium) have refractive indexes above 2, while silicon and germanium have even higher refractive indexes.

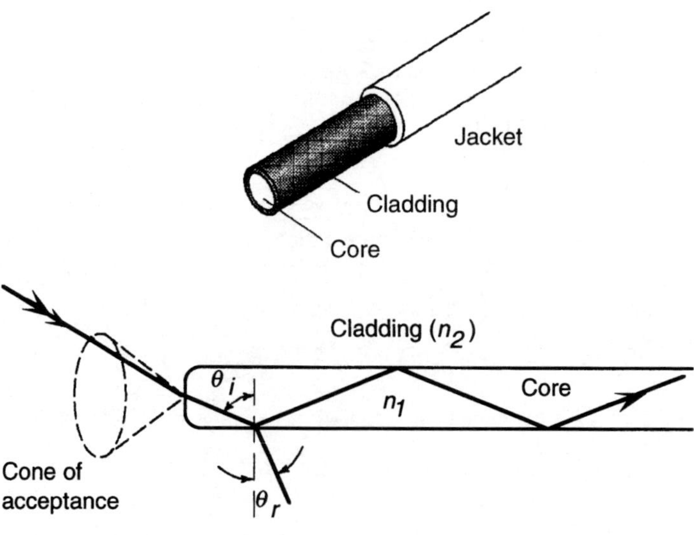

Fig. 1.4. Optical fiber construction and principle of operation.

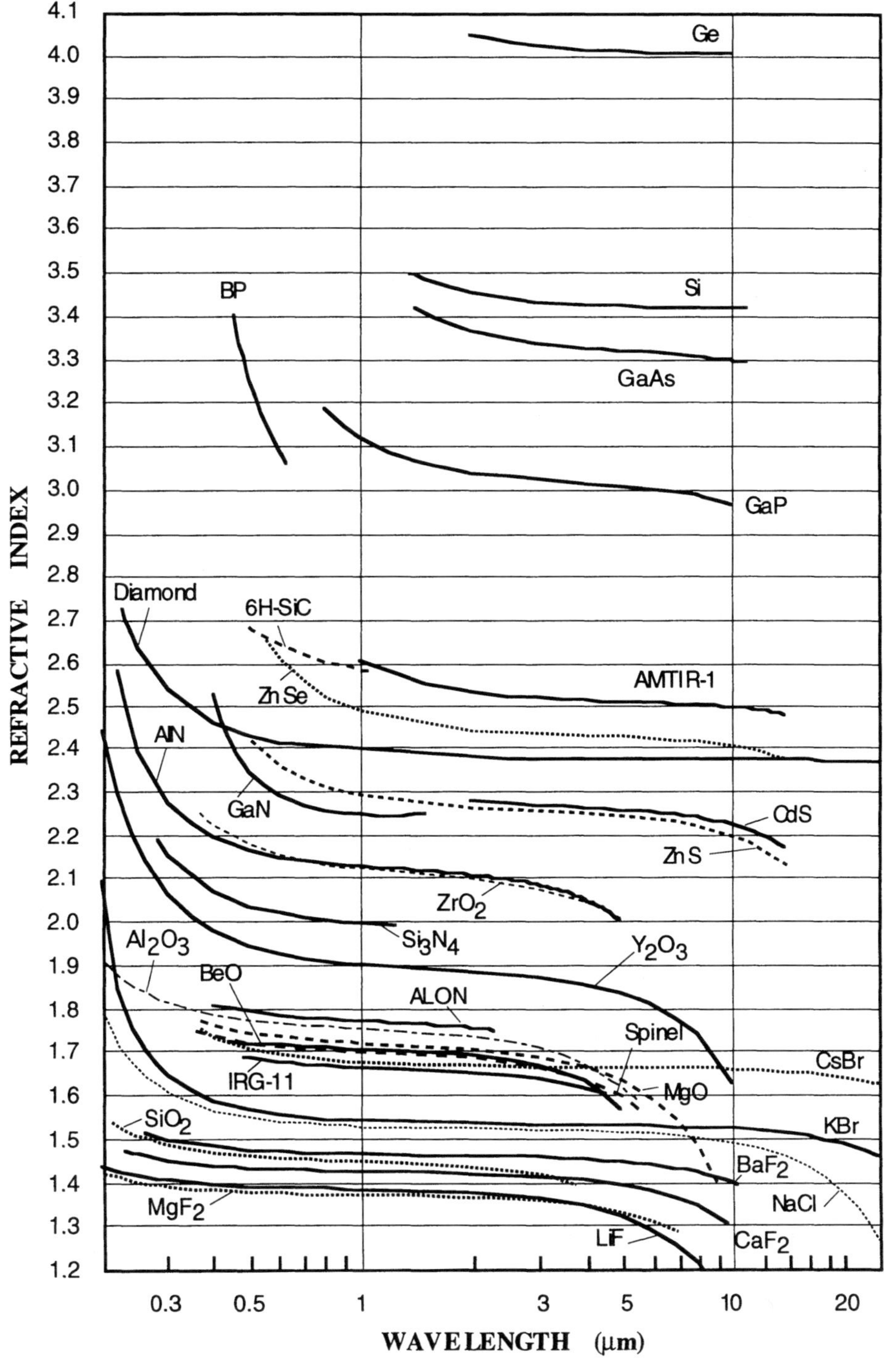

Fig. 1.5. Refractive index of infrared window materials as a function of wavelength. Curves were calculated with dispersion formulas in Table C.2 in Appendix C. The refractive index n_o is plotted for the noncubic materials AlN, BeO, Al_2O_3 and 6H-SiC. For other noncubic materials, the refractive index is for polycrystalline materials.

For most materials, a *Sellmeier equation* of the form (1-9) accurately describes the variation of refractive index, *n*, with wavelength, which is called *dispersion*:

$$n^2 - 1 = \sum_i \frac{A_i \lambda^2}{\lambda^2 - \lambda_i^2} = \frac{A_1 \lambda^2}{\lambda^2 - \lambda_1^2} + \frac{A_2 \lambda^2}{\lambda^2 - \lambda_2^2} + \cdots \tag{1-9}$$

where λ is wavelength and A_i and λ_i are empirical constants. Two or three terms on the right-hand side of Eq. (1-9) are typically required. Equation (1-9) is valid in regions where there are no strong absorption bands. Table C.2 in Appendix C gives dispersion equations used to prepare the graph in Fig. 1.5.

Table C.1 in Appendix C lists the refractive index for many infrared window materials and indicates how the index varies with temperature. For example, in zinc sulfide the variation is close to $dn/dT = 4 \times 10^{-5}$ K^{-1} in the wavelength range 1 - 10 µm and over the temperature range -100°C to +200°C. Thus if the temperature is increased by 100 K, the refractive index increases by (100 K) (4×10^{-5} K^{-1}) = 0.004.

Table C.4 shows how the refractive index changes when a material is subjected to a uniform (isostatic) pressure from all directions. For fused silica glass (SiO_2), the change is $dn/dP = +9.2 \times 10^{-4}$ kbar^{-1} at a wavelength of 0.589 nm. (One bar is 10^5 Pa, which is 0.987 atmospheres.) When subjected to a uniform pressure of 2 kbar (approximately 2000 atm), the refractive index increases by (9.2×10^{-4} kbar^{-1})(2 kbar) = 0.0018.

In many handbooks on optical materials you will encounter the name "Irtran," which is a trade name that refers to hot pressed materials formerly manufactured by Eastman Kodak. Table 1.1 lists some commonly encountered trade names of infrared materials.

Table 1.1. Trade names of some infrared optical materials

Trade name	Composition	Chemical name
Irtran 1*	MgF_2	magnesium fluoride
Irtran 2*	ZnS	zinc sulfide
Irtran 3*	CaF_2	calcium fluoride
Irtran 4*	ZnSe	zinc selenide
Irtran 5*	MgO	magnesium oxide
Irtran 6*	CdTe	cadmium telluride
KRS-5	Tl(Br,I)	thallium bromide iodide
KRS-6	Tl(Br,Cl)	thallium bromide chloride
AMTIR-1†	33wt%Ge·12wt%As·55wt%Se	germanium-arsenic-selenium glass
IRG-11‡		calcium aluminate glass
BS37A§		calcium aluminate glass
BS39B§		calcium aluminate glass
Corning 9754¶	33wt%GeO_2·37.3wt%Al_2O_3 19.7wt%CaO·5wt%BaO·5wt%ZnO	germanate glass[5]
BK-7‡		borosilicate crown glass

*Eastman Kodak ‡Schott Glass Technologies ¶Corning Glass Works
†Amorphous Materials §Barr & Stroud

1.2.1 Birefringence

We say that a glass is *isotropic* because its optical (and other) properties are the same in every direction. Crystalline materials have a regular structure in which different directions may be physically distinct. In crystals with a cubic structure, such as the sphalerite form of zinc sulfide, optical properties are the same in every direction. Crystals with lower symmetry than cubic, such as trigonal sapphire in Fig. 1.6, are *anisotropic*. That is, they have two or more distinct directions with different optical properties.

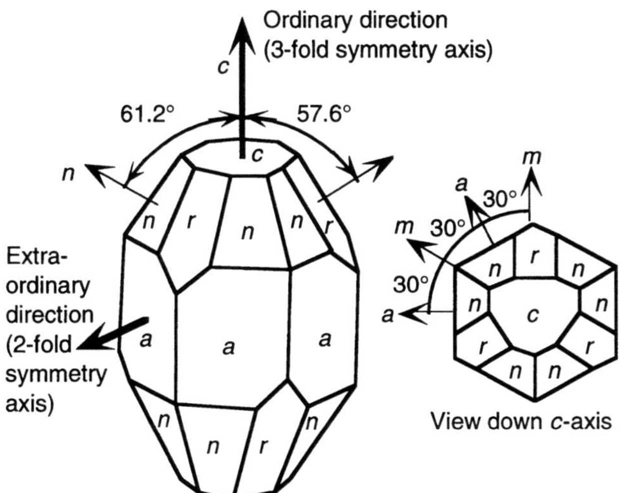

Fig. 1.6. Sapphire crystal showing 3-fold-symmetric *optical axis* (ordinary direction) and 2-fold-symmetric extraordinary direction. Sapphire crystal cleavage faces are customarily designated *c*, *n*, *r* and *a*, as indicated. When viewed down the 3-fold *c*-axis, directions designated *a* and *m* alternate every 30°.

The 3-fold symmetry axis in Fig. 1.6 is designated as the *c*-axis in the sapphire crystal. It is also called the *optical axis* or the *ordinary* direction. The 2-fold symmetry axis perpendicular to the 3-fold axis is called the *a*-axis or the *extraordinary* direction. Table 1.2 lists the refractive index for the ordinary (n_o) and extraordinary (n_e) directions for several crystals. The index of refraction of any of these materials varies continuously, depending on the direction through the crystal. We say that a material with two limiting values of the index of refraction is *birefringent.*

Table 1.2. Refractive index of birefringent materials at 589.3 nm (sodium D line)

Crystal	n_o	n_e
Calcite	1.6584	1.4864
Quartz	1.5443	1.5534
Ice	1.309	1.313
Magnesium fluoride	1.378	1.389
Sapphire (at sodium D line)	1.768	1.760
Sapphire (at 3 GHz microwave frequency)	3.06	3.40

Calcite (calcium carbonate) is a noteworthy example of a birefringent crystal with a large difference between the two optical directions. Figure 1.7 shows that light transmitted through a calcite crystal is split into two rays, each of which comes out of the crystal with a different polarization. By *polarization*, we refer to the direction of oscillation of the electric field of the electromagnetic radiation.

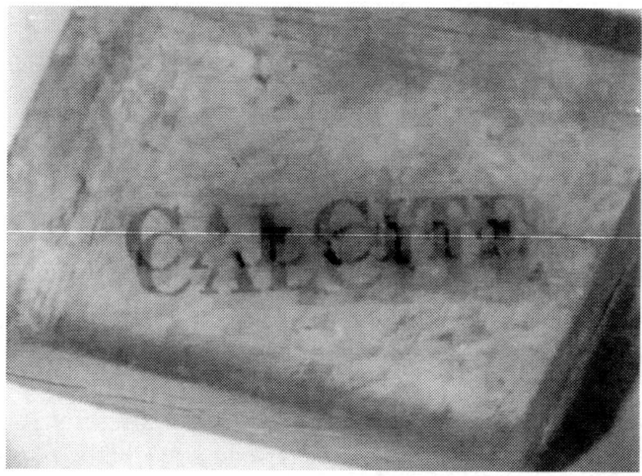

Fig. 1.7. Birefringence of calcite crystal.

An optical window that creates multiple images, depending on the viewing angle, would be disastrous. Nonetheless, the birefringent materials sapphire and magnesium fluoride in Table 1.2 are both common infrared windows. In the case of single-crystal sapphire, the quality of the optical image is limited by the slightly different refraction of the ordinary and extraordinary rays. A sufficiently high quality imaging system might not tolerate the birefringence. Magnesium fluoride, on the other hand, is usually used as a *polycrystalline* material consisting of microscopic, randomly oriented crystals compacted together. (See Box on next page) The crystal size is much smaller than the wavelength of light being viewed. As a result, the polycrystalline material is effectively isotropic, and two distinct images are not formed.

1.2.2 Preference for cubic materials

Single-crystal windows with a cubic crystal structure are preferred over noncubic crystals so that multiple images are not formed (birefringence) as light passes through the window. In polycrystalline windows, the cubic crystal structure is also preferred to reduce optical scatter. As light passes from one randomly oriented grain to another inside a polycrystalline window containing noncubic crystals, the light is refracted and reflected at each grain boundary (Fig. 1.8). The net result is that some (or most) light is diverted from the incident direction, and this is called scatter. In cubic materials, the refractive index is constant, regardless of crystal orientation. As light passes from grain to grain of a polycrystalline cubic material, no refraction or reflection (*i.e.*, no scatter) occurs. (However, stresses in a polycrystalline material could create birefringence and scatter — even with cubic crystal structures.)

Optical scattering by noncubic polycrystalline materials is negligible if the crystallite size is much smaller (< 1/20) than the wavelength of light. Otherwise, noncubic materials generally produce severe scatter and cannot be used in polycrystalline windows.

Polycrystalline materials

A perfect *single crystal* has a regular arrangement of atoms from one end of the crystal to the other. A *glass* has an amorphous structure in which there is no long range order. Most ceramic materials, including most infrared window materials, are *polycrystalline.* This means that they are composed of individual, randomly oriented, microscopic, single crystals arranged in a compact mass. The photograph below shows *grain boundaries* between individual crystals in optically clear yttria (Y_2O_3, pronounced **it**-tree-yuh). Grains in optical window materials typically range from less than one micrometer up to several hundred micrometers in size. Grain boundaries are amorphous regions that are usually a few to tens of atoms in thickness. Impurities tend to be excluded from the crystalline grains, so they aggregate at the grain boundaries. Therefore, the grain boundary is likely to have a different refractive index from the surrounding grains. Grain boundaries in the photo below are rich in silica, which was an impurity in the starting yttria powder.

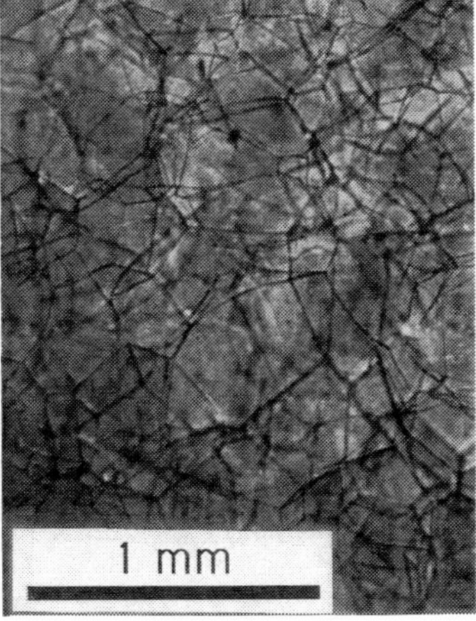

Grain structure of Raytheon yttria. Micrograph taken with crossed polarizers using first order red plate. (Courtesy Marian Hills, Naval Air Warfare Center.)

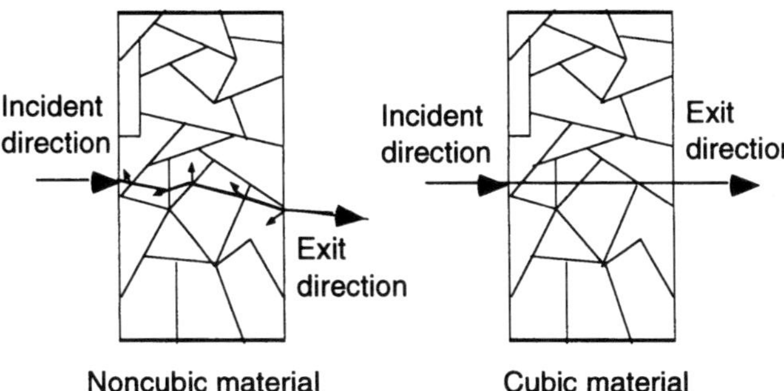

Fig. 1.8. Refraction and reflection at grain boundaries of a noncubic polycrystalline material cause optical scatter. There is no refraction or reflection in a cubic material.

Polycrystalline materials are subject to other scattering mechanisms that are not present in single crystals. Tiny voids and roughness at the grain boundaries lead to scattering (see Fig. 2.8). Impurities that aggregate at grain boundaries can give rise to scatter and absorption. Hot pressed materials may contain carbon residues from incomplete burnout of the binders used in their fabrication.

1.2.3 Reproducibility of the refractive index

Designers want to know the refractive index of optical materials very accurately. Optical glasses can be supplied with certificates stating the refractive index to the 5th decimal place with an accuracy of ~3×10^{-5}. Unfortunately, infrared window materials tend not to be so reproducible. Figure 1.9 shows that the refractive indexes of sapphire manufactured by two different methods (designated HEM and EFG) differ by ~0.001.

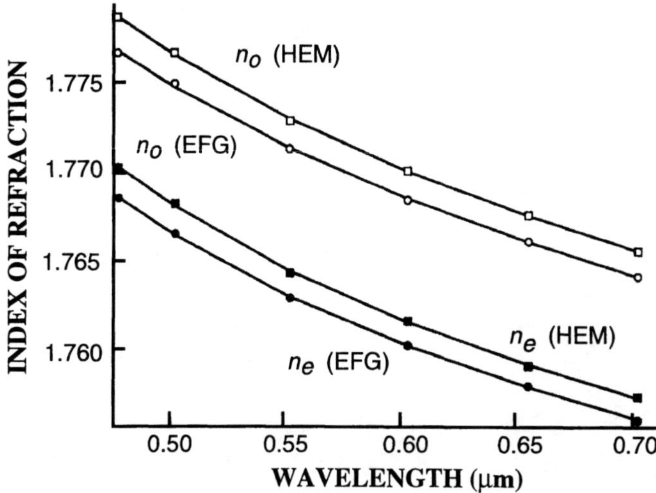

Fig. 1.9. Refractive index at 295 K of sapphire grown by different crystal growth methods designated HEM and EFG (described in Section 5.4.2).[6] The refractive index of the differently grown materials differs by ~0.001.

1.3 Reflection and transmission

When reflection from a smooth surface, called *specular reflection*, occurs, the angle of incidence is equal to the angle of reflection (Figs. 1.3 and 1.10). Light reflected from a rough surface, called *diffuse reflection*, may emerge at almost any angle because it strikes the jagged surface at various angles (Fig. 1.10). Diffuse reflection is another term for *surface scatter.*

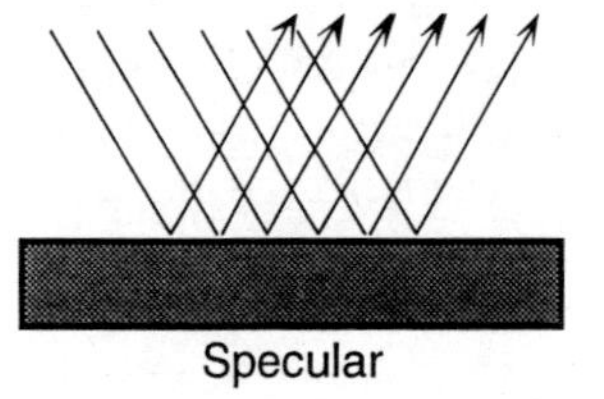

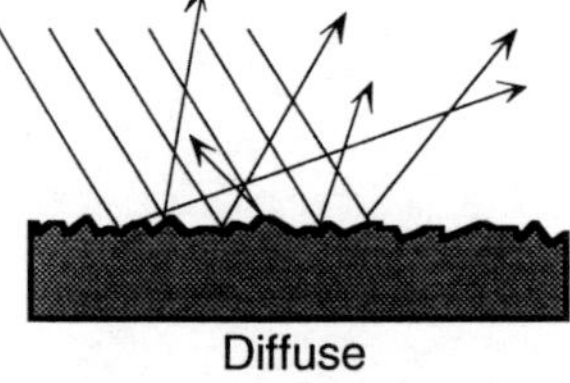

Fig. 1.10. Specular and diffuse reflection.

We restrict our discussion to perpendicular incidence of light onto a smooth surface, such as an optical window (Fig. 1.11). If the window has a refractive index n_2, and the surrounding medium has refractive index n_1 (usually air, for which $n_1 = 1$), then the fraction of incident radiant power reflected at each surface is

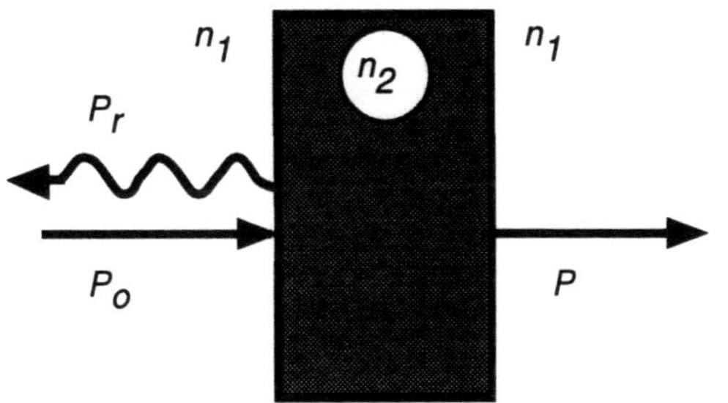

Fig. 1.11. Transmission and specular reflection with perpendicular incidence.

$$\text{Reflectance from one surface} = R = \left(\frac{n_1 - n_2}{n_1 + n_2}\right)^2 . \quad (1\text{-}10)$$

In the case where the surrounding medium is air, we can set $n_1 = 1$ and simply call $n_2 = n$ and rewrite Eq. (1-10) in the form

$$R = \left(\frac{1 - n}{1 + n}\right)^2 \quad \text{(one-surface reflection from window in air).} \quad (1\text{-}11)$$

Equation (1-11) applies to a window that has negligible absorption.

The total reflectance (fraction of incident light reflected back toward the source, P_r/P_o in Fig. 1.11) arises from light reflected from both entry and exit surfaces, and includes contributions from multiple internal reflections. The total reflectance turns out to be

$$\text{Total reflectance} = r = \frac{2R}{1 + R} . \quad (1\text{-}12)$$

If there is no absorption or scatter, the transmittance (fraction of incident light transmitted, P/P_o in Fig. 1.9), is 1 minus the reflectance:

$$\text{Transmittance} = t = 1 - r = \frac{2n}{n^2 + 1} . \quad (1\text{-}13)$$

Example: Reflection and transmission by a nonabsorbing window. Let's calculate the reflectance and transmittance for perpendicular incidence of light onto smooth windows of glass (n = 1.5), ZnS (n = 2.2) or Ge (n = 4.0) in air. To do this, we use Eqns. (1-11), (1-12) and (1-13). For glass, the calculations look like this:

$$R = \left(\frac{1 - 1.5}{1 + 1.5}\right)^2 = 0.040 \Rightarrow r = \frac{2 \times 0.040}{1 + 0.040} = 0.077 .$$

$$t = 1 - 0.077 = 0.923 .$$

That is, 4.0% of incident light is reflected from one glass surface, the total reflection is 7.7%, and the net transmittance is 92.3%. Similarly, for ZnS and Ge we can find the values in Table 1.3. *As the index of refraction of the window increases, the reflection increases and the transmission decreases.*

Table 1.3. Reflection and transmission for flat windows

Material	Refractive index (n)	Single-surface reflection (R)	Total reflectance (r)	Transmittance (t)
Glass	1.5	0.040	0.077	0.923
ZnS	2.2	0.141	0.247	0.753
Ge	4.0	0.360	0.529	0.471

Figure 1.12 is a graph of Eq. (1-13) showing the positions of several infrared window materials. Based on its low refractive index and consequent low reflection, magnesium fluoride would have a transmittance of 96% at a wavelength of 4 μm. In contrast, a germanium window would transmit only 47% because of its high refractive index. For materials like germanium, an *antireflection coating* (Section 6.1) is critical to obtain adequate transmittance.

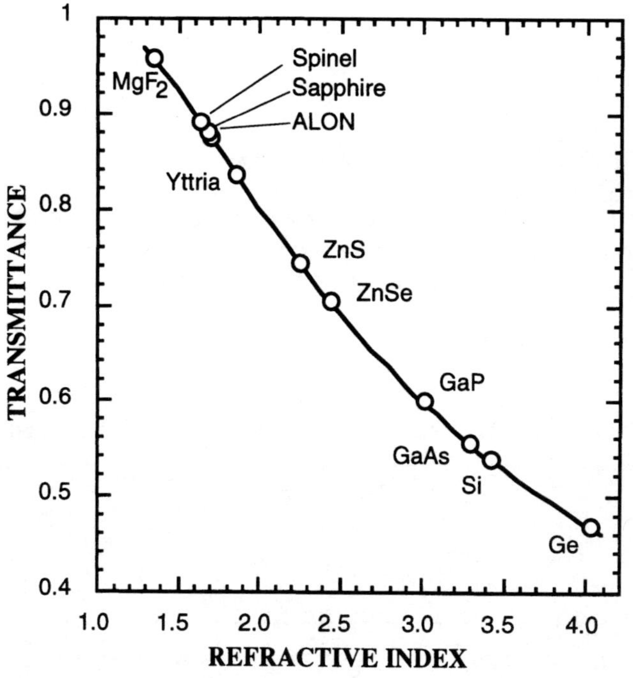

Fig. 1.12. Transmittance of perpendicular incident ray based on Eq. (1-13), assuming no absorption or scatter. The refractive index of each material is shown for a wavelength of 4 μm.

1.3.1 Transmission of an absorbing window

When a material absorbs light, as well as reflecting light, Eq. (1-13) must be modified to find the transmittance. If the absorption coefficient is α, the transmittance is

$$\text{Transmittance} = t = \frac{(1 - R)^2 \, e^{-\alpha b}}{1 - R^2 \, e^{-2\alpha b}} \tag{1-14}$$

where b is the sample thickness and R is the single-surface reflectance given by Eq. (1-11). If there is bulk scatter, it should be included in the value of α used in Eq. (1-14).

Example: Transmission through an absorbing window. Zinc sulfide has an absorption coefficient of 0.077 cm^{-1} at 10 μm wavelength.[7,8] In Table 1.3 we calculated that zinc sulfide should have a transmittance of 0.753 for unabsorbed light with perpendicular incidence. Taking into account the absorption, what is the expected transmittance of a 1.0-mm-thick or 1.0-cm-thick zinc sulfide window? Using the value R = 0.141 from Table 1.3 and the absorption coefficient $\alpha = 0.077$ cm^{-1} in Eq. (1-14), we find

$$\text{For } b = 0.10 \text{ cm:} \quad t = \frac{(1 - 0.141)^2 \, e^{-(0.077 \text{ cm}^{-1})(0.10 \text{ cm})}}{1 - 0.141^2 \, e^{-2(0.077 \text{ cm}^{-1})(0.10 \text{ cm})}} = 0.747 \,.$$

$$\text{For } b = 1.0 \text{ cm:} \quad t = \frac{(1 - 0.141)^2 \, e^{-(0.077 \text{ cm}^{-1})(1.0 \text{ cm})}}{1 - 0.141^2 \, e^{-2(0.077 \text{ cm}^{-1})(1.0 \text{ cm})}} = 0.695 \,.$$

The transmittance is reduced by only 0.006 because of absorption by the 1.0-mm window, but by 0.058 for the 1.0-cm window. The greater the absorption coefficient and the greater the thickness, the lower will be the transmittance. Optical window materials typically have absorption coefficients in the range 10^{-4} to 10^{-1} cm^{-1}.

1.3.2 Etalon effect

Optical windows with well polished, parallel surfaces occasionally have a "wavy" transmission spectrum instead of a constant, flat transmission. An extreme example is shown in Fig. 1.13. The oscillations result from constructive and destructive interference of light waves bouncing back and forth between the two surfaces. The waviness is sometimes called an *etalon effect* because an etalon is a device with two parallel plates whose transmittance is determined by the interference of waves bouncing between the plates. Figure 1.14 shows that a reflected ray passing through a window of thickness b traverses an additional distance $2b$, compared to a nonreflected ray. The reflected and nonreflected rays reinforce each other when the pathlength difference, $2b$, is an integer multiple of the wavelength λ (measured inside the window) The rays interfere destructively when the pathlength difference is a half-integral multiple of the wavelength.

The transmittance of a window that behaves as an etalon is given by a more complete version of Eq. (1-14):

$$\text{Transmittance with etalon effect} = \frac{(1 - R)^2 \, e^{-\alpha b}}{1 + R^2 \, e^{-2\alpha b} - 2R \, e^{-\alpha b} \cos\phi} \tag{1-15}$$

where $\phi = 4\pi n b / \lambda_o$. Here n is the refractive index of the window and λ_o is the wavelength of radiation measured in vacuum.

If you measure the change in the spacing between peaks in Fig. 1.13 as a function of temperature, and if the thermal expansion coefficient of the material is known independently, then you can deduce the change in refractive index (dn/dT) from the change in spacing of the fringes.[9,10] Let the mean distance between peaks in Fig. 1.13 be Δ. The change in Δ. with respect to temperature (T) is given by

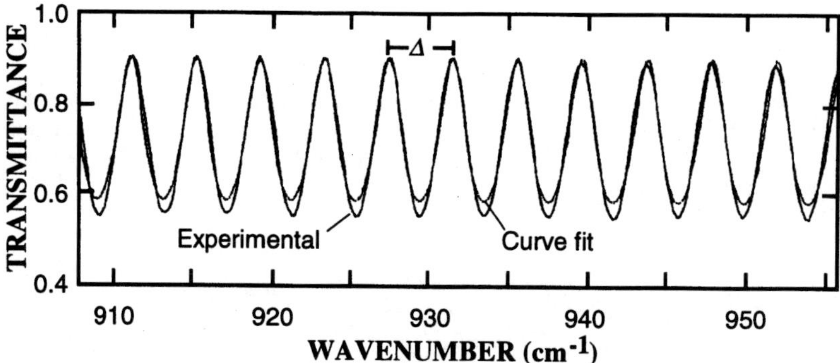

Fig. 1.13. Transmittance of diamond plate showing interference fringes from the parallel polished faces. The curve fit uses a fraction of the peak-to-valley transmittance difference predicted by Eq. (1-15). (From M. Thomas, Applied Physics Laboratory.[9,10])

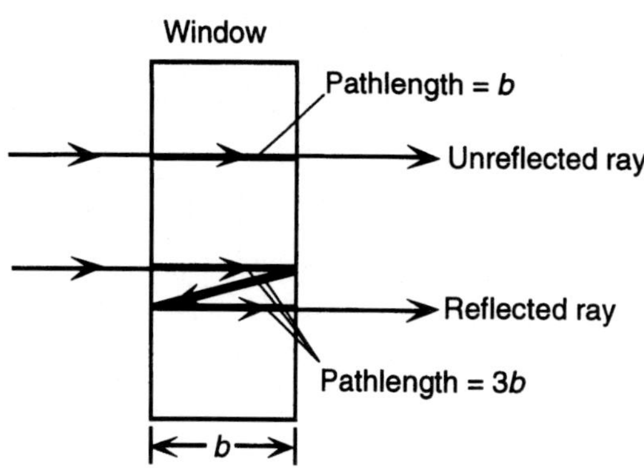

Fig. 1.14. Etalon effect. A light ray traverses a distance b through a window. A reflected ray traverses a distance $3b$. (The reflected waves are offset for clarity. They really traverse the same path back and forth through the window.) The two waves reinforce each other when the difference in pathlength, $2b$, is an integer multiple of the wavelength λ. The wavelength λ inside the window is equal to the wavelength in vacuum divided by the refractive index of the window: $\lambda = \lambda_o/n$.

$$\frac{d\Delta}{dT} = -\Delta\left[\frac{1}{n}\frac{dn}{dT} + \frac{1}{b}\frac{db}{dT}\right] \tag{1-16}$$

where n is the refractive index and b is the thickness of the window. The quantity $(1/b)(db/dT)$ is the *thermal expansion coefficient*, usually designated α.

Example: Measuring dn/dT from the etalon effect. The mean spacing between interference peaks for diamond in Fig. 1.13 is $\Delta = 4.083$ cm^{-1} at 300 K.[10] (The spacing between peaks is sometimes called the "free spectral range.") The spacing was measured as a function of temperature up to 784 K and a graph of Δ versus T had a slope $d\Delta/dT = -35.9 \times 10^{-6}$ cm^{-1} K^{-1}. The refractive index of diamond in the spectral region in Fig. 1.13 is $n = 2.38$ and the mean expansion coefficient for the range 300 - 784 K is $\alpha = 2.7 \times 10^{-6}$ K^{-1}. Therefore, the mean value of dn/dT for the temperature range 300 - 784 K is

$$\frac{d\Delta}{dT} = -\Delta\left[\frac{1}{n}\frac{dn}{dT} + \alpha\right]$$

$$-35.9 \times 10^{-6}\ \mathrm{cm^{-1}\ K^{-1}} = -\ (4.083\ \mathrm{cm^{-1}})\left[\frac{1}{2.38}\frac{dn}{dT} + 2.7 \times 10^{-6}\ \mathrm{K^{-1}}\right]$$

$$\frac{dn}{dT} = 14.5 \times 10^{-6}\ \mathrm{K^{-1}}$$

The conventional, and generally most accurate, means to measure refractive index is to carefully construct a prism of the material of interest and measure the angle of refraction of monochromatic radiation passing through the prism. By placing the prism in a furnace and varying its temperature, dn/dT can be measured directly. Figure 1.15 shows the result of such an experiment with gallium phosphide.

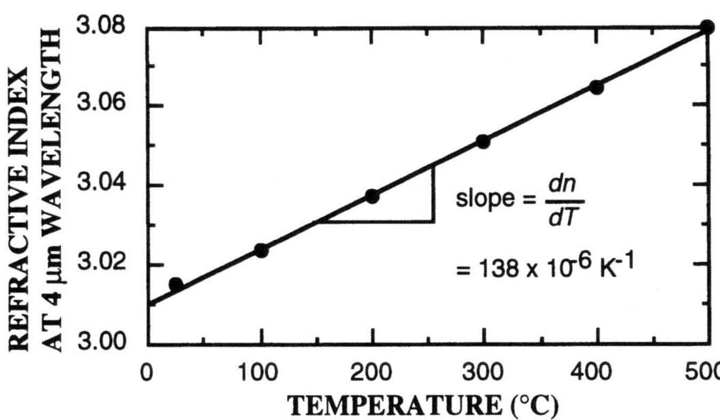

Fig. 1.15. Refractive index of gallium phosphide from Raytheon Systems Co.[13]

1.4 Optical constants: *n* and *k*

A full description of an electromagnetic wave that is attenuated by absorption requires a complex refractive index, N, defined as

$$N = n - ik \qquad (i = \sqrt{-1}) \tag{1-17}$$

where n is the real part of the refractive index, with which we are familiar, and k (the *extinction coefficient*) is the imaginary part. The absorption coefficient, α in Eq. (1-1), is proportional to k:

$$\alpha = \frac{4\pi k}{\lambda}. \tag{1-18}$$

Some authors tabulate absorption in terms of k instead of α.[7,11,12] Figure 1.16 shows the complex components of the refractive index of diamond plotted as a function of wavelength. Using the complex refractive index, the single-surface reflection in air, analogous to Eq. (1-11), is

$$R = \frac{(1 - n)^2 + k^2}{(1 + n)^2 + k^2} \quad \text{(normal incidence reflection from surface in air).} \tag{1-19}$$

Equation (1-19) gives the correct expression for R that should be used in Eqns (1-12), (1-14) and (1-15). However, in its "window" region, an infrared material is nearly transparent and k is small compared to the other terms in Eq. (1-19). Therefore it is usually fine to use Eq. (1-11) for the single-surface reflection in the window region.

Example: Relation between absorption coefficient, α, and extinction coefficient, k. The maximum absorption for diamond in the infrared region in Fig. 1.16 is at 4.96 μm, where $k = 5.2 \times 10^{-4}$. Let's find the absorption coefficient and calculate the transmittance of a 1.0-mm-thick diamond whose refractive index is 2.38. The absorption coefficient is related to k by Eq. (1-18):

$$\alpha = \frac{4\pi k}{\lambda} = \frac{4\pi(5.2 \times 10^{-4})}{(4.96\ \mu\text{m})(10^{-4}\ \text{cm}/\mu\text{m})} = 13\ \text{cm}^{-1}$$

and the transmittance can be calculated with Eqns. (1-11) and (1-14):

$$R = \left(\frac{1-n}{1+n}\right)^2 = \left(\frac{1-2.38}{1+2.38}\right)^2 = 0.167$$

$$t = \frac{(1-R)^2\, e^{-\alpha b}}{1 - R^2\, e^{-2\alpha b}} = \frac{(1-0.167)^2\, e^{-(13\ \text{cm}^{-1})(0.10\ \text{cm})}}{1 - 0.167^2\, e^{-2(13\ \text{cm}^{-1})(0.10\ \text{cm})}} = 0.189\ .$$

A 1.0-mm-thick diamond will transmit 18.9% at the absorption maximum.

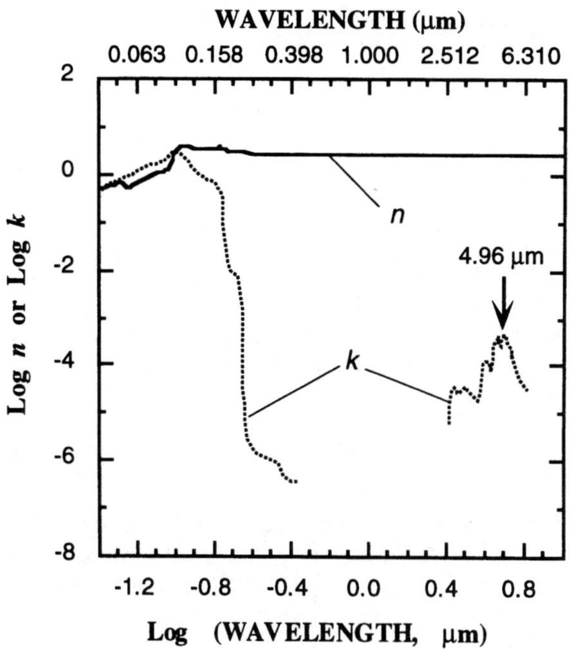

Fig. 1.16. Optical constants, n and k, for diamond.[7] Between 0.4 and 2.5 μm, k is too low to measure.

1.5 General behavior of absorption coefficient and refractive index

The shapes of the curves for n and k in Fig. 1.16 generally apply to any optical material. This behavior is illustrated for yttria in Fig. 1.17, which shows n and α, the

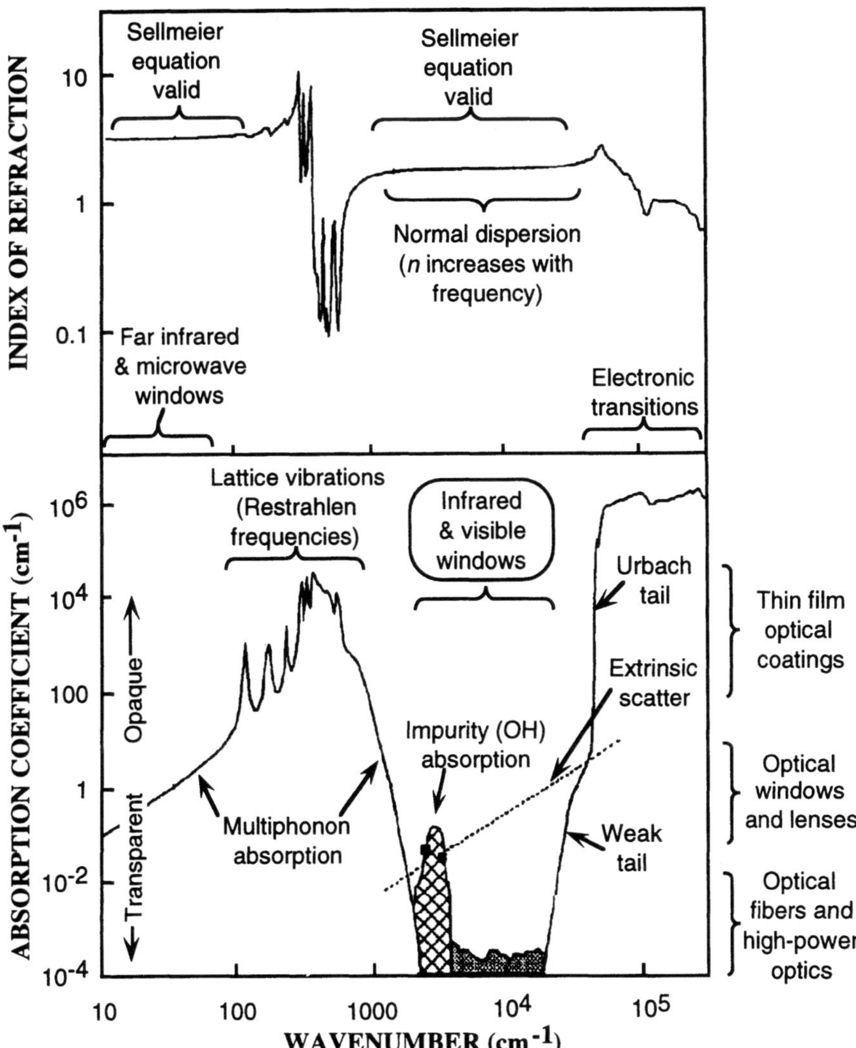

Fig. 1.17. Refractive index (n) and absorption coefficient (α) of yttria. In general, all optical materials have regions of strong absorption in the infrared due to lattice vibrations and in the ultraviolet (and sometimes visible) due to electronic transitions. Brackets at the lower right indicate approximate ranges of absorption coefficients of materials used for thin coatings, windows and optical fibers. Adapted from Thomas.[14-16]

absorption coefficient. In the lower curve in Fig. 1.17, the absorption coefficient is high in the infrared region (in the wavenumber range ~100 to 1000 cm^{-1}) and the ultraviolet region (~10^5 cm^{-1}). Infrared radiation is absorbed because it has the right energy to excite vibrational modes of the material, causing atoms to oscillate with greater amplitude about their equilibrium lattice positions. Ultraviolet radiation is absorbed because it has the right energy to promote electrons from low lying electronic energy levels (bands) to excited energy levels.

Between the two strong absorption regions is a window in which the material is transparent. The infrared edge of the window is called the *multiphonon region* because the transitions responsible for the weak absorption arise from multiple excitations of one or more vibrational modes (*phonons*). The ultraviolet edge of the window is called the *Urbach tail*. Equations describing the infrared multiphonon absorption and the Urbach tail for several infrared materials are available.[17,18]

In the window region between strong infrared and ultraviolet absorption, the very weak absorption is said to be *extrinsic*, because it is not a property of the pure, perfect material. The weak absorption arises from impurities (such as –OH groups in yttria) and crystal defects which, in principle, could be decreased to give lower absorption. Much of the transmission loss in the window region can arise from scattering by crystal defects, voids, and surfaces. In a more perfect material, this scatter could be reduced. Properties of the pure, perfect material are said to be *intrinsic*. Observed properties arising from impurities and imperfections are extrinsic.

As an example of the extrinsic nature of absorption in the window region, the absorption coefficient of sapphire fibers from different sources varied from 3×10^{-6} to 2×10^{-4} cm^{-1} at a wavelength near 1 μm.[19] If absorption were intrinsic, it would be the same in all specimens. In another study, different sapphire fibers had attenuations of 0.016 to 2.16 dB/cm at a wavelength of 0.83 μm.[20] These attenuations include contributions from both absorption and scatter.

Example: Converting dB/cm to an absorption coefficient. What absorption coefficient (α) corresponds to an attenuation coefficient (γ) of 0.016 dB/cm? Equation (0-9) expressed transmittance in terms of the attenuation coefficient: transmittance = t = $10^{-\gamma b/10}$. If the pathlength, b, is expressed in cm, then the units of γ are dB/cm. But we also know from Eq. (1-1) that $t = e^{-\alpha b}$. Equating the two expressions for t gives

$$t = 10^{-\gamma b/10} = e^{-\alpha b} \, .$$

Writing the number 10 as $e^{\ln 10}$ lets us solve for α in terms of γ:

$$(e^{\ln 10})^{-\gamma b/10} = e^{-\alpha b} \quad \Rightarrow \quad -\frac{\gamma b}{10} \ln 10 = -\alpha b \quad \Rightarrow \quad \alpha = \frac{\gamma}{10} \ln 10 \, .$$

Substituting in γ = 0.016 dB/cm gives α = (0.016/cm)(ln10)/10 = 0.0037 cm^{-1}.

The refractive index in Fig. 1.17 is nearly constant in the window region of any material, but varies wildly in the regions of strong absorption. The refractive index in the visible/infrared window region decreases slightly as the wavelength increases, which is described as "normal dispersion." In the far infrared/millimeter wavelength region, the refractive index also decreases slightly with increasing wavelength.

1.6 Transmission spectra of infrared materials

Figures 1.18 and 1.19 show infrared and ultraviolet/visible transmission spectra of many infrared window materials.[21] Most of the materials are nearly opaque ($t \approx 0$) at

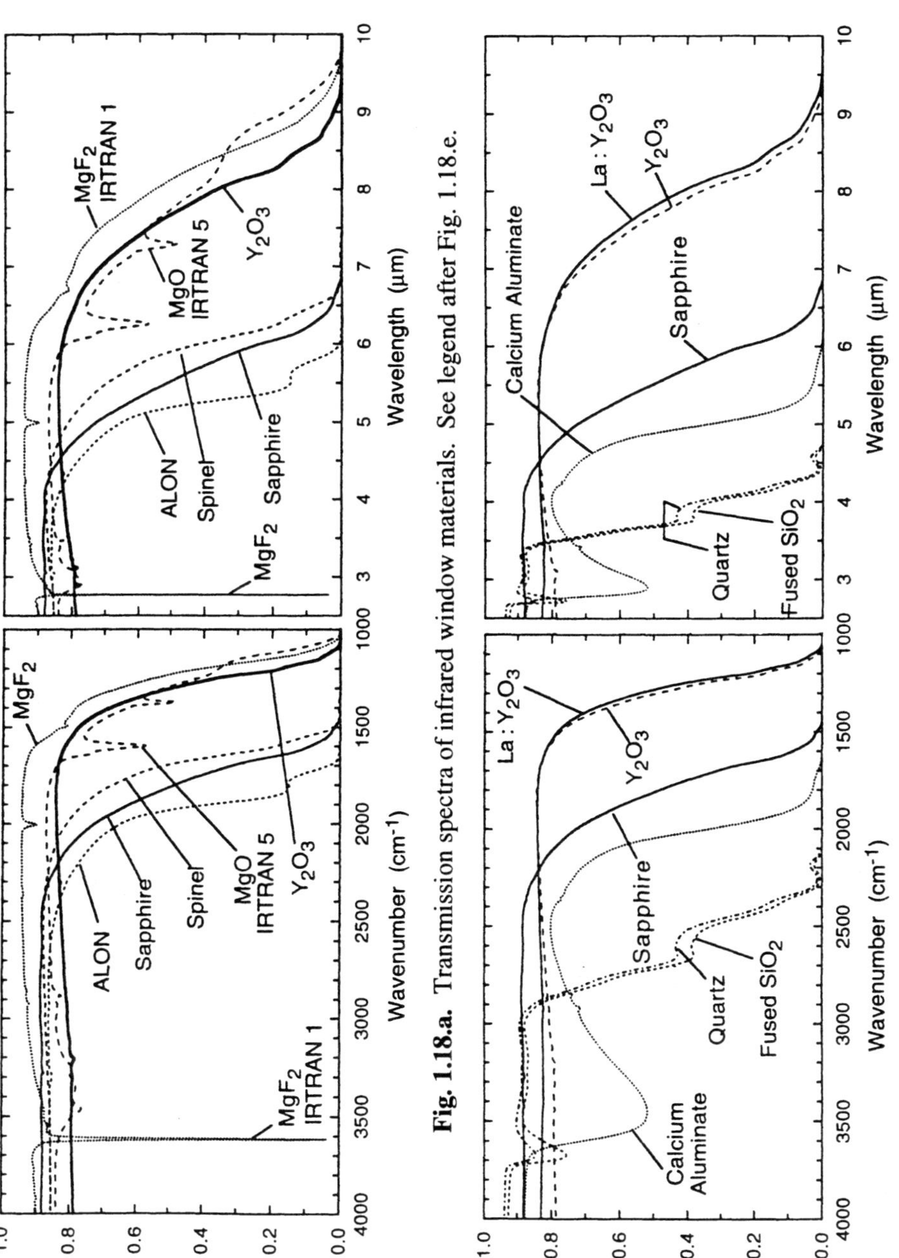

Fig. 1.18.a. Transmission spectra of infrared window materials. See legend after Fig. 1.18.e.

Fig. 1.18.b. Transmission spectra of infrared window materials. See legend after Fig. 1.18.e.

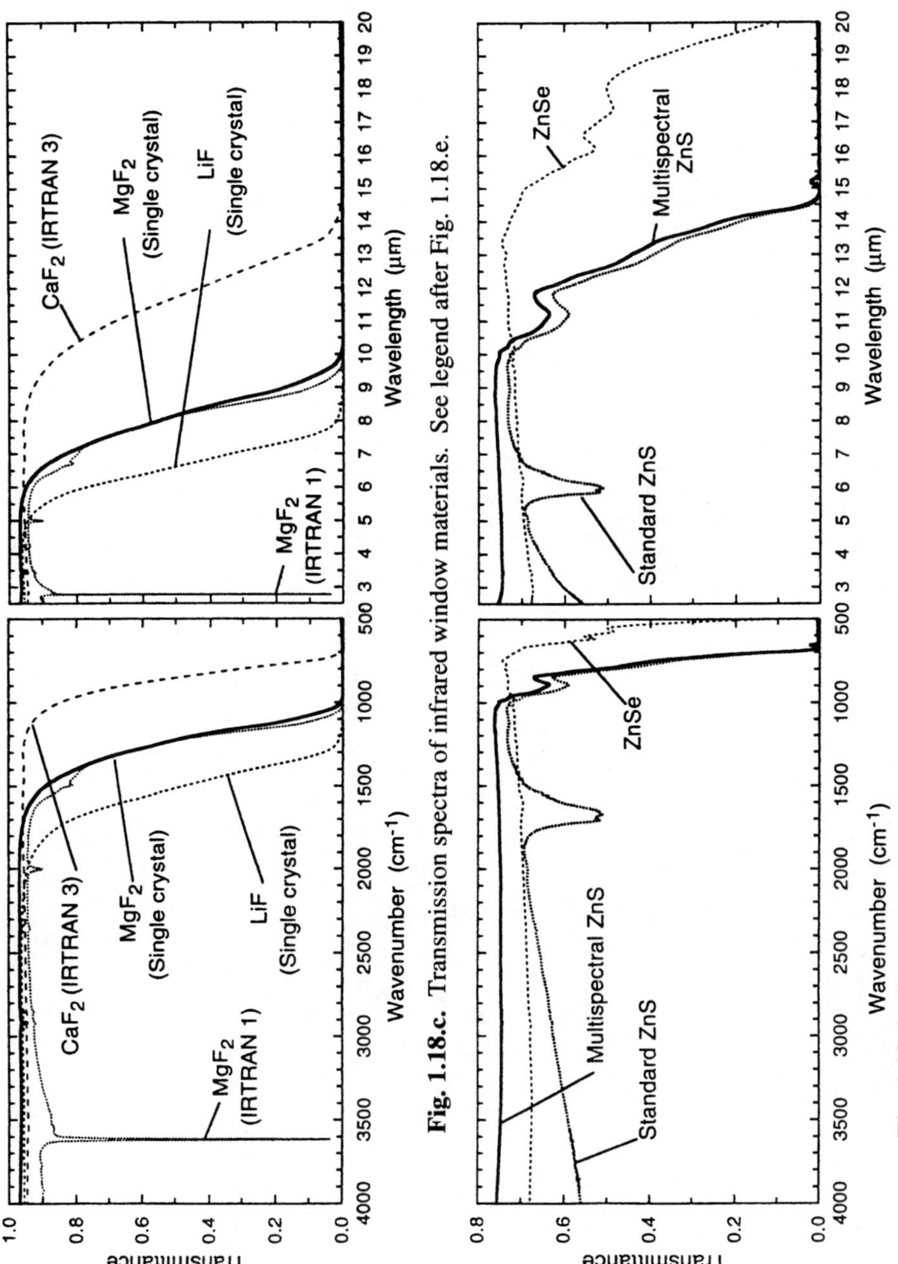

Fig. 1.18.c. Transmission spectra of infrared window materials. See legend after Fig. 1.18.e.

Fig. 1.18.d. Transmission spectra of infrared window materials. See legend after Fig. 1.18.e.

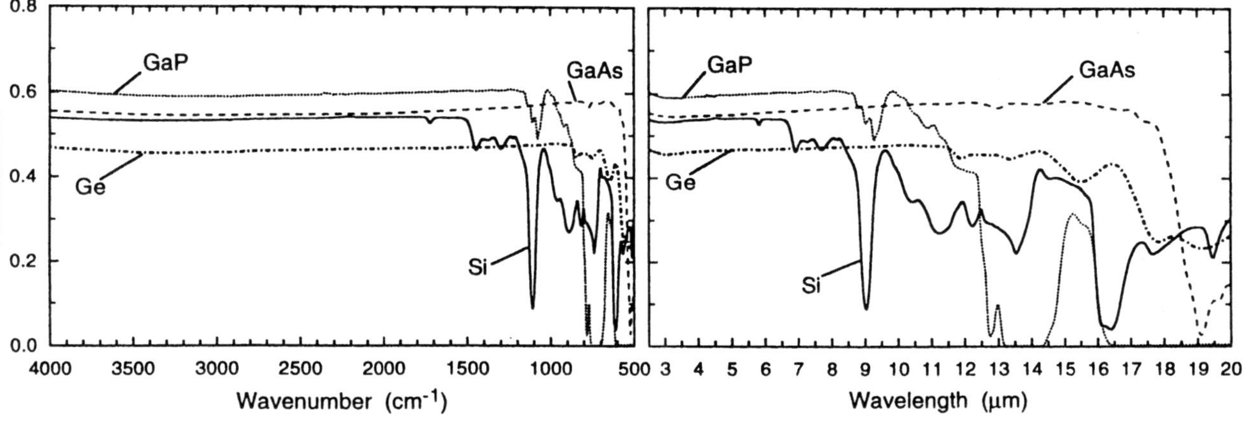

Fig. 1.18.e. Transmission spectra of infrared window materials.

Legend: **(a)** Magnesium fluoride: Kodak hot pressed Irtran 1®, 2.0 mm thick. Aluminum oxynitride: Raytheon polycrystalline ALON, 1.9 mm thick. Sapphire: Union Carbide single crystal, 60° cut, 2.6 mm thick. Spinel: Coors polycrystalline $MgAl_2O_4$, 1.7 mm thick. Magnesium oxide: Kodak hot pressed Irtran 5®, 2.1 mm thick. Yttrium oxide: Raytheon polycrystalline yttria, 2.1 mm thick. **(b)** Calcium aluminate glass: unknown source, 2.6 mm thick. Quartz: single crystal, 4.7 mm thick. Fused silica: unknown source, 5.3 mm thick. Sapphire: Union Carbide single crystal, 60° cut, 2.6 mm thick. Yttria: Raytheon polycrystalline Y_2O_3, 2.1 mm thick. Lanthana-doped yttria: GTE polycrystalline $0.9La_2O_3 \cdot 0.91Y_2O_3$, 1.9 mm thick. **(c)** Magnesium fluoride: Kodak hot pressed Irtran 1®, 2.0 mm thick. Magnesium fluoride: John H. Ransom Laboratories single crystal, 2.7 mm thick. Lithium fluoride: Harshaw single crystal, 5.4 mm thick. Calcium fluoride: Kodak hot pressed Irtran 3®, 1.1 mm thick. **(d)** Standard grade zinc sulfide: Raytheon polycrystalline material, 6.0 mm thick. Multispectral grade zinc sulfide: Raytheon polycrystalline material, 5.2 mm thick. Zinc selenide: Raytheon polycrystalline material, 7.1 mm thick. **(e)** Germanium: 2.8 mm thick. Silicon 2.8 mm thick. Gallium phosphide: Raytheon polycrystalline material, 3.0 mm thick. Gallium arsenide: (100) single crystal, 0.65 mm thick.

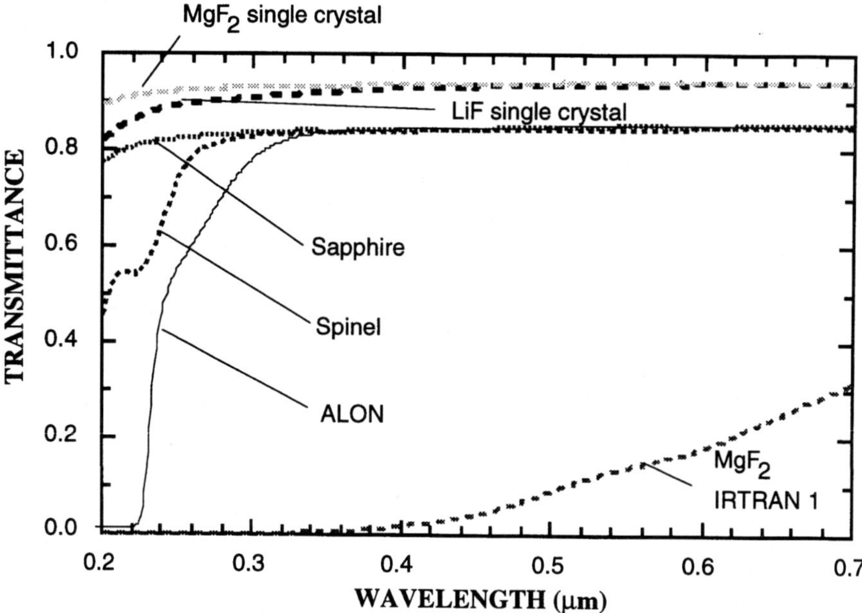

Fig. 1.19.a. Ultraviolet/visible transmittance of infrared window materials. See legend after Fig. 1.19.c. Spectra of sapphire, ALON and spinel have been interpreted in terms of band gaps of 9.1, 6.5, and 8.0 eV, respectively.[22]

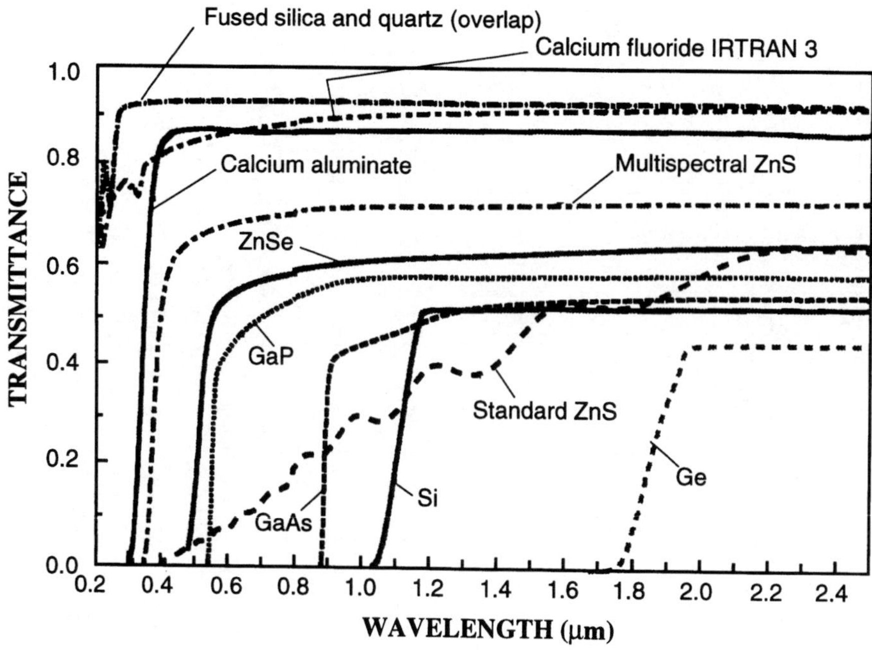

Fig. 1.19.b. Ultraviolet/visible transmittance of infrared window materials.

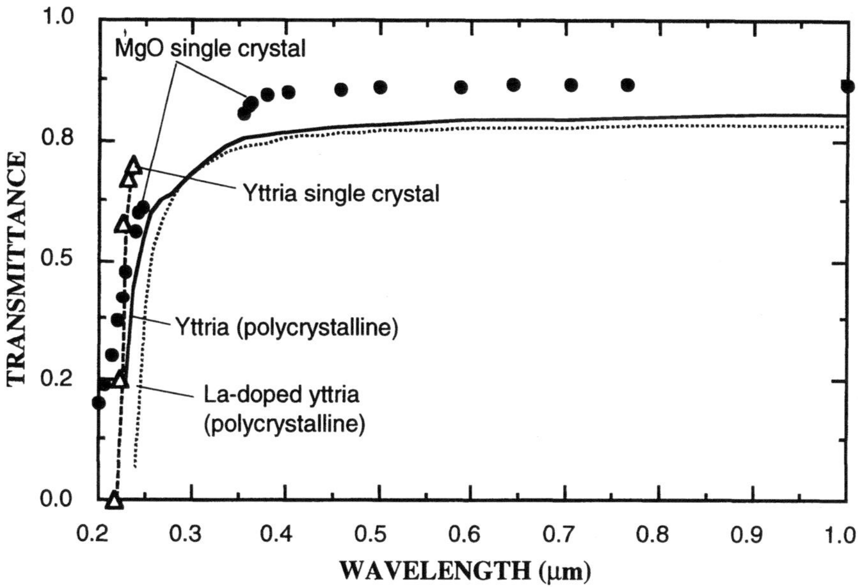

Fig. 1.19.c Ultraviolet/visible transmittance of infrared window materials.

Legend: (a) Single-crystal MgF_2: John H. Ransom Laboratories, 2.7 mm thick. Polycrystalline MgF_2: Kodak hot pressed Irtran 1®, 2.0 mm thick. Single-crystal LiF, 5.4 mm thick. Sapphire: Union Carbide single crystal, 60° cut, 2.6 mm thick. Spinel: Coors polycrystalline $MgAl_2O_4$, 1.7 mm thick. Polycrystalline ALON: Raytheon, 1.9 mm thick. **(b)** Quartz: 4.7 mm thick. Fused silica: 5.3 mm thick. CaF_2: Kodak Irtran 3®, 1.1 mm thick. Calcium aluminate glass: 2.6 mm thick. Multispectral ZnS: Raytheon, 5.2 mm thick. ZnSe: Raytheon, 7.1 mm thick. GaP: Raytheon, 3.0 mm thick. Standard ZnS: Raytheon, 6.0 mm thick. GaAs: single crystal, 0.65 mm thick. Si: 2.8 mm thick. Ge: 2.8 mm thick. **(c)** Polycrystalline yttria: Raytheon Y_2O_3, 2.0 mm thick. Polycrystalline lanthana-doped yttria: GTE $0.9La_2O_3{\cdot}0.91Y_2O_3$, 2.0 mm thick. Single-crystal MgO, 2.0 mm thick. Single-crystal Y_2O_3, 2.0 mm thick. Spectra of single-crystal MgO and Y_2O_3 were computed from the optical constants n and k given in Palik[11] using Eqns. (1-19), (1-18) and (1-14).

sufficiently long wavelength in the infrared region and at sufficiently short wavelength in the ultraviolet region. Opaque regions correspond to lattice and electronic absorptions in Fig. 1.17. Between these two opaque regions is the useful transparent infrared/visible range for each window material. Transmittance in the window region is governed by extrinsic scattering and impurities.

For example, the sharp absorption band near 3615 cm^{-1} in polycrystalline MgF_2 (Irtran 1) at the left in Fig. 1.18a arises from an impurity, which might be HF. The absorption is absent in the spectrum of single-crystal MgF_2 in Fig. 1.18c. Single-crystal MgF_2 is transparent in the visible and ultraviolet regions in Fig. 1.19a. The ultraviolet absorption of single-crystal MgF_2 becomes strong at wavelengths below 0.11 μm, reaching a maximum at 0.06 μm, which is not shown in Fig. 1.19. By contrast, polycrystalline MgF_2 (Irtran 1) in Fig. 1.19a is nearly opaque at visible and ultraviolet wavelengths because of strong scattering from the polycrystalline material (Fig. 1.8).

1.7 Measuring the absorption coefficient

If sufficiently thick (5 - 20 mm), polished samples of a material with absorption coefficient $\alpha > 0.05$ cm^{-1} are available, there are two relatively easy ways to measure the absorption coefficient from the transmittance. For lower absorption coefficients and/or thinner samples, laser calorimetry becomes the method of choice.

1.7.1 Direct transmittance measurements

The simplest method applicable to a single thick sample with sufficiently large absorption coefficient is to rearrange Eq. (1-14) and solve for α:

$$\alpha = \frac{1}{b} \ln\left(\frac{(1-R)^2}{2t} + \sqrt{R^2 + \frac{(1-R)^4}{4t^4}}\right) \tag{1-20}$$

where b is the sample thickness, t is the measured transmittance, and R is the single-surface reflectance calculated with Eq. (1-11), which requires knowledge of the index of refraction. Equation (1-20) is the least accurate, but easiest, way to find α for a relatively transparent material. The main problem is that when α is small, a small uncertainty in t gives a large uncertainty in α. If the material has significant absorption, then Eq. (1-20) is accurate. Another problem with this method is that there is no distinction between scatter losses and absorption losses. The value of α measured in this way includes contributions from both bulk and surface scatter.

Example: Calculating the absorption coefficient from a single sample. A 1.59-cm-thick sample of zinc sulfide exhibits a transmittance of 66.5% at a wavelength of 9.0 μm, where the refractive index is 2.211. The single-surface reflectance, R, is calculated with Eq. (1-11):

$$R = \left(\frac{1 - 2.211}{1 + 2.211}\right)^2 = 0.1422\ .$$

Now we will use Eq. (1-20) to find α:

$$\alpha = \frac{1}{1.59\text{ cm}} \ln\left(\frac{(1 - 0.1422)^2}{2(0.665)} + \sqrt{0.1422^2 + \frac{(1 - 0.1422)^4}{4(0.665)^2}}\right) = 0.074\text{ cm}^{-1}\ .$$

This value of α includes contributions from both bulk and surface scatter.

A better, but still simple, way to find α uses two samples of different thickness (Fig. 1.20).[18] With the assumptions that reflection and surface scatter are the same from both samples, then the only difference in external transmittance is due to absorption in the bulk.* In this case the absorption coefficient is given by

$$\alpha = \frac{\ln(t_1/t_2)}{b_2 - b_1} \tag{1-21}$$

---

*To help guarantee consistent surfaces, a single prism may be used instead of two samples. Transmission is measured through thick and thin regions of the prism.

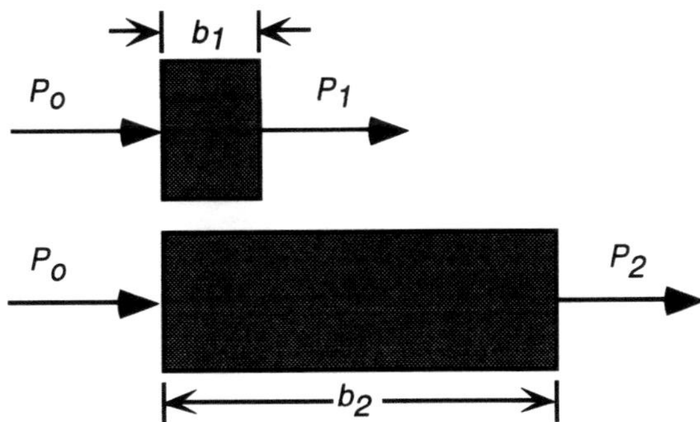

Fig. 1.20. Use of two equivalent samples of different thickness to measure the absorption coefficient.

where t_1 is the transmittance through sample 1 (= P_1/P_o), t_2 is the transmittance of sample 2, and b_1 and b_2 are the two thicknesses. If the surface scatter from both samples is the same, it will cancel in this measurement. However, bulk scatter contributes to the difference between the two transmittances. Therefore the value of α determined with Eq. (1-21) includes a contribution from bulk scatter, but not surface scatter.

1.7.2 Laser calorimetry

A sensitive way to measure low absorption coefficients of transparent materials is by *laser calorimetry*.[23] In this method, a laser is transmitted through the thermally insulated material under study and its temperature is measured as a function of time. From the rise in temperature, we can calculate how much laser power was absorbed, and, hence, the absorption coefficient. This method is useful for values of α as low as 10^{-5} cm^{-1}. Bulk scatter decreases the accuracy of laser calorimetry because it increases the effective pathlength of light through the specimen.

Figure 1.21 shows a laser calorimetry experiment and typical results.[24] Two crystals of an optical window material were suspended on thin nylon fibers in a vacuum chamber and one was irradiated with a laser whose power was carefully stabilized. The temperatures of both crystals were monitored by low-mass thermistors labeled 1 - 4. The responses of thermistors 1 and 2 were essentially identical, and the responses of thermistors 3 and 4 were also nearly identical. The graph displays the temperature difference between the sample crystal and the reference crystal.

The laser was turned on at time 0 and left on until the time t_s (~2100 s) on the graph. The temperature of the sample rose during irradiation, but in a nonlinear manner. The hotter the sample became, the faster it lost heat by radiation (and some conduction) to the surroundings. After time t_s, the sample cooled off.

Analysis of the heating and cooling curves in Fig. 1.21 proceeds as follows.[24] If the absorption coefficient is α and the pathlength is b, the power absorbed from the laser is

$$\text{Power absorbed} = aP_o = \frac{(1 - R)(1 - e^{-\alpha b})}{(1 - Re^{-\alpha b})} P_o \tag{1-22}$$

where a is the absorptance — the fraction of incident laser power, P_o, absorbed by the sample. The one-surface reflection coefficient, R, was given by Eq. (1-11).

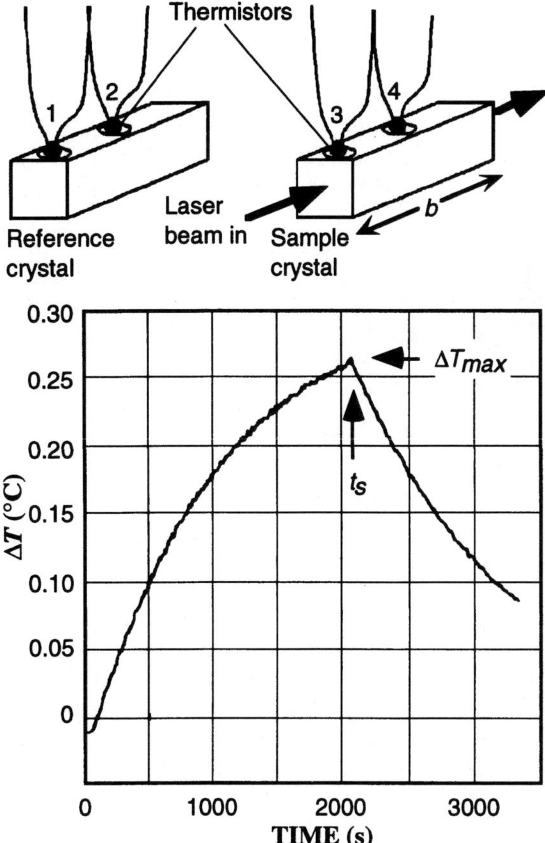

Fig. 1.21. Laser calorimetry experiment.[24]

If the sample were not losing heat, the rate at which its temperature, T, would rise would be

$$\text{Heating rate} = \frac{dT}{dt} = \frac{aP_o}{mC} \tag{1-23}$$

where t is time, m is the mass of the sample and C is the heat capacity of the sample (units = J/(g·K)). The numerator on the right side of Eq. (1-23) is the power absorbed by the sample. The denominator is the energy required to raise the sample temperature by 1 degree. In the absence of heating, the rate at which the sample cools is proportional to the difference in temperature between the sample and its surroundings (ΔT).

$$\text{Cooling rate} = \frac{dT}{dt} = -\frac{\rho}{mC}\,\Delta T \tag{1-24}$$

where ρ is an empirical constant that accounts for radiative and conductive heat losses. In Fig. 1.21, ΔT is the temperature difference between the sample and reference crystals.

The curve in Fig. 1.21 is described by the sum of Eqns. (1-23) and (1-24), which can be integrated to obtain

$$\text{Combined heating and cooling: } \Delta T = \frac{aP_o}{\rho}(1 - e^{-\rho t/mC}) . \tag{1-25}$$

To analyze the curve in Fig. 1.21, we first find the constant ρ by a least-squares fit to the cooling portion after time t_s. The integrated form of Eq. (1-24) which describes the cooling curve is $\Delta T = \Delta T_{max}\, e^{-\rho(t-t_s)/mC}$. Once the constant ρ is known, the rising portion of the curve is fit to Eq. (1-25) to find the absorptance, a. From the absorptance, we can find the absorption coefficient, α, by using Eq. (1-22).

Table 1.4. Absorption coefficient at 10.6 μm of one diamond specimen (0.75 mm thick) measured by laser calorimetry in different laboratories[25]

Lab 1	0.07 ± ?
Lab 2	0.11 ± 0.03
Lab 3	0.23 ± 0.04

Accurate measurements of absorption coefficients require extreme attention to detail. Caution is indicated by the example in Table 1.4 in which one optical sample was sent to three different laboratories for measurements. Reported absorption coefficients vary by a factor of 3 and do not agree within the estimated uncertainties of the measurements.

By studying samples that have different thicknesses, but are otherwise identical, it is possible to decompose the measured absorption coefficient into bulk and surface components, as in Fig. 1.2. For example,[26] it was found that the surface absorption coefficient of zinc selenide at 10.6 μm was constant near 6×10^{-4} cm^{-1} in the temperature range 20-140°C. Between 140 and 200°C the surface absorption coefficient decreased by 50%, perhaps as a surface impurity baked off. Above 250°C, the surface absorption coefficient increased as a new surface species was created by heating.

1.8 Emissivity

In Section 0.2 we saw that a *blackbody* is a perfect absorber and emitter of light. The *emissivity* of a material tells us the fraction of the radiation of a blackbody that is emitted:*

$$\text{Emissivity} = \varepsilon = \frac{\text{radiant power emitted by material}}{\text{radiant power emitted by blackbody}} . \tag{1-26}$$

Emissivity is a number that varies between 0 (for a transparent material) and 1 (for a black material with zero reflection). It varies with wavelength and temperature. In general, *a good absorber is a good emitter.*

Figure 1.22 gives examples of emission spectra.[28,29] Notice the mirror image relationship between transmittance and emittance. When the transmittance is high, the emittance is low, and *vice versa.* In its transparent region, an optical window is a low emitter. In its opaque region, it is a strong emitter.

---

* The terms "emissivity" and "emittance" are used interchangeably in the literature. One possible distinction is that "emissivity" is an ideal property of a pure, perfect material and "emittance" is a property of an actual specimen of material (which is never pure), including the radiation derived from intentional or unintentional surface coatings (such as naturally occurring oxide coatings on metals).[27] We will tend to use the word "emittance" for real specimens of window materials.

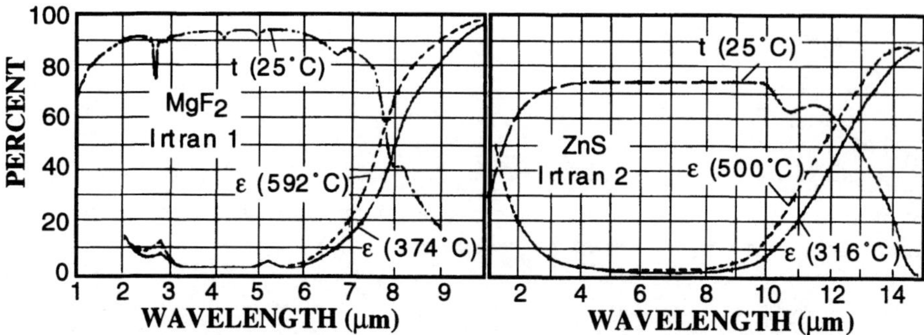

Fig. 1.22. Transmittance and emittance of hot pressed MgF_2 (1.0 mm thick) and ZnS (3.8 mm thick).[28] Reference 29 shows emittance spectra of other common infrared window materials. The following emittance values were reported[45] for 0.305-cm-thick Bausch & Lomb hot pressed MgF_2 averaged over the wavelength range 3 - 5 μm:

Temperature (°C)	21	201	356	542
Average emittance (±0.005)	0.007	0.017	0.017	0.022

For a window material with a single-surface reflection, R, given by Eq. (1-19), the emissivity normal to the surface is given by[30]

$$\varepsilon = \frac{(1 - R)(1 - e^{-\alpha b})}{1 - Re^{-\alpha b}} \tag{1-27}$$

where α is the absorption coefficient and b is the sample thickness.* For fairly transparent windows with low emissivity, it is usually true that $\alpha b \ll 1$. In this case, Eq. (1-27) can be simplified (using the approximation $e^{-x} \approx 1 - x$) to the form

Estimating emittance of a window: $\varepsilon \approx \alpha b$ (for $\alpha b \ll 1$). (1-28)

Example: Estimating emittance. To estimate the emittance normal to the surface of a window with a thickness of $b = 0.20$ cm and an absorption coefficient of $\alpha = 0.10$ cm^{-1}, we simply plug these values into Eq. (1-28):

$\varepsilon \approx \alpha b = (0.10 \text{ cm}^{-1})(0.20 \text{ cm}) = 0.020 = 2.0\%$ of emittance from a blackbody

---

Figure 1.23 shows experimental and calculated emittance of four midwave infrared materials.[32,33] Calculations are based on a multiphonon model[14,15] available in software called *OPTIMATR*®.[34] The software predicts the refractive index and absorption coefficient versus temperature, and emissivity is calculated with Eq. (1-27).

**Hemispheric emissivity* is the radiant emission into a hemisphere adjacent to a point on a surface. Calculation of hemispheric emissivity for a transparent body is rather more complicated than Eq. (1-27).[31]

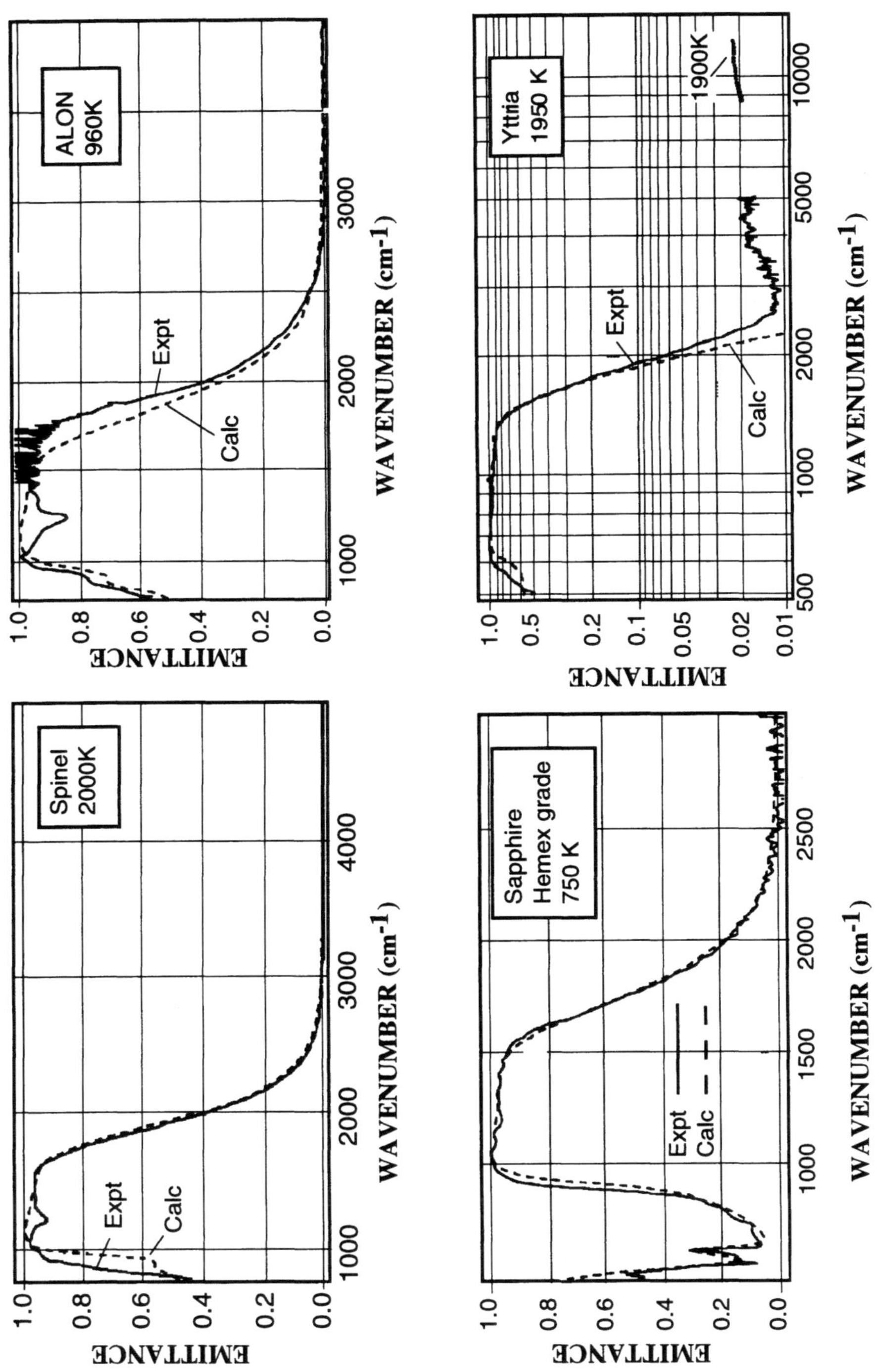

Fig. 1.23. Experimental and calculated emittance of spinel, ALON, sapphire and yttria normal to the surface of 1.0-mm-thick specimens.[32,33] Calculations are based on a multiphonon model[14,15] embodied in commercial software called *OPTIMATR*®.[34]

Mathematical models are unreliable for the window region of any material, where emittance and absorbance are governed by extrinsic impurities and defects which are not in the model. In Fig. 1.23, yttria exhibits a broad band with ~0.02% emittance at 4000 to 10 000 cm^{-1} which does not arise from pure, perfect yttria. We cannot use a model based on intrinsic properties to predict the behavior of a real material in the window region. Models are useful up to — but not beyond — the edges where absorption or emission decrease to low levels.

1.9 Effect of temperature on absorption and emission

The absorption edge of an infrared window material shifts to shorter wavelength as temperature increases (Fig. 1.24).[35-37] What really happens is that the weak absorptions present in this region gain intensity. When measuring transmission at elevated temperature, it is necessary to subtract the emission from the sample, which makes the apparent transmission greater than the actual transmission.[38]

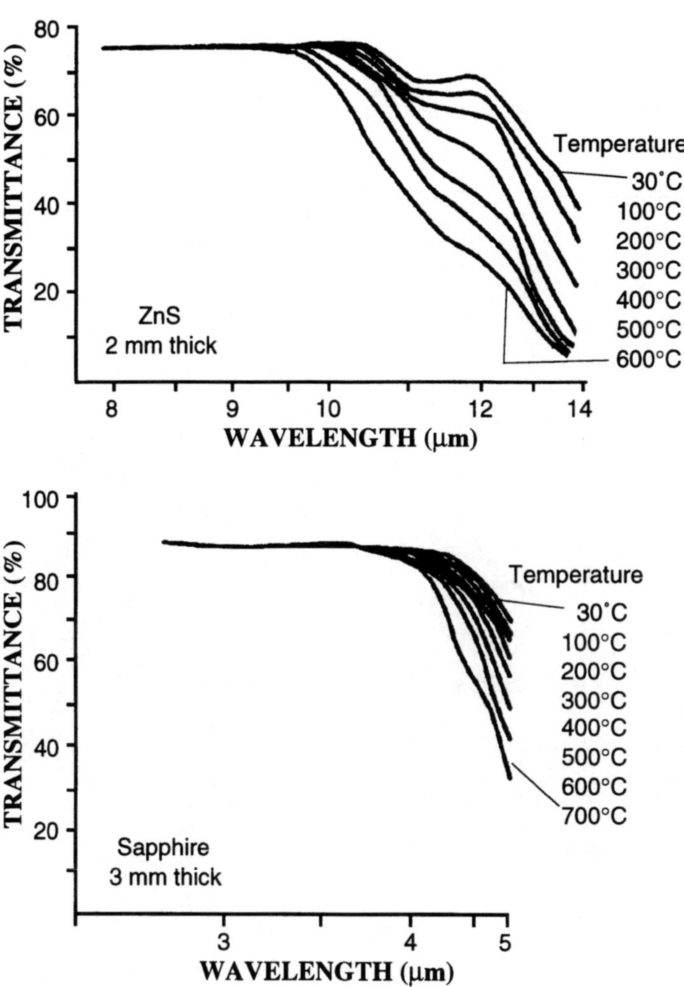

Fig. 1.24. Temperature dependence of transmission of zinc sulfide and sapphire.[36]

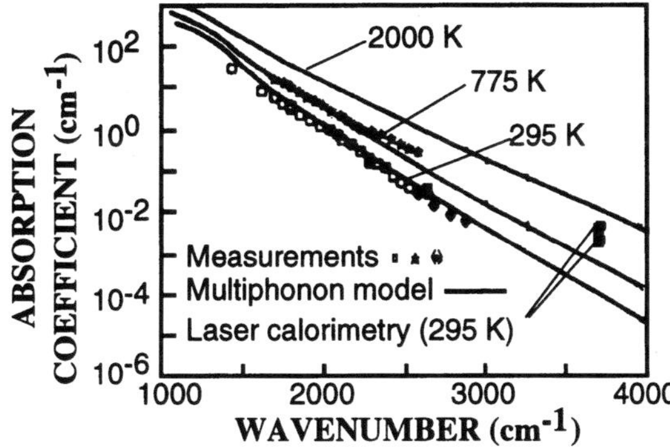

Fig. 1.25. Sapphire absorption coefficient as a function of temperature.[39]

Figure 1.25 is another way of plotting the sapphire absorption of Fig. 1.24. Data points in Fig. 1.25 are measured at 295 or 775 K and the solid lines are fit to model equations.[39] The two data points represented by squares at the right side of Fig. 1.25 were measured by laser calorimetry and are thought to be accurate, even though they lie far above the theoretical line for 295 K. In the window region where the squares were measured, absorption is dominated by impurities and/or defects in the material. If the sapphire could be made more perfect, the measured absorption ought to decrease and approach the theoretical line.

When the absorbance of a material increases, so does its emittance. Figure 1.22 showed the actual increase of emittance for magnesium fluoride and zinc sulfide, while Fig. 1.26 shows the predicted behavior of spinel, ALON, sapphire and yttria.

Recall from Eq. (0-4) that radiant emission from a blackbody is proportional to the fourth power of absolute temperature. If the temperature triples from 300 K to 900 K, emission increases by a factor of $3^4 = 81$ (and the peak emission moves to shorter wavelength). A sapphire window at 1000°C (1273 K) emits 7.4 times as much radiant energy as the same window at 500°C (773 K) if the emissivity is the same. Near the absorption edge the emissivity increases with temperature, so the emission is even more than 7.4 times greater at a given wavelength.

At elevated temperature, it is emission, not absorption, that limits the performance of an optical window.

There are two severely detrimental effects of increased emission from a window or dome. One is that radiation emitted from the window can be so great as to obscure radiation from the object being observed. The signal-to-noise ratio becomes too small to detect signal. This problem is discussed in Section 2.4. Another possible effect is that if emission from the window or dome is great enough, the detector may be saturated by photons and no longer respond to signals.

How hot do infrared windows and domes become in military systems? Figure 1.27 shows the stagnation temperature at the front tip of an object traveling through the atmosphere. This is the steady state temperature that is reached as a result of frictional heating of the object by air. This temperature is typically attained by the front of the

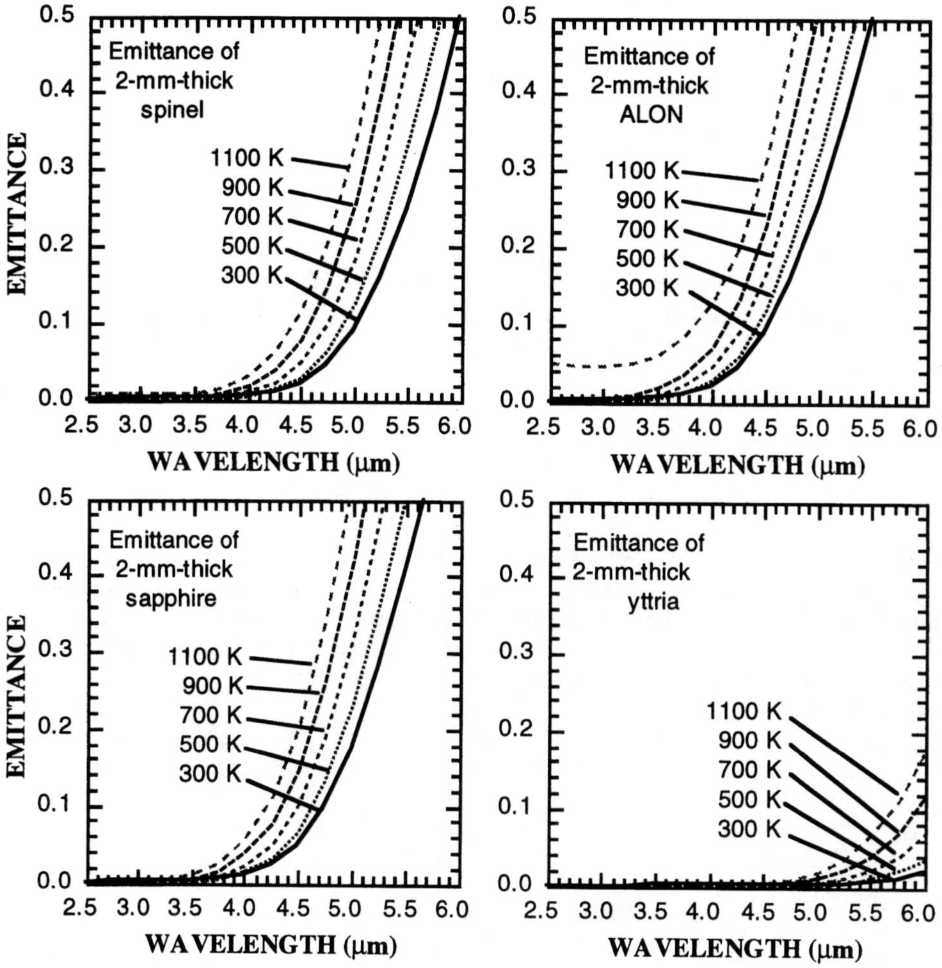

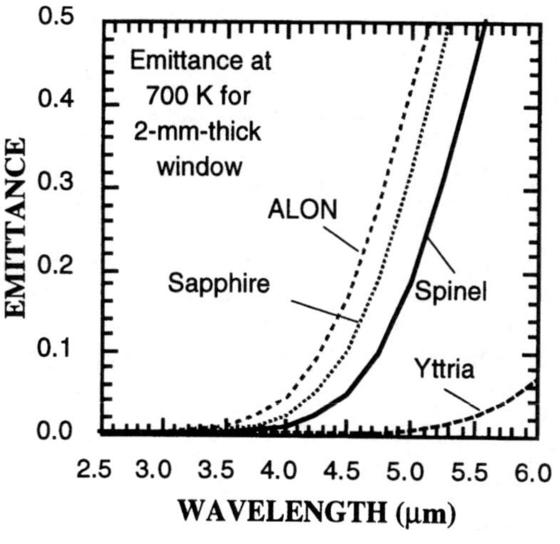

Fig. 1.26. Temperature dependence of the emittance of spinel, ALON, sapphire and yttria calculated with absorption coefficients from the program *OPTIMATR*.[34] Table C.3 in Appendix C provides representative absorption coefficients from *OPTIMATR*. The bottom graph compares all four materials. Transparent yttria was developed in the 1980's because of its low midwave emissivity. Unfortunately, the thermal shock resistance of yttria is less than that of sapphire, so yttria has not replaced sapphire in applications involving rapid heating of the window.

Table 1.5. Stagnation temperature as a function of speed at 10 km altitude*

Mach number	Stagnation temperature (°C)	Mach number	Stagnation temperature (°C)
1	-5	4	660
2	130	5	1060
3	350	6	1560

*Stagnation temp. (K) = (ambient temp.) × (1+0.2 M^2), where M is Mach number and ambient temperature is 288 K at sea level, 223 K at 10 km, 216 K at 20 km and 231 K at 30 km.

dome on a missile (Fig. 0.1) at low altitude* after a few seconds of flight. Table 1.5 shows that domes reach significantly high temperatures for speeds above Mach 3. Emission of light from the dome of a missile traveling at Mach 4 or higher is a very important consideration.

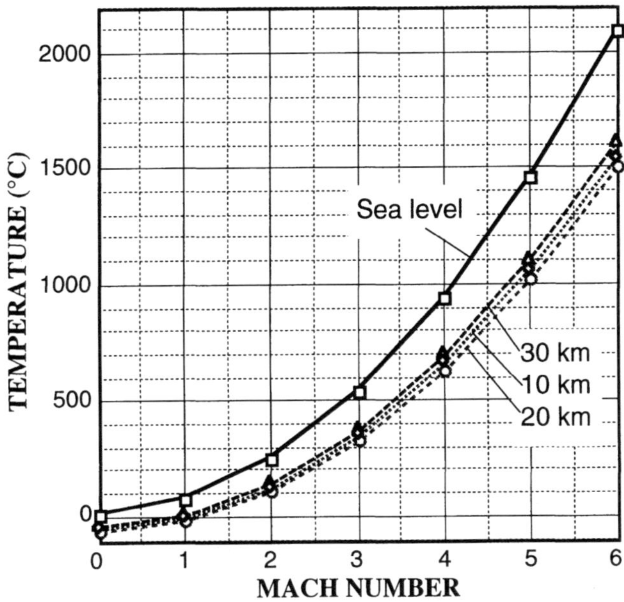

Fig. 1.27. Stagnation temperature as a function of Mach number and altitude.

*At high altitude a dome takes longer to reach the stagnation temperature, and may not reach that temperature if the heat transfer coefficient is too small. The heat transfer coefficient is the rate of transfer of energy (W/[m^2·K]) from the air to the dome. Heat transfer becomes smaller as the air density decreases at increasing altitude.

1.10 Free carrier absorption in semiconductors

Semiconductors have a relatively low energy *band gap* separating the *valence band* that is filled with electrons and the *conduction band* that is empty, or nearly so (Fig. 1.28). *Interband absorption* occurs when light whose energy is greater than or equal to the band gap promotes an electron from the valence band to the conduction band. If the band gap is small enough and the temperature is high enough, there will be some thermal population of electrons in the conduction band and holes in the valence band. Electrons in the conduction band can absorb virtually any small increment of energy to be promoted to a higher state within the same band. This is called *free carrier absorption*, or *intraband absorption.*

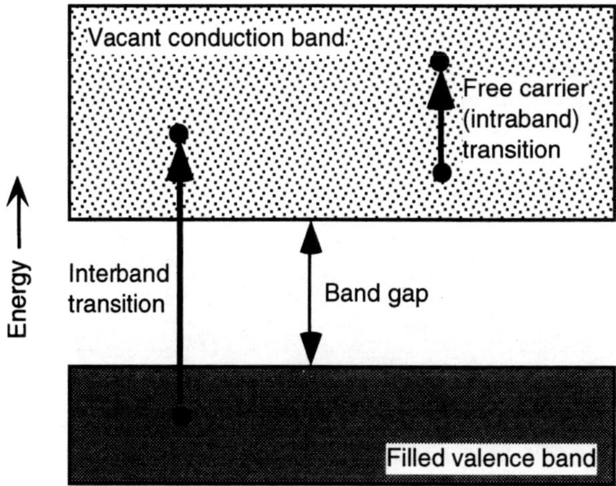

Fig. 1.28. Energy bands and types of transitions for a semiconductor.

Table 1.6 shows that germanium has a relatively small band gap and will, therefore, have a relatively high population of free carriers. Figure 1.29 shows that as the band gap increases in different materials, free carrier concentration decreases.[40] As the temperature increases, the free carrier concentration increases.

Table 1.6. Band gaps and approximate upper temperature limits for transmission

Material	Band gap (electron volts, eV)	Estimated upper useful temperature (°C)*
Ge	0.7	~100
Si	1.1	~260
GaAs	1.4	~460
GaP	2.2	~640
ZnS	3.6	-

*The rough estimate of upper operating temperatures is based on drawing a horizontal line near 10^{15} carriers/cm^3 in Fig. 1.29. Doped semiconductors with higher carrier concentrations will have lower useful operating temperatures.

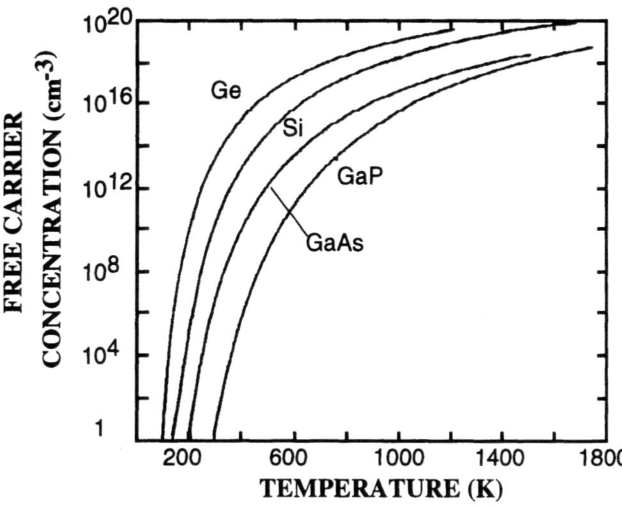

Fig. 1.29. Free carrier concentration as a function of temperature for optical semiconductor materials.[40]

Since the free carrier absorption coefficient is proportional to the concentration of free carriers, absorption by a semiconductor increases with temperature. The band gap of germanium is so low that this increase in absorption is quite significant by 150°C in Fig. 1.30. For semiconductors with a wider band gap, increased absorption requires higher temperature. In the case of silicon in Fig. 1.30, the absorption is significant at 300°C.

Gallium arsenide and gallium phosphide in Table 1.6 have even higher band gaps than silicon and therefore can be used up to higher temperatures without significant absorption. Figures 1.31 and 1.32 show the behavior of these materials.

We see in Fig. 1.30 that free carrier absorption increases at longer wavelength (λ). In theory, the free carrier absorption coefficient increases as the square of λ. Doubling the wavelength should increase the free carrier absorption coefficient by a factor of 4. Figure 1.33 shows the calculated behavior of the absorption coefficient of silicon as a function of temperature at two different wavelengths.

Electrons in the conduction band in Fig. 1.28 can absorb infrared radiation over a wide range of frequencies. Heating a window promotes electrons from the valence band to the conduction band and decreases infrared transmission. Another way to populate the conduction band is illustrated in Fig. 1.34 in which germanium was irradiated with near infrared radiation (0.808 μm wavelength) having enough energy (1.41 eV) to promote electrons from the valence band to the conduction band.[44] The infrared response can be turned on and off rapidly by external irradiation.

1.11 What makes a window midwave or long wave?

The lower part of Fig. 1.17 showed the behavior of the absorption coefficient of optical window materials. Infrared radiation is absorbed because it stimulates vibrations of atoms in the crystal. Ultraviolet radiation is absorbed because it promotes electrons to higher energy levels. The transparent window region is located between the two regions of strong absorption. Figure 1.24 highlighted the infrared absorption edge between the transparent window region and opaque lattice vibration region in the spectra of two materials. For zinc sulfide, the cutoff is near 11 μm, whereas for sapphire the cutoff is near 5 μm. What determines the position of this cutoff edge?

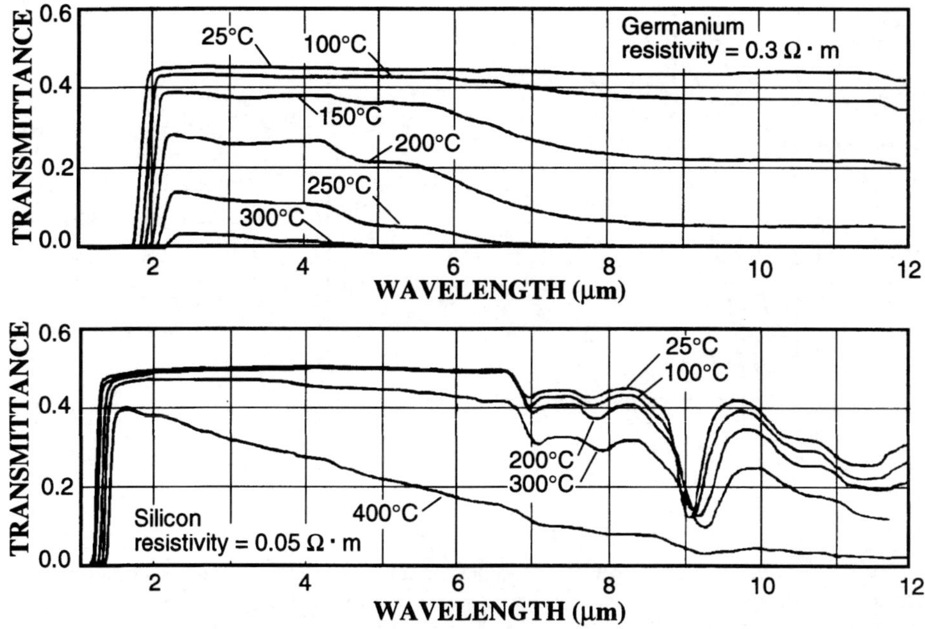

Fig. 1.30. Infrared transmission of 2.80-mm-thick single-crystal *p*-type germanium (resistivity = 0.3 ohm·m) and *n*-type silicon (resistivity = 0.05 ohm·m).[38] The prominent absorption band in most specimens of silicon near 9 μm is due to an oxygen impurity that can be nearly eliminated if the silicon is grown by techniques such as the float zone method.[41,42]

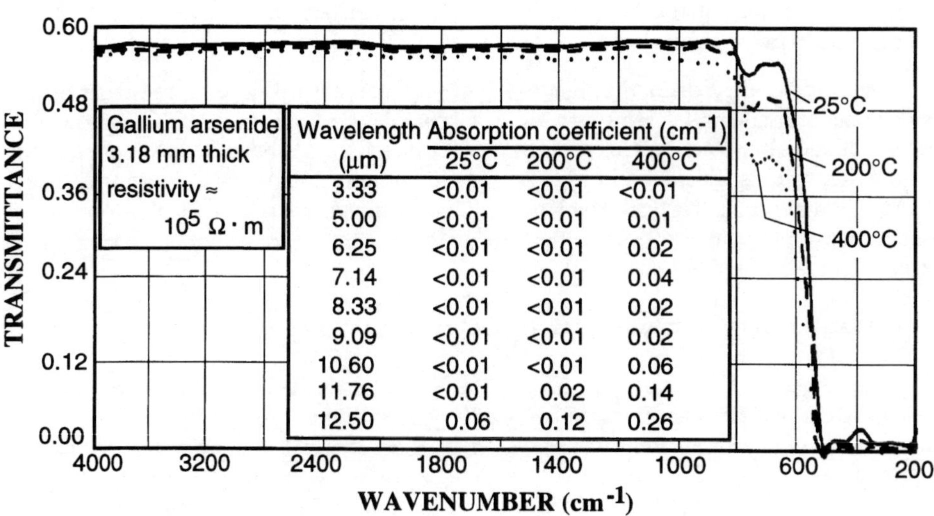

Wavelength (μm)	Absorption coefficient (cm^{-1}) 25°C	200°C	400°C
3.33	<0.01	<0.01	<0.01
5.00	<0.01	<0.01	0.01
6.25	<0.01	<0.01	0.02
7.14	<0.01	<0.01	0.04
8.33	<0.01	<0.01	0.02
9.09	<0.01	<0.01	0.02
10.60	<0.01	<0.01	0.06
11.76	<0.01	0.02	0.14
12.50	0.06	0.12	0.26

Fig. 1.31. Temperature dependence of the infrared transmission of 3.18-mm-thick polycrystalline gallium arsenide with a resistivity of ~10^5 ohm·m.[43]

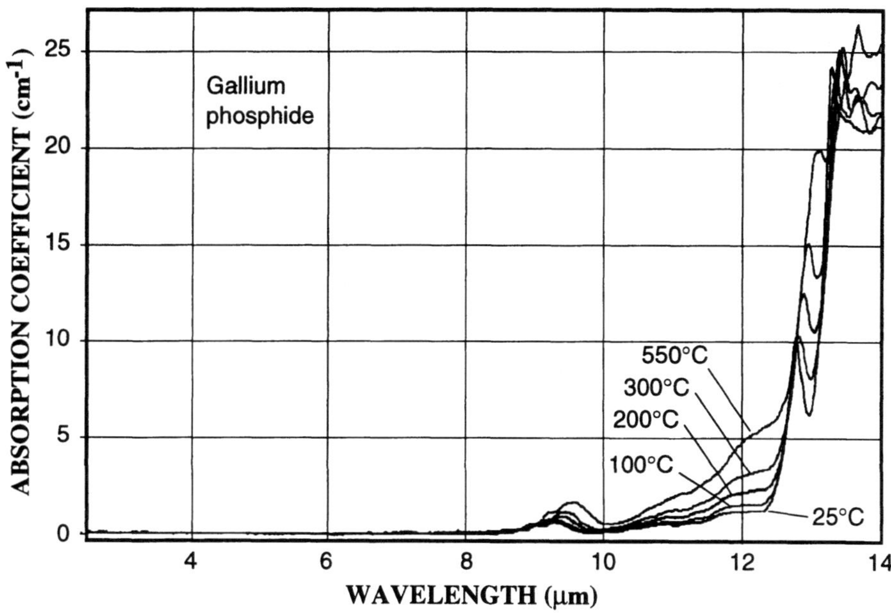

Fig. 1.32. Temperature dependence of the infrared absorption coefficient of gallium phosphide.[13] The absorption coefficient in the 3-5 μm region at temperatures up to 400°C is reported to be ≤0.01 cm^{-1}.[43]

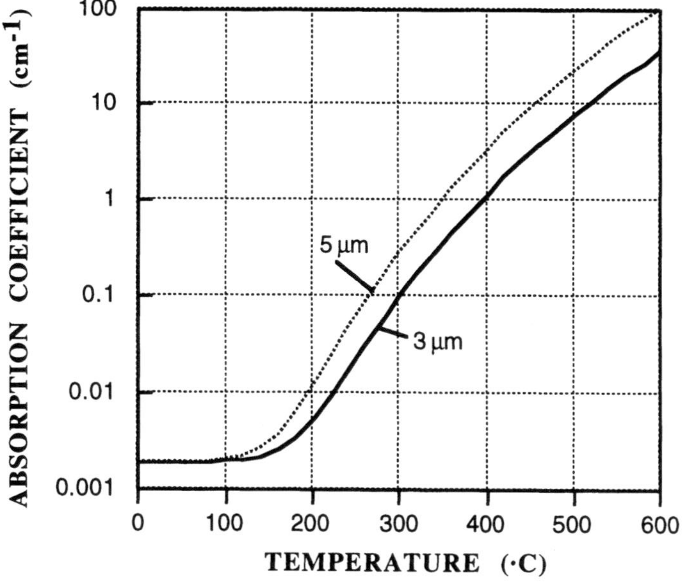

Fig. 1.33. Predicted behavior of the absorption coefficient of pure silicon at specific wavelengths. (Personal communication from C. Klein, C.A.K. Analytics.)

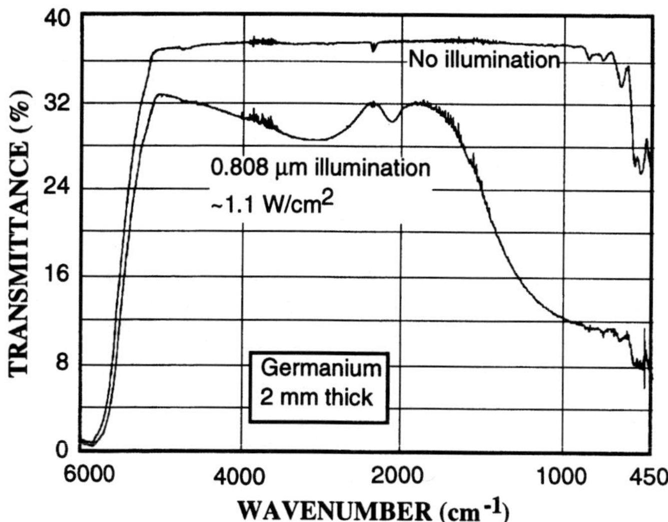

Fig. 1.34. Illuminating germanium with light having enough energy to promote electrons from the valence band to the conduction band increases infrared absorption.[44]

Figure 1.35 shows a simple model for the vibration of a diatomic molecule in which two atoms of masses m_1 and m_2 are connected by a spring of *force constant* k_f. When the bond (the spring) is stretched from its equilibrium length, r_e, to the length $r = r_e + q$, the restoring force is proportional to the displacement, q:

$$\text{Restoring force} = -k_f q \qquad (k_f = \text{force constant}). \tag{1-29}$$

The stiffer (stronger) the chemical bond, the greater is the force constant. The vibrational energy levels turn out to be

$$\text{Vibrational energy levels} = (\mathrm{v} + \frac{1}{2})\, h\omega \tag{1-30}$$

where v is the *vibrational quantum number* (0, 1, 2, 3,...), h is Planck's constant and ω is given by

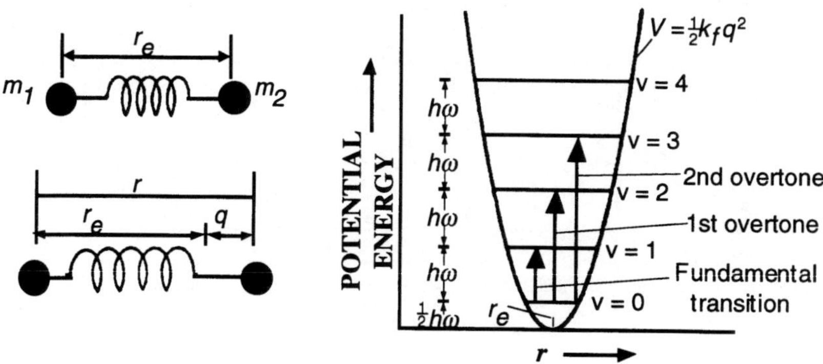

Fig. 1.35. *Left*: Model of vibrating diatomic molecule. *Right*: Potential well for harmonic oscillator.

$$\omega = \frac{1}{2\pi}\sqrt{\frac{k_f}{\mu}} . \quad (1\text{-}31)$$

The quantity μ, called the *reduced mass*, is

$$\text{Reduced mass} = \mu = \frac{m_1 m_2}{m_1 + m_2} . \quad (1\text{-}32)$$

The greater the masses of the atoms, the greater the reduced mass and the smaller the vibrational energies.

On the right side of Fig. 1.35, we see that the lowest vibrational energy level of a diatomic molecule (the *ground state*) has an energy of $^1/_2 h\omega$ above the bottom of the well. That is, even at absolute zero temperature, the molecule is vibrating with the energy $^1/_2 h\omega$. The first excited state lies at an energy $h\omega$ above the ground state, and successive excited states each lie at equally spaced intervals. The lowest infrared energy that the molecule can absorb upon going from v=0 to v=1 is $h\omega$, and this absorption is called the *fundamental transition.* The molecule could also go from v=0 to the v=2 excited state, which would be called the *first overtone*.

Two key points that apply to crystals as well as gaseous diatomic molecules are:

1. Vibrational energies increase as the force constant of the atomic bond increases. That is, *stronger bonds give higher vibrational energies.*
2. Vibrational energies increase as the masses of the atoms in the molecule decrease. *Heavier atoms have lower vibrational energies.*

In vibrations of crystals, whole planes of atoms are displaced from their equilibrium positions. Figure 1.36 shows two types of crystal vibrations designated *longitudinal* and *transverse.* As the longitudinal oscillation propagates from left to right in the diagram, the atoms move to the left or right from their equilibrium positions in a sinusoidal manner. As the transverse oscillation propagates from left to right, the atoms move up or down from their equilibrium positions. As in diatomic molecules, we can associate a force constant with each type of vibration. The vibrational energy increases as the force constant increases and as the masses of the vibrating atoms decrease.

Crystal vibrations are further divided into higher energy *optical modes* and lower energy *acoustic modes*, shown for a transverse vibration in Fig. 1.37. The designations TO, LO, TA and LA, standing for transverse optical, longitudinal optical, transverse acoustic and longitudinal acoustic, respectively, describe crystal vibrational modes. A quantum unit of crystal vibrational energy is called a *phonon.*

Absorption coefficients for fundamental transitions are typically in the range 10^3-10^5 cm^{-1}. That is, infrared window materials are opaque in the 1-phonon energy region where the fundamental vibrational transitions occur. Spectroscopic investigation of this region usually relies upon reflection, rather than transmission, to identify phonon energies.[46] Figure 1.38 shows that absorption coefficients for zinc sulfide in the 2-phonon region are roughly in the range 10^1-10^3 cm^{-1}. This is the region where overtones (*e.g.*, 2TO and 2LO) of the fundamental vibrations are observed, as well as *combination bands*, such as TO + LO, TO + LA, *etc.* Combination bands represent simultaneous excitation of two different crystal vibrations by a single photon of infrared radiation. The energy of the

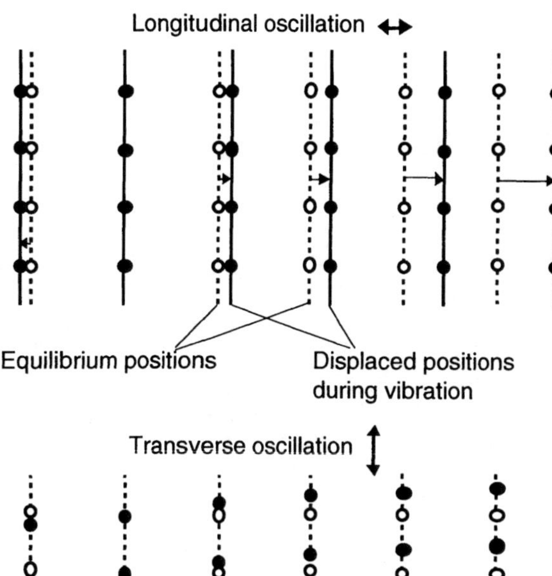

Fig. 1.36. Longitudinal and transverse oscillations of planes of atoms in a crystal.

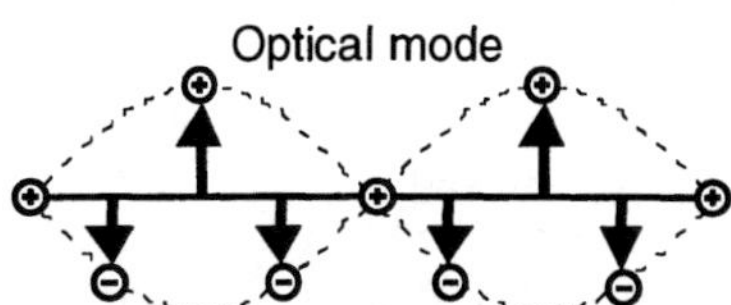

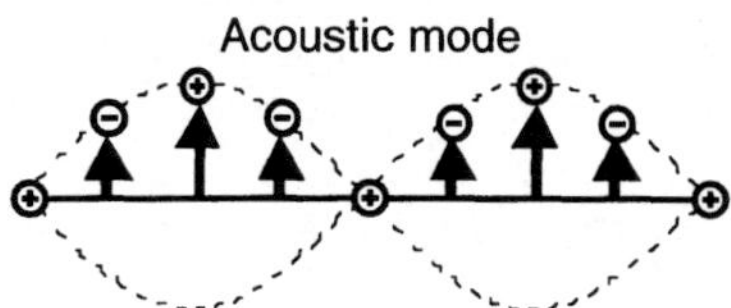

Fig. 1.37. Transverse optical and acoustic vibrations, in which adjacent cations and anions in a lattice are displaced out of phase (optical) or in phase (acoustic) with each other. The vibrational wave is propagating from left to right.

photon is the sum of the energies of the two phonons involved. The right side of Fig. 1.38 shows how all of the weak features, including every shoulder, can be assigned in terms of overtones and combination bands of four fundamental phonon frequencies.[47]

Infrared window materials with a thickness of several millimeters normally exhibit their transmission cutoff in the 3-phonon region at room temperature. They are quite opaque in the 2-phonon region and generally transparent in the 4-phonon region. At elevated temperature, absorption in the 4- and 5-phonon regions can be significant.[16] Figures 1.39 and 1.40 give two more examples of window materials whose transmission cutoffs are in the 3-phonon region.

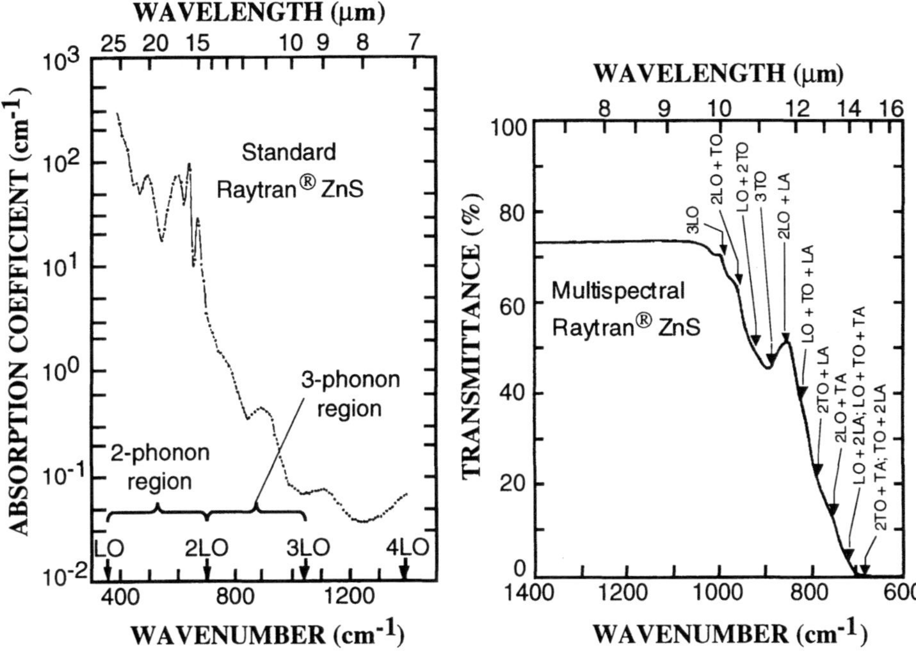

Fig. 1.38. *Left*: Absorption features of zinc sulfide in the 2- and 3-phonon regions.[48,8] Notice that the ordinate is logarithmic. *Right*: Assignment of weak features in the 3-phonon region of zinc sulfide.

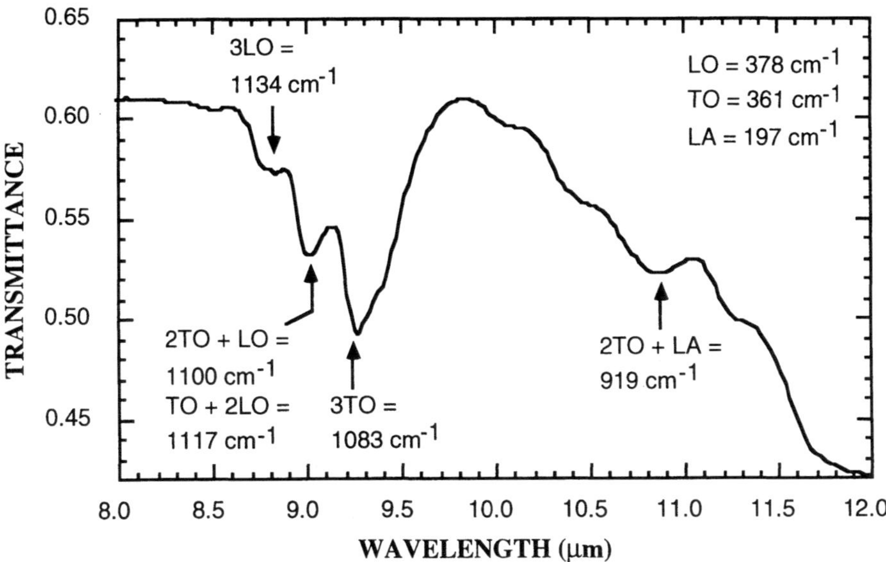

Fig. 1.39. Fine structure in the 3-phonon region of gallium phosphide (3.0-mm-thick specimen from Raytheon Systems Co.). Figure 1.18e shows that there are no significant absorptions below 8.5 μm. Assignments are from Klein[49] based on literature values of the fundamental lattice vibrations.[50]

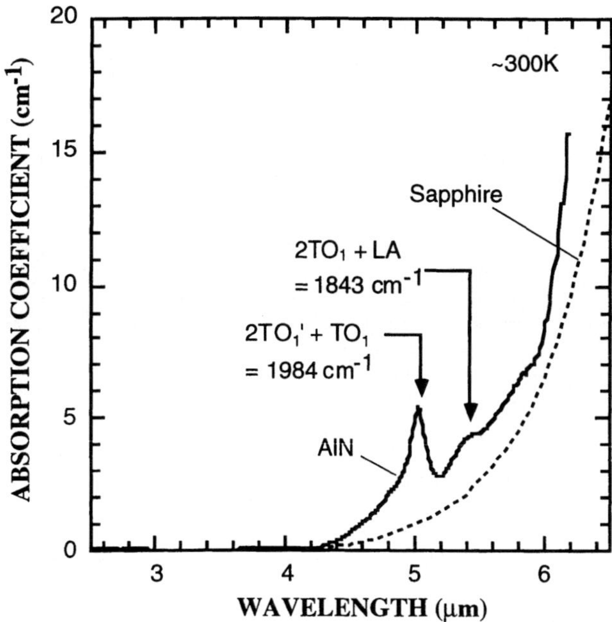

Fig. 1.40. Comparison of the infrared cutoff regions of aluminum nitride (AlN) and sapphire (Al_2O_3). AlN data come from a 4.29-mm-thick (0001) crystal from Crystal IS (Latham, NY), courtesy L. Schowalter and G. A. Slack. Transmittance was measured at Rensselaer Polytechnic Institute by I. Bhat. Assignments of the 3-phonon peaks for AlN are based on the following fundamental frequencies:[51] LO = 738, TO_1 = 668, TO_1' = 658, TO_2 = 627, LA = 507, TA_1 = 444, TA_2 = 413 cm^{-1}.

Table 1.7. Fundamental vibrational frequencies of infrared window materials*

Midwave materials

Sapphire (*o*-ray)[52]		ALON[52]		Spinel[52]		Yttria[52]	
cm^{-1}	Strength	cm^{-1}	Strength	cm^{-1}	Strength	cm^{-1}	Strength
385	0.3	346	1.0	307	0.4	240	0.16
442	2.7	395	0.4	488	2.48	304	2.45
569	3.0	497	3.0	510	1.07	336	1.75
635	0.3	631	1.55	589	0.07	372	2.70
806	0.004	758	0.19	671	0.70	415	0.043
		922	0.035	808	0.022	461	0.048
						490	0.005
						555	0.094

Long wave materials

Zinc Sulfide[48]		Zinc Selenide[48]	
cm^{-1}	Assignment	cm^{-1}	Assignment
330	LO	222	LO
295	TO	204	TO
193	LA	131	LA
89	TA	65	TA

*For the upper part of the table, cm^{-1} refers to the wavenumber of the vibration and strength refers to the relative intensity of infrared absorption.

We are now in a position to appreciate why different materials have different infrared cutoff frequencies. Figure 1.18a showed the cutoffs for the midwave window materials, sapphire, ALON, spinel and yttria[53] and Fig. 1.18d showed the long wave windows, zinc sulfide and zinc selenide. Table 1.7 lists fundamental vibrational frequencies for each of these materials. Notice that the vibrational frequencies of zinc sulfide and zinc selenide are significantly lower than those of the midwave materials. The 3-phonon cutoffs for zinc sulfide and zinc selenide occur at correspondingly lower energy, or longer wavelength. Among midwave materials, yttria has lower energy vibrations than sapphire, ALON or spinel, so yttria has the longest wavelength (lowest energy) cutoff among these materials in Fig. 1.18a. Zinc selenide has lower energy vibrations than zinc sulfide, so zinc selenide has the longer wavelength cutoff in Fig. 1.18d.

Recall that lower energy vibrations result from weaker chemical bonds and heavier atoms. Both of these factors operate in Table 1.7. Among the midwave (3-5 μm) materials, the yttrium atoms in yttria (Y_2O_3) are heavier than the aluminum and magnesium atoms in sapphire (Al_2O_3), ALON ($9Al_2O_3 \cdot 5AlN$) and spinel ($MgAl_2O_4$). (As we go down lower in the periodic table of the elements, Fig. 1.41, atoms become heavier.) Yttria therefore has the lowest vibrational energies and longest wavelength cutoff. The long wave (8-14 μm) materials have sulfur (S) or selenium (Se) anions instead of oxygen anions. Sulfur and selenium are heavier than oxygen and form weaker bonds. Both factors lower the vibrational frequencies in zinc sulfide and zinc selenide and make them useful long wave window materials. Selenium is heavier than sulfur and forms weaker bonds than sulfur. Therefore zinc selenide has a longer wavelength cutoff than zinc sulfide.

Li	Be				B	C	N	**O**	F	Ne
Na	**Mg**				**Al**	Si	P	**S**	Cl	Ar
K	Ca	Sc	···	**Zn**	Ga	Ge	As	**Se**	Br	Kr
Rb	Sr	**Y**	···	Cd	In	Sn	Sb	Te	I	Xe
Cs	Ba	La	···	Hg	Tl	Pb	Bi	Po	At	Rn

Fig. 1.41. Partial view of the periodic table of the elements. Atoms become heavier and tend to form weaker bonds as we move down the table.

Figure 1.42 shows the transmission windows of numerous optical materials. Those with heavier atoms have longer wavelength cutoffs. In general, oxides (containing the oxygen anion) never transmit in the long wave region because oxygen is too light and forms strong bonds. Its vibrational frequencies are too high for transmission in the long wave region. Sulfur and selenium come below oxygen in the periodic table (Fig. 1.41) and form compounds with long wave transmission. To the left of sulfur in the periodic table is phosphorus (P). To the left of selenium is arsenic (As). Since arsenic is heavier than phosphorus, we predict that gallium arsenide (GaAs) transmits to longer wavelength than gallium phosphide (GaP), and this is indeed the case. Gallium arsenide cuts off near 17 μm in Fig. 1.31, while gallium phosphide cuts off near 12 μm in Fig 1.32.

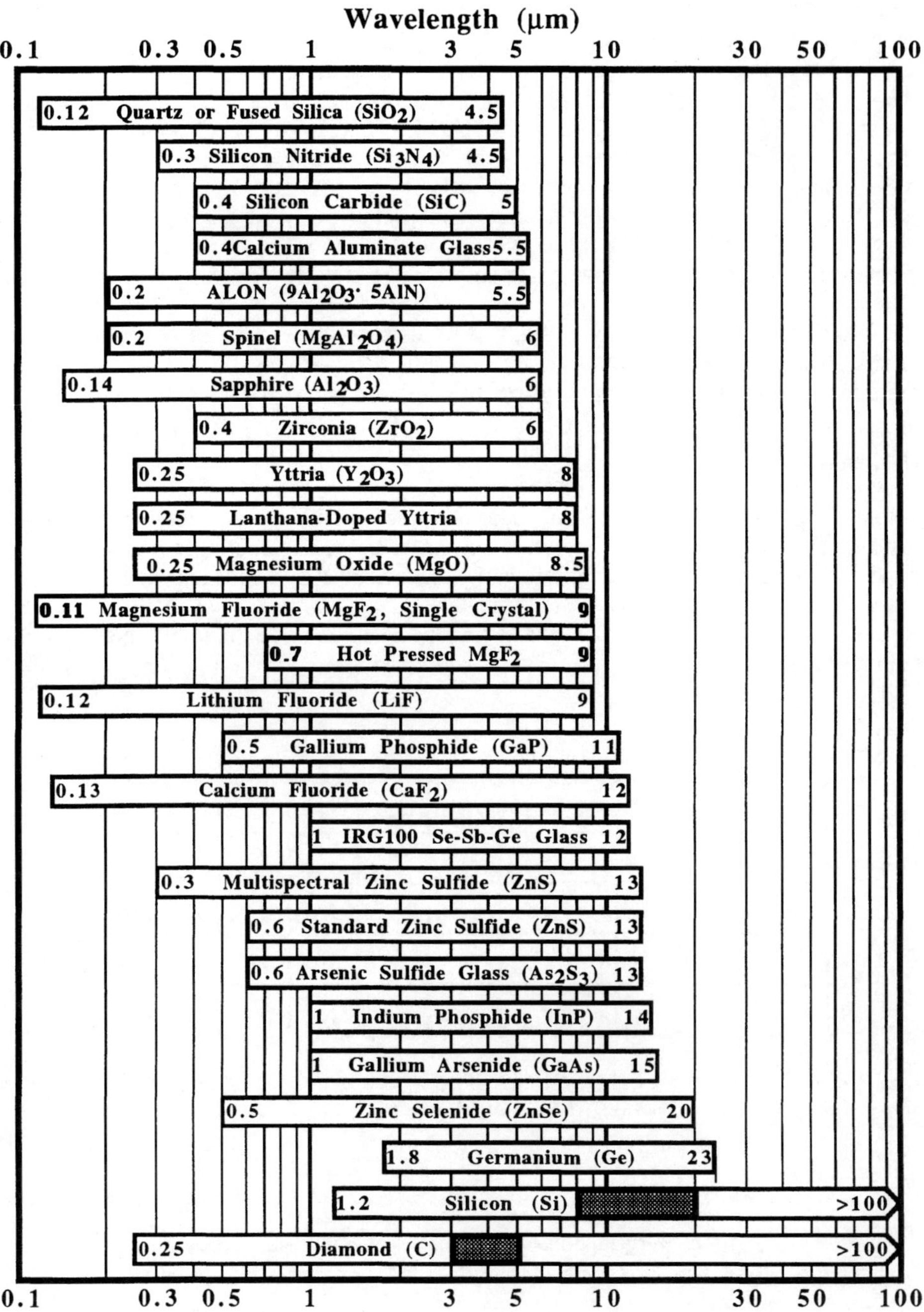

Fig. 1.42. Transmission windows for selected materials. Numbers give cutoff edge, taken as the wavelength (µm) at which a 2 mm thick window has 10% transmittance. Cutoff edges are only approximate, and may vary with the quality of the material. Shaded regions for diamond and silicon have low transmission. (Figure continued on next page.)

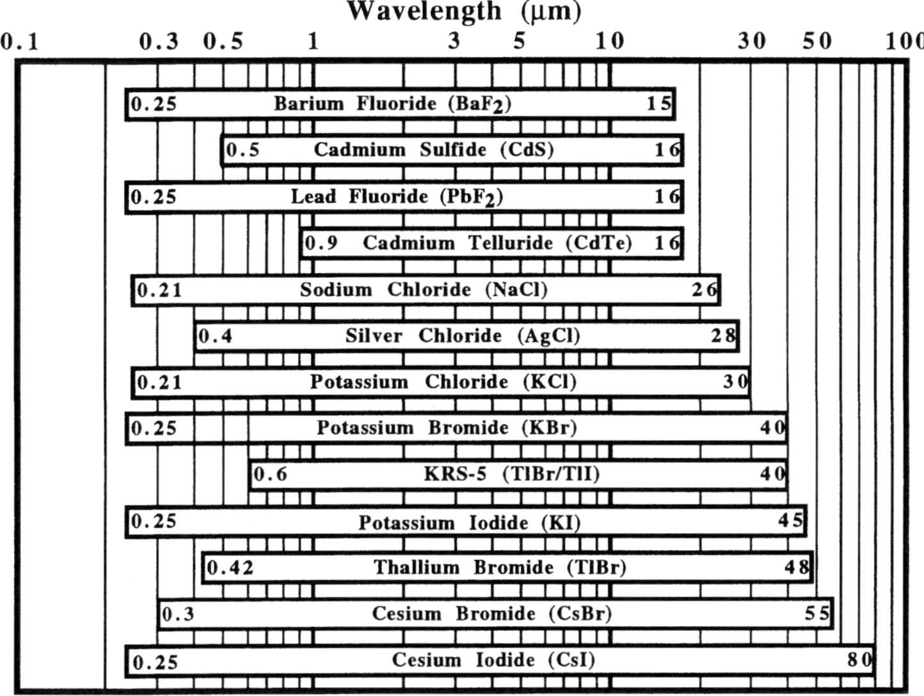

Fig. 1.42 (continued). Transmission windows for selected materials.

If we want long wave transmitting materials, we look for compounds with heavy atoms.[54] Cesium iodide (CsI), for example, transmits all the way to 80 μm. The problem is that chemical bonds within the crystal are weaker for heavier atoms, so materials made of the heavier elements have mechanical properties that are inferior to those of materials made of lighter atoms. The long wave transmitting materials are generally weaker, softer and less tough than midwave transmitting materials. *There is almost always a tradeoff between gaining long wave transmission and losing desirable mechanical properties.*

An exception to this tradeoff rule is diamond, which combines the best possible mechanical performance with excellent long wave transmission (Fig. 1.43). Diamond consists of carbon atoms, each of which forms strong bonds to its four nearest neighbors arranged at the corners of a regular tetrahedron (see Fig. 9.2). The fundamental vibrational frequencies of diamond are high because the atoms are light and the bonds are strong. However, the symmetry of the diamond crystal is such that the vibrating atoms cannot absorb energy from the electromagnetic field in an infrared experiment.* As Fig.

*A requirement for infrared absorption is that the dipole moment of the crystal must change as the atoms vibrate. The high symmetry of diamond is such that the dipole moment does not change and no energy can be absorbed, even though the energy of the electromagnetic radiation matches that of the crystal vibrations.

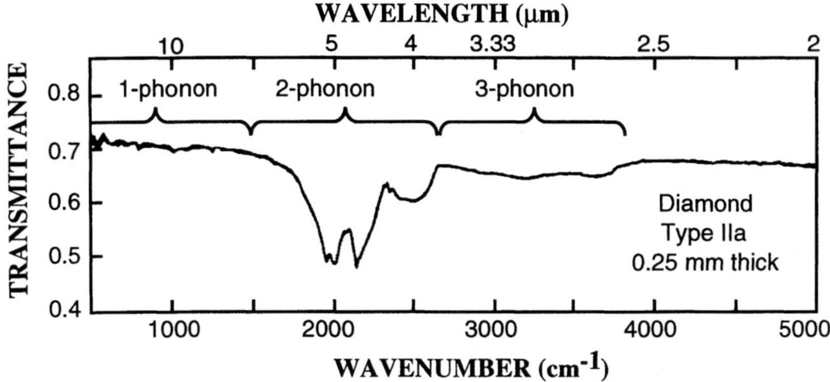

Fig. 1.43. Infrared transmission spectrum of natural gem quality diamond.[55]

1.43 shows, the 2- and 3-phonon transitions are allowed, but they are relatively weak.[55] Diamond does not exhibit the characteristic long wavelength cutoff seen in other optical windows. If the fundamental transition were not forbidden by symmetry, diamond would absorb throughout the infrared region and be useless as an infrared window.

1.12 "Two-color" materials

"Two-color" window materials transmit radiation in two different spectral regions. In this section we consider a 2-color material to be one that is transparent in both the midwave (3-5 µm) and long wave (8-14 µm) infrared regions. Depending on the application, different people might refer to 2-color performance as combining transmission in the ultraviolet, visible, near-infrared, midwave infrared, or long wave infrared regions.

All of the window materials in the second part of Fig. 1.42 (from barium fluoride to cesium iodide) are 2-color materials, but none are generally useful for external exposure to rain, blowing particles, rapid heating, or mechanical stress. Among the commonly available materials for external use, only germanium, silicon, gallium arsenide, gallium phosphide, zinc sulfide and zinc selenide are candidates for 2-color applications.

Transmission spectra of the candidate materials are shown in Fig. 1.44. We now give a brief critique of each candidate:

Germanium has low absorption in both the midwave and long wave windows, but its high index of refraction gives it high reflection and less than 50% transmission. It therefore requires antireflection coatings for most applications. With its small band gap, Ge becomes conductive and absorptive above 100°C (Fig. 1.30). Germanium has only modest resistance to erosion by rain and sand particles and usually requires a durable, protective coating when it is used as an infrared window in aircraft.

Silicon is a good midwave material with marginal capabilities in the long wave region. The strongest absorption band near 9 µm in Fig. 1.44 is due to an oxygen impurity that is present in most specimens of Si, but can be almost eliminated by very careful crystal growth.[41,42] Other bands in the long wave infrared region are intrinsic to the material. Like Ge, Si is highly reflective and requires antireflection coatings. It becomes absorptive from free carriers above ~250°C. Silicon has modest rain erosion resistance and poor sand erosion resistance.

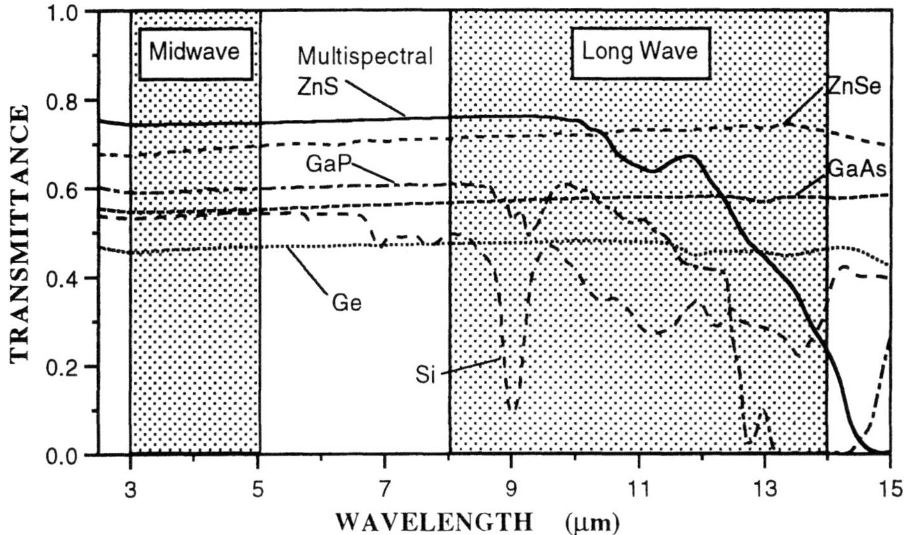

Fig. 1.44. Candidate 2-color infrared window materials for external use. Specimens are the same as in Fig. 1.18. The thicknesses are: ZnS, 5.2 mm; ZnSe, 7.1 mm; GaP, 3.0 mm; GaAs, 0.65 mm; Si, 2.8 mm; and Ge, 2.8 mm. A thicker sample of GaAs is shown in Fig. 1.31.

Gallium arsenide has excellent optical properties in both the midwave and long wave windows, but its upper operating temperature is limited to ~460°C by free carrier absorption. Its high refractive index demands good antireflection coatings and its poor erosion resistance requires good protective coatings for external use.

Gallium phosphide is an excellent midwave material and has 2-color capability at room temperature because emittance is not significant. At elevated temperature, emission from weak absorption bands obliterates the long wave region. At best, GaP has a small usable window from about 8 - 8.5 µm at elevated temperature. The upper operating temperature of GaP is estimated at ~640°C, limited by free carrier absorption. Gallium phosphide has outstanding thermal shock resistance — better than sapphire — for rapid heating environments. Gallium phosphide, like GaAs, Si, and Ge, has a large change of refractive index with temperature (*dn*/*dT* in Table 1.8), which causes optical distortions when there is a temperature gradient across the window. The erosion resistance of GaP is similar to that of ZnS, which is only modest. Gallium phosphide requires antireflection coatings and good protective coatings for external use.

Zinc sulfide comes in at least three varieties called *standard grade*, *multispectral* (or Cleartran®* or Waterclear®) and *elemental*®. All three types, which are discussed in Section 5.3.1, have good transmittance from 8 - 10 µm, beyond which absorption

*"Cleartran" is a trademark of Morton CVD Materials, Woburn, MA. "Waterclear" is a trademark of Sassoon Advanced Materials, Dumbarton, Scotland. "Elemental" is a trademark of Raytheon Systems Co., Lexington, MA.

Table 1.8. Change in refractive index with temperature (*dn/dT*) near 4 μm wavelength*

Material	*dn/dT* (10^{-6} / K)	Material	*dn/dT* (10^{-6} / K)
ALON	2.8	Sapphire	6 - 12
Diamond	16	Silicon	159
Fused silica	10	Sodium chloride	-36
Gallium arsenide	150	Spinel	3
Gallium phosphide	137	Yttria	~30
Germanium	424	Zinc sulfide	43
Lithium fluoride	-15	Zinc selenide	63
Magnesium fluoride	1		

*Data from Appendix C.

sets in. Although standard grade material has too much optical scatter to be used in the midwave region (Fig. 1.18d), multispectral and elemental materials perform well in the midwave range. Zinc sulfide can be used up to ~800°C[56] and has good thermal shock resistance (but not as good as GaP or sapphire). The erosion resistance is poor, so ZnS requires protective coatings.

Zinc selenide has outstanding 2-color optical characteristics, with an absorption coefficient of ~10^{-3} to 10^{-4} cm^{-1}. However, its erosion and thermal shock resistance are poor. Thin coatings (~10 μm) are insufficient to protect ZnSe from exposure to rain and dust at aircraft speeds, so thick claddings (~200-1000 μm) are required for protection.

References

1. C. A. Klein, R. P. Miller and D. L. Stierwalt, "Surface and Bulk Absorption Characteristics of Chemically Vapor-Deposited Zinc Selenide in the Infrared," *Appl. Opt.*, **33**, 4304-4313 (1994).
2. S. Allen and J. A. Harrington, "Optical Absorption in KCl and NaCl at Infrared Laser Wavelengths," *Appl. Opt.*, **17**, 1679-1680 (1978).
3. G. Mann and G. Phillipps, "Rough Surface Absorption of Nd:YAG Laser Rods," *Opt. Mater.*, **4**, 811-814 (1995).
4. P. Klocek and G. H. Sigel, Jr., *Infrared Fiber Optics*, SPIE Press (Tutorial Text TT 2), Bellingham, Washington (1989).
5. W. H. Dumbaugh, "Infrared Transmitting Germanate Glasses," *Proc SPIE*, **297**, 80-85 (1981).
6. A. C. DeFranzo and B. G. Pazol, *Appl. Opt.*, **32**, 2224 (1993).
7. E. D. Palik, ed., *Handbook of Optical Constants of Solids*, Academic Press, Orlando, Florida (1985).
8. C. A. Klein and R. N. Donadio, "Infrared-Active Phonons in Cubic Zinc Sulfide," *J. Appl. Phys.*, **51**, 797-800 (1980).
9. M. E. Thomas, W. J. Tropf and A. Szpak, "Optical Properties of Diamond," *Diamond Films & Technol.*, **5**, 159-180 (1995).
10. M. E. Thomas and W. J. Tropf, "Optical Properties of Diamond," *Proc SPIE*, **2286**, 144-151 (1994).
11. E. D. Palik, ed., *Handbook of Optical Constants of Solids II*, Academic Press, San Diego, California (1991).
12. E. D. Palik, ed., *Handbook of Optical Constants of Solids III*, Academic Press, San

Diego, California (1998).
13. J. Trombetta, Raytheon Systems Co., to be published.
14. W. J. Tropf and M. E. Thomas, "Models of the Optical Properties of Solids," *Proc. SPIE*, **1760**, 318-328 (1992).
15. M. E. Thomas, "A Computer Code for Modeling Optical Properties of Polar Crystals," *Proc. SPIE*, **1112**, 260-267 (1989).
16. M. E. Thomas and R. I. Joseph, "The Infrared Properties of Window Materials," *Johns Hopkins Appl. Phys. Lab. Tech. Digest*, **9**, 328-338 (Oct. - Dec. 1988).
17. W. J. Tropf, M. E. Thomas and T. J. Harris, "Properties of Crystals and Glasses," in *Handbook of Optics* (M. Bass, E. W. van Stryland, D. R. Williams and W. L. Wolfe, eds.), Vol II, Chap 33, McGraw-Hill, New York (1995). For the Urbach tail and the infrared multiphonon cutoff, see Eqns. 15 and 17 and Tables 46 and 48.
18. T. F. Deutsch, "Absorption Coefficient of Infrared Laser Window Materials," *J. Phys. Chem. Solids*, **34**, 2091-2104 (1973).
19. D. Blair, F. Cleva and C. N. Man, "Optical Absorption Measurements in Monocrystalline Sapphire at 1 μm," *Optical Mater.*, **8**, 233-236 (1997).
20. A. E. Barnes, R. G. May, S. Gollapudi and R. O. Claus, "Sapphire Fibers: Optical Attenuation and Splicing Techniques," *Appl. Opt.*, **34**, 6855-6861 (1995).
21. D. C. Harris, "Durable 3-5 μm-Infrared Transmitting Window Materials, *Infrared Phys. Technol.*, **39**, 185-201 (1998).
22. M. E. Thomas, W. J. Tropf, and S. L. Gilbert, "Vacuum-Ultraviolet Characterization of Sapphire, ALON, and Spinel Near the Band Gap," *Opt. Eng.*, **32**, 1340-1343 (1993).
23. P. A. Temple, "Thin-Film Absorptance Measurements Using Laser Calorimetry," in *Handbook of Optical Constants of Solids* (E. D. Palik, ed.), Academic Press, Orlando, Florida (1985).
24. R. G. Schlecht, C. I. Zanelli, A. M. Schlecht, P. .F. Bordui, C. D. Bird and R. Blachman, *Single Crystal Growth Optimization of Magnesium-Doped Lithium Niobate*, Army Materials Technology Laboratory Report MTL TR 92-24 (Contract DAAL04-88-C0029), April 1992.
25. D. C. Harris, *Development of Chemical-Vapor-Deposited Diamond for Infrared Optical Applications. Status Report and Summary of Properties*, Naval Air Warfare Center Weapons Division Report TP 8210, China Lake, California, July 1994, p 28.
26. P. A. Miles, "Temperature Dependence of Multiphonon Absorption in Zinc Selenide," *Appl. Opt.*, **16**, 2891-2896 (1977).
27. J. M. Palmer, "The Measurement of Transmission, Absorption, Emission, and Reflection," in *Handbook of Optics* (M. Bass, E. W. van Stryland, D. R. Williams and W. L. Wolfe, eds.), Vol II, Chap 25, McGraw-Hill, New York (1995).
28. S. E. Hatch, "Emittance Measurements on Infrared Windows Exhibiting Wavelength Dependent Diffuse Transmittance," *Appl. Opt.*, **1**, 595-601 (1962).
29. D. L. Stierwalt, "Infrared Spectral Emittance Measurements of Optical Materials," *Appl. Opt.*, **5**, 1911-1915 (1966).
30. M. A. Bramson, *Infrared Radiation: A Handbook for Applications*, Plenum Press, New York (1968).
31. R. Gardon, "The Emissivity of Transparent Materials," *J. Am. Ceram. Soc.*, **39**, 278-287 (1956).
32. R. M. Sova, M. J. Linevsky, M. E. Thomas and F. F. Mark, "High Temperature Infrared Properties of Sapphire, ALON, Fused Silica, Yttria, and Spinel," *Infrared Phys. Technol.*, **39**, 251-261 (1998).
33. R. M. Sova, M. J. Linevsky, M. E. Thomas and F. F. Mark, "High Temperature Optical Properties of Oxide Dome Materials," *Proc. SPIE*, **1760**, 27-40 (1992).
34. *OPTIMATR®* (A Computer Program to Calculate Optical Properties of Materials),

ARSoftware, 8201 Corporate Drive, Suite 1110, Landover MD 20785 (Phone: 301-459-3773).

35. C. A. Klein, B. diBenedetto and J. Pappis, "ZnS, ZnSe, and ZnS/ZnSe Windows: Their Impact on FLIR System Performance," *Opt. Eng.*, **25**, 519-531 (1986).
36. J. A. Cox, D. Greenlaw, G. Terry, K. McHenry and L. Fielder, "Comparative Study of Advanced IR Transmissive Materials," *Proc. SPIE*, **683**, 49-62 (1986).
37. C. M. Freeland, "High Temperature Transmission Measurements of IR Window Materials," *Proc. SPIE*, **929**, 79-86 (1988).
38. D. T. Gillespie, A. L. Olsen and L. W. Nichols, "Transmittance of Optical Materials at High Temperatures in 1-μ to 12-μ Range," *Appl. Opt.*, **4**, 1488-1493 (1965).
39. M. E. Thomas, R. I. Joseph and W. J. Tropf, "Infrared Transmission Properties of Sapphire, Spinel, Yttria and ALON as a Function of Temperature and Frequency," *Appl. Opt.*, **27**, 239-245 (1988).
40. C. D. Thurmond, "The Standard Thermodynamic Functions for the Formation of Electrons and Holes in Ge, Si, GaAs, and GaP," *J. Electrochem. Soc.*, **122**, 1133-1141 (1975).
41. C. R. Poznich and J. C. Richter, "Silicon for Use as a Transmissive Material in the Far IR?" *Proc. SPIE*, **1760**, 112-120 (1992).
42. K. V. Ravi, "Diamond Technology for Infrared Seeker Windows," *AIAA J. Spacecraft and Rockets*, **30**, 79-86 (1993).
43. P. Klocek, J. Hoggins, T. McKenna, J. Trombetta and M. W. Boucher, "Optical Properties of GaAs, GaP, and CVD Diamond," *Proc. SPIE*, **1498**, 147-157 (1991).
44. H. Rutt, "Effect of Visible and Near-Infrared Illumination on the Mid-Infrared Transmission of Silicon and Germanium," *Proc. SPIE*, **2286**, 100-107 (1994).
45. M. E. Thomas, S. K. Andersson, T. M. Cotter and K. T. Constantikes, "Infrared Properties of Polycrystalline Magnesium Fluoride," *Infrared Phys. Technol.*, **39**, 213-222 (1998).
46. M. E. Thomas, "Temperature Dependence of the Complex Index of Refraction," in *Handbook of Optical Constants of Solids II* (E. D. Palik, ed.), Academic Press, San Diego, California (1991).
47. C. A. Klein and R. N. Donadio, "Infrared-Active Phonons in Cubic Zinc Sulfide," *J. Appl. Phys.*, **51**, 797-800 (1980).
48. C. A. Klein, *Compendium of Property Data for Raytran Zinc Selenide and Raytran Zinc Sulfide*, Report RAY/RD/T-1154, Raytheon Co., Lexington, Massachusetts (31 Aug. 1987).
49. C. A. Klein, "Infrared Transmittance of Gallium Phosphide," Raytheon Memorandum RAY/RD/M-4879 (25 June 1990); J. M. Wahl and R. W. Tustison, "Optical, Mechanical and Water Drop Impact Characteristics of Polycrystalline GaP," *J. Mater. Sci.*, **29**, 5765-5772 (1994).
50. D. Kleinman and W. Spitzer, "Infrared Lattice Absorption of GaP," *Phys. Rev.*, **118**, 110-117 (1960).
51. J. Pastrňák and B. Hejda, "Infrared Absorption of AlN Single Crystals in the Two-Phonon Region," *Phys. Stat. Sol.*, **35**, 953-958 (1969),
52. M. E. Thomas and R. I. Joseph, "A Comprehensive Model for the Infrared Transmission Properties of Optical Windows," *Proc. SPIE*, **929**, 87-93 (1988).
53. R. L. Gentilman, "Current and Emerging Materials for 3-5 Micron IR Transmission," *Proc. SPIE*, **683**, 2-11 (1986).
54. M. P. Nadler, C. K. Lowe-Ma and T. A. Vanderah, "Single-Crystal Infrared Characterization of Ternary Sulfides," *Mater. Res. Bull.*, **28**, 1345-1354 (1993).
55. M. E. Thomas and R. I. Joseph, "Optical Phonon Characteristics of Diamond, Beryllia, and Cubic Zirconia," *Proc. SPIE*, **1326**, 120-126 (1990).
56. R. W. Tustison and R. L. Gentilman, "Current and Emerging Materials for LWIR External Windows," *Proc. SPIE*, **968**, 25-34 (1988).

Chapter 2

OPTICAL PERFORMANCE OF INFRARED WINDOWS

This chapter discusses several measures of the performance of infrared optical systems with emphasis on how properties of the window material affect performance. We conclude by considering the ability of infrared windows to serve as microwave windows.

2.1 Resolution

When a point source of light is viewed through an optical system, diffraction of light at the system aperture broadens the image to a finite width and creates a series of faint concentric rings around the central spot. This pattern of rings shown at the right in Fig. 2.1 is called an *Airy disk*. In the optical system at the left of Fig. 2.1, the lens diameter is the aperture, D, and the focal length is L. The f number is $f\# = L/D$.

In a *diffraction limited* optical system, the optical wavefront does not deviate from its ideal shape by more than one-fourth of the wavelength, λ. Suppose that we look at two equal intensity point sources of light through an ideal, diffraction limited optical system whose aperture has diameter, D. The *resolution* of the system is the minimum angular separation between the two sources such that it is barely possible to discern two separate sources. With less separation, the two sources appear to be one (Fig. 2.2). The *Rayleigh criterion* for resolution is that the central spot of one Airy disk overlaps the first dark ring of the adjacent Airy disk. This means that the trough between the two bright spots is 74% as bright as the peaks.

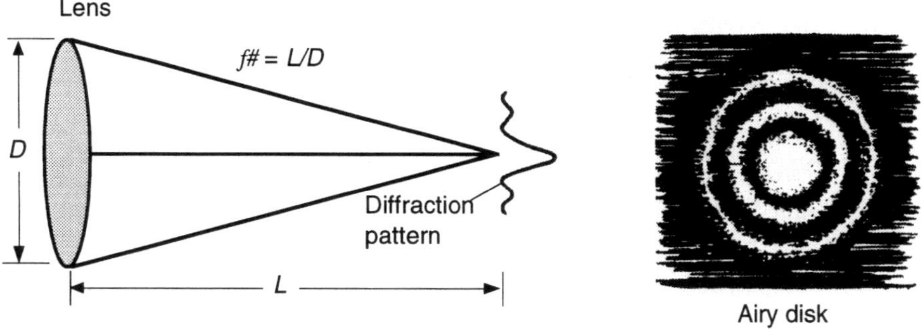

Fig. 2.1. Imaging a point source through an optical system produces a diffraction pattern called an Airy disk.

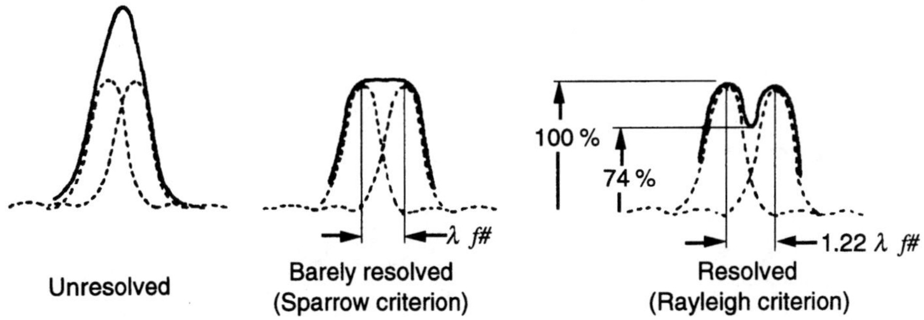

Fig. 2.2. Resolution of two closely spaced peaks.

The *resolution* of a diffraction limited system is given by

$$\text{Resolution (radians)} = \frac{1.22\lambda}{D} . \tag{2-1}$$

For example, at a wavelength of 10 μm, the resolution of a system with an 8.0-cm-diameter aperture is $(1.22)(10 \times 10^{-6}\ \text{m})/(0.080\ \text{m}) = 1.5 \times 10^{-4}$ radians. If two point sources are separated by more than 1.5×10^{-4} radians, they will be seen as separate objects. If they have less separation, they will appear to be a single object.

Example: Aperture requirements for resolution. What aperture size is required for 0.25 milliradian resolution in a 3-5 μm midwave infrared system and in an 8-10 μm long wave system? Using $\lambda = 4$ μm for the midwave system, we use Eq. (2-1) to calculate

$$D = \frac{1.22\lambda}{\text{resolution}} = \frac{(1.22)\ (4 \times 10^{-6}\ \text{m})}{2.5 \times 10^{-4}} = 1.95\ \text{cm} .$$

For the long wave system ($\lambda = 9$ μm), the same calculation gives $D = 4.39$ cm. The longer the wavelength of light, the larger the aperture needed for the same resolution.

2.2 Scatter

Scattering of light by an optical window degrades the sharpness of the image that is observed. If scatter is great enough, the outline of an object may be obscured or the ability to distinguish two closely spaced objects may be lost. For systems operated in the outdoors in daylight, sunlight scattered into the field of view of the detector by the window may badly obscure the intended target.

When two equally intense objects are viewed in the presence of background scatter, the resolution is not affected very much by low levels of scatter. However, a critical point is reached at which resolution rapidly falls to zero when the scatter is 2.8 times as great as the intensity of light from the objects (Fig. 2.3).[1] Low levels of scatter can reduce the resolution between a faint object and a bright object, such as the sun, to zero.

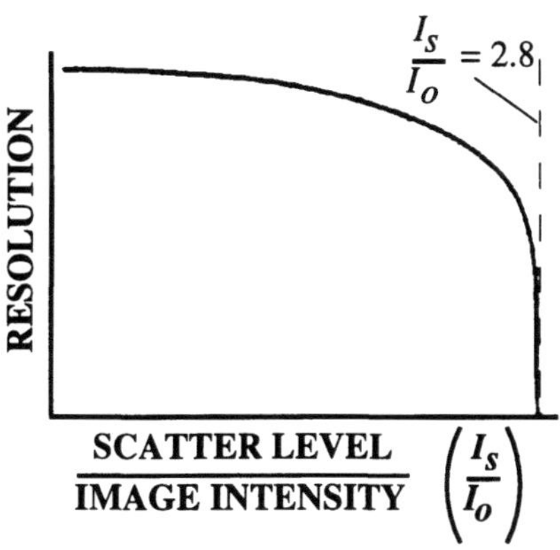

Fig. 2.3. Effect of scatter on resolution of two equally bright, closely spaced objects.[1]

Figure 2.4 shows how scatter from a dome can be measured.[2] A 1- to 2-mm-diameter laser beam of the desired frequency is passed through various locations on the dome and the transmitted light is measured by the detector at the top of the diagram. The Coblentz sphere reflects scattered light that is measured by a second detector. This particular arrangement measures what is called *total integrated scatter* in the forward hemisphere. An alternate arrangement can measure scatter in the backward hemisphere, which is typically from 10 - 100% as great as the forward scatter in infrared window materials.

Figure 2.5 shows visible (0.647 μm) and infrared (3.39 μm) scatter measured at various points on magnesium fluoride domes. The used dome, which had experienced 30 to 40 aircraft carrier landings in the Pacific Ocean, had about twice as much total integrated forward scatter at 3.39 μm as the unused dome (1.7% scatter vs 0.8% scatter), and the used dome had lost the characteristic glossy surface finish of a new dome. The visible scatter of this polycrystalline magnesium fluoride is very high (23% for the unused dome, 44% for the used dome), and this material is unsuitable for visible applications.

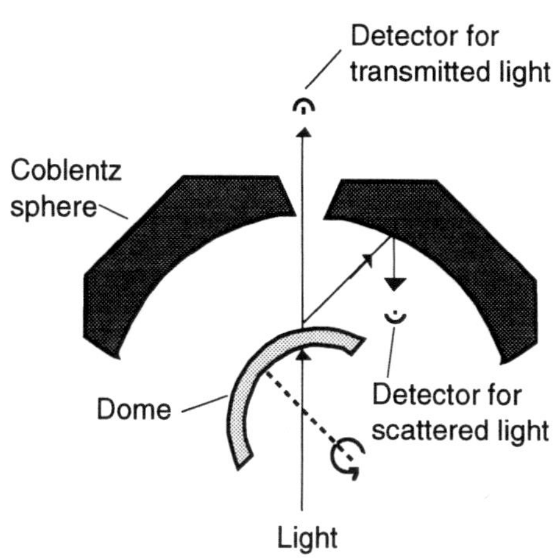

Fig. 2.4. Schematic diagram of apparatus[2] used to measure total integrated scatter in the forward hemisphere for a missile dome. The Coblentz sphere collects light scattered between 2.5° and 70° off the incident axis.

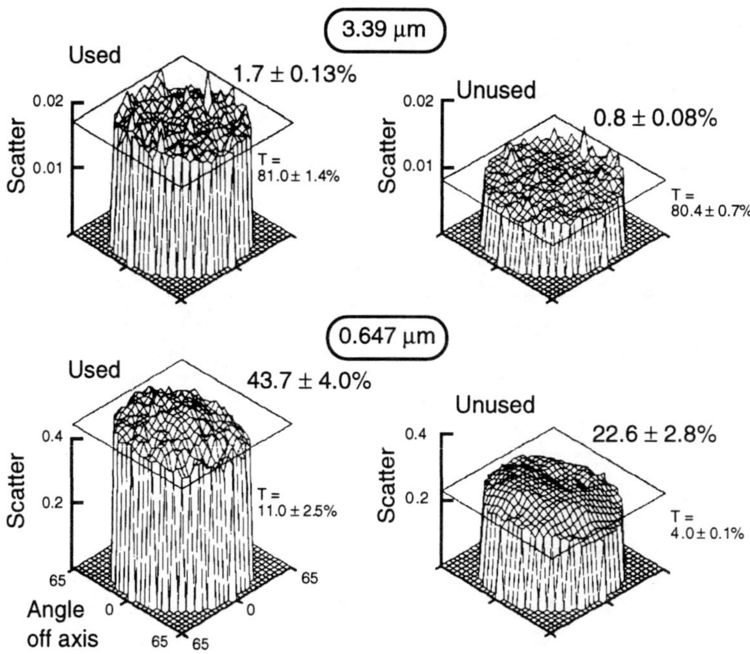

Fig. 2.5. Total integrated scatter measured in the forward hemisphere for new and used polycrystalline magnesium fluoride missile domes.[2] Horizontal axes measure angular position on the dome. The average scatter of the top left dome measured at many positions is 1.7 (± 0.13)% and the average transmittance is 81.0 (±1.4)%.

Total integrated scatter in the forward hemisphere is typically in the range 0.5 - 5% for polycrystalline optical materials at infrared wavelengths. Glasses are more typically near 0.1% and single-crystal materials such as sapphire are in the 0.01 - 0.1% range.

While the experiment in Fig. 2.4 measures total integrated scatter, the experiment in Fig. 2.6 measures scatter as a function of angle from the incident axis.[3-5] The *bidirectional transmittance distribution function* (BTDF) measured in Fig. 2.6 is

$$\text{BTDF} = \frac{P_s/P_i}{\Omega \cos \theta_s} \tag{2-2}$$

where P_s is the scattered power (watts), P_i is the incident power (watts), Ω is the solid angle of scatter viewed by the detector and θ_s is the scatter angle. This definition is

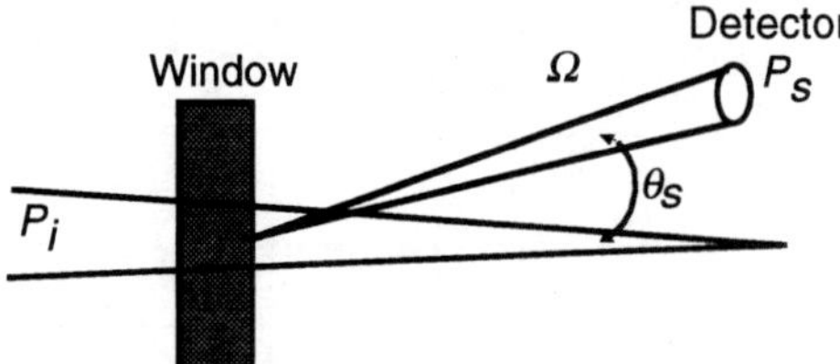

Fig. 2.6. Experiment used to measure bidirectional transmittance distribution function (BTDF).[3]

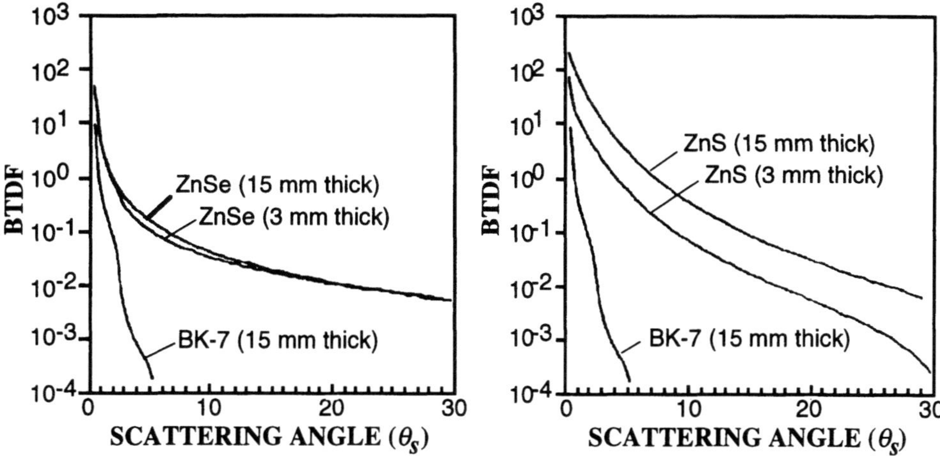

Fig. 2.7. Bidirectional transmittance distribution function (BTDF) of zinc selenide (left) and multispectral zinc sulfide (right) at 0.633 μm wavelength.[6] The lower curve at the left shows very low scattering from BK-7 glass (Table 1.1).

equivalent to saying that BTDF is the ratio of transmitted *radiance* to incident *irradiance*, terms which are defined in Appendix D.

Figure 2.7 shows examples of BTDF measurements. Note that the ordinate is logarithmic because the scattered light intensity falls off rapidly with increasing angle from the incident direction. BK-7 glass has much less visible scatter than zinc selenide, which, in turn, has less scatter than multispectral zinc sulfide. Furthermore, because the scatter from ZnSe is essentially independent of specimen thickness, the scatter must arise almost entirely from the surface, not from the bulk. The significant increase in scatter from ZnS with increasing thickness demonstrates that the scatter is predominantly from the bulk, not from the surface.

Because optical scatter typically falls off rapidly with increasing angle from the incident direction, total integrated scatter measured by the apparatus in Fig. 2.4 depends on the cutoff angle used to define the difference between "scattered" and "unscattered" light. In Fig. 2.4 light scattered less than 2.5° is considered to be "unscattered." If an angle of 1.5° were used instead of 2.5°, the total integrated scatter would be somewhat greater.* The scatter integrated between 2.5° and 70° in a BTDF measurement in Fig. 2.7 should be equal to the total integrated scatter measured by the method in Fig. 2.4. Indeed, when the scatter from a sample of lanthana-doped yttria was measured by both methods, the results were close:[7]

	Wavelength (μm)	
	~0.64	3.39
Total integrated scatter measured with Coblentz sphere:	3.9%	0.5%
Scatter integrated from BTDF measurement:	4.2%	0.6%

*The contribution of angles below 2.5° to total integrated scatter is not as great as it would seem from Fig. 2.7, because the solid angle subtended by such small cone angles is much less than the solid angle subtended by larger cone angles.

Scatter is caused by variations in refractive index as light passes through a material. Causes include foreign particles or voids, gradual changes of composition, second phases at grain boundaries, and strains in the material. Optical scatter from lanthana-doped yttria was attributed to microscopic amounts of a second crystalline phase left from the fabrication process.[7] Annealing for increasing time decreased the quantity of second phase and decreased the scatter.

Rayleigh scattering occurs when the size of the scattering centers is much smaller than the wavelength of light, λ. In this case, the scattering intensity is proportional to $1/\lambda^4$. Rayleigh scattering increases rapidly at short wavelengths. *Mie scattering*, which occurs when the size of the scattering centers is comparable to the wavelength of light, has a complex dependence on wavelength. When the scattering center is much larger than λ, there is almost no dependence of scattering intensity on wavelength. Both Rayleigh and Mie scattering are very sensitive to scattering angle.

Figure 2.8 shows the calculated effect of 0.1-, 0.5- and 1.0-μm pores (voids) on transmission through alumina.[8] Without pores, the expected transmittance of alumina is 87%, based on 13% reflection for a refractive index of 1.70 (using Eqns. [1-11] and [1-12]). Figure 2.8 shows that this transmittance is approached for 0.1-μm pores at a wavelength of 5 μm. As the wavelength approaches the pore size, scatter increases and transmittance decreases. For 0.1-μm pores, scatter is significant at wavelengths below 2.5 μm, which is 25 times greater than the size of scattering centers. This is the reason we said in Section 1.2.2 that particle size should be 20 times smaller than the wavelength to be transmitted for a noncubic (birefringent) polycrystalline material.

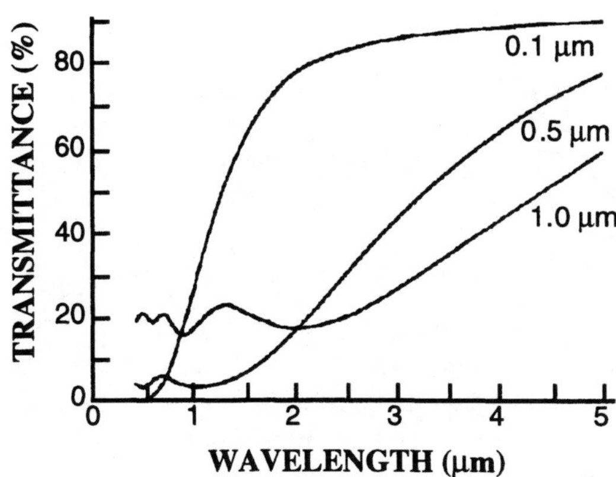

Fig. 2.8. Calculated effect of pore size (0.1, 0.5 and 1.0 μm) on transmission of 0.5-mm-thick alumina.[8] It is assumed that the pores (voids) occupy 0.2% of the volume of the material.

2.3 Modulation transfer function: a measure of imaging quality

Scatter in a window is one factor that can degrade the sharpness of an image formed by an optical system. A quantitative description of image degradation is the *modulation transfer function*, abbreviated MTF. Figure 2.9(a) shows a typical bar pattern used to characterize the imaging capability of an optical system.[9,10] The intensity of dark bars (the *object* being viewed) at a regular interval is described by the square wave at the right

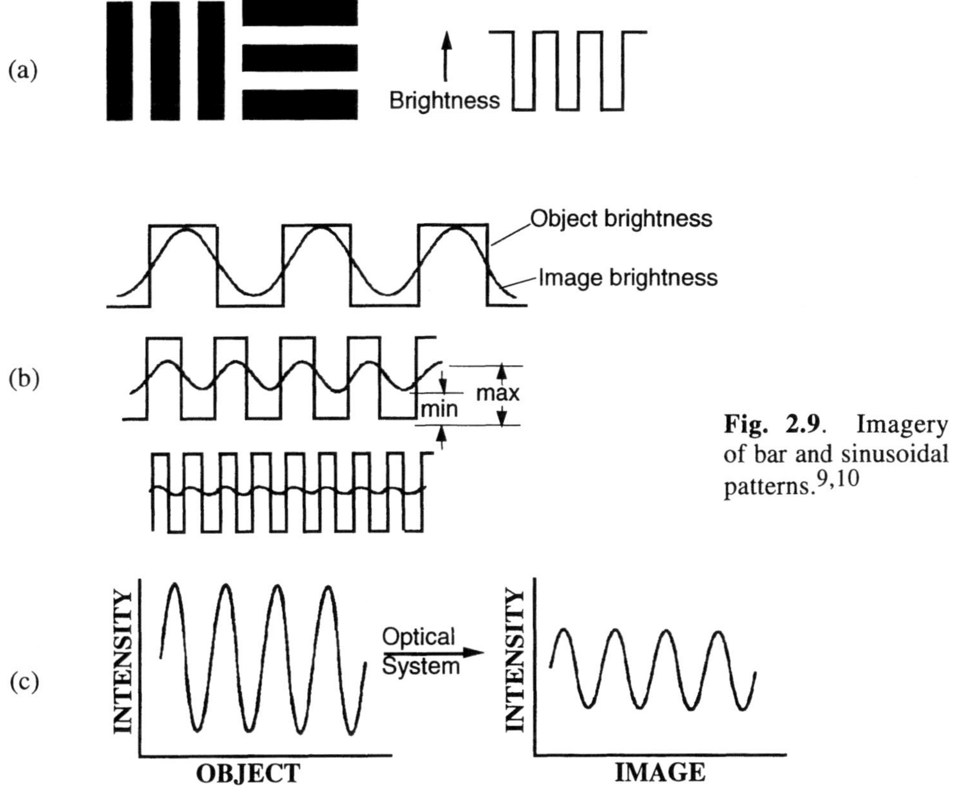

Fig. 2.9. Imagery of bar and sinusoidal patterns.[9,10]

of Fig. 2.9(a). The top of Fig. 2.9(b) shows that an optical system forms an *image* whose edges are never as sharp as those of the original object. In the best possible diffraction limited optical system, blurring occurs because of diffraction of light at the edge of each optical element.

In most real systems, additional blurring occurs. Figure 2.9(b) shows that as the *spatial frequency* (number of lines per millimeter) increases, the image resolution decreases. That is, the more closely spaced the lines in the object, the less contrast between maximum and minimum brightness in the image. In Fig. 2.9(c) we see that sinusoidally varying object intensity gives rise to a sinusoidally varying image with decreased contrast.

Referring to Fig. 2.9(b), the contrast *modulation* is defined as

$$\text{Modulation} = \frac{\text{max - min}}{\text{max + min}} \tag{2-3}$$

where max and min are the maximum and minimum image intensities formed by the optical system. A modulation of unity corresponds to perfect reproduction of the contrast, while a modulation of zero corresponds to no contrast in the image.

At the bottom of Fig. 2.10, we see that as the spatial frequency of the object intensity increases, the modulation of the image decreases. The curves in the upper part of Fig. 2.10, called modulation transfer functions, show how modulation decreases with increasing spatial frequency for perfect (diffraction limited) and real optical systems. The

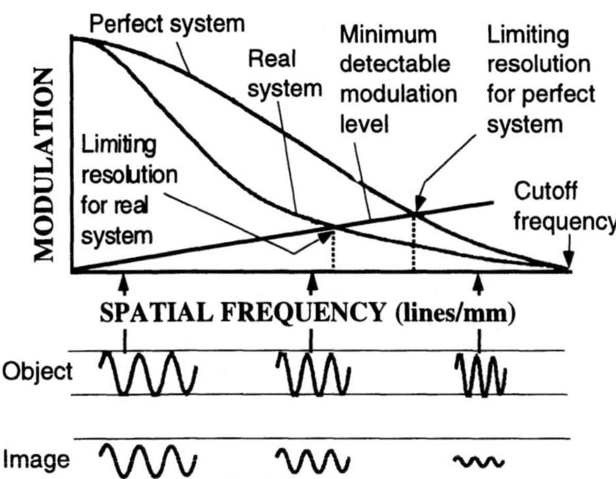

Fig. 2.10. Modulation transfer function.[9,10] The cutoff frequency (limiting resolution, lines/mm) at which the MTF of a diffraction limited system drops to zero is $1/(\lambda f\#)$, where λ is wavelength and $f\#$ is the f number of the optical system ($f\#$ = focal length/aperture diameter, Fig. 2.1).

straight line through the origin shows how the minimum detectable contrast level might vary for a particular detector. The intersection between the straight line and the modulation transfer function corresponds to the limiting resolution for the optical system.

MTF specifications for an optical system are based on how much qualitative image degradation is allowable. The MTF of a window is determined by measuring the MTF of a system without the window and then measuring the MTF with the window in place. The quotient (MTF with window)/(MTF without window) is equal to the MTF of the window. In a high-quality imaging system, an MTF of 0.95 can give a discernibly fuzzy or grainy image and might be considered unacceptable. The higher the quality of the system, the greater the MTF must be to avoid image degradation.

Figure 2.11 shows modulation transfer functions for ALON (aluminum oxynitride) and yttria (yttrium oxide) windows.[11] The higher modulation transfer function at greater spatial frequency indicates that this particular ALON window would be better for imaging than this particular yttria window. Be aware that data for Fig. 2.11 were obtained while these optical materials were under development. The modulation transfer functions of more recent materials might be better. Figure 2.12 shows the modulation transfer function for a gallium phosphide dome that gives near-diffraction-limited performance.[12]

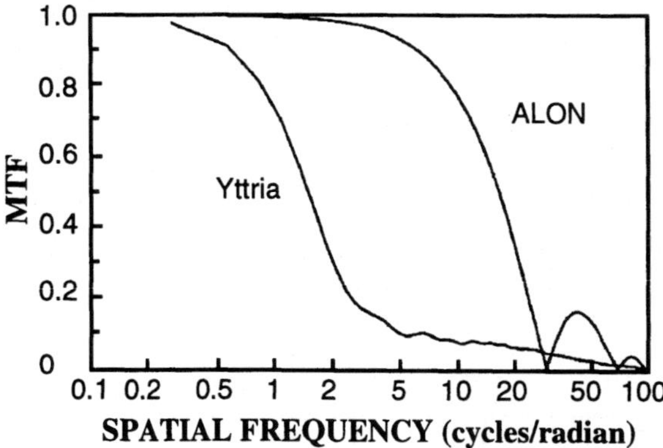

Fig. 2.11. Modulation transfer functions for ALON and yttria at a wavelength of 0.633 μm.[11] The object being viewed was a radial contrast pattern with a spatial frequency expressed as cycles/radian. Note logarithmic abscissa. Nodes in the ALON modulation transfer function near 30 and 70 cycles/radian correspond to phase reversal frequencies where the contrast goes to 0.

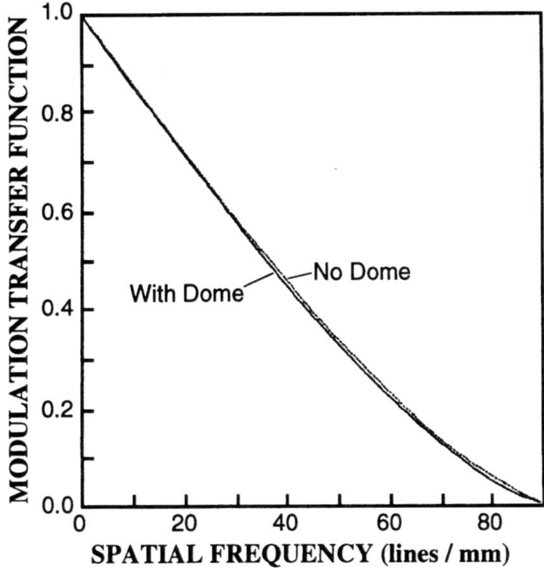

Fig. 2.12. Modulation transfer function for Raytheon Systems Co. 76-mm-diameter gallium phosphide dome at a wavelength of 9.3 μm.[12] The upper curve shows the response of the system without the dome.

Modulation transfer functions are multiplicative. If a window with a modulation transfer function of 0.80 at a particular spatial frequency is placed in front of an optical detection system with a modulation transfer of 0.60 at the same frequency, the combined modulation transfer function of the whole system is $0.80 \times 0.60 = 0.48$.

2.4 Degradation of infrared sensing by a hot window

A hot window emits photons that create a randomly fluctuating signal in the detector behind the window. Window emittance can set the high-temperature limit for signal-to-noise ratio of an infrared system. Thermal gradients in a window create refractive index gradients that distort an optical image. This section discusses both effects.

2.4.1 Emittance from a hot window

For the experiment in Figure 2.13, a magnesium fluoride window is placed in front of a 3-5 μm lead telluride infrared detector. The graph shows that the minimum detectable signal rises as the window is heated.[13] Near room temperature, the minimum detectable signal is nearly constant because the noise level in the detector is almost constant. Near room temperature, the noise is governed by electrical characteristics of the detector itself. Noise created by photons from the window is insignificant compared to the electrical noise in the detector. When the window becomes hot enough, photons from the window striking the detector create additional noise. The higher the noise level, the greater must be the signal to be detected. With the window at 700°C, the minimum detectable signal is more than 7 times greater than the signal that could be detected with the window at 0°C. Photon noise is proportional to the square root of the photon flux density at the detector. The right side of the curve in Fig. 2.13 rises in proportion to the square root of the flux density.

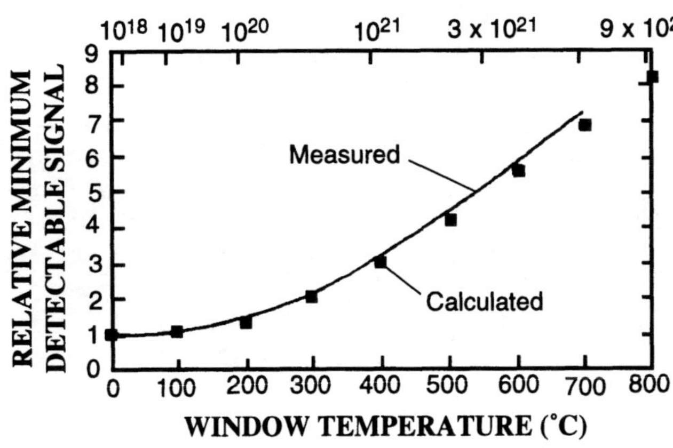

Fig. 2.13. Effect of MgF_2 (1.75 mm thick) window temperature on the minimum signal that can be detected by a 3-5 µm lead telluride infrared sensor with f/1 optics.[13]

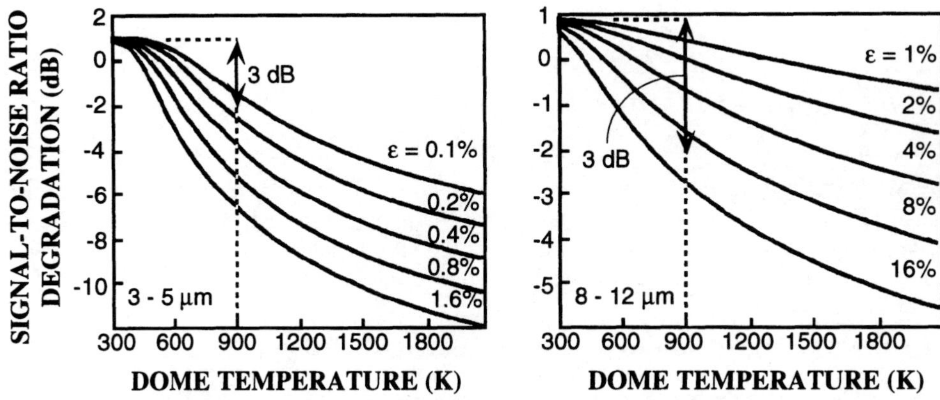

Fig. 2.14. Theoretical degradation of infrared seeker signal-to-noise ratio by a hot dome in midwave (*left*) and long wave (*right*) regions calculated with Eqns. (2-4), (2-5) and (2-6).[14] The ordinate is $10 \log (\Phi_o/\Phi)^{1/2}$. Each curve corresponds to a different dome emissivity. For examples of calculations of the power reaching the detector or the absolute (not relative) signal-to-noise ratio, see Refs. 18-21.

Figure 2.14 shows how the signal-to-noise ratio of a midwave or long wave infrared seeker is degraded by a hot dome.[14-17] The important conclusion that we will discuss is:

> *A long wave infrared seeker can tolerate much greater emittance from the dome than can a midwave seeker.*

Let's see how this comes about. The calculations that underlie Fig. 2.14 assume that the dominant sources of noise are photon flux from the background of the scene being viewed, from the seeker optics and from the hot dome. Other possible sources of noise, such as the electronics, are not included. If other sources of noise are significant, the effect of the hot dome in Fig. 2.14 would be somewhat attenuated and the relative difference between the two wavebands would not be as great.

For Fig. 2.14, the signal-to-noise ratio, S/N, is calculated for the dome at various temperatures. The reference signal-to-noise ratio, $(S/N)_O$, is calculated for a temperature of T_o = 295 K. The signal-to-noise ratio is taken to be inversely proportional to the square root of the photon flux (photons/m^2/s) striking the detector:[14]

$$\frac{S/N}{(S/N)_O} = \sqrt{\frac{\Phi_o}{\Phi}} \tag{2-4}$$

where Φ is the photon flux within each waveband (3-5 or 8-12 μm) striking the detector when the dome is at the indicated temperature. Φ_o is the photon flux when the sensor and scene are both at the temperature T_o = 295 K. The flux ratio is given by

$$\frac{\Phi_o}{\Phi} = \frac{Q_o}{\underbrace{(1-\varepsilon)Q_b\tau_o}_{\text{Scene background}} + \underbrace{(1-\tau_o)Q_s}_{\text{Seeker optics}} + \underbrace{\varepsilon Q_d\tau_o}_{\text{Dome emittance}}} . \tag{2-5}$$

In Eq. (2-5), Q_o is the photon flux from a blackbody at temperature T_o = 295 K. Q_b is the flux from a blackbody at the effective temperature of the background of the scene being viewed, which was assumed to be T_b = 227 K for Fig. 2.14. Q_s is the flux from a blackbody at the seeker temperature assumed to be T_s = 300 K. Q_d is the flux from a blackbody at the dome temperature. The emissivity of the dome, ε, is assumed to be constant. The transmittance of the seeker optics, excluding the dome, is denoted τ_o, which is taken as 0.5 for Fig. 2.14.

The three terms in the denominator of Eq. (2-5) account for photons from the background of the scene being viewed, the inside of the seeker, and the hot dome. The term $\varepsilon Q_d\tau_o$ is easiest to understand: εQ_d is the photon flux from the dome and τ_o gives the fraction of this flux transmitted through the seeker optics to the detector. The term $(1-\varepsilon)Q_b\tau_o$ gives the flux of photons from the background of the scene being viewed that is transmitted to the detector. The term $(1-\varepsilon)$ gives the transmittance of the dome because the emittance is equal to the absorptance, which is the fraction of light absorbed. [It was assumed that reflection and scatter are zero in this simplified calculation.] The middle term in the denominator is the contribution of blackbody radiation from inside the seeker (from the optics and housing) that reaches the detector. You can rationalize the factor $(1-\tau_o)$ by noting that the entire expression for Φ_o/Φ must equal unity if the temperatures of the background, seeker and dome were all at T_o = 295 K. Then $Q_b = Q_s = Q_d = Q_o$ and there must be a factor of $(1-\tau_o)$ before Q_s to make the entire fraction unity.

The blackbody photon flux terms in Eq. (2-5) are computed from the equation

$$Q_T = \int_{\lambda_1}^{\lambda_2} \frac{2\pi c\, d\lambda}{\lambda^4(e^{hc/\lambda kT} - 1)} = \int_{\lambda_1}^{\lambda_2} \frac{1.88365157\times 10^8\, d\lambda}{\lambda^4(e^{1.43876866\times 10^4/\lambda T} - 1)} \tag{2-6}$$

where T is temperature (K), c is the speed of light, h is Planck's constant, and k is Boltzmann's constant. Eq. (2-6) is derived from the Planck distribution by dividing Eq. (0-5) by hc/λ, the energy of one photon. The numbers on the right side of Eq. (2-6) give

Q_T in photons/m^2/s when wavelength is expressed in micrometers. For the midwave seeker, $\lambda_1 = 3$ µm and $\lambda_2 = 5$ µm. For the long wave seeker, $\lambda_1 = 8$ µm and $\lambda_2 = 12$ µm.

For the moment, let's forget about the details of Eqns. (2-4) - (2-6) and examine the important qualitative conclusions from Fig. 2.14. The curves show the relative loss of signal-to-noise ratio as the dome temperature rises. A 3-dB change corresponds to 50% loss in signal-to-noise ratio. Consider a dome heated to 900 K, which is the stagnation temperature of a vehicle flying near Mach 4 in Fig. 1.27. For a midwave seeker, the left side of Fig. 2.14 shows that there will be a 3-dB loss if the emissivity of the dome is approximately 0.1%. For a long wave seeker, the 3-dB loss occurs when the dome emissivity is approximately 10%. That is, *a long wave infrared seeker can tolerate a dome with much greater emissivity than can a midwave seeker.*

Why should this be? The answer lies in the three terms in the denominator of Eq. (2-5). The 300 K dome with an emissivity of 0.1% provides only 0.1% as many photons as do the other two terms. That is, photon flux from the dome in the midwave region at 300 K is negligible in comparison with photon flux from the seeker optics and the background of the scene being viewed. When the dome is heated to 900 K, its midwave photon flux increases by a factor of 2000 and becomes twice as great as the other two sources combined.

In the long wave region, the situation is different. At 300 K, a dome with 10% emittance produces approximately the same number of photons as the seeker optics and scene background. When the dome temperature is raised to 900 K, the photon flux from the dome goes up only by a factor of 30.

The essential difference between the midwave and long wave behavior in Fig. 2.14 is that midwave emittance from the dome is negligible at 300 K and rises by a factor of 2000 at 900 K. Long wave emittance from the dome is quite significant at 300 K and only rises by a factor of 30 at 900 K. Therefore, degradation of signal-to-noise ratio of a midwave seeker is strongly dependent on dome emissivity as the dome temperature rises. A long wave seeker is much more tolerant of dome emissivity because the dome is already emitting a high photon flux at 300 K and this flux does not rise rapidly with increasing temperature.

2.4.2 Temperature gradients in windows

Figure 2.15 shows an edge view of a water-cooled silicon window which has been proposed for use in high-temperature environments.[22] Coolant is forced through tiny channels near the hot surface of the window. One of the performance issues for such a window is the optical pathlength difference for light waves passing through the warmer and cooler regions between the channels.

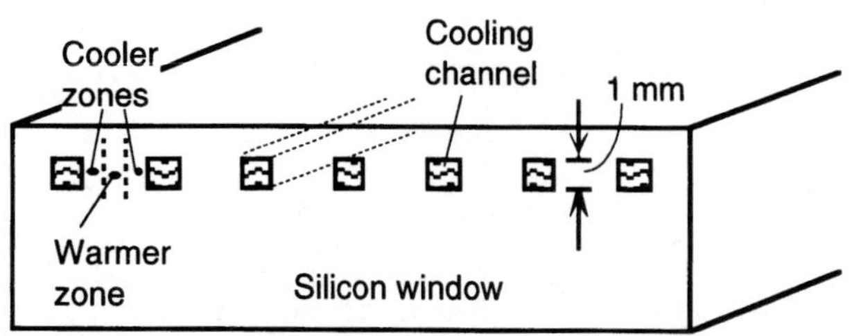

Fig. 2.15. Side view of a water-cooled silicon window.

Consider two light rays in Fig. 2.16 with wavelength $\lambda_1 = \lambda_2 = 4$ μm striking the upper surface in phase with each other. One passes through the warmer zone midway between two cooling channels and the other passes through the cooler zone adjacent to a channel. After the rays have passed through the regions of different temperature, they will be out of phase with each other for two reasons: (1) The index of refraction (n) of the warmer region is different from the index of refraction of the cooler region. Therefore the wavelength of light, which is λ/n, is different in each region. (2) The warmer region expands relative to the cooler region, so the actual distance traveled through the warm region is greater.

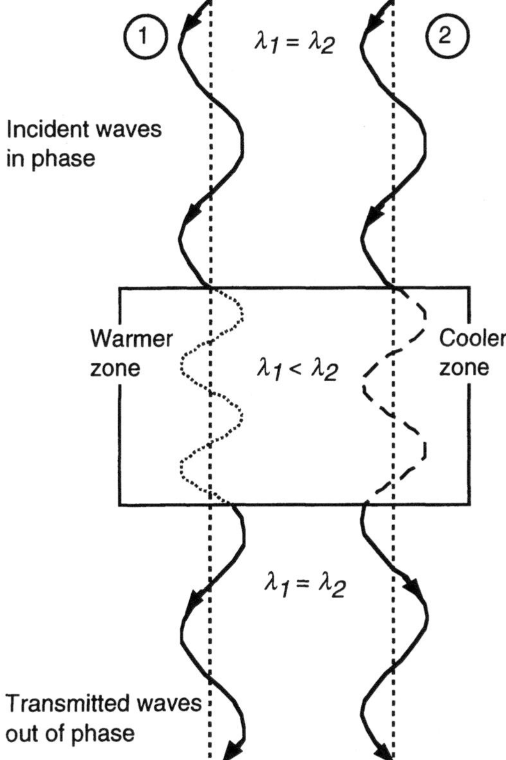

Fig. 2.16. Incident waves that are in phase are out of phase when they emerge from the region with a temperature gradient.

The *optical pathlength* is defined as nb, where n is the refractive index and b is the actual distance traversed by a light wave. If the refractive index is 2, then the optical pathlength is twice as great as the actual pathlength. The change in optical pathlength with respect to temperature (T) is

$$\frac{d(nb)}{dT} = n\frac{db}{dT} + b\frac{dn}{dT} .$$

Using the definition of the thermal expansion coefficient, $\alpha = (1/b)db/dT$, we can write

$$\frac{d(nb)}{dT} = nb\alpha + b\frac{dn}{dT} = nb\left[\alpha + \frac{1}{n}\frac{dn}{dT}\right].$$

The *optical pathlength difference* for waves traversing warm and cool regions differing by ΔT is therefore

$$\text{Optical pathlength difference} = \frac{d(nb)}{dT}\Delta T = nb\Delta T\left[\alpha + \frac{1}{n}\frac{dn}{dT}\right]. \qquad \text{(2-7)}$$

Example: Optical pathlength difference for a cooled silicon window. Suppose that there is a temperature difference of 5 K between the warmer and cooler zones in Fig. 2.15. The actual pathlength of the region is 1 mm. What is the optical pathlength difference for 4 μm infrared radiation passing through the window? Appendix C tells us that the refractive index of silicon is 3.425 and dn/dT is approximately $159 \times 10^{-6}\ \text{K}^{-1}$. The thermal expansion coefficient is $2.6 \times 10^{-6}\ \text{K}^{-1}$ in Table 4.1. (The expansion coefficient is temperature dependent. The value cited applies near 300 K.) We find the optical pathlength difference with Eq. (2-7):

$$\text{Optical pathlength difference} = nb\Delta T\left[\alpha + \frac{1}{n}\frac{dn}{dT}\right]$$

$$= (3.425)(1\ \text{mm})(5\ \text{K})\left[(2.6 \times 10^{-6}\ \text{K}^{-1}) + \frac{1}{3.425}(159 \times 10^{-6}\ \text{K}^{-1})\right]$$

$$= 0.84\ \mu\text{m}.$$

An optical pathlength difference of 0.84 μm means that two 4-μm infrared rays that are in phase at the top of Fig. 2.16 would be out of phase by 0.84 μm / 4 μm = 0.21 wavelengths after traversing the regions of different temperature. A phase difference is equivalent to refraction of the light as it passes through the window. *The window distorts the image being viewed.*

In the preceding example, the thermal expansion term is small in comparison to the refractive index term. The distortion of an image as it passes through a window with a temperature gradient is especially great for germanium, silicon, gallium arsenide and gallium phosphide, which have exceptionally large values of dn/dT in Table 1.8. A method for measuring actual wavefront distortion of the cooled window in Fig. 2.15 by laser heating has been described.[23]

2.5 Frequency doubling

Intense sources of radiation, such as laser beams, give rise to effects whose magnitude is proportional to the square or cube of the oscillating electric field of the beam. These effects are collectively called "nonlinear" optical effects because they depend on more than the first power of the electric field. One of these effects is called second harmonic generation, or *frequency doubling*. Zinc selenide and zinc sulfide give rise to observable nonlinear effects.

For example, when a 1.06-μm near infrared laser beam with an energy of 70 mJ is passed through a 1-mm-thick laminate made of layers of ZnS and ZnSe, a tiny fraction of the transmitted light is doubled in frequency to give 0.53-μm green light (Fig. 2.17).[24] Only one green photon is created for every 10^6 - 10^7 incident infrared photons, but the green light is strong enough to see with the naked eye in laboratory lighting.

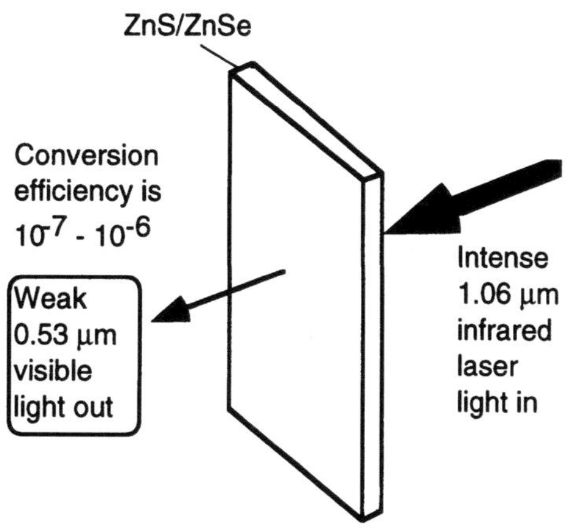

Fig. 2.17. When a near infrared laser beam passes through a ZnS/ZnSe window, enough frequency-doubled green light is created to be seen by eye with laboratory lighting.[24]

2.6 Microwave transmission properties of infrared materials

In systems requiring both infrared and microwave (or millimeter wave) sensors, there is interest in using a single window or dome for both wavelengths. The property that characterizes microwave absorption is called the *loss tangent.* If the microwave radiant power entering a window of thickness b in Fig. 1.1 is P_1 watts per square meter, and the power reaching the second surface is P_2, the internal transmittance is

$$\text{Internal microwave transmittance} = \frac{P_2}{P_1} = e^{-[(2\pi/\lambda)\tan\delta]b} \qquad (2\text{-}8)$$

where λ is the wavelength of microwave radiation within the window and $\tan\delta$ is the loss tangent. The quantity in brackets in Eq. (2-8) is equivalent to the absorption coefficient, α, for infrared radiation in Eq. (1-1). The smaller the loss tangent, the smaller is the absorption coefficient for microwave radiation. Small loss tangents are therefore desired. Figure 2.18 compares the millimeter-wave frequency dependence of the absorption coefficients of several infrared window materials.

Microwave transmission through a material is decreased by absorption or by reflection. We saw in Eq. (1-11) that infrared reflection increases as the refractive index of the window material increases. The same is true for microwave windows, but we customarily speak of the *dielectric constant,* ε, instead of the refractive index, n. For materials that are relatively transparent to microwave radiation, the dielectric constant is just the square of the refractive index:

$$\varepsilon \approx n^2 . \qquad (2\text{-}9)$$

A low dielectric constant is preferred in a microwave transmitting material to reduce reflection losses.

Microwave absorption in a window or radome is reduced if the loss tangent is low, and reflection is reduced if the dielectric constant is low. It is possible to reduce reflection to near zero if the window thickness is an integral multiple of $\lambda/2$. The wavelength, λ, of radiation in the window or radome material is equal to the vacuum wavelength, λ_o, divided by the refractive index:

$$\lambda = \frac{\lambda_o}{n} \approx \frac{\lambda_o}{\sqrt{\varepsilon}} \ . \tag{2-10}$$

Infrared domes are hemispheres, and systems are designed so each ray that reaches the detector passes through the dome at normal incidence. Radomes have a pointed aerodynamic shape. Radiation that is transmitted or received passes through the radome at many angles of incidence. Since the optical pathlength of radiation through the radome is not constant, it is not possible to design a radome for zero reflection under all conditions. The dome is designed to minimize reflection at the most common angle of incidence. Since reflection cannot be reduced to zero for all angles of incidence, it is important that the dielectric constant of the radome be as small as possible to minimize reflection losses.

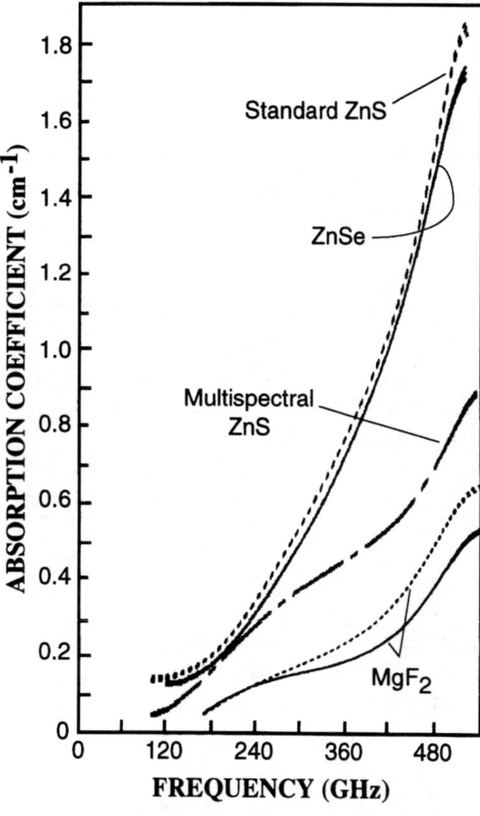

Fig. 2.18. Frequency dependence of absorption coefficient of infrared materials in the millimeter wave region.[30] The two hot pressed magnesium fluoride samples were prepared by different methods.

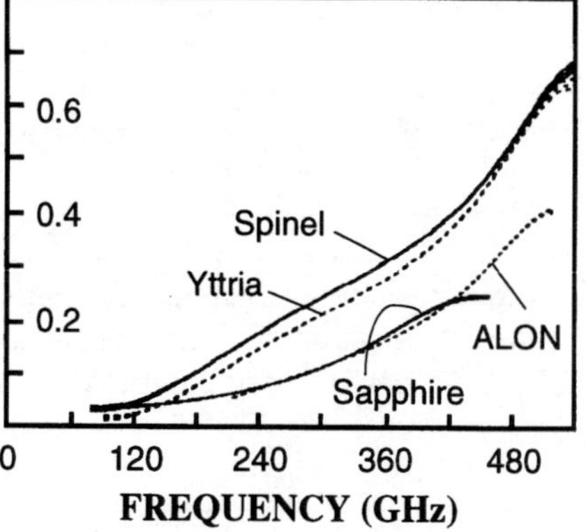

Table 2.1 lists the dielectric constant and loss tangent for common radome materials and some infrared window materials.[25-27] In general, loss tangents of infrared materials are suitably low for microwave transmission. However, the dielectric constants of infrared materials are generally much higher than those of radome materials, giving much more reflection loss for microwaves passing through infrared materials. The two infrared materials with the lowest microwave dielectric constants are magnesium fluoride and diamond, both of which would be excellent microwave windows.

Example: Microwave absorption and reflection. Let's compare the transmission and reflection of 10 GHz microwave radiation impinging on 1.0-cm-thick zinc sulfide or magnesium fluoride at normal incidence. Equation (0-1) tells us that a frequency of 10 GHz corresponds to a wavelength of $\lambda_o = c/v = (3.00 \times 10^8 \text{ m/s})/(10 \times 10^9 \text{ s}^{-1}) = 3.0$ cm in air. Inside ZnS, where the dielectric constant is 8.35 (Table 2.1), the wavelength is $\lambda = \lambda_o/\sqrt{\varepsilon} = 3.0 \text{ cm} / \sqrt{8.35} = 1.04$ cm. With the value of $\tan\delta$ from Table 2.1, we can find the internal transmittance of zinc sulfide with Eq. (2-8):

$$\text{Transmittance} = e^{-[(2\pi/\lambda)\tan\delta]b} = e^{-[(2\pi/(1.04 \text{ cm}))(0.0024)](1.0 \text{ cm})} = 0.986 \ .$$

Only 1 - 0.986 = 1.5% of the radiation is absorbed. For magnesium fluoride, the absorption is even less (0.1%), because its loss tangent is lower. These computations indicate that *microwave absorption losses for the materials in Table 2.1 are negligible.*

Now let's calculate the single-surface reflectance for zinc sulfide using Eq. (1-11). Recall that the refractive index, n, is the square root of the dielectric constant. For zinc sulfide we find $n = \sqrt{8.35} = 2.89$, and

$$\text{One-surface reflectance} = \left(\frac{1 - n}{1 + n}\right)^2 = \left(\frac{1 - 2.89}{1 + 2.89}\right)^2 = 24\% \ .$$

Nearly one-fourth of the microwave energy striking a zinc sulfide surface at normal incidence is reflected. For less-than-normal angles of incidence, the reflection is even greater. You can see that it is critical to design high-dielectric-constant radomes to minimize reflection, or little radiation would be transmitted through the dome. Carrying out the same calculation for magnesium fluoride with a refractive index of $n = \sqrt{5.1} = 2.26$ gives a one-surface reflectance of 15%. An organic composite radome with a dielectric constant of 3 has a one-surface normal-incidence reflectance of just 7%.

Missile radomes become very hot during flight, so it is desirable that the change in dielectric constant with temperature be small. Figure 2.19 shows the temperature dependence of dielectric constant and loss tangent for several infrared materials.[25] One of the factors that increases the dielectric constant is the thermal expansion of the material. Materials with lower thermal expansion tend to have a smaller change in dielectric constant with temperature. Table 2.2 shows the change in microwave refractive index with respect to temperature for some infrared materials.

Table 2.1. Microwave dielectric properties[25,26] of infrared and radome materials*

Material	Dielectric constant (ε)	Loss tangent ($\tan\delta$)
Infrared materials		
Zinc sulfide, ZnS	8.35	0.0024 (35 GHz)
Zinc selenide, ZnSe	8.98	0.0017 (35 GHz)
Magnesium fluoride, MgF_2	5.1	0.0001
Magnesium fluoride, MgF_2, single-crystal[29]	4.87 ($E \parallel c$)	--- (0.1-40 MHz)
	5.45 ($E \perp c$)	--- (0.1-40 MHz)
Calcium fluoride, CaF_2	6.5	0.00015
ALON, $9Al_2O_3 \cdot 5AlN$	9.28	0.00027 (35 GHz)
Spinel, $MgAl_2O_4$	9.19	0.00022 (35 GHz)
Sapphire, Al_2O_3	9.39 ($E \parallel c$)	0.00005 (35 GHz)
	11.58 ($E \perp c$)	0.00006 (35 GHz)
Yttria, Y_2O_3	11.8	0.0005
Lanthana-doped yttria, $0.09La_2O_3 \cdot 0.91Y_2O_3$	12.2	0.0005
Calcium aluminate glass	9.0	0.0025
Diamond	5.7	<0.0004
Gallium arsenide, GaAs (10^5 $\Omega \cdot$m)	~12	~0.003
Silicon (10 $\Omega \cdot$m)	~12	~0.009
Radome materials		
Organic composites	2 - 4	0.0001 - 0.01
Quartz-polyimide	3.2	0.008
DI-100/200	3	0.01
Pyroceram 9606	5.58	0.0008
Rayceram 8 (magnesium aluminum silicate)	4.72	0.003
Nitroxyceram	5.5	0.001
IRBAS	7.75	0.0012
ZPBSN	5.6	0.002
Duroid (Teflon)	2.65	0.003
Fused silica, SiO_2	3.33	0.001
Silicon nitride, Si_3N_4 (reaction sintered)	7.90	0.0017
Silicon nitride, Si_3N_4 (hot pressed)	8.14	0.0006
Boron nitride, BN	5.0	0.0005

*Values in this table generally apply over a wide range of frequencies, from hundreds of megahertz to tens of gigahertz. Two values for sapphire refer to the ordinary and extraordinary directions (Fig. 1.6). Numbers in parentheses for GaAs and Si are low frequency resistivity. DI-100 and DI-200 are quartz-reinforced materials based on silicon polymers manufactured by Textron. Nitroxyceram is particulate-reinforced silicon oxynitride produced by LORAL. IRBAS (in-situ reinforced barium aluminosilicate) and ZPBSN (zirconium-based phosphate-bonded silicon nitride) are products of Lockheed-Martin. Absorption coefficients increase at higher frequencies as shown in Fig. 2.19.[28] Absorption coefficients for sapphire, yttria, spinel, ALON, zinc sulfide and sodium chloride in the 50-200 cm^{-1} far infrared region have been measured.[27] Data for some radome materials are from J. M. Wright, Lockheed Martin Vought Systems.

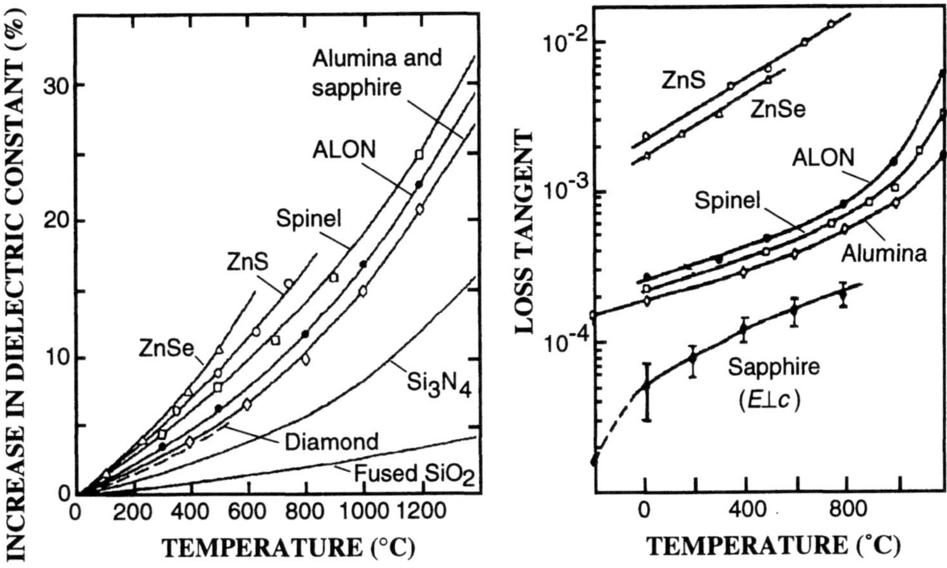

Fig. 2.19. Temperature dependence of dielectric constant and loss tangent for some infrared window materials measured at 35 GHz.[25]

Table 2.2. Microwave refractive index at 3 GHz

Material	Refractive index (n)	dn/dT (10^{-6} K^{-1})
Zinc sulfide, ZnS	2.89	280
Diamond, C	2.38	46
Sapphire, Al_2O_3	3.056 (n_o)	190
	3.400 (n_e)	210
ALON, $9Al_2O_3{\cdot}5AlN$	3.046	210
Spinel, $MgAl_2O_4$	2.88	20
Yttria, Y_2O_3	3.43	100
Magnesium oxide, MgO	3.401	130
Beryllium oxide, BeO	2.59 (n_o)	---
Fused silica, SiO_2	1.944	---

Data from M. E. Thomas, Applied Physics Laboratory.

References

1. H. E. Bennett and D. W. Ricks, "Effects of Surface and Bulk Defects in Transmitting Materials on Optical Resolution and Scattered Light," *Proc. SPIE*, **683**, 153-159 (1986).

2. P. C. Archibald and H. E. Bennett, "Scattering from Infrared Missile Domes," *Opt. Eng.*, **17**, 647-651 (1978).
3. J. C. Stover, "Practical Measurement of Rain Erosion and Scatter from IR Windows," *Proc. SPIE*, **1326**, 321-330 (1990).
4. J. C. Stover, *Optical Scattering: Measurement and Analysis*, McGraw-Hill, New York (1990).
5. F. O. Bartell, E. L. Dereniak and W. L. Wolfe, "The Theory and Measurement of BRDF and BTDF," *Proc. SPIE*, **257**, 154 (1980).
6. M. Melozzi, A. Mazzoni and G. Curti, "Bidirectional Transmittance Distribution Function Measurements on ZnSe and ZnS Cleartran," *Proc. SPIE*, **1512**, 178-188 (1991).
7. G. C. Wei, A. Hecker and W. H. Rhodes, "Scattering in Lanthana-Strengthened Yttria (LSY) Infrared Transmitting Materials," *Proc. SPIE*, **1760**, 14-26 (1992).
8. J. G. J. Peelen and R. Metselaar, "Light Scattering by Pores in Polycrystalline Materials: Transmission Properties of Alumina," *J. Appl. Phys.*, 45, 216-220 (1974).
9. W. J. Smith, *Modern Optical Engineering*, McGraw-Hill, New York (1990).
10. R. E. Fischer, *Optical Design and Engineering: Visible and IR*, SPIE Short Course Notes, SPIE, Bellingham, Washington (1991).
11. D. D. Duncan, C. H. Lange and D. G. Fischer, "Imaging Performance of Crystalline and Polycrystalline Oxides," *Proc. SPIE*, **1326**, 59-70 (1990).
12. J. Trombetta, Raytheon Systems Co. to be published.
13. C. A. Klein, "On Photosaturation of Intrinsic Infrared Detectors," *Appl. Opt.*, **8**, 1897-1900 (1969).
14. C. A. Klein, "Diamond Windows for IR Applications in Adverse Environments," *Diamond and Related Materials*, **2**, 1024-1032 (1993).
15. C. A. Klein, "Infrared Systems and Hot IRdomes: I. The Noise-Equivalent Irradiance," *Proc. SPIE*, **15**, 29-42 (1970).
16. C. A. Klein, "Hot Infrared Domes: A Case Study," *Proc. SPIE*, **1326**, 217-230 (1990).
17. C. A. Klein, "How Missile Windows Degrade the Noise-Equivalent Irradiance of Infrared Seeker Systems," *Proc. SPIE*, **2286**, 458-470 (1994).
18. J. M. Lloyd, "Fundamentals of Electro-Optical Imaging Systems Analysis," in *The Infrared and Electro-Optical Systems Handbook* (M. C. Dudzik, ed.), Vol. 4, pp. 1-53, Environmental Research Institute of Michigan, Ann Arbor, Michigan, and SPIE Press, Bellingham, Washington (1993).
19. P. Klocek, T. McKenna and J. Trombetta, "Thermo-Optic, Thermo-Mechanical, and Electromagnetic Effects in IR Windows and Domes, and the Rationale for GaAs, GaP, and Diamond," *Proc. SPIE*, **2286**, 70-90 (1994).
20. E. F. Cross, "Window Heating Effects on Airborne Infrared System Calibration," *Proc. SPIE*, **1762**, 576-583 (1992).
21. E. F. Cross, "Analytical Method to Calculate Window Heating Effects on IR Seeker Performance," *Proc. SPIE*, **2286**, 493-499 (1994).
22. D. A. Kalin, L. C. Brooks, C. J. Wojciechowski and G. W. Jones, "Performance Characterization of an Internally Cooled Window in a Nonuniform High Heat Flux Environment," *Proc. SPIE*, **3060**, 296-305 (1997).
23. J. Simoneau, R. Feller, C. Peatfield, G. Mackenzie, S. Wirth and D. Zarkh, "New Technique Using a CO_2 Laser for High Heating and Simultaneous Optical Characterization of IR Windows," *Proc. SPIE*, **3060**, 320-328 (1997).
24. J. M. Hall, "Army Applications for Multi-Spectral Windows," *Proc. SPIE*, **3060**, 330-334 (1997).

25. W. W. Ho, "Millimeter-Wave Dielectric Properties of Infrared Window Materials," *Proc. SPIE*, **750**, 161-165 (1987).
26. E. B. Joy, "MMW Radomes," in *Principles and Applications of Millimeter-Wave Radar* (N. C. Currie and C. E. Brown, eds.), Artech House, Norwood, Massachusetts (1987).
27. M. E. Thomas, "Optical Properties of IR Materials," *Thin Solid Films*, **206**, 241-247 (1991).
28. M. Stead and G. Simonis, "Near Millimeter Wave Characterization of Dual Mode Materials," *Appl. Opt.*, **28**, 1874-1776 (1989).
29. A. A. Duncanson and R. W. H. Stevenson, "Some Properties of Magnesium Fluoride Crystallized from the Melt," *Proc. Phys. Soc. London*, **72**, 1001-1006 (1958).

Chapter 3

MECHANICAL PROPERTIES

Infrared window and dome materials are classified as *ceramics*, which is a broad term encompassing inorganic, nonmetallic materials. Ceramics are typically brittle, stable at high temperature, and poor conductors of heat and electricity. In this chapter we discuss the mechanical behavior of brittle materials, which are somewhat different from ductile metals.

For example, Fig. 3.1 shows specimens of yttria fabricated for a test of tensile strength. Each piece was gripped at the ends and pulled apart until it broke. If these had been metal, they would have broken in the narrow section near the middle, and all failures would have occurred at nearly the same load. Instead, three of five samples broke in the thick end region, and the load at failure varied over a range of >20%. It is difficult to make meaningful measurements of tensile strength of optical ceramics with specimens similar to those in Fig. 3.1.

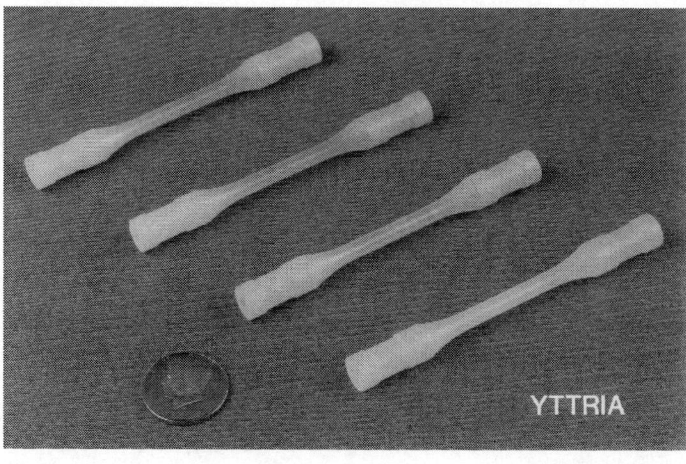

Fig. 3.1. Yttria specimens with 0.4-cm-diameter central gauge section, prepared for unsuccessful tensile strength measurement. (Courtesy Raytheon Co.)

3.1 Elastic constants

Consider a cylinder of solid material that is gripped at the ends and pulled apart until it breaks. The force per unit area pulling on the object is called the *stress*, σ:

$$\text{Stress} = \sigma = \frac{\text{force}}{\text{area}} . \tag{3-1}$$

The area in Eq. (3-1) is the cross sectional area of the cylinder. Units of stress are newtons per square meter, known as *pascals* (Pa = N/m^2). The English units of stress are pounds per square inch (psi). A common abbreviation for 1000 psi is *ksi*. The relation between the English and metric units is ksi × 6.895 = MPa (M = mega = 10^6). For example, a stress of 5.3 ksi is equivalent to 5.3 × 6.895 = 37 MPa.

If the cylinder being pulled apart from the ends has a length ℓ before force is applied, and length $\ell + \Delta\ell$ when stretched, we say that the *strain*, ε, is the fractional increase in length:

$$\text{Strain} = \varepsilon = \frac{\Delta\ell}{\ell} . \tag{3-2}$$

Figure 3.2 shows the behavior of a brittle material and a ductile material when tensile stress is applied. The brittle material stretches slightly, with strain proportional to stress: If the stress is doubled the strain doubles also. Eventually the stress exceeds the strength of the material and it fractures. The *strength* of the material is equal to the stress at failure. The slope of the stress *vs*. strain curve is called *Young's modulus, E:*

$$\text{Young's modulus} = E = \frac{\text{stress}}{\text{strain}} = \frac{\sigma}{\varepsilon} . \tag{3-3}$$

The greater the value of Young's modulus for a material, the stiffer it is. That is, a greater stress is required to produce a given strain. The behavior of the brittle material in Fig. 3.2, prior to rupture, is said to be *elastic:* When the stress is removed, the material returns to its original shape.

In contrast to the brittle material, the ductile material in Fig. 3.2 is easier to stretch. The initial slope of the stress *vs*. strain curve is smaller for the ductile material than for the brittle material. Furthermore, after a certain amount of stretching, the slope of the stress *vs*. strain curve decreases, and the ductile material becomes even easier to stretch. If the stress were removed at this point, the ductile material would not return to its original shape. Eventually, the ductile material fails (breaks) also, but it has stretched much more than the brittle material at a similar stress.

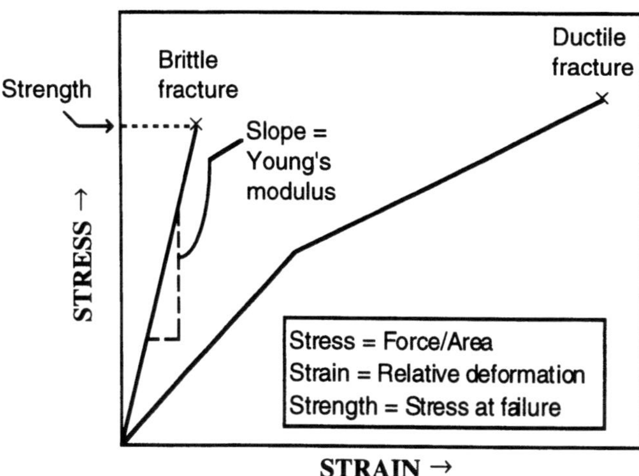

Fig. 3.2. Response of brittle and ductile materials to tensile stress.

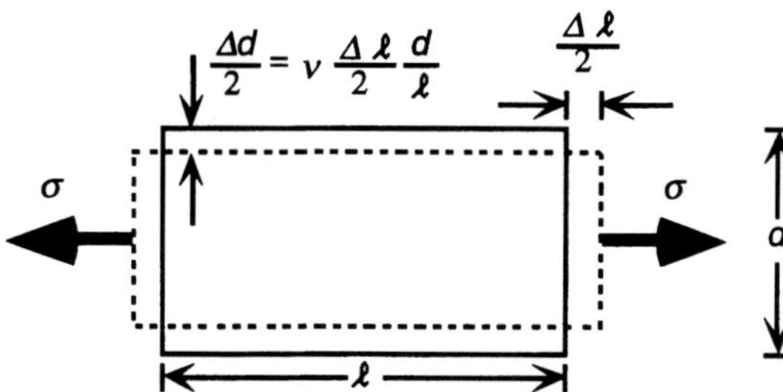

Fig. 3.3. The length of a solid object increases by $\Delta\ell$ and the diameter shrinks by Δd when the stress σ is applied.

Figure 3.3 shows a cross section of a cylindrical solid with initial length ℓ and initial diameter *d*. After a tensile stress σ is applied, the length increases to $\ell + \Delta\ell$ and the diameter shrinks to $d - \Delta d$. *Poisson's ratio*, ν, (pronounced **pwa**-san) is defined as the fractional change in diameter divided by the fractional change in length:

$$\text{Poisson's ratio} = \nu = \frac{\Delta d/d}{\Delta\ell/\ell} . \tag{3-4}$$

Ceramic materials undergo an elastic deformation for which Poisson's ratio is typically in the range 0.2 - 0.3. A material that exhibits plastic (viscous) flow or creep maintains a constant volume as it distorts. In such case Poisson's ratio is 0.5.

Obtaining stress-strain data by a common method

"Obtaining stress-strain data by a common method." So said the caption of this photograph from A. A. Somerville, J. M. Ball and L. A. Edland, "Autographic Stress-Strain Curves of Rubber at Low Elongations," *Industrial and Engineering Chemistry, Analytical Edition*, **2,** 289-293 (1930). Anthropologists who have studied this behavior generally believe that the method was harder for the man at the bottom.

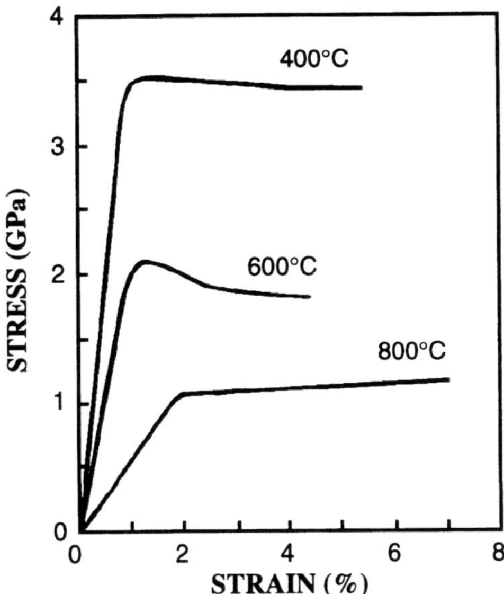

Fig. 3.4. Illustration of plastic deformation of a ceramic.[1] Stress-strain curves for sapphire under uniaxial compression along the *c*-axis of the crystal (see Fig. 1.6). In most other kinds of mechanical tests, sapphire behaves elastically at the temperatures in this graph. Plastic behavior is not generally observed until higher temperatures are reached.

Figure 3.4 shows the response of a ceramic material which exhibits a transition from elastic behavior to plastic behavior at high stress. When the stress is reduced to zero, the elastic portion of the strain is recovered, but the plastic portion remains. In a tensile test, the *yield strength* is usually defined as the stress that leaves a residual strain of 0.2% when the stress is removed. In Fig. 3.4, the yield strength would be close to the bend in each curve. Plastic behavior is common in metals at room temperature, and in ceramics at sufficiently high temperature.

If a body is compressed by a uniform stress of P N/m^2 in all directions (Fig. 3.5, *left*), and its volume shrinks from V to $V - \Delta V$, we define the *bulk modulus* as

$$\text{Bulk modulus} = K = \frac{P}{\Delta V / V} . \tag{3-5}$$

Similarly, if the body at the right of Fig. 3.5 is subjected to a shear stress, τ, and distorts by the angle γ (measured in radians), we define the *shear modulus* as

$$\text{Shear modulus} = \mu = \frac{\tau}{\gamma} . \tag{3-6}$$

Our discussion of elastic constants applies to *isotropic* solids, such as glasses or polycrystalline materials, whose properties are the same in all directions. The behavior of single crystals is *anisotropic* (not the same in all directions) and is discussed in Appendix E. Elastic constants can be measured from stress-strain curves or by measuring the speed of sound in a solid. As a concrete example, the elastic constants for Corning 7940 fused silica (a glass) are: E (Young's modulus) = 73 GPa, μ (shear modulus) = 31 GPa, and K (bulk modulus) = 36.9 GPa. In general the shear modulus is approximately 40% as great as Young's modulus, with the relation between them being $\mu = E/(2 + 2\nu)$, where ν is Poisson's ratio. The bulk modulus is given by $K = E/(3 - 6\nu)$.

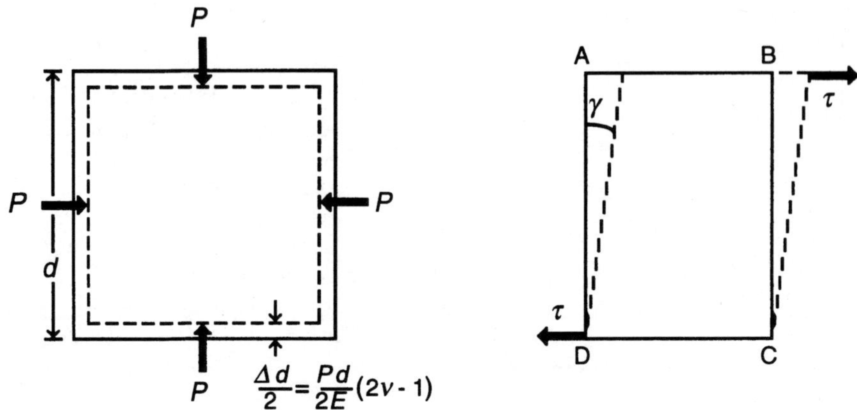

Fig. 3.5. Definition of elastic constants for bulk compression (*left*) and shear (*right*). The shear stress, τ, is defined as the shear force in the direction of τ divided by the area of the plane ABCD. For more on shear, see Fig. E.3 in Appendix E.

Most ceramics maintain their mechanical properties over a wide temperature range, before becoming weaker and more compliant at high temperature. For example, the temperature dependence of different elastic moduli of polycrystalline lanthana-doped yttria is shown in Fig. 3.6.

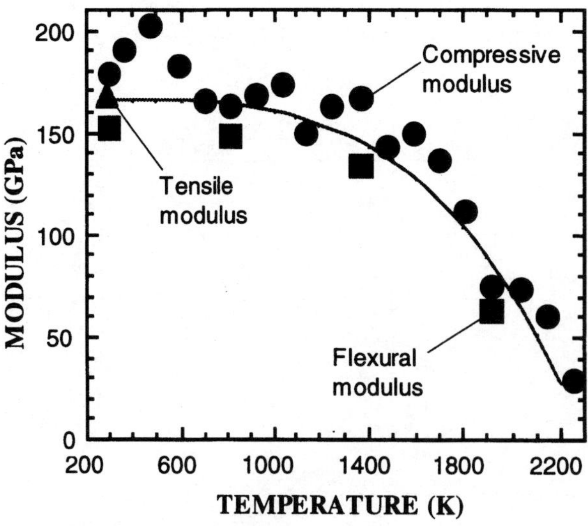

Fig. 3.6. Temperature dependence of moduli of lanthana-doped yttria.[2] The triangle shows the tensile modulus. Circles refer to the experiment in Fig. 3.3 with compression instead of tension. Squares come from an experiment in which bars of material are bent. The solid line is an approximate fit by the equation $E = (166 \text{ GPa}) \times \{1 - [T/(2300 \text{ K})]^4\}$, where E is modulus and T is temperature (K).

3.2 Measuring the strength of brittle materials

We saw in Fig. 3.2 that the strength of a brittle material is defined as the stress at which the material fails. The strength is also called the *modulus of rupture*, or MOR.

For metals, the strength of similar pieces is reproducible. For ceramics, the strength of seemingly identical pieces is highly variable. It is not unusual for the strongest piece in a set to be twice as strong as the weakest piece. We will discuss how strength is measured, how it is characterized statistically, and why it varies so much.

3.2.1 3-point and 4-point flexure tests

Pulling apart the samples in Fig. 3.1 to measure tensile strength does not work well because the pieces break anywhere, not just in the thin central section in which the tensile stress is easily calculated. A more successful approach is to bend a ceramic bar until it breaks, and thereby measure the *flexure strength*.[3]

The simplest approach to flexure testing is the 3-point bending test in Fig. 3.7, in which a load is applied at the top center of a test specimen with a rectangular cross section. When the bar flexes, the bottom surface is in tension and the top surface is in compression. Fracture normally originates at the tensile surface. The tensile stress directly below the load on the bottom surface of the specimen is

$$\text{Maximum tensile stress} = \sigma = \frac{3PL}{2bd^2} \qquad (3\text{-}7)$$

where P is the applied force (newtons, N), L is the length between supports, b is the width of the specimen and d is the thickness of the specimen. The stress decreases linearly from the center until it reaches zero at the positions of the supports. A major problem with this method of measuring strength is that the exact position at which failure originates must be measured in order to calculate strength.

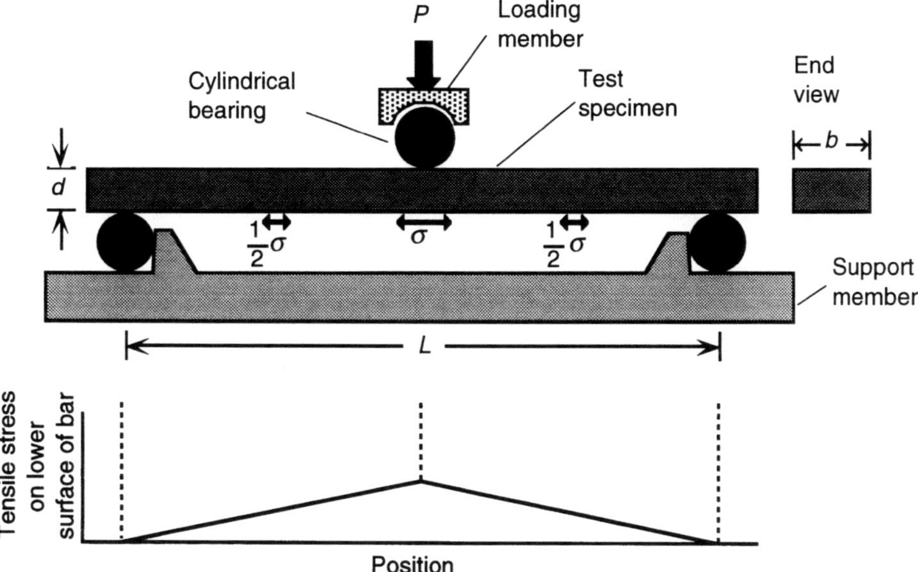

Fig. 3.7. Flexure strength measurement using 3-point bending. Load is applied at the middle of the specimen. Black circles are end views of cylinders.

Example: Modulus of rupture in 3-point bend test. Suppose that the length, L, in Fig. 3.7 is 40.0 mm, the width, b, is 4.00 mm, and the thickness, d, is 3.00 mm. Let's find the modulus of rupture (the strength) if the specimen breaks 8.3 mm from the right hand support when a load of $P = 352$ N (79.1 pounds, N × 0.2248 = pounds) is applied. The maximum tensile stress beneath the load is given by Eq. (3-7):

$$\sigma = \frac{3\ (352\ \text{N})\ (0.0400\ \text{m})}{2\ (0.00400\ \text{m})\ (0.00300\ \text{m})^2} = 5.87 \times 10^8\ \text{N/m}^2 = 587\ \text{MPa}.$$

The horizontal distance from the support to the load point is 20.0 mm. The fraction of this distance at which failure occurred is 8.3/20.0 = 0.415, so the stress at the failure point is (0.415) (587 MPa) = 244 MPa. The strength of this specimen is 244 MPa.

A better measurement of strength is the 4-point flexure test in Fig. 3.8.[4,5,6] In this case, *the tensile stress on the bottom (tensile) surface of the bar is constant between the two load points*. The stress on the tensile face between the load points is

$$\text{Constant tensile stress between load points} = \sigma = \frac{3PD}{bd^2} \tag{3-8}$$

where D is the distance between the outer and inner cylindrical bearings and the other symbols have the same meaning as in Eq. (3-7). The length D is commonly equal to L/4 so that the span between the upper load points is $L/2$. As long as the sample breaks anywhere between the two load points, Eq. (3-8) gives the correct strength. As in the 3-point test, the stress decreases linearly from the maximum value down to zero in the span between the load and support points.

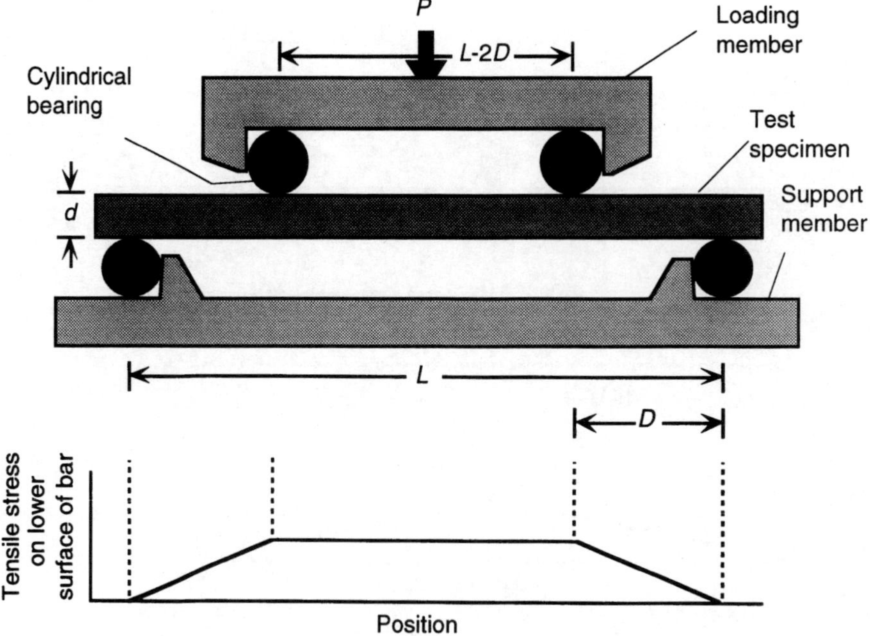

Fig. 3.8. Flexure strength measurement by 4-point bending.

The deflection of the lower surface of the bend bars in Figs. 3.7 and 3.8 can be used to compute Young's modulus, which is called the *flexural modulus* in Fig. 3.6. The maximum vertical deflection at the center of the bar, measured with respect to the support points, is

$$\text{Maximum deflection in 3-point bending} = \frac{PL^3}{4Ebd^3} . \tag{3-9}$$

$$\text{Maximum deflection in 4-point bending} = \frac{PD}{4Ebd^3}(3L^2 - 4D^2) . \tag{3-10}$$

If a flexure specimen is not stressed to failure, and if its deflection is elastic, it will return to its original shape when the load is removed. Figure 3.9 shows a ceramic specimen that experienced plastic deformation at 1500°C and did not return to its original shape when the stress was removed, even though the specimen did not break.

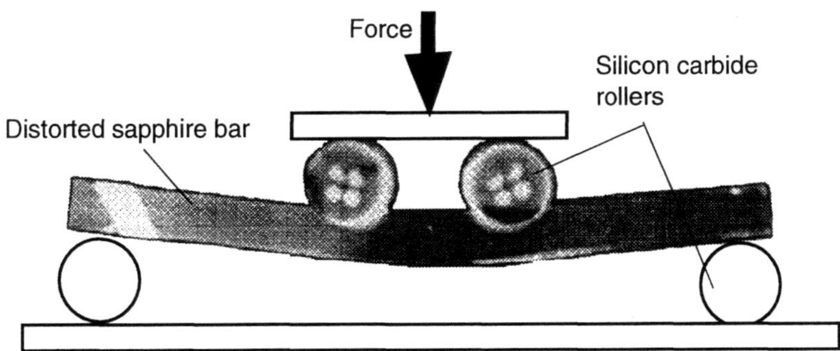

Fig. 3.9. Plastic deformation: Digital image (200% of actual size) of a permanently distorted sapphire bar after flexure testing at 1500°C. The image was made after the load was removed and the sample cooled to room temperature. The inner load cylinders were impressed into the surface of the sapphire. Lower circles show approximate positions of the outer support cylinders, which did not stick to the sapphire. The diagonal feature near the left is a crystal twin plane seen with crossed polarizers. [Courtesy G. A. Graves, University of Dayton Research Institute.]

The *strain rate*, $\dot{\varepsilon}$, in the flexure test in Fig. 3.8 is the rate of change of strain ($\Delta\ell/\ell$ in Fig. 3.3) in the region of highest stress on the tensile surface. If the testing machine is perfectly rigid and has a constant displacement speed, s, then the strain rate in the flexure bar is[7]

$$\text{Strain rate in flexure test} = \dot{\varepsilon} = \frac{3sd}{D\,(3L - 4D)} . \tag{3-11}$$

Equation (3-11) also applies to 3-point flexure in Fig. 3.7 with $D = L/2$.

Example: Strain rate in 4-point bend test. Suppose that L = 40.0 mm, D = 15.0 mm and d = 3.00 mm in Fig. 3.8. Let's find the strain rate if the testing machine has a constant vertical displacement rate (called the *crosshead speed*) of s = 0.508 mm/min. Using Eq. (3-11) we find

$$\dot{\varepsilon} = \frac{3sd}{D\,(3L - 4D)} = \frac{3(0.508 \text{ mm/min})(3.00 \text{ mm})}{(15.0 \text{ mm})\,[3(40.0 \text{ mm}) - 4(15.0 \text{ mm})]} = 0.00508 \text{ min}^{-1}\,.$$

If the same crosshead speed were used for a pure tensile test, as in Fig. 3.3, and if the length of the tensile specimen were 40.0 mm, the strain rate would be (s = 0.508 mm/min)/(40.0 mm) = 0.0127 min^{-1}, or 2.5 times greater than in the flexure test.

Fracture usually originates on the tensile surface. The fracture origin in a ceramic is a strength-limiting flaw such as a microscopic scratch or a tiny inclusion of foreign material, or a void. The strength depends on the size and distribution of flaws, which depends on the quality of the optical finish. The better the polished surface, the smaller are the microscopic scratches, and the stronger the sample becomes. *To get a meaningful measurement of strength, the tensile surface of test specimens should have the same finish as a window or dome.*

A common problem with both the 3-point and 4-point bend tests is that fracture originates where stress is concentrated on the long edges of the tensile face, instead of on the face. Therefore, the long edges should be uniformly *chamfered* (beveled) at a 45° angle, or rounded. In either case, the edge should be polished to the same quality finish as the face. Even with these precautions, edge failures of optically polished materials are common.

3.2.2 Equibiaxial disk flexure test

The preferred measure of strength of an optical window or dome material is the ring-on-ring equibiaxial flexure test in Fig. 3.10 because the face of the disk can be polished to similar specifications as an optical window.* A flat metal ball or a metal ring with radius a is used to apply a load, P, to the top of the ceramic disk test specimen that is supported on a metal ring of radius b. The radius of the ceramic disk is c and its thickness is d. The lower surface of the disk is in tension and the upper surface is in compression during the test. The two components of stress parallel to the surface are called radial and hoop (or tangential) stress, as shown in Fig. 3.10. This test is said to be "equibiaxial" because the magnitude of the radial stress is equal to the magnitude of the hoop stress for all points on the tensile surface within the load radius, a. As long as disk failure occurs within the load radius, we can be sure that we know the stress.

The hoop and radial tensile stress inside the load radius on the tensile surface of the disk in the ring-on-ring flexure test are given by

$$\text{Stress within load radius} = \frac{3P\,(1-\nu)}{4\pi d^2}\left(\frac{b^2-a^2}{c^2} - 2\,\frac{1+\nu}{1-\nu}\ln\frac{a}{b}\right) \tag{3-12}$$

---

*The preferred strength test should closely simulate the stress state in the intended application. If a window or dome will see maximum stress on its polished faces, the ring-on-ring test is preferred. If the window will see maximum stress at the edge, a 4-point bending test may be preferred.

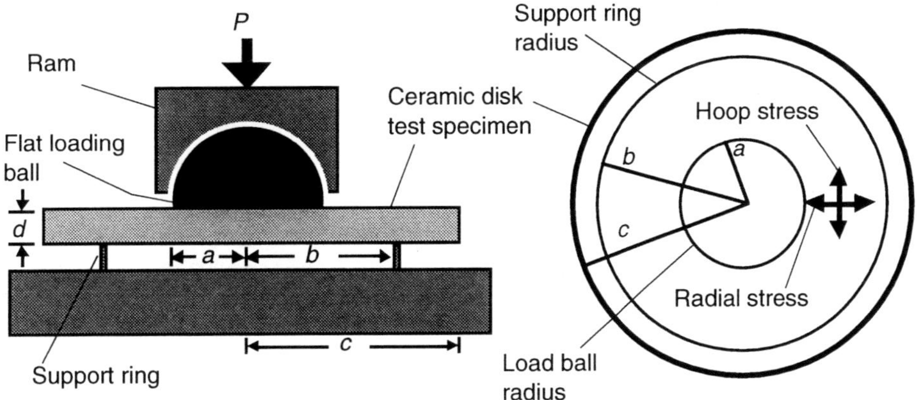

Fig. 3.10. Ring-on-ring equibiaxial flexure test. Radial stress is along the radius and hoop stress is tangential.

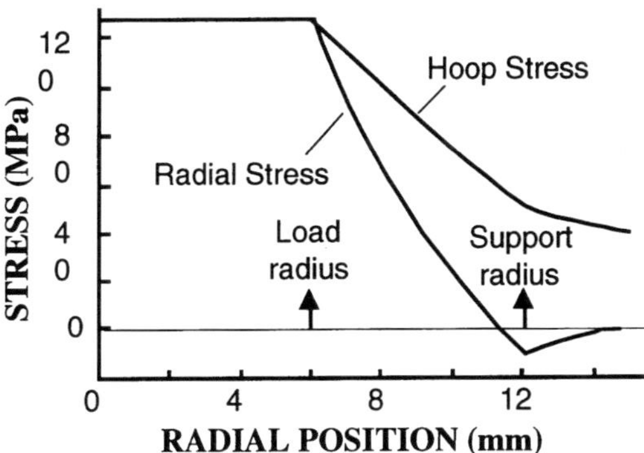

Fig. 3.11. Radial and hoop stresses on lower surface of disk in ring-on-ring flexure test in Fig. 3.10 (a = 6 mm, b = 12 mm, c = 15 mm, d = 2 mm, ν = 0.3). The negative sign of the radial stress near 12 mm indicates that it is compressive, rather than tensile. Formulas to calculate stresses throughout the entire volume of the disk can be found in References 8 and 9.

where ν is Poisson's ratio for the material being tested. For any radius greater than a in Fig. 3.10, the hoop and radial stresses are not equal, as illustrated in Fig. 3.11. The low values of both stresses near the edge of the disk reduce the probability of failure originating at the edge instead of the face. The maximum deflection of the tensile surface of the disk, relative to the support ring, is

$$\text{Maximum disk deflection} = \frac{3P\,(1-\nu^2)}{2\pi E d^3}\left(a^2 \ln\frac{a}{b} + (b^2 - a^2)\left[1 + \frac{1-\nu}{2(1+\nu)}\,\frac{b^2}{c^2}\right]\right) \quad (3\text{-}13)$$

Example: Stresses and deflection in ring-on-ring flexure test. Let's find the stress in the central region of a disk bearing a load of 1000 N (224.8 pounds). Suppose that the disk radius is c = 15.0 mm and thickness is d = 2.00 mm. Let the load ring radius be a = 6.0 mm and the support ring radius be b = 12.0 mm. Suppose also that the disk is made of yttria whose Poisson's ratio is 0.30 in Table E.3 in Appendix E. Plugging these values into Eq. (3-12) gives

$$\sigma = \frac{3\,(1000\text{ N})\,(1-0.30)}{4\pi(0.00200\text{ m})^2}\left(\frac{12.0^2-6.0^2}{15.0^2} - 2\,\frac{1+0.30}{1-0.30}\ln\frac{6.0}{12.0}\right) = 128\text{ MPa}\,.$$

What is the deflection at the center of the tensile surface? We need to know Young's modulus, which is 173 GPa (173 × 10^9 N/m) for yttria in Table E.3. Eq. 3-13 gives:

$$\text{Maximum deflection} = \frac{3\,(1000\text{ N})(1-0.30)^2}{2\pi\,(173\text{ GPa})(2.00\text{ mm})^3}\times$$

$$\left((6.0\text{ mm})^2\ln\frac{6.0}{12.0} + [(12.0\text{ mm})^2 - (6.0\text{ mm})^2]\left[1 + \frac{1-0.30}{2(1+0.30)}\,\frac{12.0^2}{15.0^2}\right]\right)$$

$$= 17\ \mu\text{m}$$

The deflection is 17 μm, which is just 0.86% of the thickness of the disk.

Equations in this text are based on elastic bending of plates with the approximation that the only important stresses are bending stresses. Satisfying the approximation places upper and lower bounds on the allowed thickness of a flexure disk.[10] The disk should not be thicker than 1/5 of the diameter of the support ring ($2b$ in Fig. 3.10) and the disk should not be so thin that the center deflection exceeds 1/2 of the disk thickness.

Several experimental and theoretical studies have investigated the stresses in the ring-on-ring equibiaxial flexure test more closely.[10-12] In different studies, the radial or hoop stress on the tensile surface beneath the load ring was 20-50% higher than the stress at the center of the disk where Eq. (3-12) is valid. High fidelity finite element calculations for the loading of a zinc sulfide disk by a steel load ring with a curved contact surface show that nearly 10% of the disk radius experiences a stress at least 30% larger than that calculated by Eq. (3-12).[11] Amplification of the stress beneath the load ring explains the common observation that many fracture origins in disk tests occur near the load radius. Substitution of a compliant, flat load ring made of Delrin for the steel load ring decreased the excess stress beneath the load ring by a factor of 5.[11]

Figure 3.12 shows two ceramic disks that were broken in a ring-on-ring flexure test. The disk on the right broke at 1.4 times as much load as the disk on the left. It is typical that the stronger sample broke into more pieces because it stored 40% more energy than the weaker sample prior to shattering. An experienced person can look at the fracture patterns in Fig. 3.12 and make an educated guess as to the failure origin of each sample. However, it normally requires detailed investigation with a scanning electron microscope to identify the failure origin unequivocally. The dashed lines in Fig. 3.12 show the positions of the load and support rings. You can see in the disk on the right that a good deal of the damage is concentrated beneath the load ring, probably because the stress at the load radius is higher than the stresses elsewhere in the disk. In an ideal test, the stress would not be amplified beneath the load ring.

3.3 Ceramics fracture at pre-existing flaws

A close-up look at the fracture origin in disks such as those in Fig. 3.12 may reveal a scene such as that in Fig. 3.13. Distinct "river marks" (a pattern of lines) lead back toward a "fracture mirror" within which the failure began. The approximate dimensions

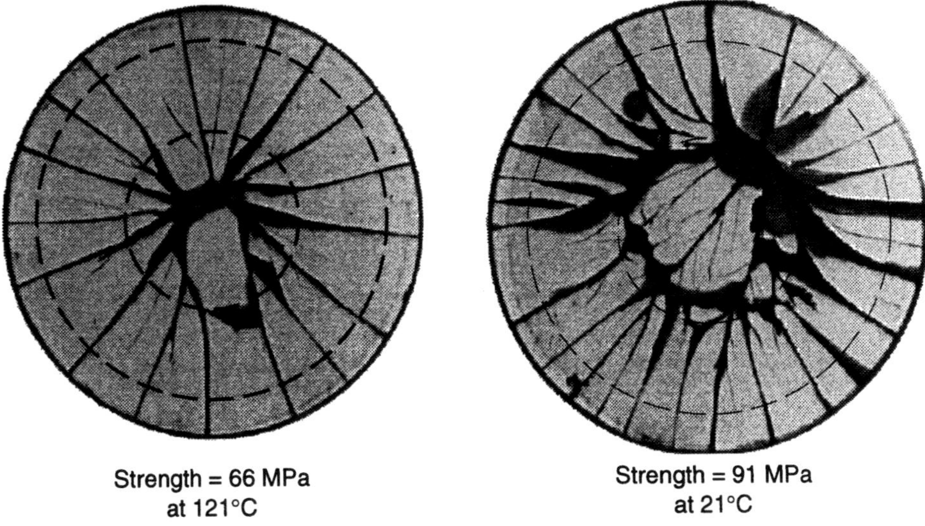

Fig. 3.12. Zinc sulfide disks from a ring-on-ring flexure test illustrating that stronger specimens tend to break into more fragments. Positions of the load and support rings are indicated by dashed lines. Note the high concentration of damage at the load ring radius in the specimen at the right.

of the mirror are designated *a* and *2b* in Fig. 3.13. Examination of the tensile surface of the sample near the fracture mirror might show polishing scratches or mechanical damage at which stress was concentrated to initiate fracture. Alternatively, the origin might be a defective grain or grain boundary or a void. Occasionally the fracture origin is internal, but most often it is at the surface of the specimen. The "critical flaw" is the one at which fracture begins when the flaw strength is exceeded by the applied stress. The larger the size of the flaw, the easier it is for the ceramic to fracture. Examination of the fracture origin is called *fractography*.

Fig. 3.13. Fracture origin in a ceramic disk seen under a scanning electron microscope. Horizontal bar at the bottom has a length of 100 μm (0.1 mm). The label *2b* marking the length of the fracture mirror is drawn at the surface of the specimen. The depth of the critical flaw is *a*. "River marks" are the series of lines radiating down into the sample from the fracture mirror. [Photograph courtesy Jack Mecholsky, University of Florida.]

Ceramic failure originates at pre-existing flaws. When the applied stress exceeds the strength of the critical flaw, the specimen shatters.

Figure 3.14 shows the edge of an optically polished ceramic disk prepared for strength measurement. Compared to the polished face, the edge is very rough and, therefore, weaker. If the stress were as high at the edge of the disk as at the center, most samples would fracture at the edge. The measurement of strength would be meaningless.

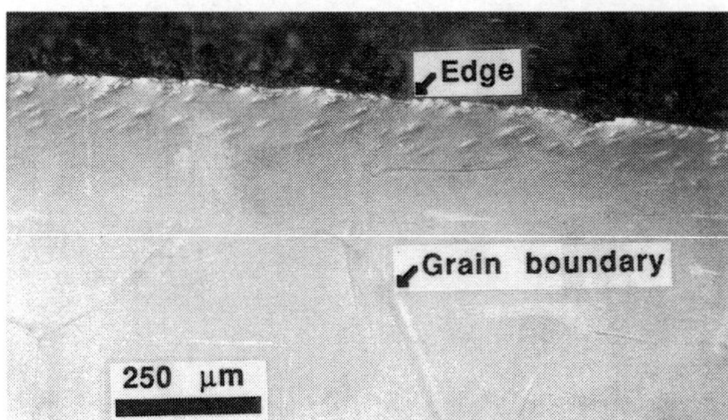

Fig. 3.14. Irregular edge of optically polished ceramic disk showing numerous flaws at which fracture might originate. Lines on the flat, polished surface are grain boundaries. (Photograph courtesy Marian Hills, Naval Air Warfare Center.)

3.3.1 Stress concentration by cracks

The theoretical strength of a defect-free material can be estimated from the forces needed to pull planes of atoms apart from each other.[13] The result is

$$\text{Theoretical strength} = \sqrt{\frac{E\gamma_{SE}}{a_o}} \approx \frac{E}{10} \qquad (3\text{-}14)$$

where E is Young's modulus, γ_{SE} is the surface energy (the energy needed to create a unit area of new surface, J/m^2), and a_o is the spacing between rows of atoms in the crystal.

Although fine whiskers of nearly perfect crystals can achieve strengths near the theoretical limit, bulk ceramics are typically weaker by two orders of magnitude. For polycrystalline zinc sulfide, for example, E = 74 GPa, so its theoretical strength is ~7 GPa = 7000 MPa. The actual strength of zinc sulfide is ~100 MPa, which is around 1% of the theoretical strength.

We can understand why most real materials are so weak by considering the concentration of stress at the tip of a flaw in the real material. Figure 3.15 shows a simple approximation for a flaw in a plate of material whose dimensions are assumed to be much larger than the flaw. The flaw is an elliptical opening with axis lenghs $2b$ and $2c$. The stress in the material in the y direction (parallel to the applied stress) at the end of the flaw is

$$\sigma_y = \sigma_{applied}\left(1 + \frac{2c}{b}\right) = \sigma_{applied}\left(1 + 2\sqrt{\frac{c}{\rho}}\right) \qquad (3\text{-}15)$$

where ρ is the radius of curvature ($= b^2/c$) of the ellipse at the left and right ends.

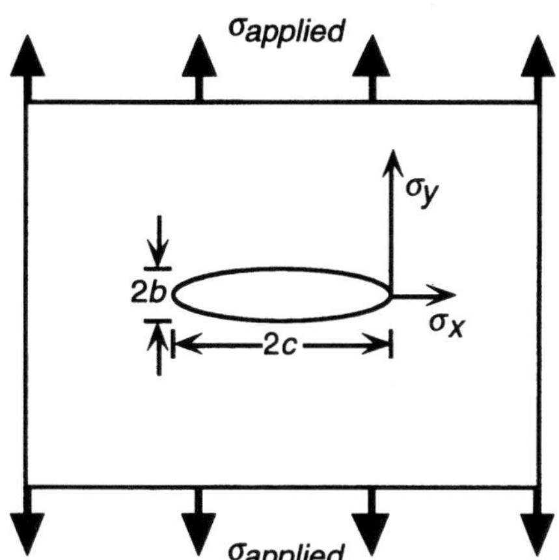

Fig. 3.15. Elliptical cavity with semi-axes b and c in large plate with applied stress $\sigma_{applied}$ in the y direction.

Equation (3-15) says that the more elongated the crack, the smaller its radius of curvature at the tips and the more amplified the stress at the crack tips. If $c/b = 5$, then $\sigma_y = 11\sigma_{applied}$: That is, the stress at the tip of the crack is 11 times greater than the applied stress. Thin cracks amplify the applied stress and can bring the material to the point of mechanical failure.

Most of our current thinking about the fracture of brittle materials is based on energy considerations first introduced by Griffith.[13,14] He considered a thin crack of length c_o penetrating a material from the edge, as in Fig. 3.16. Under the applied stress, $\sigma_{applied}$, the crack might propagate by an infinitesimal distance, dc.

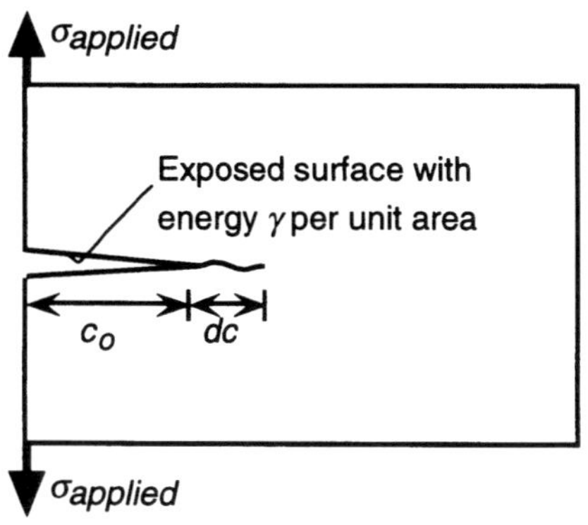

Fig. 3.16. Thin crack of length c at the surface of a material.

Griffith equated the total energy of the system to the mechanical potential energy stored in the strained ceramic (which behaves as a stretched spring) plus the mechanical work associated with moving apart the two loading points plus the surface energy of the exposed crack. The total energy reaches a maximum when the applied stress is

Griffith strength relation: $$\sigma_{critical} = \sqrt{\frac{2E\gamma_{SE}}{\pi c_o}} \qquad (3\text{-}16)$$

where E is Young's modulus and γ_{SE} is the surface energy per unit area. For an applied stress less than $\sigma_{critical}$, the crack does not propagate because the energy needed to expose new surface is greater than the mechanical energy applied to the system. For an applied stress greater than $\sigma_{critical}$, the crack will open abruptly. Ideally, nothing happens to the crack as stress is applied until a critical stress is reached. Then the crack opens abruptly. The strength of the material is $\sigma_{critical}$. The Griffith equation predicts that the strength of a material decreases inversely as the square root of the size of the critical flaw from which failure originates.

3.3.2 Strain rate dependence of strength

Contrary to the ideal behavior predicted by Eq. (3-16), it is possible for a crack of subcritical size to grow slowly under an applied stress. When the crack reaches the critical size, it cannot resist the load and failure occurs abruptly. Therefore, the strength measured in various mechanical tests can depend on the rate at which stress is applied. If the strain rate is slow, subcritical cracks have time to grow and the material will ultimately fail at a lower stress than would have been observed if the cracks had not grown. Figure 3.17 shows that the faster the strain rate used for strength measurement, the stronger the ceramic will be.[15,16] From measurements such as those in Fig. 3.17, we can deduce the growth rates of subcritical cracks during the mechanical test.[17-19]

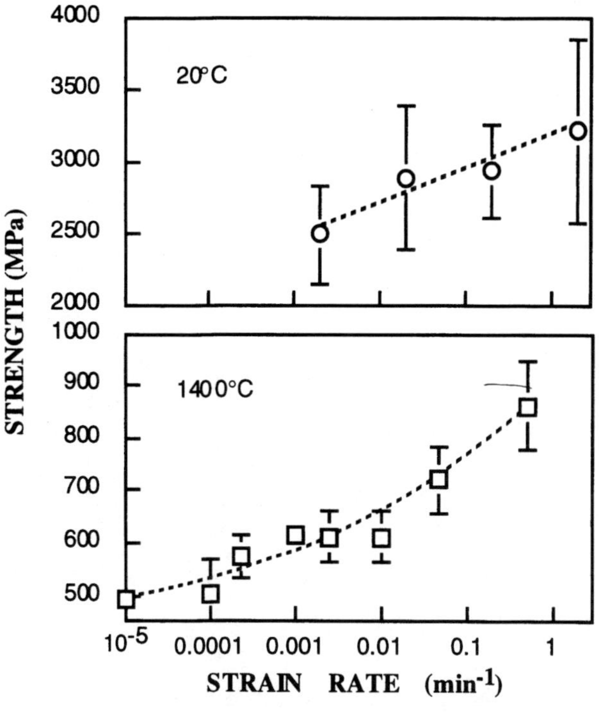

Fig. 3.17. Dependence of rupture strength on strain rate for c-axis sapphire fibers (~200 μm diameter) at 20°C[15] and 1400°C.[16] Dashed curves are guides for the eye. Error bars are one standard deviation.

3.4 Weibull statistics

Ceramic strength is statistical in nature. What is the probability that the right flaw with the right orientation is present to initiate failure? While there are elaborate approaches to dealing with this question,[9,20] we introduce only the simplest and most common description of ceramic failure statistics – the Weibull distribution.[21,22] Appendix F provides more detail on Weibull statistics and explains how to predict the behavior of a material in one kind of mechanical test from its behavior in a different kind of mechanical test.

3.4.1 The Weibull distribution

As a concrete example, consider the flexure strengths of a set of zinc sulfide disks in Table 3.1. The samples are numbered $i = 1$ to 13, with strengths ordered from weakest to strongest in the second column. We define the *probability of failure* as[23]

$$\text{Probability of failure} = P = \frac{i - {}^1/_2}{N} \tag{3-17}$$

where N is the total number of samples tested ($N = 13$). Weibull related the probability of failure to strength, S, with the empirical equation

$$\text{Weibull distribution:} \quad P = 1 - e^{-(S/S_o)^m} \tag{3-18}$$

where m and S_o are constants called the *Weibull modulus* and *scaling factor*, respectively.

More complete forms of the Weibull distribution discussed in Appendix F take into account the area (A) or volume (V) of the specimen, depending on whether fracture originates on the surface or in the volume. The more complete equations are

$$P = 1 - e^{-(A/A_o)(S/S_o)^m} \quad \text{or} \quad P = 1 - e^{-(V/V_o)(S/S_o)^m} \tag{3-19}$$

where A_o and V_o are arbitrarily chosen unit areas or volumes such as 1 cm^2 or 1 cm^3. In

Table 3.1. Data for Weibull plot of zinc sulfide strength*

Sample number, i	Strength, S (MPa)	Probability of failure $P = (i - {}^1/_2)/N$	$\ln\{\ln[1/(1-P)]\}$	$\ln S$
1	62	0.038	-3.239	4.127
2	69	0.115	-2.099	4.234
3	73	0.192	-1.544	4.290
4	76	0.269	-1.159	4.331
5	87	0.346	-0.856	4.466
6	89	0.423	-0.598	4.489
7	90	0.500	-0.367	4.500
8	93	0.577	-0.151	4.533
9	100	0.654	0.059	4.605
10	107	0.731	0.272	4.673
11	110	0.808	0.500	4.700
12	125	0.885	0.770	4.828
13	126	0.962	1.181	4.836

*Values were measured with standard grade zinc sulfide disks of radius 12.7 mm and thickness 1.96 mm using a ring-on-ring test fixture with a load radius of 5.36 mm and support radius of 10.13 mm.

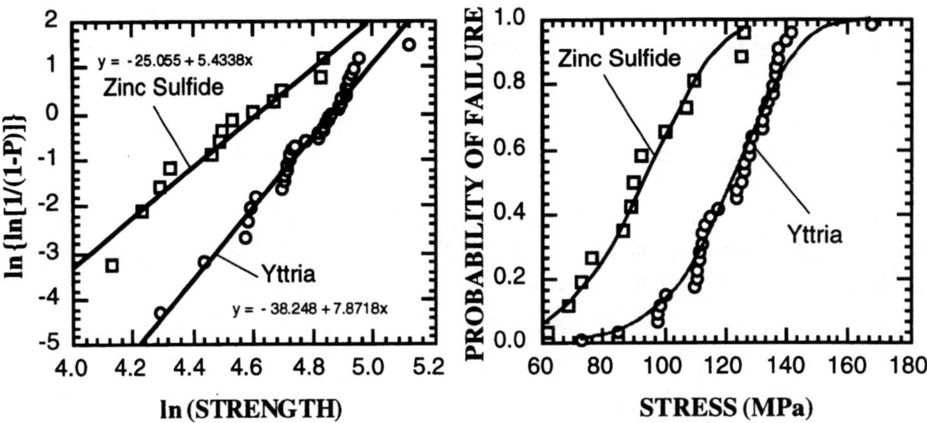

Fig. 3.18. *Left:* Weibull plot for zinc sulfide and yttria disks tested in ring-on-ring flexure. *Right:* Probability of failure *vs.* stress for the same samples shown at the left. Smooth curves are best-fit Weibull curves given by Eq. (3-18) with $m = 5.4338$ and $S_o = 100.6$ MPa for zinc sulfide and $m = 7.8718$ and $S_o = 128.9$ MPa for yttria.

Equation (3-18) the unit area or volume is taken as that of the test specimen. That is, $A = A_o$ or $V = V_o$.

Equation (3-18) can be cast in linear form by rearranging and taking the natural logarithm twice. The result is:

$$\ln\left(\ln\frac{1}{1-P}\right) = m \ln S - m \ln S_o . \qquad (3\text{-}20)$$

To find m and S_o, we prepare a graph of ln {ln [1/(1-P)]} *vs.* ln S. The slope is m, and the intercept is $m \ln S_o$. The upper left curve in Fig. 3.18 is derived from the data in Table 3.1. The slope (5.4338*) of the straight line drawn through the points is the Weibull modulus. The scaling factor is obtained from the intercept:

$$S_o = e^{(-\text{intercept}/m)} = e^{(25.055/5.4338)} = 100.6 \text{ MPa} . \qquad (3\text{-}21)$$

The scaling factor is somewhat greater than the mean strength of the disks, which is 93 MPa. Equation (3-24) states how the mean strength is related to the scaling factor.

At the right side of Fig. 3.18 is a graph of probability of failure *vs.* stress. The smooth curve is the Weibull function in Eq. (3-18) with the values $m = 5.4338$ and $S_o = 100.6$ MPa. We associate the lowest probability of failure with the lowest stresses and the highest probability of failure with the highest stresses. That is, by the time the applied stress is 120 MPa, the probability that the zinc sulfide disk will break is high.

---

*We retain many digits to avoid roundoff errors in subsequent calculations. The spread in the data rarely justifies more than one decimal place in the Weibull modulus.

A higher Weibull modulus implies that the strength is more reproducible. The lower curve at the left in Fig. 3.18 is for a set of 37 yttria disks. The steeper slope tells us that the Weibull modulus is higher for this set of yttria disks than for this set of zinc sulfide disks. The relative spread of strength is less for yttria than for zinc sulfide. The average strength of the yttria samples is 121 MPa, with a standard deviation of 18 MPa, or 15%. The average strength of the zinc sulfide samples is 93 MPa, with a standard deviation of 20 MPa, or 22%.

Equation (3-18) is a 2-parameter Weibull distribution with the empirical parameters m and S_o. This equation adequately describes the data in Fig. 3.18. Sometimes the behavior in Fig. 3.19 is observed and a third parameter is required to fit the data. The 3-parameter Weibull distribution includes the factor S_u, which is the critical stress below which failure does not occur:

$$\text{3-Parameter Weibull distribution: } P = 1 - e^{-(S/[S_o - S_u])^m} . \tag{3-22}$$

The 2-parameter equation is mathematically easier to deal with, so we will use the 2-parameter Weibull distribution in subsequent discussions.

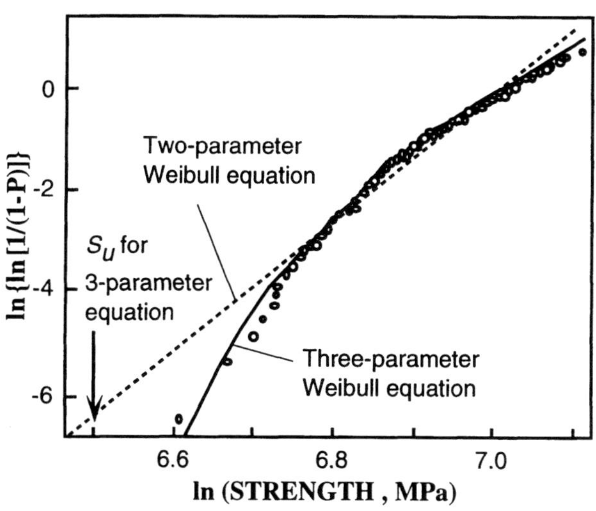

Fig. 3.19. Weibull plots for tensile strength of silicon nitride dogbone specimens with the shape shown in Fig. 3.1. The 2-parameter fit gives m = 13 and S_o = 1102 MPa. The 3-parameter fit gives m = 4, S_o = 1109 MPa and S_u = 665 MPa.[24]

3.4.2 Safety factors

With a standard deviation of 15%, the yttria strengths in Fig. 3.18 span a range in which the strongest sample (168 MPa) is 2.3 times stronger than the weakest sample (73 MPa). The strength of the weakest sample is 60% of the average strength of the set. If you were designing a system with this yttria, it would be wise to add a conservative *safety factor* to avoid the likelihood of catastrophic failure.

$$\textit{Safety factor:} \quad \text{Maximum allowed stress} = \frac{\text{mean strength}}{\text{safety factor}} . \tag{3-23}$$

Customarily, a safety factor of 4 is used for ceramic windows. For an average strength of 121 MPa, we would not design a system to exceed a stress of 30 MPa.

Although a safety factor of 4 is commonly used, it is sensible to base the safety factor on the Weibull modulus, which typically varies over the range 2 to 20 for different materials. A material with a Weibull modulus of 20 should safely handle stresses much closer to the average strength than one with a Weibull modulus of 2.

Before going further, we note that if the strengths of a set of identical coupons follow a 2-parameter Weibull distribution, then the mean strength, $\bar{S}$, is a fraction Γ of the Weibull scaling factor, S_o:

$$\bar{S} = S_o \times \Gamma . \tag{3-24}$$

Γ is a function of the Weibull modulus and is typically in the range 0.90 to 0.95. Numerical values of Γ are listed in Table F.2 in Appendix F. For $m = 5$, $\Gamma = 0.918$, so the mean strength is about 92% of S_o.

Now let's look at an example of how to choose a safety factor. Suppose that a set of ceramic samples has a Weibull modulus of $m = 5$ and a scaling factor of $S_o = 100$ MPa. At what stress would the probability of failure be 50%? At what stress would it be 0.1%? To answer these questions, rearrange Eq. (3-20) to the form

$$\ln\left(\frac{S}{S_o}\right) = \left(\frac{1}{m}\right) \ln\left(\ln\frac{1}{1-P}\right) . \tag{3-25}$$

For the 50% probability of failure, insert $P = 0.50$ and $m = 5$ on the right side to get

$$\ln\left(\frac{S}{S_o}\right) = \left(\frac{1}{5}\right) \ln\left(\ln\frac{1}{1-0.5}\right) = -0.0733$$

$$\frac{S}{S_o} = e^{-0.0733} \quad \Rightarrow \quad S = (0.929)S_o = 92.9 \text{ MPa} \quad (\text{since } S_o = 100 \text{ MPa}).$$

For the 0.1% probability of failure, insert $P = 0.001$ into Eq. (3-25) to find $S = 25.1$ MPa.

The example shows why a safety factor of 4 is reasonable. For $S_o = 100$ MPa and $m = 5$, Eq. (3-24) tells us that the mean strength is $\bar{S} \approx 91.8$ MPa. If we take 0.1% as the desired probability of failure, then the maximum allowed stress is 25.1 MPa. The safety factor from Eq. (3-23) is

$$\text{Safety factor} = \frac{\text{mean strength}}{\text{maximum allowed stress}} = \frac{91.8 \text{ MPa}}{25.1 \text{ MPa}} = 3.7 .$$

A safety factor of 4 puts us in the right region for a reliability of 99.9%.

The example we just considered was based on a Weibull modulus of 5. If the material were more reproducible in its behavior, perhaps it would have a Weibull modulus of 20. In this case, for the same scaling factor of $S_o = 100$ MPa, the mean strength is 97.4 MPa (using Γ from Table F.2), the 50% probability of failure comes at 98.1 MPa and the 0.1% probability of failure comes at 70.8 MPa. For this material with $m = 20$, a safety factor of 97.4/70.8 = 1.4 gives the same reliability as a safety factor of 3.7 if the Weibull modulus were 5.

A higher safety factor is required when reliability must be greater. If failure of a window could lead to loss of an airplane, perhaps you would want a probability of failure of 10^{-7} instead of 10^{-3}. Table 3.2 indicates that for a probability of failure of 10^{-7}, the safety factor needs to be 23 if $m = 5$ and 2.2 if $m = 20$.

Table 3.2. Safety factors from Weibull probability of survival for $S_o = 100$ MPa

Weibull modulus	Mean strength	Stress for 0.1% failure ($P = 0.001$)	Safety factor	Stress for $P = 10^{-7}$	Safety factor
5	91.8 MPa	25.1 MPa	3.7	4.0 MPa	23
20	97.4 MPa	70.8 MPa	1.4	44.7 MPa	2.2

3.5 Strength scales with area (or volume) under stress

Our discussion so far only describes the behavior of identical test specimens. If you test one kind of coupon, such as a small disk, and want to find the failure probability of another kind of object, such as a large window, differences in volume and stress state must be considered. We consider simple volume or area scaling in this section.

Since the strength of a ceramic is governed by the presence of defects, it is plausible that a given stress operating over a large area is more likely to encounter a strength-limiting defect than the same stress operating over a small area. That is, *large objects tend to fail at lower stress than small objects.* In the following discussion, we consider the case in which failure originates at the surface. If failure can originate anywhere inside the component, we would substitute volume for area in the discussion.

Suppose that a ceramic manufactured by a particular method has a Weibull modulus, m. Suppose also that we test a series of disks in ring-on-ring flexure with a load area of A_1 ($= \pi a^2$ in Fig. 3.10) and find an average strength S_1. If we test a second set of the same kind of disks using a load area A_2 ($>A_1$), the strength S_2 will be lower:[25,26]

$$\frac{S_2}{S_1} = \left(\frac{A_1}{A_2}\right)^{1/m} . \qquad (3\text{-}26)$$

Equation (3-26) is derived for volume scaling in Eq. (F-17) in Appendix F.

Example: Predicting flexure strength for parts with different size. The zinc sulfide disks in Table 3.1 have an average strength of $S_1 = 93$ MPa with a Weibull modulus of 5.4 when the load radius is 5.36 mm. If larger disks were tested with a load radius of 10.0 mm, what is the expected average strength, S_2? We answer this with Eq. (3-26):

$$\frac{S_2}{93 \text{ MPa}} = \left(\frac{\pi(5.36 \text{ mm})^2}{\pi(10.0 \text{ mm})^2}\right)^{1/5.4} = 0.79 \Rightarrow S_2 = (0.79)\,(93 \text{ MPa}) = 74 \text{ MPa}.$$

The disks with the larger load area are expected to be only 79% as strong as the disks with the smaller load area, even though they are made of the same material.

To compare parts with complicated geometries, a finite element stress analysis should be carried out. Appendix F explains that the effective area under stress is

$$\text{Effective area} = \int \left(\frac{\sigma}{\sigma_{max}} \right)^m dA \tag{3-27}$$

where σ is the stress in a given surface area element, dA, and σ_{max} is the maximum stress found in any surface area element. The integration in Eq. (3-27) is carried out numerically by summing over all surface elements with a tensile stress. Surface elements in compression are ignored. This method has been used* to predict strengths of components with different sizes and shapes under different types of loads.[3,27]

Different tests of the same size samples give different strengths if the loading geometry is different. For example, high quality sintered alumina disks tested in ring-on-ring flexure exhibit a strength of 154 ± 6 MPa. A set of identical disks tested in a similar ball-on-ring test exhibit a strength of 292 ± 14 MPa, since much less area is loaded by the ball than by a ring (Fig. 3.20). There are many experimental examples in which ball-on-ring disk flexure strength is greater than ring-on-ring flexure strength[12] and 3-point bar flexure strength is greater than 4-point flexure strength.[4,30]

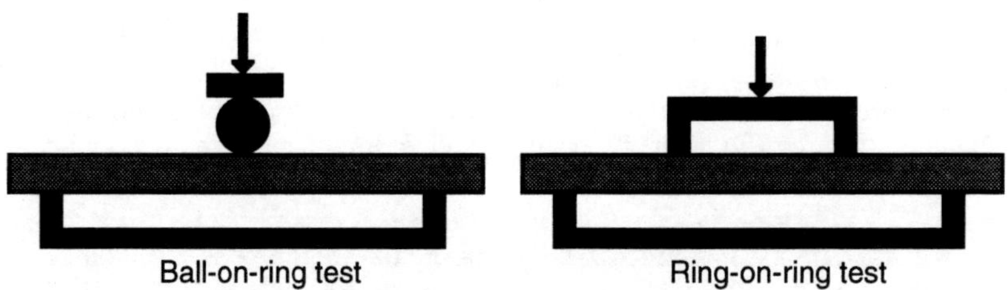

Strength of high purity sintered alumina:
Ball-on-ring: 292 ± 14 MPa
Ring-on-ring: 154 ± 6 MPa

Fig. 3.20. Ball-on-ring and ring-on-ring flexure tests. The ball loads a smaller area (with an effective radius of ~1/3 of the disk thickness) than the ring. (An equation to calculate the stress beneath the ball is found in Refs. 10 or 31.)

---

*Although we have examples in which Eqns. (3-26) and (3-27) are successful, they may not always work. In one series of experiments,[28] the strength of silicon nitride was measured in tension, 4-point bending, 3-point bending and ring-on-ring equibiaxial flexure. Mean strengths measured in the four types of tests were 526, 629, 660 and 721 MPa, respectively, with Weibull moduli between $m = 8.0$ and $m = 9.6$. A graph of ln(mean strength) vs ln(effective volume) for the four types of tests gave a straight line with a slope of -0.040, instead of the expected slope of $-1/m \approx -1/9 = -0.11$. In another case, as specimen size was increased, the predominant type of critical flaw changed from surface damage to internal flaws characterized by different Weibull parameters.[29]

3.6 Strengths of optical ceramics

For a given material manufactured in a consistent manner with a consistent surface finish having consistent flaw characteristics, mechanical strength depends on the kind of test that is performed and on the size of the test coupon. Rarely do we have such perfect control of manufactured parts because the nature and quality of machining and polishing different kinds of test specimens and real windows and domes are highly variable.

The moral of this story is:

> *It is dangerous to quote the strength of a material, since it depends on the type and quality of surface finish, material fabrication method, material purity, test method and specimen size.*

Nonetheless, virtually every application requires us to know the "strength" of the material. So we will go out on a limb and present strengths drawn from numerous sources in Table 3.3.

Table 3.3. Approximate room-temperature strengths of infrared window materials*

Material	Strength (MPa)	Material	Elastic limit (MPa)†
Silicon carbide	600	CaF_2 (hot forged¶)	55
Silicon nitride	600	MgF_2 (single-crystal)	50
Sapphire[32,33]	300-1000	CaF_2 (single-crystal)	37
Diamond (CVD‡)[34-40]	100-800	BaF_2 (single-crystal)	27
ALON	300	KRS-5	26
Spinel	190	LiF (single-crystal)	11
Yttria (doped/undoped)	160	KCl (hot forged)	11
Silicon	120	NaCl (hot forged)	10
MgF_2 (hot pressed)[41,42]	100-150	CsI (single-crystal)	6
CaF_2 (single-crystal)[43]	100-150	NaCl (single-crystal)	2
Gallium phosphide§[44]	100-120	KCl (single-crystal)	2
Gallium arsenide[45]	60-130	KBr (single-crystal)	1
ZnS (standard)	100		
ZnS (multispectral)	70		
Germanium	90		
SrF_2 (single-crystal)[43]	70-110		
Fused silica	60		
Zinc selenide	50		

*Material strength should always be considered approximate. It varies with the quality of surface finish, fabrication method, material purity, type of test and size of the sample.

†Elastic limit is the load at which the rate of deformation is twice as great as the initial rate. The elastic limit is used in place of the strength, S, in Eq. (3-28). Values are quoted from the Harshaw/Filtrol *Crystal Optics Catalog*, Solon, Ohio.

‡Chemical vapor deposited diamond.

¶Hot forging is described in Section 5.4.3.

§Strength ≈ 195 MPa at 500°C.[44]

If the strength of a window material is marginal for your particular application, it is important to seek primary data from the manufacturer of the specific material you plan to use, with the same surface finish you plan to use. The data should include all of the information on the geometry of the test so that you can try to account for the effects of size and stress state with Eqns. (3-26) and (3-27) to predict the probability of failure for the ceramic component of interest to you. Alas, usually the data you want do not exist and you may need to conduct your own mechanical test program with parts manufactured to your specifications.

In some cases, a window is required to be exposed to stresses that will fracture an unacceptable fraction of the window population. For example, the required probability of survival might be 99.9% and the predicted probability of survival might be 75%. It is still possible to make reliable windows if each one is subjected to a *proof test* (described in Chapter 8) in which it is stressed somewhat beyond the operational stress. The windows that fail end up in the garbage and the windows that pass are known to be reliable *if it can be shown that the proof test does not weaken the window.* The simplest way to ensure the reliability of the proof test is to stress the same part again and again to show that if it passes once it passes repeatedly. It may not be trivial to devise a proof test that mimics the stress that will be encountered in actual operation of the window.

3.6.1 Strength is not an intrinsic property of a material

It cannot be stated too strongly that strength depends on the quality of the material being tested and how you choose to make the measurement. Strength is not an intrinsic, invariable property of an ideal material.

Difficulties in reporting the "strength" of a material are illustrated in Table 3.4 for hot pressed MgF_2 from the 1970's and single-crystal MgF_2 from the 1990's. Consider the ring-on-ring disk flexure data. The hot pressed material has a mean strength of 88 MPa and single-crystal material with three different surface finishes has a mean strength of 142 MPa. Is the hot pressed material really weaker than the single-crystal material? This sounds like a dumb question, but the answer is probably "no."

If failure originates at the surface within the load ring, we can use Eq. (3-26) to predict the strength of the hot pressed material from the single-crystal data. The load diameter for the single crystals was 5.79 mm and the load diameter of the polycrystalline material was 38.1 mm. The ratio of their areas is $5.79^2/38.1^2$. Eyeballing a mean Weibull modulus of $M \approx 5$, we calculate

$$\frac{S_{polycrystal}}{S_{single\ crystal}} = \left(\frac{A_{single\ crystal}}{A_{polycrystal}}\right)^{1/m}$$

$$\frac{S_{polycrystal}}{142\ \text{MPa}} = \left(\frac{5.79^2}{38.1^2}\right)^{1/5} \quad \Rightarrow \quad S_{polycrystal} = 67\ \text{MPa}\ .$$

Based on area scaling, the hot pressed disks should have a mean strength of 67 MPa. The observed strength of 88 MPa is *higher* than the value predicted from the single crystals. It appears that the polycrystalline material is stronger — not weaker — than the single-crystal material. This is not obvious from the strengths reported in Table 3.4 because the load areas are so different.

Table 3.4. Strength of magnesium fluoride

Material	Strength (MPa) ± standard deviation (n = number of specimens)	Temperature (°C)	Weibull parameters *m*	S_o (MPa)
Ring-on-ring disks (85.1 mm diameter with unstated thickness) (Load diameter = 38.1 mm; support diameter = 79.4 mm)				
Hot pressed	88 ± 21 (n = 15)	24	5.0	96
Ring-on-ring disks with 3 different polishes (38.1 mm diameter × 2.5 mm thick)[41] (Load diameter = 5.79 mm; support diameter = 25.4 mm)				
Single crystal, polish A	137 ± 29 (n = 21)	~24	5.7	148
Single crystal, polish B	167 ± 32 (n = 19)	~24	5.7	181
Single crystal, polish C	122 ± 33 (n = 20)	~24	4.4	134
4-point flexure bars (length = 25.4, thickness = width = 1.78, load span = 8.38 mm)				
Hot pressed	129 ± 16 (n = 20)	24	9.7	136
Hot pressed	129 ± 14 (n = 10)	121	9.9	135
Hot pressed	140 ± 44 (n = 23)	260	3.5	152
Hot pressed	139 ± 36 (n = 20)	399	4.8	151
Hot pressed	111 ± 32 (n = 17)	538	4.0	122

The 4-point flexure samples have a much higher strength than the hot pressed disks, and much less surface is in tension. Which strength should you use for magnesium fluoride? A good choice would be to measure the strength of the same kind of material with the same kind of surface finish placed in the same kind of stress state (uniaxial or biaxial) that will be encountered in practice. Then a scaling by area or volume needs to be made. *Strength is not an intrinsic property of a material.*

3.6.2 Temperature dependence of strength

Ceramics lose strength at sufficiently high temperature. Figure 3.21 shows that spinel gradually weakens between 600° and 1200°C, while lanthana-doped yttria does not lose strength up to 1600°C (although a precipitous loss occurs between 1600° and 1700°C).

Many studies of sapphire (single-crystal Al_2O_3) with different crystal orientations, different finishes, and from different growth methods show significant strength loss above room temperature.[33,48-52] In the work of Wachtman and Jackman in Fig. 3.22, the strength reached a broad minimum at 300-600°C. Gentilman found no minimum up to 1000°C. By contrast, the strength of polycrystalline alumina, which has the same chemical composition as sapphire, is essentially constant up to 800°C and then begins to fall.[49,50,53] The right side of Fig. 3.22 shows that the tensile strength of 0.2-mm-diameter sapphire filaments is much greater than that of bulk crystals. Filaments are stronger than bulk samples of many materials because filaments have a small surface or volume in which flaws can be found.

Zinc sulfide is unusual in that its strength increases with increasing temperature.[54,55] In Fig. 3.23 the strength is almost twice as great at 700°C as at 20°C.

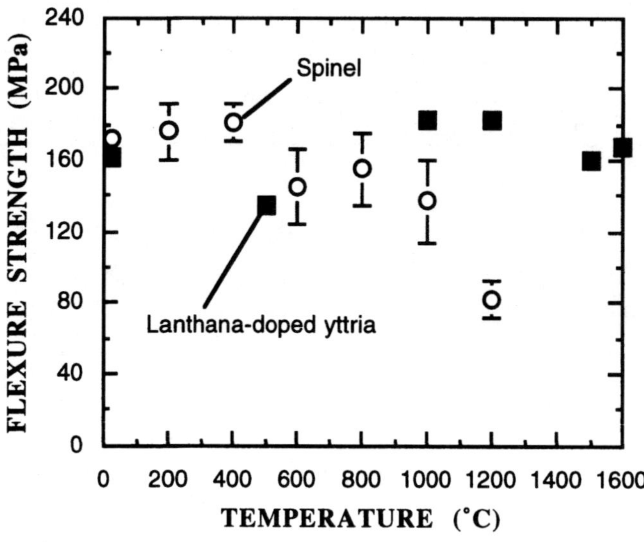

Fig. 3.21. Temperature dependence of strength of polycrystalline spinel[46,47] and lanthana-doped yttria (data from GTE Laboratories).[2] Standard deviation error bars are shown for spinel, but not for yttria.

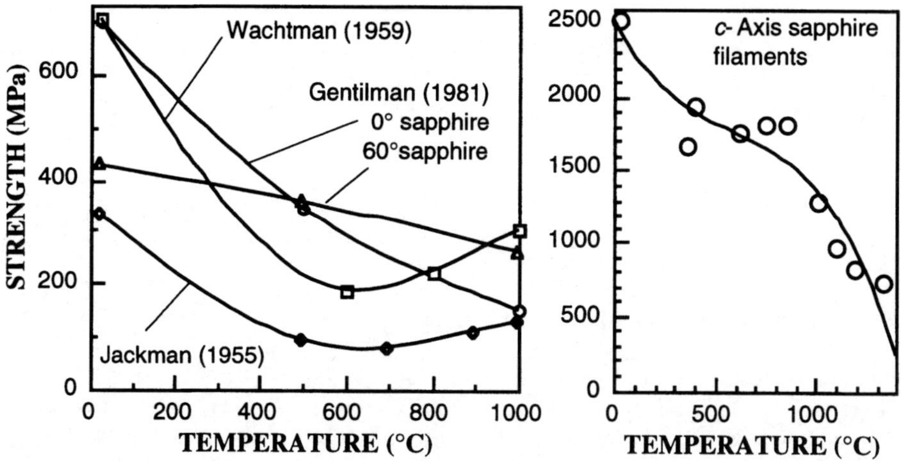

Figure 3.22. *Left:* Flexure strength of sapphire. Wachtman:[48] 4-point flexure of flame-polished 2.5 × 38 mm rods with rod axis ∥ *c* (0° rods). Jackman:[49,50] 4-point flexure of 2 × 2 × 58 mm ground prisms with random orientation. Gentilman:[51] Biaxial flexure of 2.5 × 51 mm polished disks with surface normal to *c* or 60° from *c*. *Right:* Tensile strength of *c*-axis filaments.[52]

Although the strength of ZnS increased with temperature when measured in air,[54,55] measurements under nitrogen showed little change in strength from 25 to 600°C.[56] A possible explanation is that when ZnS is heated in air, oxygen diffuses to the microscopic crack tips where it chemically reacts and the product blunts the crack tips.[56] Blunting the crack tips increases the strength of the material.

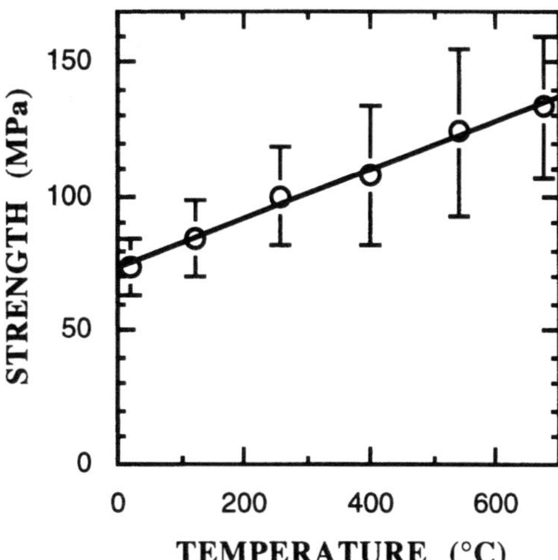

Fig. 3.23. Ring-on-ring biaxial flexure strength of standard grade ZnS.[54] Twenty disks (3.2 mm thick × 38.1 mm diameter) were tested at each temperature with a 15.9-mm-diameter load ring and a 31.8-mm-diameter support ring.

°C	Strength (MPa) ± std. dev.	Weibull modulus
21	73 ± 11	7.7
121	84 ± 14	7.0
260	100 ± 18	5.9
399	108 ± 26	4.6
538	124 ± 31	4.6
677	134 ± 27	5.6

Gallium phosphide also has an unusual increase of strength with increasing temperature. The strength of 25-mm-diameter × 2-mm-thick polycrystalline disks increased from 118 ± 24 MPa at 25°C (18 disks with Weibull modulus = 5.3) to 195 ± 30 MPa at 500°C (15 disks with Weibull modulus = 6.5).[44]

3.7 Window and dome design

The more pressure a window or dome must withstand, the thicker it should be. We now consider examples of window and dome design.

3.7.1 Designing a circular window

Suppose that a circular window with thickness d and exposed diameter L (Fig. 3.24) is subjected to a pressure of P MPa. Let the strength of the window material be S MPa and the design safety factor be f. We can choose any number we desire for f, but a factor

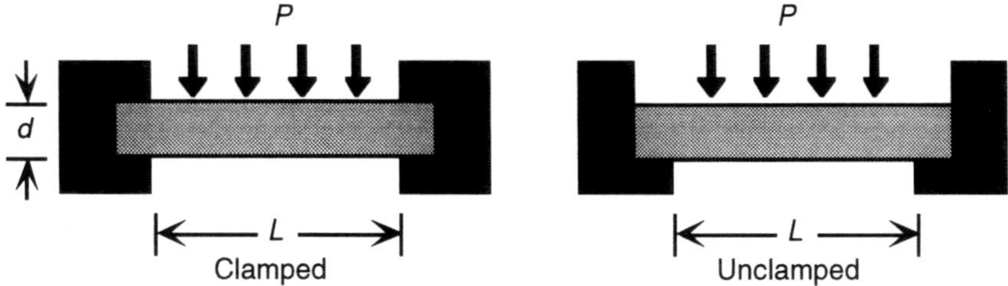

Fig. 3.24. Clamped and unclamped circular windows. The clamped window experiences maximum tensile stress at the edge of the clamped area, while the unclamped window experiences maximum stress at the center.

of 4 is customary. That is, we choose not to exceed a stress greater than one-fourth of the strength of the material. The required ratio of thickness to diameter is given by

$$\frac{d}{L} = \frac{1}{2}\sqrt{\frac{k\,f\,P}{S}} \tag{3-28}$$

where k is 0.75 for a clamped window and 1.125 for an unclamped window (Fig. 3.24).

Example: Thickness of a circular window. How thick should a 10-cm-diameter sapphire window be if it is clamped at the edge and must withstand a pressure of 100 atm? One atmosphere equals 0.101 MPa, so 100 atm = 10.1 MPa. If the average strength of sapphire is taken as S = 400 MPa, then Eq. (3-28) gives

$$\frac{d}{L} = \frac{1}{2}\sqrt{\frac{(0.75)\,(4)\,(10.1\text{ MPa})}{400\text{ MPa}}} = 0.14\ .$$

That is, the thickness should be 14% of the exposed diameter, or (0.14)(10 cm) = 1.4 cm. If we had used a calcium fluoride window instead, with a strength of 37 MPa, the required thickness would be 4.5 cm. Figure 3.25 shows the ratio of thickness-to-diameter for clamped circular windows as a function of pressure and material strength.

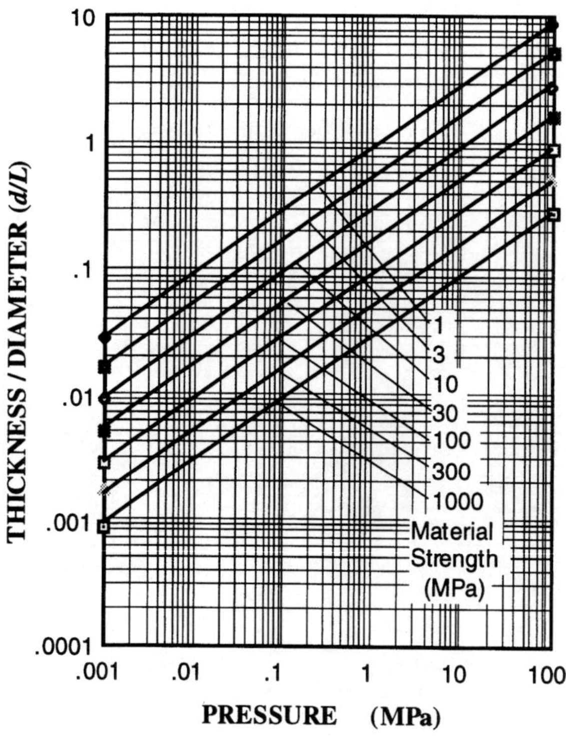

Fig. 3.25. Thickness-to-diameter ratio required for the clamped window in Fig. 3.24 as a function of applied pressure and material strength. A safety factor of f = 4 was used for the calculations.

3.7.2 Designing a dome

Now we discuss how thick the dome in Fig. 3.26 must be to withstand a pressure load. Section 4.3.3 discusses thermal considerations in dome design. Let the dome have a thickness d, a spherical radius r, a base diameter D, and an included angle 2θ. If a pressure, P, is applied across the dome, and if the dome is simply supported so that it is free to bend, then the maximum tensile stress, σ, induced by the bending is[57]

$$\sigma = \frac{rP}{2d}\left[\cos\theta\left(1.6 + 2.44\ \sin\theta\sqrt{\frac{r}{d}}\right) - 1\right]. \qquad (3\text{-}29)$$

The geometry in Fig. 3.26 tells us that $\sin\theta = D/2r$ and, always, $\cos\theta = \sqrt{1\text{-}\sin^2\theta}$. Equation (3-29) is valid if d/r is in the range $(^1/_{12})(\sin^2\theta) \le d/r \le (^1/_{1.2})(\sin^2\theta)$. If the dome is not simply supported, but is clamped at its base, then the maximum stress is compressive and is located near the base of the dome. In this case the maximum compressive stress is $\sigma = -1.2r/d$ for $(^1/_{12})(\sin^2\theta) \le d/r \le (^1/_3)(\sin^2\theta)$. We are only concerned with the simply supported dome under tensile stress.

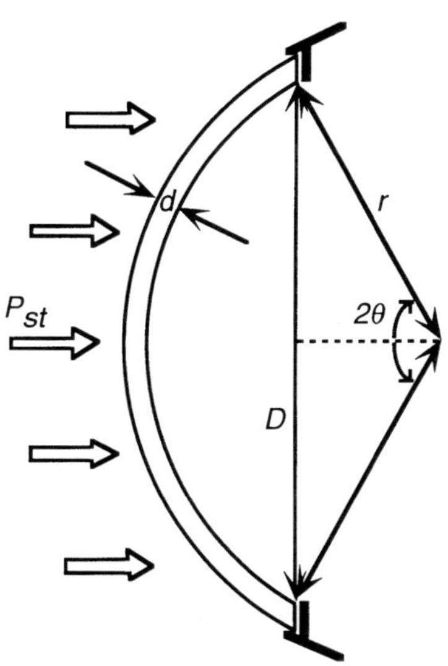

Fig 3.26. Missile dome with simply supported base. For the example in the text, r = 50 mm, D = 50 mm, and θ = 30°.

We will consider an instructive example from Klein[58] of a dome on a missile that must be capable of the low, medium, and high altitude flights in Table 3.5. The three flights require cruising at Mach 3 at an altitude of 1 km, Mach 4 at an altitude of 3 km, or Mach 6 at an altitude of 30 km. We designate the external atmospheric pressure far away from the missile as the free-stream pressure, P_∞, and the stagnation pressure at the front of the dome as P_{st}. If the dome cavity is vented so its internal pressure is P_∞, then the pressure difference across the nose of the dome is

$$\text{Pressure across dome} = P = P_{st} - P_\infty = P_\infty\left(\frac{P_{st}}{P_\infty} - 1\right). \qquad (3\text{-}30)$$

Table 3.5. Airstream characteristics for three different missile flights*

	Low altitude	Medium altitude	High altitude
Flight altitude (km)	1	3	30
Peak Mach number	3	4	6
Free stream pressure (kPa)	89.9	70.1	1.20
Stagnation pressure ratio, Eq. (3-31)	12.1	21.1	46.8
Pressure across dome (kPa)	998	1410	55.0
Free-stream temperature (K)	282	269	227
Stagnation temperature (K)	790	1130	1861

*Stagnation temperature (K) = (free-stream temperature) × $[(1 + (^1/_2)(\gamma-1)M_\infty{}^2]$, where M_∞ is Mach number and $\gamma = 1.4$. Free-stream temperature is the temperature of the atmosphere far away from the missile. Temperature and pressure are from Table 4.4, U.S. Standard Atmosphere, 1962 [*American Institute of Physics Handbook*, 3rd ed, McGraw-Hill, New York (1972)].

The ratio of stagnation pressure to free-stream pressure is

$$\frac{P_{st}}{P_\infty} = \left(\frac{\gamma+1}{2}M_\infty{}^2\right)^{\gamma/(\gamma-1)}\left(\frac{2\gamma}{\gamma+1}M_\infty{}^2 - \frac{\gamma-1}{\gamma+1}\right)^{-1/(\gamma-1)} \quad (3\text{-}31)$$

where M_∞ is the Mach number at which the missile is traveling and γ is the heat capacity ratio of air, taken as 1.4. With the free-stream pressures in Table 3.5 for each altitude, Eqns. (3-30) and (3-31) give us the pressure difference across the dome on the fifth line of Table 3.5. The worst-case pressure of 1410 kPa (14 times standard atmospheric pressure) occurs in the medium altitude flight.

To find the thickness of the dome in Fig. 3.26 needed to survive a pressure of 1410 kPa, we use Eq. (3-29). With P = 1.41 MPa, r = 50 mm, and $\theta = 30°$, we can plot the

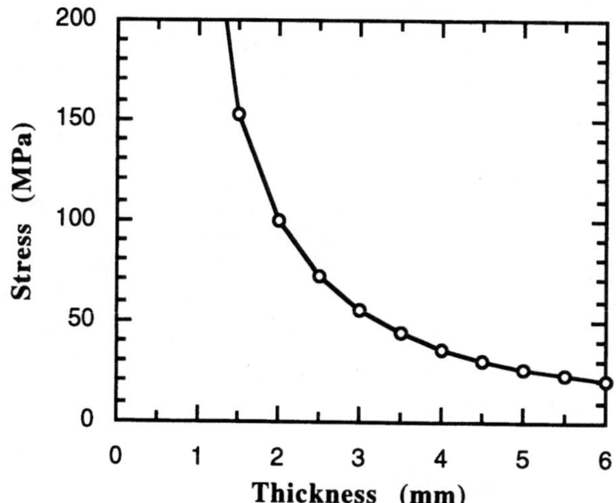

Fig 3.27. Graph of Eq. (3-29) for P = 1410 kPa, r = 50 mm, and $\theta = 30°$.

maximum tensile stress on the dome as a function of dome thickness, which is shown in Fig. 3.27. Suppose that the dome material has a mean strength of 100 MPa and we want to use a safety factor of 4. Then the maximum allowed stress would be 100/4 = 25 MPa. Figure 3.27 says that the dome thickness must be d = 5.1 mm to keep the stress below 25 MPa. Is Eq. (3-29) valid for this dome thickness? The caveat beneath the equation was that d/r must be in the range $(^1/_{12})(\sin^2\theta)$ to $(^1/_{1.2})(\sin^2\theta)$, which is 0.0208 to 0.208 for $\theta = 30°$. The required value of d/r is (5.1 mm)/(50 mm) = 0.102, which is in the valid range. We therefore need to specify a dome thickness of 5.1 mm.

We can use Eq. (3-29) to evaluate the capabilities of an optical dome on a cannon-launched projectile. The effective pressure on the dome arises from the rapid acceleration when the cannon is fired:

$$\text{Pressure from acceleration} = \text{density} \times \text{thickness} \times \text{acceleration} . \qquad (3\text{-}32)$$

Example: Pressure on cannon-launched projectiles. What is the effective pressure on sapphire and zinc sulfide domes with a thickness of 2 mm if they are launched from a cannon with an acceleration of 15 000 times the acceleration of gravity (15 000 g's)? The density of sapphire in Table 3.6 is 3.98 g/mL. Since 1 mL = 1 cm^3, there are 10^6 cm^3 in a cubic meter. The density is 3980 kg/m^3. The acceleration of gravity ("1 g") is 9.81 m/s^2, so 15 000 g's are (15 000)(9.81 m/s^2) = 1.47×10^5 m/s^2. Plugging into Eq. (3-32) gives the pressure:

$$\begin{aligned}\text{Pressure} &= \text{density} \times \text{thickness} \times \text{acceleration}\\ &= (3980\ \text{kg/m}^3) \times (0.002\ \text{m}) \times (1.47 \times 10^5\ \text{m/s}^2) = 1.17\ \text{MPa} .\end{aligned}$$

For zinc sulfide, we would use a density of 4.08 g/mL to find a pressure of 1.20 MPa.

Example: Stress in domes on cannon-launched projectiles. Given the pressures from the previous example, what are the maximum tensile stresses on the sapphire and zinc sulfide domes? For sapphire, we use Eq. (3-29) with d = 2 mm, r = 50 mm, $\theta = 30°$, and P = 1.17 MPa to find

$$\sigma = \frac{rP}{2d}\left[\cos\theta\left(1.6 + 2.44\sin\theta\sqrt{\frac{r}{d}}\right) - 1\right] = 83\ \text{MPa} .$$

If sapphire has a mean strength of 400 MPa and we use a safety factor of 4, then we can tolerate a stress of 100 MPa. Since the estimated stress is 83 MPa, the sapphire dome is predicted to survive the cannon launch.

For zinc sulfide, the calculated stress is 85 MPa. However, the mean strength of zinc sulfide is around 100 MPa and a safety factor of 4 would limit the allowed stress to 25 MPa. The zinc sulfide dome design is not adequate for the cannon shell. If zinc sulfide domes were *proof tested* (Chapter 8) to throw away domes with a strength less than, say, 100 MPa, the material could be used in a reliable design.

Table 3.6. Density of dome materials

Material	Density (g/mL)	Material	Density (g/mL)
ALON ($9Al_2O_3{\cdot}5AlN$)	3.69	Lanthana-doped yttria ($0.09La_2O_3{\cdot}0.91Y_2O_3$)	5.13
Aluminum nitride (AlN)	3.26	Lithium fluoride (LiF)	2.64
AMTIR-1 (Ge/As/Se glass)	4.40	Magnesium fluoride (MgF_2)	3.18
Barium fluoride (BaF_2)	4.89	Magnesium oxide (MgO)	3.58
Cadmium sulfide (CdS)	4.82	Potassium bromide (KBr)	2.75
Calcium aluminate glass	2.9-3.1	Quartz (SiO_2)	2.65
Calcium fluoride (CaF_2)	3.18	Sapphire (Al_2O_3)	3.98
Calcium lanthanum sulfide ($CaLa_{2.7}S_{5.05}$)	4.61	Silicon (Si)	2.33
Cesium bromide (CsBr)	4.44	Silicon carbide (SiC)*	3.21
Diamond (C)	3.51	Sodium chloride (NaCl)	2.16
Fused silica (SiO_2)	2.20	Silicon nitride (Si_3N_4)	3.24
Gallium arsenide (GaAs)	5.32	Spinel ($MgAl_2O_4$)	3.58
Gallium phosphide (GaP)	4.13	Yttria (Y_2O_3)	5.03
Germanate glass (Corning 9754)	3.58	Zinc selenide (ZnSe)*	5.27
Germanium (Ge)	5.35	Zinc sulfide (ZnS)*	4.08
KRS-5 ($TlI_{0.543}Br_{0.457}$)	7.37		

*Chemical vapor deposited SiC, ZnSe and ZnS.

3.8 Hardness and fracture toughness

Hardness is a measure of resistance to indentation.[59] Figure 3.28 shows the indentation pattern left by a Vickers indentor, which has a square pyramidal point made from a crystal of diamond. The primary indentation has a diagonal length of $2a$ and the cracks radiating from the corners span a length $2c$. If we could look at a cross section of an ideal indented sample, we would see a semicircular "half-penny" radial crack of diameter $2c$, with a median crack extending down into the ceramic from the center of the indentor.

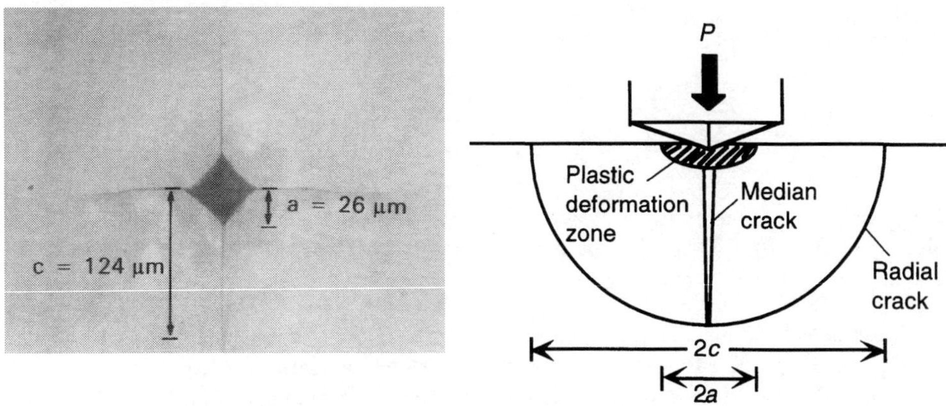

Fig. 3.28. *Left*: Micrograph of Vickers indentation pattern in yttria. *Right*: Idealized cross section of crack system.

If the indentation is made lightly enough so that cracks do not extend beyond the primary diamond-shaped impression in Fig. 3.28, the hardness is defined as

$$\text{Hardness} = H = \frac{P}{\alpha_o a^2} \qquad (3\text{-}33)$$

where P is the indentor load (newtons, N), $2a$ is the length defined in Fig. 3.28, and α_o is a constant that depends on the indentor geometry. For the Vickers indentor, $\alpha_o = 2$. *The harder the material, the less of an indentation will be made.*

The units of hardness are N/m^2, which is Pa. Frequently, the units kg/mm^2 are encountered. In such case, the indentation length is expressed in mm and the load is expressed as the mass required to produce the force ($P = mg$, where m is mass [kg] and g is the acceleration of gravity [9.806 65 m/s^2]). A load of 9.807 N is equivalent to a mass of 1.000 kg. You can think of hardness as being proportional to the number of kg of load required to make a 1-square-mm indentation.

The Knoop indentor in Fig. 3.29, which leaves an elongated impression, rather than the square impression in Fig. 3.28, is commonly used to measure hardness. Hardness measured with this indentor is called *Knoop hardness*. The hardness of single crystals is different for different crystal planes and even for one crystal plane with the indentor rotated in different directions. A Berkovich indentor[60] makes the triangular impression in Fig. 3.29 that is most useful for testing single-crystal materials with trigonal or hexagonal symmetry.

The measured hardness is a function of the load applied to the indentor.* At low loads, hardness increases. Normally, measurements are made at increasing loads and the hardness that is reported is the value in the plateau region (between 2 and 4.5 N in Fig. 3.30).

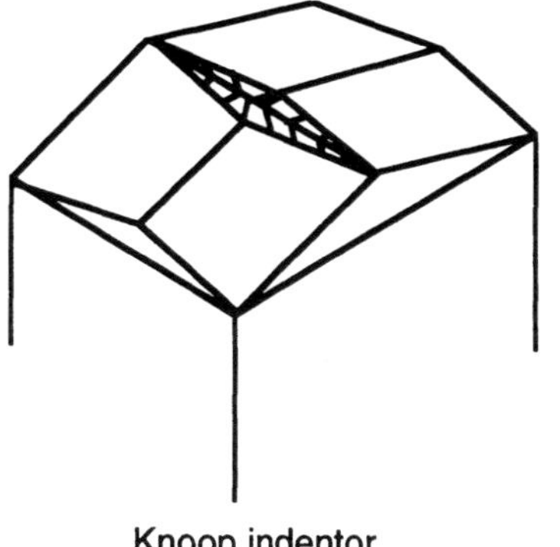

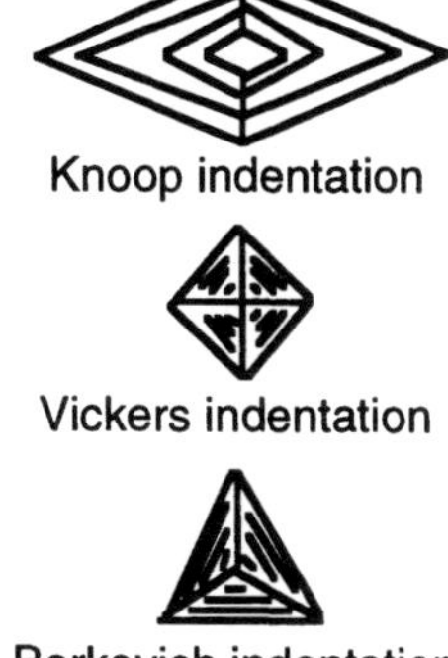

Fig. 3.29. *Left:* Knoop indentor showing elongated indenting surface. *Right:* Impressions left by three common indentors for measuring hardness.

*Hardness should be measured with light loading that does not produce the crack of length c in Fig. 3.28. The apparent hardness measured under heavier loading that does produce cracking is not very different from hardness measured with lighter loads.

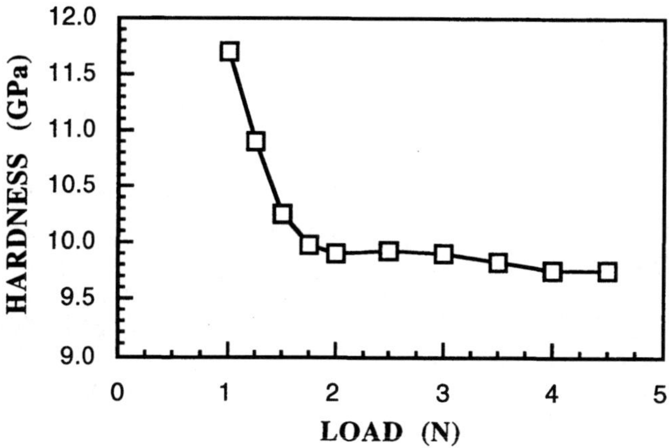

Fig. 3.30. Dependence of hardness on indentor load for a polycrystalline ceramic.[61] Normally, the reported value is taken where the curve is flat.

Geologists use the Mohs hardness scale in Table 3.7, in which selected minerals are given values from 1 to 10. Any material that is harder than a given mineral can scratch the softer mineral. Any material softer than the mineral can be scratched by the mineral.

Table 3.8 gives the hardness of many infrared window materials. In general, the harder a material, the more scratch resistant it will be. Hardness should correlate with resistance to sand and dust erosion.

Table 3.7. Comparison of Mohs and Knoop hardness scales

Material	Formula	Mohs hardness	Knoop hardness (kg/mm^2)
Talc	$3MgO{\cdot}4SiO_2{\cdot}H_2O$	1	–
Gypsum	$CaSO_4{\cdot}2H_2O$	2	32
Calcite	$CaCO_3$	3	135
Fluorite	CaF_2	4	163
Apatite	$CaF_2{\cdot}3Ca_3(PO_4)_2$	5	430
Feldspar	$K_2O{\cdot}Al_2O_3{\cdot}6SiO_2$	6	560
Quartz	SiO_2	7	820
Topaz	$Al_2F_2SiO_4$	8	1340
Sapphire (corundum)	Al_2O_3	9	2200
Silicon carbide	SiC	–	2540
Boron carbide	B_4C	–	2750
Boron nitride	BN	–	4500
Diamond	C	10	9000

Table 3.8. Knoop hardness of infrared window materials

Material	Hardness kg/mm^2	Hardness GPa	Material	Hardness kg/mm^2	Hardness GPa
KBr	6	0.06	MgO	640	6.3
KCl	8	0.08	GaAs (single-crystal)	700	6.9
AgCl	10	0.10	Y_2O_3	720	7.1
NaCl	17	0.17	La-doped Y_2O_3	760	7.5
CsBr	20	0.20	GaP (single-crystal)	840	8.2
KRS-5	40	0.39	Ge	850	8.3
BaF_2	80	0.78	Si	1150	11.3
ZnSe	105	1.03	$MgAl_2O_4$ (spinel)	1600	16
LiF	110	1.1	ALON	1800	18
CaF_2	160	1.6	Si_3N_4 (silicon nitride)	2200	22
ZnS (multispectral)	160	1.6	Al_2O_3 (sapphire)	2200	22
ZnS (standard)	250	2.5	SiC (silicon carbide)	2540	24.9
Fused silica	460	4.5	Diamond [(111) face]	9000	88
MgF_2	580	5.7			

Multiply kg/mm^2 by 0.009807 to convert to GPa.

Solution hardening is sometimes observed when two materials form a solid solution. For example, PbS has a hardness of 93 kg/mm^2 in Fig. 3.31, and PbTe has a hardness of 37 kg/mm^2. Yet when a solid solution is made by gradually adding PbTe to PbS, the hardness increases to a maximum value near 150 kg/mm^2 at 30 mole percent PbTe.[62] The explanation is based on the fact that Te^{2-} is 22% larger than S^{2-}. As Te^{2-} is substituted into the S^{2-} lattice of PbS, the larger Te^{2-} ions introduce a compressive stress into the crystal lattice, resulting in a harder material. Eventually the lattice becomes predominately Te^{2-} and takes on the inferior hardness of PbTe.

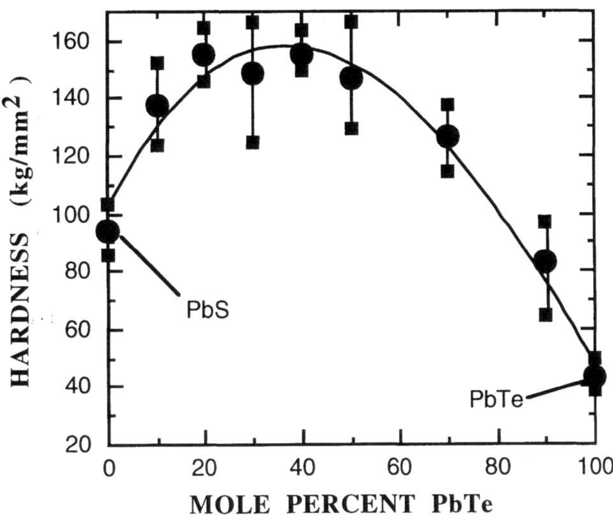

Fig. 3.31. Illustration of solution hardening, showing Vickers hardness of PbS/PbTe solid solution as a function of composition.[62] Bars give range of 10 measurements.

Fracture toughness, also called the *critical stress intensity*, K_{Ic}, is a measure of resistance to crack extension in a material.[63-65] From the indentation experiment in Fig. 3.28, the fracture toughness is calculated with the equation

$$\text{Fracture toughness} = K_{Ic} = \delta\sqrt{\frac{E}{H}}\frac{P}{c^{3/2}} \tag{3-34}$$

where δ is a material-independent constant equal to 0.016 ± 0.004, E is Young's modulus, H is hardness from Eq. (3-33), P is indentor load and c is the crack length in Fig. 3.28. The units of fracture toughness are usually $\text{MPa}\sqrt{\text{m}}$ or, equivalently, MN $\text{m}^{-1.5}$. *The tougher a material is, the shorter will be the cracks radiating out from the indentation in Fig. 3.28.* Fracture toughness for some infrared window materials is given in Table 3.8.

Table 3.8. Fracture toughness of infrared window materials

Material	Fracture toughness ($\text{MPa}\sqrt{\text{m}}$)	Material	Fracture toughness ($\text{MPa}\sqrt{\text{m}}$)
ZnSe	0.5	$MgAl_2O_4$ (spinel)	1.9
ZnS (standard)	1.0	ALON	1.4
GaAs [(100) face]	0.4	Y_2O_3 (doped/undoped)	0.7
GaP (single-crystal)	0.8	Al_2O_3 (sapphire)	2.0
GaP (polycrystalline)*	1.0	Si_3N_4 (silicon nitride)	4
Ge (single-crystal)	0.7	SiC (silicon carbide)	4
Si	0.9	Diamond (single-crystal)	3.4
Fused SiO_2	0.8	Diamond (CVD)*	2-8
BK-7 glass	0.6		

*Chemical vapor deposited GaP and diamond.

The subscript "I" in K_{Ic} designates mode I opening of a fracture in Fig. 3.32. We usually assume that opening of a crack is the dominant failure mode for a brittle material.

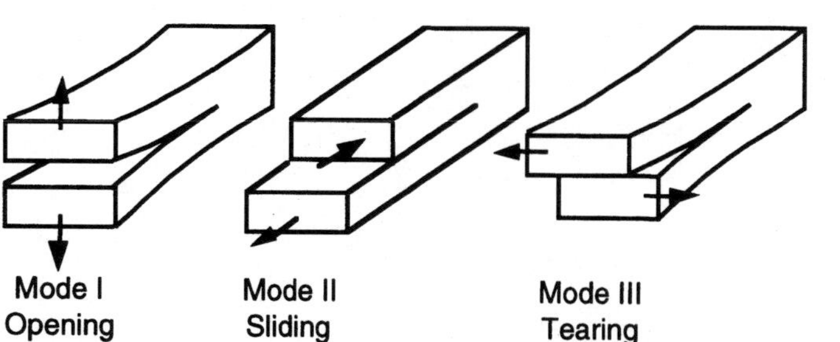

Fig. 3.32. Three modes of fracture.

3.8.1 Relation of strength to fracture toughness and grain size

The strength of a brittle material is proportional to the fracture toughness and inversely proportional to the square root of the radius, r, of the fracture-initiating flaw:[66-68]

$$\text{Strength} = S = \frac{K_{Ic}}{1.24\sqrt{r}} \tag{3-35}$$

where K_{Ic} is fracture toughness. Usually the flaw has more of an elliptical shape than a circular shape (as in Fig. 3.13), in which case we use the effective radius $r = \sqrt{ab}$, where a and b are the depth and half-width of the flaw in Fig. 3.13.

Fracture toughness is a measure of the intrinsic strength of a material. That is, for the same flaw population, strength is directly proportional to fracture toughness. Equation (3-35) says that strength is inversely proportional to the square root of flaw size. For a given fracture toughness, strength decreases in proportion to $1/\sqrt{r}$. Figure 3.33 shows the application of Eq. (3-35) to a wide range of flaw sizes for flexure and tension specimens of sintered silicon nitride.

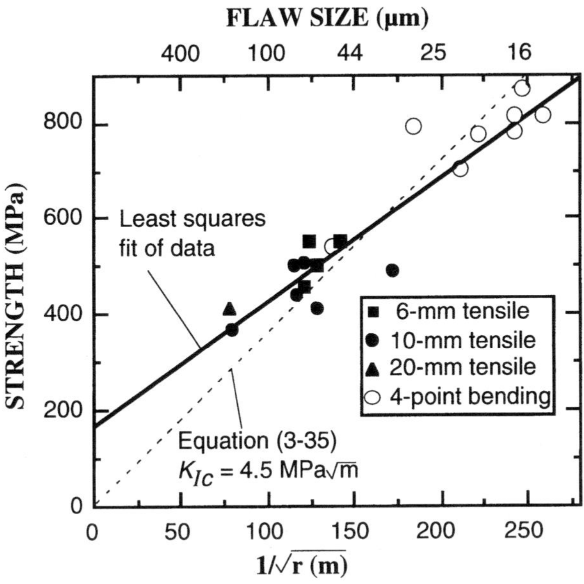

Fig. 3.33. Graph of strength *vs.* $1/\sqrt{r}$ for flexure and tensile specimens of Si_3N_4, showing application of Eq. (3-35). Data from Ref. 3.

Example: Relation of strength and critical flaw size. Zinc sulfide has a strength of 100 MPa and a fracture toughness of 1.0 MPa$\sqrt{\text{m}}$. What is the radius, r, of the critical flaw? To answer this question, we rearrange Eq. (3-35) to solve for c:

$$r = \left(\frac{K_{Ic}}{1.24\,S}\right)^2 = \left(\frac{1.0\ \text{MPa}\sqrt{\text{m}}}{(1.24)\ (100\ \text{MPa})}\right)^2 = 6.5 \times 10^{-5}\ \text{m} = 65\ \mu\text{m}\ .$$

If ZnS could be fabricated with a critical flaw size of 10 μm, how strong would it be?

$$\text{Strength} = \frac{1.0 \text{ MPa}\sqrt{\text{m}}}{1.24\sqrt{10 \times 10^{-6}\text{ m}}} = 255 \text{ MPa} .$$

Zinc sulfide would be much stronger than 100 MPa if the critical flaw size could be reduced by improved fabrication.

The *Petch equation* is an empirical relationship between strength and grain size, G, for polycrystalline materials:

$$\text{Strength} = \sigma_o + \frac{k}{\sqrt{G}} \quad (3\text{-}36)$$

where σ_o and k are constants.[69] If the flaw size of a polycrystalline material is similar to the grain size, then Eq. (3-35) predicts a relationship of the form (3-36), with $\sigma_o = 0$. Figure 3.34 shows that alumina and CsI follow Eq. (3-36) with $\sigma_o \approx 0$.[70,71] Magnesium oxide and zinc selenide have nonzero intercepts, so $\sigma_o \neq 0$. The strength of lanthana-doped yttria is independent of grain size over the limited range that was studied, so $k \approx 0$ for this material. Hardness also tends to be proportional to $1/\sqrt{G}$ for small grain sizes, but for larger grain sizes hardness tends to go through a minimum and then to increase again.[72,43]

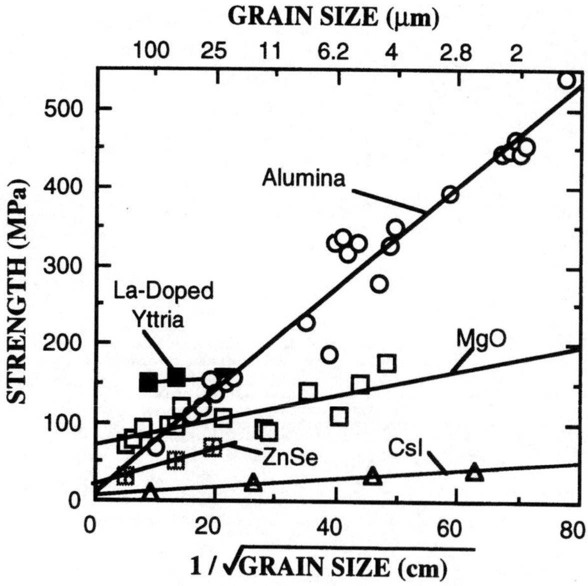

Fig. 3.34. Graph of strength *vs.* $1/\sqrt{G}$ for alumina,[70] magnesium oxide,[70] cesium iodide,[71] zinc selenide,[43] and lanthana-doped yttria (GTE data), illustrating Petch equation (3-36).

3.8.2 Temperature dependence of hardness and fracture toughness

Just as modulus and strength are temperature dependent, so are fracture toughness and hardness. Figure 3.35 illustrates this effect.[73-77] Some materials, such as gallium arsenide, deform plastically at elevated temperature. In such case, the fracture toughness increases with increasing temperature. For oxide ceramics, it is possible to estimate the hardness and elastic modulus as a function of temperature based on the change in separation between atoms in the crystal as a function of temperature.[78]

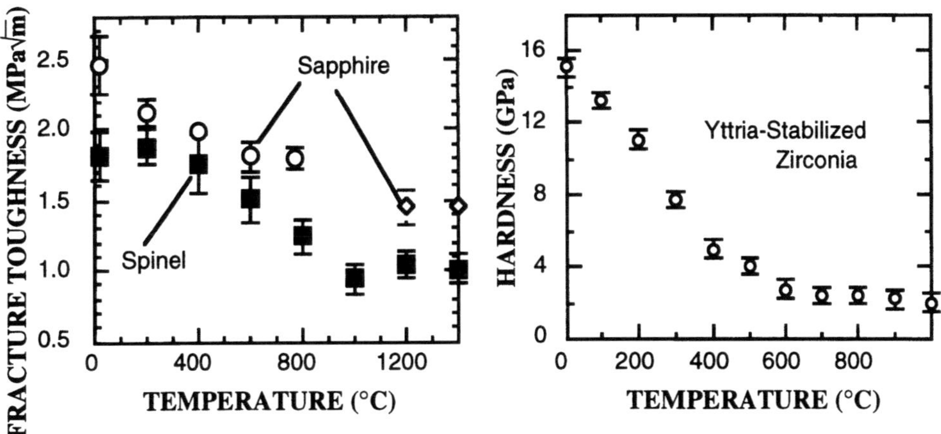

Fig. 3.35. Effect of temperature on fracture toughness of single-crystal sapphire,[73,74] and polycrystalline spinel,[75] and hardness of single crystals of yttria-stabilized zirconia.[76] The minimum in the fracture toughness curve for spinel is attributed to the sintering aid, LiF, which melts at 842°C.[77] LiF is used in the fabrication of polycrystalline spinel and accumulates at the grain boundaries.

Example: Unit conversion for hardness. Hardness is often expressed in GPa. What is the equivalent hardness in kg/mm^2? The load produced by 1 kg is P = mass × acceleration of gravity = (1 kg) × (9.807 m/s^2) = 9.8 N. One mm = 10^{-3} m, so 1 mm^2 = 10^{-6} m^2. The conversion of kg/mm^2 to N/m^2 looks like this:

$$1\,\frac{\cancel{\text{kg}}}{\cancel{\text{mm}^2}} \times \frac{9.807\ \text{N}/\cancel{\text{kg}}}{10^{-6}\ \text{m}^2/\cancel{\text{mm}^2}} = 9.807 \times 10^6\,\frac{\text{N}}{\text{m}^2} = 9.807\ \text{MPa}\,.$$

A hardness of 10 GPa (= 10^{10} Pa) is equivalent to x kg/mm^2, where x is found from the ratio

$$\frac{x\ \text{kg/mm}^2}{1\ \text{kg/mm}^2} = \frac{10^{10}\ \text{Pa}}{9.807 \times 10^6\ \text{Pa}} \Rightarrow x = 1020\,\frac{\text{kg}}{\text{mm}^2}\,.$$

References

1. K. P. D. Lagerlöf, A. H. Heuer, J. Castaing, J. P. Rivière and T. E. Mitchell, "Slip and Twinning in Sapphire (α-Al_2O_3)," *J. Am. Ceram. Soc.*, **77**, 385-397 (1994).
2. W. J. Tropf and D. C. Harris, "Mechanical, Thermal, and Optical Properties of Yttria and Lanthana-Doped Yttria," *Proc. SPIE*, **1112**, 9-19 (1989).
3. G. D. Quinn and R. Morrell, "Design Data for Engineering Ceramics: A Review of the Flexure Test," *J. Am. Ceram. Soc.*, **74**, 2037-2066 (1991).
4. G. D. Quinn, "Flexure Strength of Advanced Structural Ceramics: A Round Robin," *J. Am. Ceram. Soc.*, **73**, 2374-2384 (1990).
5. MIL-STD-1942, Department of the Army, Washington, D. C. (1983).
6. A 4-point loading test in which the upper and lower loading points are staggered, instead of nested, creates a pure shear stress in the middle of a rectangular bar. It can be used to measure the shear strength of a ceramic-to-ceramic bond. See O. Ünal, I. E. Anderson and S. I. Maghsoodi, *J. Am. Ceram. Soc.*, **80**, 1281-1284 (1997).
7. F. I. Baratta, W. T. Matthews and G. D. Quinn, *Errors Associated with Flexure Testing of Brittle Materials*, Army Materials Laboratory Report MTL TR 87-35, July 1987, p. 29.
8. H. Fessler and D. C. Fricker, "A Theoretical Analysis of the Ring-on-Ring Loading Disk Test," *J. Am. Ceram. Soc.*, **67**, 582-588 (1984). This paper also computes the stresses that result from friction between the disk and the test fixture.
9. T. Thiemeier, A. Brückner-Foit and H. Kölker, "Influence of the Fracture Criterion on the Failure Prediction of Ceramics Loaded in Biaxial Flexure," *J. Am. Ceram. Soc.*, **74**, 48-52 (1991).
10. D. K. Shetty, A. R. Rosenfield, P. McGuire, G. K. Bansal and W. H. Duckworth, "Biaxial Flexure Tests for Ceramics," *Ceram. Bull.*, **59**, 1193-1197 (1980).
11. W. F. Adler and D. J. Mihora, "Biaxial Flexure Testing: Analysis and Experimental Results" in *Fracture Mechanics of Ceramics* (R. C.Bradt, D. P. H. Hasselman, D. Munz, M. Sakai and V. Y. Shevchenko, eds.), Plenum Publishing Corp., New York (1992).
12. J. E. Ritter, Jr., K. Jakus, A. Batakis and N. Bandyopadhyay, "Appraisal of Biaxial Strength Testing," *J. Non-Crystalline Solids*, **38** & **39**, 419-424 (1980).
13. J. B. Wachtman, *Mechanical Properties of Ceramics*, Wiley, New York (1996).
14. B. Lawn, *Fracture of Brittle Solids*, 2nd ed., Cambridge University Press, Cambridge (1993).
15. J. T. A. Pollock and G. F. Hurley, "Dependence of Room Temperature Fracture Strength on Strain-Rate in Sapphire," *J. Mater. Sci.*, **8**, 1595-1602 (1973).
16. S. A. Newcomb and R. E. Tressler, "Slow Crack Growth in Sapphire Fibers at 800° to 1500°C," *J. Am. Ceram. Soc.*, **76**, 2505-2512 (1993).
17. S. M. Wiederhorn, A. G. Evans, E. R. Fuller and H. Johnson, "Application of Fracture Mechanics to Space-Shuttle Windows," *J. Am. Ceram. Soc.*, **57**, 319-323 (1974).
18. A. G. Evans, "Slow Crack Growth in Brittle Materials Under Dynamic Loading Conditions," *Int. J. Fracture Mech.*, **10**, 251-259 (1974).
19. E. R. Fuller, Jr., S. W. Freiman, J. B. Quinn, G. D. Quinn and W. C. Carter, "Fracture Mechanics Approach to the Design of Glass Aircraft Windows: A Case Study," *Proc. SPIE*, **2286**, 419-430 (1994).
20. J. Lamon, "Statistical Approaches to Failure for Ceramic Reliability Assessment," *J. Am. Ceram. Soc.*, **71**, 106-112 (1988).
21. W. Weibull, "A Statistical Distribution Function of Wide Applicability," *J. Appl. Mech.*, **13**, 293-297 (1951).
22. R. B. Abernethy, J. E. Breneman, C. H. Medlin and G. L. Reinman, *Weibull Analysis Handbook*, Air Force Report AFWAL-TR-83-2079, Pratt & Whitney

Aircraft, West Palm Beach, Florida (1983).
23. J. D. Sullivan and P. H. Lauzon, "Experimental Probability Estimators for Weibull Plots," *J. Mat. Sci. Lett.*, **5**, 1245-1247 (1980).
24. V. K. Pujari, D. M. Tracey, M. R. Foley, N. I. Paille, P. J. Pelletier, L. C. Sales, C. A. Willkens and R. L. Yeckley, "Reliable Ceramics for Advanced Heat Engines,"*Ceram. Bull.*, **74** [4], 86-90 (1995).
25. O. M. Jadaan, D. L. Shelleman, J. C. Conway, Jr., J. J. Mecholsky, Jr., and R. E. Tressler, "Prediction of the Strength of Ceramic Tubular Components: Part I - Analysis," *J. Testing. Eval.*, **19**, 181-191 (1991).
26. D. G. S. Davies, "The Statistical Approach to Engineering Design in Ceramics," *Proc. Brit. Ceram. Soc.*, **22**, 429-452 (1973).
27. D. L. Shelleman, O. M. Jadaan, J. C. Conway, Jr., and J. J. Mecholsky, Jr., "Prediction of the Strength of Ceramic Tubular Components: Part II - Experimental Verification," *J. Testing. Eval.*, **19**, 192-200 (1991).
28. F. Hild, E. Amar and D. Marquis, "Stress Heterogeneity Effect on the Strength of Silicon Nitride," *J. Am. Ceram. Soc.*, **75**, 700-702 (1992).
29. R. W. Rice, "Specimen Size-Tensile Strength Relations For a Hot-Pressed Alumina and Lead Zirconate/Titanate," *Ceram. Bull.*, **66**, 794-798 (1987).
30. M. N. Giovan and G. Sines, "Biaxial and Uniaxial Data for Statistical Comparisons of a Ceramic's Strength," *J. Am. Ceram. Soc.*, **62**, 510-515 (1979).
31. B. J. de Smet, P. W. Bach, H. F. Scholten, L. J. M. G. Dortmans and G. de With, "Weakest-Link Failure Predictions for Ceramics III: Uniaxial and Biaxial Bend Tests on Alumina," *J. Eur. Ceram. Soc.*, **10**, 101-107 (1992).
32. M. R. Borden and J. Askinazi, "Improving Sapphire Window Strength," *Proc. SPIE*, **3060**, 246-249 (1997).
33. F. Schmid and D. C. Harris, "Effects of Crystal Orientation and Temperature on the Strength of Sapphire," *J. Am. Ceram. Soc.*, **81**, 885-893 (1998).
34. D. S. Olson, G. J. Reynolds, G. F. Virshup, F. I. Friedlander, B. G. James and L. D. Partain, "Tensile Strength of Synthetic Chemical-Vapor-Deposited Diamond," *J. Appl. Phys.*, **78**, 5177-5179 (1995).
35. T. J. Valentine, A. J. Whitehead, R. S. Sussmann, C. J. H. Wort and G. A. Scarsbrook, "Mechanical Property Measurements of Bulk Polycrystalline CVD Diamond," *Diamond and Related Materials*, **3**, 1168-1172 (1994).
36. K. L. Jackson, D. L. Thurston, P. J. Boudreaux, R. W. Armstrong and C. CM. Wu, "Fracturing of Industrial Diamond Plates," *J. Mater. Sci.* **32**, 5035-5045 (1997).
37. J. E. Field and C. S. J. Pickles, "Strength, Fracture and Friction Properties of Diamond," *Diamond and Related Materials*, **5**, 625-634 (1996).
38. T. E. Steyer, K. T. Faber and M. D. Drory, "Fracture Strength of Free-Standing Chemically Vapor-Deposited Diamond Films," *Appl. Phys. Lett.*, **66**, 3105-3107 (1995).
39. J. A. Savage, C. J. H. Wort, C. S. J. Pickles, R. S. Sussmann, C. G. Sweeney, M. R. McClymont, J. R. Brandon, C. N. Dodge and A. C. Beale, " Properties of Free-Standing CVD Diamond Optical Components," *Proc. SPIE*, **3060**, 144-159 (1997).
40. C. B. Willingham, T. M. Hartnett, R. P. Miller and R. B. Hallock, "Bulk Diamond for IR/RF Windows and Domes," *Proc. SPIE*, **3060**, 160-168 (1997).
41. R. W. Sparrow, H. Herzig, W. V. Medenica and M. J. Viens, "Influence of Processing Techniques on the VUV Transmittance and Mechanical Properties of Magnesium Fluoride Crystal," *Proc. SPIE*, **2286**, 33-45 (1994).
42. M. D. Herr and W. R. Compton, *Evaluation of Statistical Fracture Criteria for Magnesium Fluoride Seeker Domes*, Naval Weapons Center Report TP 6226, China Lake, California, December 1980.
43. P. Miles, "High Transparency Infrared Materials — A Technology Update," *Opt. Eng.*, **15**, 451-459 (1976).

44. J. Trombetta, Raytheon Systems Co. to be published.
45. P. Klocek, M. W. Boucher, J. M. Trombetta and P. A. Trotta, "High Resistivity and Conductive Gallium Arsenide for IR Optical Components," *Proc. SPIE*, **1760**, 74-85 (1992).
46. D. W. Roy, "Hot-Pressed $MgAl_2O_4$ for Ultraviolet (UV), Visible, and Infrared (IR) Optical Requirements," *Proc. SPIE*, **297**, 13-18 (1981).
47. C. Baudín, R. Martínez and P. Pena, "High-Temperature Mechanical Behavior of Stoichiometric Magnesium Spinel," *J. Am. Ceram. Soc.*, **78**, 1857-1862 (1995).
48. J. B. Wachtman, Jr. and L. H. Maxwell, "Strength of Synthetic Single Crystal Sapphire and Ruby as a Function of Temperature and Orientation," *J. Am. Ceram. Soc.*, **42**, 432-433 (1959).
49. E. A. Jackman and J. P. Roberts, "Strength of Synthetic Single-Crystal and Polycrystalline Corundum," *Phil. Mag.*, **46**, 809-811 (1955).
50. E. A. Jackman and J. P. Roberts, "Strength of Polycrystalline and Single-Crystal Corundum," *Trans. Brit. Ceram. Soc.*, **54**, 389-398 (1955).
51. R. L. Gentilman, E. A. Maguire, H. S. Starrett, T. M. Hartnett and H. P. Kirchner, "Strength and Transmittance of Sapphire and Strengthened Sapphire," *J. Am. Ceram. Soc.*, **64**, C116-C117 (1981).
52. G. F. Hurley, "Mechanical Behavior of Melt-Grown Sapphire at Elevated Temperature," *Appl. Polymer Symp.*, **21**, 121-130 (1973).
53. R. J. Charles and R. R. Shaw, "Delayed Failure of Polycrystalline and Single-Crystal Alumina," General Electric Research Laboratory Report 62-RL-3081M, Schenectady, New York (1962).
54. J. D. Spain and H. S. Starrett, *Biaxial Flexural Evaluations of Zinc Sulfide*, Southern Research Institute Report SRI-MME-88-259-6270, Birmingham, Alabama, October 1988.
55. A. A. Déom, D. L. Balageas, F. G. Laturelle, G. D. Gardette and G. J. Freydefont, "Sensitivity of Rain Erosion Resistance of Infrared Materials to Environmental Conditions Such as Temperature and Stress," *Proc. SPIE*, **1326**, 301-309 (1990).
56. C. S. J. Pickles and J. E. Field, "The Dependence of the Strength of Zinc Sulfide on Temperature and Environment," *J. Mater. Sci.*, **29**, 1115-1120 (1994).
57. C. A. Klein, "Infrared Missile Domes: Is There a Figure of Merit for Thermal Shock?" *Proc. SPIE*, **1760**, 338-357 (1992).
58. C. A. Klein, "Diamond Windows for Infrared Applications in Adverse Environments," *Diamond and Related Materials*, **2**, 1024-1032 (1993).
59. I. J. McColm, *Ceramic Hardness*, Plenum Press, New York (1990).
60. R. D. Dukino and M. V. Swain, "Comparative Measurement of Indentation Fracture Toughness with Berkovich and Vickers Indenters," *J. Am. Ceram. Soc.*, **75**, 3299-3304 (1992).
61. A. L. Yurkov, E. Breval and R. C. Bradt, "Cracking During Indentation in Sialon-Based Ceramics: Kinetic Microhardness and Acoustic Emission," *J. Mater. Sci. Lett.*, **15**, 987-990 (1996).
62. M. S. Darrow, W. B. White and R. Roy, "Micro-Indentation Hardness Variation as a Function of Composition for Polycrystalline Solutions in the Systems PbS/PbTe, PbSe/PbTe, and PbS/PbSe," *J. Mater. Sci.*, **4**, 313-319 (1969).
63. G. R. Anstis, P. Chantikul, B. R. Lawn and D. B. Marshall, "A Critical Evaluation of Indentation Techniques for Measuring Fracture Toughness: I, Direct Crack Measurements,"*J. Am. Ceram. Soc.*, **64**, 533-538 (1981).
64. P. Chantikul, G. R. Anstis, B. R. Lawn and D. B. Marshall, "A Critical Evaluation of Indentation Techniques for Measuring Fracture Toughness: II, Strength Method," *J. Am. Ceram. Soc.*, **64**, 539-543 (1981).

65. C. B. Ponton and R. D. Rawlings, "Vickers Indentation Fracture Toughness Test, Part I: Review of Literature and Formulation of Standardised Indentation Toughness Equations," *Mater. Sci. Tech.*, **5**, 865-872 (1989).
66. P. Chantikul, G. R. Anstis, B. R. Lawn and D. B. Marshall, "A Critical Evaluation of Indentation Techniques for Measuring Fracture Toughness: II, Strength Method," *J. Am. Ceram. Soc.*, **64**, 539-543 (1981).
67. R. W. Hertzberg, *Deformation and Fracture Mechanics of Engineering Materials*, 3rd ed., Wiley, New York (1989), pp. 281-284.
68. J. J. Mecholsky, Jr., "Quantitative Fractography: An Assessment," in *Ceramic Transactions*, Vol. 17 (V. D. Freschette and J. Varmer, eds.), American Ceramic Society, Westerville, Ohio (1991).
69. R. W. Rice, "Review: Effects of Environment and Temperature on Ceramic Tensile Strength-Grain Size Relations," *J. Mater. Sci.*, **32**, 3071-3087 (1997).
70. S. C. Carniglia, "Reexamination of Experimental Strength-vs-Grain-Size Data for Ceramics," *J. Am. Ceram. Soc.*, **55**, 243-249 (1972).
71. H.-E. Kim and A. J. Moorhead, "Effect of Doping on the Strength and Infrared Transmittance of Hot-Pressed Gallium Arsenide," *J. Am. Ceram. Soc.*, **74**, 161-165 (1991).
72. R. W. Rice, C. Cm. Wu and F. Borchelt, "Hardness-Grain Size Relations in Ceramics," *J. Am. Ceram. Soc.*, **77**, 2539-2553 (1994).
73. S. M. Wiederhorn, B. J. Hockey and D. E. Roberts, "Effect of Temperature on the Fracture of Sapphire," *Phil. Mag.*, **28**, 783-796 (1973).
74. S. A. Newcomb and R. E. Tressler, "High-Temperature Fracture Toughness of Sapphire," *J. Am. Ceram. Soc.*, **77**, 3030-3032 (1994).
75. A. Ghosh, K. W. White, M. G. Jenkins, A. S. Kobayashi and R. C. Bradt, "Fracture Resistance of a Transparent Magnesium Aluminate Spinel," *J. Am. Ceram. Soc.*, **74**, 1624-1630 (1991).
76. G. N. Morscher, P. Pirouz and A. H. Heuer, "Temperature Dependence of Hardness in Yttria-Stabilized Zirconia Single Crystals," *J. Am. Ceram. Soc.*, **74**, 491-500 (1991).
77. K. W. White and G. P. Kelkar, "Fracture Mechanisms of a Coarse-Grained, Transparent $MgAl_2O_4$ at Elevated Temperatures, *J. Am. Ceram. Soc.*, **75**, 3440-3444 (1992).
78. W. B. Hillig, "A Methodology for Estimating the Mechanical Properties of Oxides at High Temperatures," *J. Am. Ceram. Soc.*, **76**, 129-138 (1993).

Chapter 4

THERMAL PROPERTIES

If you plunge a ceramic or glass object that is sufficiently hot into a bucket of water, the object is likely to shatter. The same fate awaits a window or dome if it is heated too rapidly by a laser or on the front of a missile. Resistance to failure by thermal shock is a critical requirement for some applications. We begin this chapter by discussing thermal expansion and thermal conductivity and then proceed to consider thermal shock.

4.1 Thermal expansion and heat capacity

Most materials expand when heated.* When the temperature of the rod in Fig. 4.1 is raised by ΔT, the length increase, ΔL, is given by

$$\frac{\Delta L}{L} = \alpha \, \Delta T \tag{4-1}$$

where L is the initial length and α is called the *expansion coefficient*, or the coefficient of thermal expansion (C.T.E.).

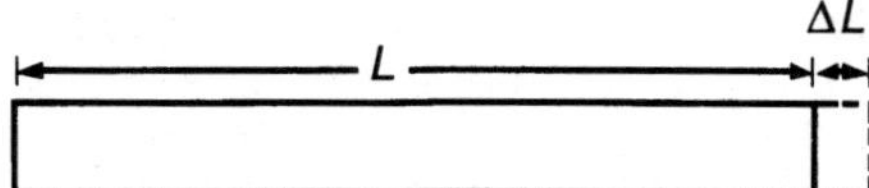

Fig. 4.1. Thermal expansion.

Expansion is not exactly a linear function of temperature. The behavior of yttria is shown in Fig. 4.2. When heated from 300 to 1000 K, $\Delta L/L = 0.00522$. Therefore the *average* expansion coefficient for yttria between 300 and 1000 K is $\alpha = (\Delta L/L)/\Delta T = 0.00522/700 = 7.46 \times 10^{-6}$ K^{-1}. The expansion coefficient at a particular temperature is the slope of the curve in Fig. 4.2 at that temperature. At 300 K the expansion coefficient of yttria is 6.63×10^{-6} K^{-1}, while at 1000 K the expansion coefficient is 8.49×10^{-6} K^{-1}. When you look up an expansion coefficient for a material, note what temperature or temperature range applies. Expansion coefficients of some infrared window materials are given in Table 4.1. More detailed thermal property data appear in Appendix G.

---

*A notable exception is the ceramic material zirconium tungstate, ZrW_2O_8, which *contracts* isotropically when heated in the temperature range 0.3-1050 K.[1,2] This material is a candidate for fabricating zero-expansion composite structural materials.

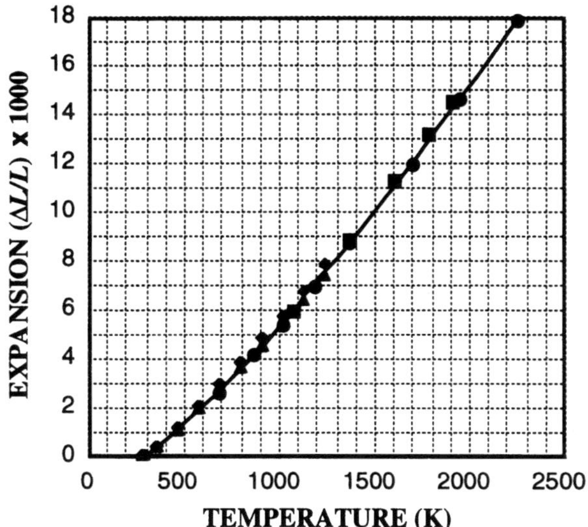

Fig. 4.2. Thermal expansion of yttria. The four different kinds of symbols correspond to four different experiments. Ordinate has been multiplied by 1000. The value 8, for example, corresponds to $\Delta L/L$ = 0.008. Smooth curve[3] is given by $\Delta L/L = -1.69277 \times 10^{-3} + 5.83245 \times 10^{-6}\,T + 1.33104 \times 10^{-9}\,T^2$.

Most materials in Table 4.1 that are suitable for external use have expansion coefficients in the range 5-10 × 10^{-6} K^{-1}. Silicon, diamond and fused silica have unusually low expansion coefficients. Zerodur® glass ceramic (Section 5.1.1) has an expansion coefficient near 10^{-7} K^{-1} and is suitable for such demanding applications as telescope mirrors. The expansion coefficient of Zerodur is slightly *negative* at 300 K.

Table 4.1 also gives *heat capacity* (also called *specific heat*), which is the energy required to raise the temperature of 1 gram of material by 1 K at constant pressure. Figure 4.3 shows how the heat capacity of yttria depends on temperature. A material with a large heat capacity changes temperature sluggishly, while one with low heat capacity changes temperature easily. The heat capacity of any material approaches zero as temperature approaches absolute zero and approaches a level value at high temperature (typically above 1000°C).

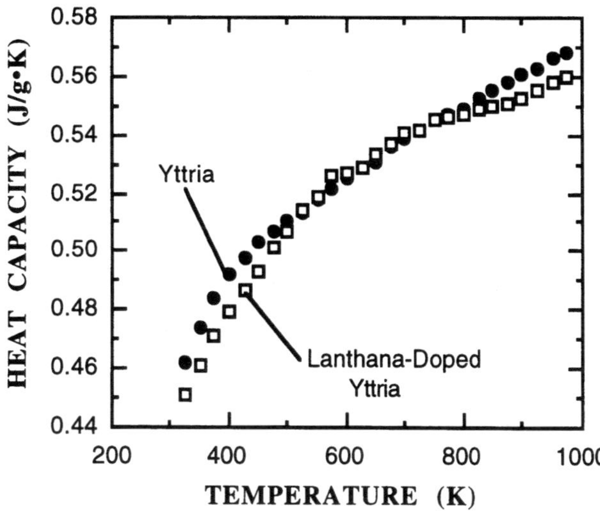

Fig. 4.3. Heat capacity of yttria and lanthana-doped yttria as a function of temperature.

Table 4.1. Thermal properties of infrared window materials near 300 K[4]

Material	Expansion coefficient (10^{-6} K^{-1})	Heat capacity (J/g·K)	Thermal conductivity (W/m·K)
LiF	34.4	1.55	11.3
KBr	38.5	0.44	4.8
KCl	36.6	0.69	6.7
AgCl	31.0	0.36	1.12
NaCl	40.0	0.84	6.5
CsBr	47	0.26	0.94
CsI	48	0.20	1.1
KRS-5	58	—	0.54
As_2S_3 glass	24.6	0.46	0.17
Schott IRG 100	15	—	0.3
Corning 9754 germanate glass	6.2	0.54	1.0
AMTIR-1 Ge-As-Se glass	12.0	0.29	0.25
BS39B calcium aluminate glass	8.4	0.86	1.23
Fused SiO_2	0.52	0.74	1.4
CaF_2	18.9	0.85	10
MgF_2 (hot pressed)	10.4	0.50	14.7
MgO	10.5	0.88	59
Sapphire	5.0 ($\parallel c$)[5] 4.4 ($\perp c$)[5]	0.75	35.1 ($\parallel c$) 33.0 ($\perp c$)
Y_2O_3	6.6	0.48	13.5
La-doped Y_2O_3	6.6	0.48	5.3
ALON	5.8	0.77	12.6
Spinel	5.6	0.88	14.6
ZnS (standard)	7.0	0.47	19
ZnS (multispectral)	7.0	0.47	27
ZnSe	7.6	0.34	16
GaP	5.3	0.84	110
GaAs	5.7	0.32	55
Ge	6.1	0.31	59
Si	2.6	0.75	163
SiC	2.8	0.69	490
Si_3N_4	2.1 ($\parallel c$) 1.1 ($\perp c$)	0.88	33
AlN	~4.9	0.80	320
BP[6,7]	3.6	0.71	400
Diamond (Type IIa)	0.8	0.52	2000

4.2 Thermal conductivity[8]

When heat is added to the left side of the object in Fig. 4.4, the energy flux from left to right is

$$\text{Heat flux}\left(\frac{\text{watts}}{\text{m}^2}\right) = \frac{k\,A\,\Delta T}{L} \tag{4-2}$$

where k is the thermal conductivity (W/m·K), A is the cross-sectional area, ΔT is the temperature difference across the sample and L is the distance between the two faces of the material. *The greater the thermal conductivity, the more rapidly heat is transferred across the sample.*

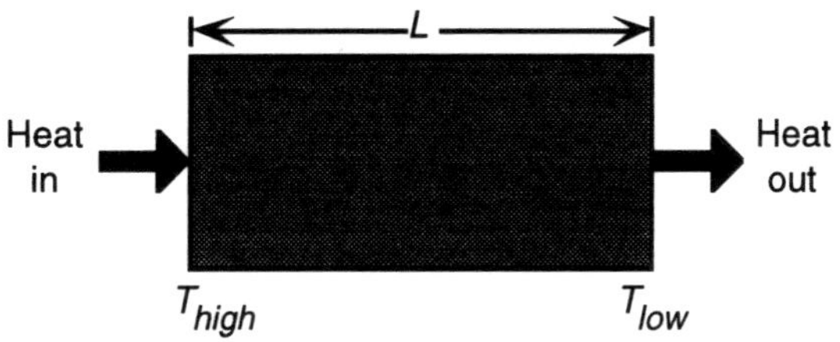

Fig. 4.4. Thermal conductivity measures the ease with which heat can be transferred across a material.

A related quantity is *thermal diffusivity*, which is the thermal conductivity of a material divided by its heat capacity per unit volume:

$$\text{Thermal diffusivity}\left(\frac{\text{m}^2}{\text{s}}\right) = \frac{k}{C_p\,\rho} \tag{4-3}$$

where k is thermal conductivity, C_p is its heat capacity per gram (J/g·K), and ρ is density (g/m^3). The rate of change of temperature of a material with time is proportional to the thermal diffusivity and the change of temperature gradient in the material. The greater the thermal diffusivity, the faster the temperature change. Thermal diffusivity characterizes the rate of "diffusion of temperature" through a material.

Figure 4.5 compares the thermal conductivity of a ceramic window material, calcium fluoride, and a metal. In general, the conductivity of a crystalline material that is not an electrical conductor goes through a maximum at cryogenic temperatures and then falls off approximately as $1/T$ at high temperature, where T is temperature in kelvins. The conductivity of an electrical conductor also goes through a maximum at low temperature, but levels off at a constant value at elevated temperature.

We can understand the thermal conductivity (k) of electrically insulating ceramics through the equation

$$k = \frac{1}{3} C_p v \ell \tag{4-4}$$

where C_p is the heat capacity, v is the mean velocity of acoustic waves (called *phonons*) through the crystal, and ℓ is the mean free path of such waves. The mean velocity is approximately equal to the speed of sound in the crystal. The mean free path is the distance required for the vibrational wave to be attenuated to 1/e of its initial amplitude.

To a good approximation, v is independent of temperature. As the temperature approaches zero, heat capacity approaches zero and therefore thermal conductivity also approaches zero. At high temperature heat capacity approaches a constant value. However, the mean free path of phonons in the crystal decreases as $1/T$ at high temperature. Therefore the thermal conductivity falls as $1/T$. In an electrical conductor such as copper, most thermal energy is carried by conduction elections which obey a different set of rules from lattice vibrations and lead to a constant thermal conductivity at high temperature.

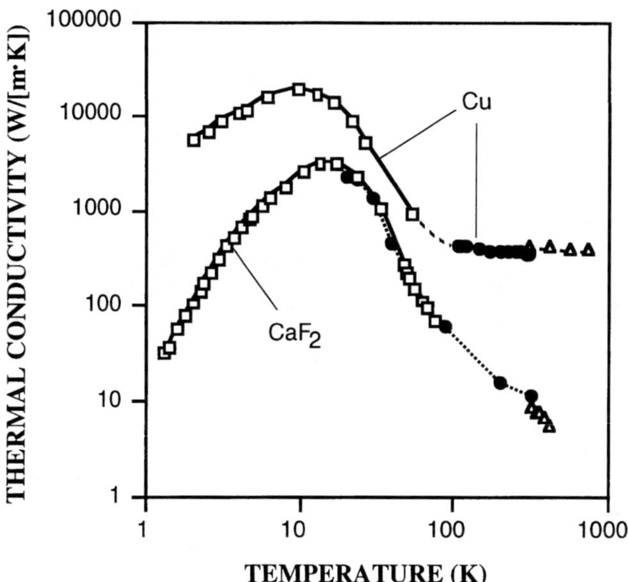

Fig. 4.5. Thermal conductivity of copper and single-crystal calcium fluoride, which has a cubic structure.[9] Different symbols correspond to measurements on different samples by different investigators.

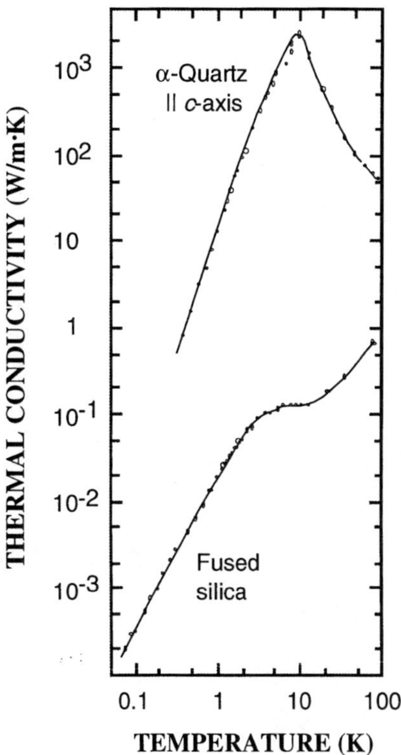

Fig. 4.6. Thermal conductivity of crystalline and amorphous SiO_2.[10]

Amorphous materials, such as optical glasses, generally have lower thermal conductivities than do crystalline materials because the mean free path of a phonon in a glass is shorter than the mean free path in a crystal. At high temperature, the conductivity of the glass approaches that of the crystal. Figure 4.6 compares the behavior of crystalline SiO_2 (quartz) and amorphous SiO_2 (fused silica). Notice also that the glasses in Table 4.1 have lower conductivities than the crystalline materials.

Thermal conductivity of most ceramics (which are poor conductors) can be measured by an experiment similar to the one in Fig. 4.4, using a water calorimeter to measure the rate at which heat reaches the second surface.[11] A notable complication for transparent materials is that *radiant heat transfer* (by photons traveling through the material) becomes significant in comparison to thermal conduction by vibrating atoms of the crystal lattice at elevated temperature. The total heat flux across a transparent sample *increases* at sufficiently high temperature because of radiant heat transfer. The circles in Fig. 4.7 show the apparent conductivity of yttria and the diamonds are corrected by subtracting the calculated radiant transfer across the test specimens.[12] The corrections are most significant above 1000 K.

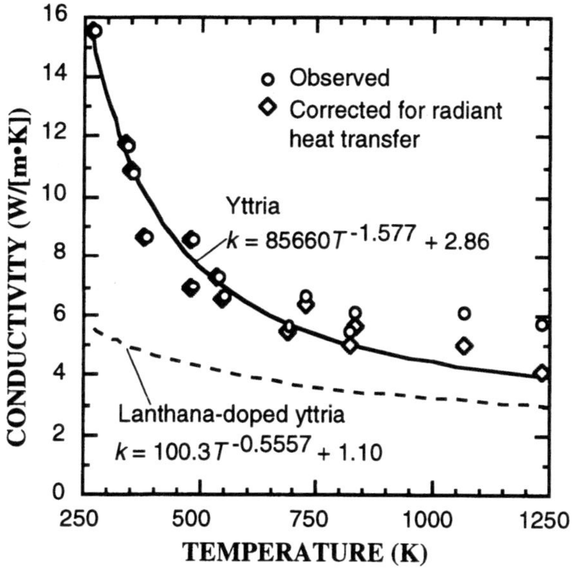

Fig. 4.7. Thermal conductivity of yttria and lanthana-doped yttria as a function of temperature. Solid and dashed lines are best fits to multiple data sets.[3] Circles show one measured data set for yttria and diamonds give corrected values after subtracting the calculated radiant contribution from the circles.

Figure 4.8 illustrates that the thermal conductivity of electrically insulating, crystalline materials decreases as $1/T$ at elevated temperature. The reciprocal of thermal conductivity is called *thermal resistivity*. A graph of thermal resistivity versus temperature is a straight line at high temperature.

The thermal conductivities of materials with very high thermal conductivity can be measured by the laser flash method (which also works for lower conductivity).[11] In this experiment, a laser provides a rapid pulse of energy at the front surface of a disk. The rise of temperature at the back surface is then measured to determine conductivity.

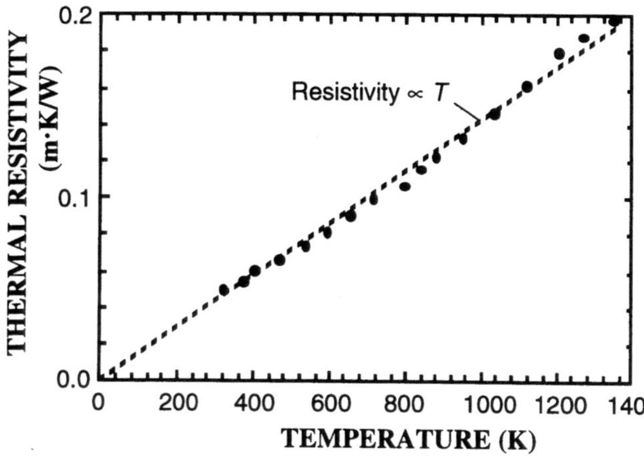

Fig. 4.8. Thermal resistivity (1/conductivity) of polycrystalline zinc sulfide is proportional to temperature at elevated temperature.[13]

Small quantitites of impurity in high-conductivity crystals greatly reduce the thermal conductivity. Figure 4.9 shows that 1 wt % oxygen atom impurity in the lattice of aluminum nitride reduces the conductivity by a factor of 3 at room temperature.[14]

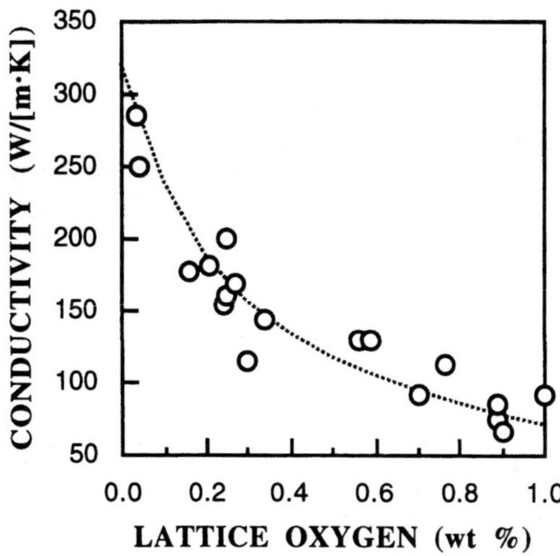

Fig. 4.9. Oxygen atom impurities in the crystal lattice of AlN decrease the thermal conductivity at constant temperature.[14]

4.3 Thermal shock

Consider a hemispheric infrared missile dome (Fig. 0.1) immediately after launch. Friction with the air heats the outside leading surface of the blunt nose more rapidly than the rest of the dome. The hot front surface expands more than the cooler inside and soon there is significant stress between the expanded and unexpanded material. If the stress exceeds the strength of a critical flaw, the dome shatters. This is *thermal shock* induced by aerothermal heating. Maximal tensile stress is on the inside surface of the dome, where failure is likely to occur. After the boost stage ends and the dome attains maximum temperature, the dome cools down from the outside during glide stage. At this time, maximum tensile stress is found on the outside surface, from which failure may originate. Thermal shock failure may occur at either surface of the dome at different stages of flight.

The proper way to model these processes for engineering design is by finite element thermal stress analysis of the entire front of the missile, including the mount. In fact, a poorly designed mount with different thermal expansion from the dome can place more stress on the dome than can aerothermal heating.

In the absence of a sophisticated analysis, a simple *thermal shock figure of merit* can be used for a semiquantitative comparison of the thermal shock resistance of different materials. *The larger the figure of merit, the greater the heat flux that can be withstood by the material without catastrophic failure.*

Two figures of merit are used, depending on whether or not heat added to a window or dome has time to diffuse away from the hot surface.[15-17] The parameter that governs the rate of thermal diffusion is called the *Biot number*, β:

$$\text{Biot number} = \beta = \frac{t\,h}{k} \tag{4-5}$$

where t is the thickness of the material, h is the surface *heat transfer coefficient*, and k is thermal conductivity of the material. The heat transfer coefficient gives the power that enters the material per unit area per unit temperature difference between the atmosphere and surface. The units are $W/m^2{\cdot}K$. A high Biot number means that heat moves slowly from one surface of the material to the other. The Biot number is low when heat moves rapidly through the material.

In the limit of severe thermal shock, $\beta >> 1$ in Eq. (4-5) and the appropriate figure of merit is designated R:

$$\text{Hasselman severe thermal shock figure of merit} = R = \frac{S\,(1-\nu)}{\alpha\,E} \tag{4-6}$$

where S is the strength of the material, ν is Poisson's ratio, α is the expansion coefficient and E is Young's modulus. Severe thermal shock is caused by rapid heating of one surface of a thick dome with low thermal conductivity.

In the limit of mild thermal shock, $\beta << 1$ and the figure of merit is designated R':

$$\text{Hasselman mild thermal shock figure of merit} = R' = \frac{S\,(1-\nu)\,k}{\alpha\,E} . \tag{4-7}$$

Mild thermal shock is caused by slow heating of a thin dome with high thermal conductivity. The difference between Eqns. (4-6) and (4-7) is the presence or absence of k, the thermal conductivity.

Which figure of merit do we use? Aerothermal heating of missile domes involves the following range of parameters:[15]

Parameter in Eq. (4-5)	Range	Typical value
t (m)	0.002-0.004	0.003
h ($W/m^2{\cdot}K$)	400-8000	1000
k ($W/m{\cdot}K$)	7-50	12
β	0.01-5	0.25

A Biot number of 0.25 puts us in the range of mild thermal shock, for which R' is the appropriate figure of merit. In severe thermal shock, the rate of heating is so great that heat cannot spread through the material and thermal conductivity drops out of consideration. We will use the R' figure of merit for much of our discussion. Recognize, however, that real systems can enter the severe thermal shock regime in which R is the appropriate figure of merit. Recognize also that one material can be in its mild thermal shock regime while another is in severe thermal shock under the same conditions because of its lower thermal conductivity.

To understand "severe" and "mild" heating, consider a dome whose temperature is initially at $T_{inner\ wall}$. Then the dome is suddenly exposed to an airstream at the temperature $T_{stagnation}$. Figure 4.10 is a schematic profile of the temperature inside the apex of the dome at the time of peak stress under laminar flow conditions.[18] "Severe heating" is synonymous with a "thermally thick" dome, which is characterized by a "large" Biot number ($\beta > 1$). "Mild heating" is synonymous with a "thermally thin" dome, which is characterized by a "small" Biot number ($\beta < 1$). Under severe heating, the outside wall of the dome reaches the stagnation temperature, which is the temperature of the air at the outside surface at the apex of the dome. In the case of severe heating, the outside of the dome warms up, but the inside remains at its initial temperature at the time of peak stress. Under mild heating conditions, there is time for heat to diffuse through the thickness of the dome before the outside wall reaches the stagnation temperature. The difference in temperature between the inner and outer surfaces at the time of peak stress is $(T_{stagnation} - T_{inner\ wall}) \times \beta$, where β is the Biot number (Eq. 4-5).

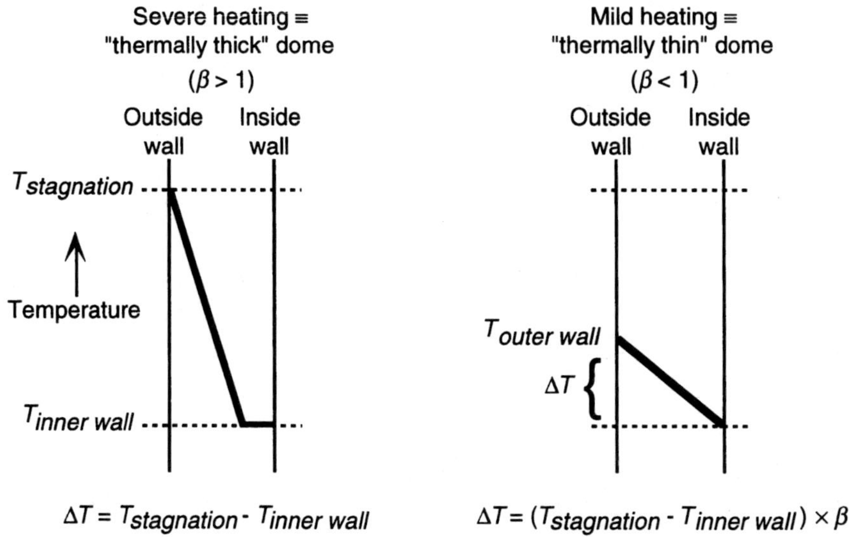

Fig. 4.10. Schematic temperature profile at the apex of a dome under "severe" or "mild" heating at peak stress under laminar flow conditions.[18]

Common experimental measures of thermal stress shock resistance are obtained by dropping hot objects into cold liquids and measuring the maximum temperature difference at which the material survives or the temperature difference at which the residual flexure strength of the material suddenly decreases.[19] A related test measures quench-induced crack extension of deliberately introduced indentation cracks.[20] Testing methods that are more relevant to aerothermal heating of a missile dome involve heating the front surface of a ceramic disk very rapidly with a lamp or laser[21-23] or with hot air.[24] Appropriate heating rates are several hundred degrees per second.

4.3.1 Hasselman figures of merit

The figure of merit for mild heating, R' in Eq. (4-7), tells us that thermal shock resistance is favored by high strength, high thermal conductivity, low thermal expansion and low Young's modulus. Making a dome or window thinner also increases its thermal shock resistance. Table 4.2 gives thermal shock parameters for some infrared window materials.

Table 4.2. Thermal shock parameters of infrared window materials

Material	Expansion coefficient α (10^{-6} K^{-1})	Young's modulus E (GPa)	Thermal conductivity k (W/m·K)	Poisson's ratio ν	Mean Strength S (MPa)	Figures of merit R' (mild)*	R (severe)*
Midwave							
MgF_2 (hot pressed)	10.4	142	14.7	0.27	125	0.9	0.06
Fused SiO_2	0.52	73	1.4	0.16	60	3.4	2.4
MgO	10.5	249	59	0.18	200?	3.7	0.06
Sapphire	5.3	344	36	0.27	300	4.3	0.12
GaP†	5.3	103	110	0.31	110	15	0.14
Y_2O_3	6.6	173	13.5	0.30	160	1.3	0.10
La-doped Y_2O_3	6.6	170	5.3	0.30	160	0.5	0.10
ALON	5.8	323	12.6	0.24	300	1.5	0.12
Spinel	5.6	193	14.6	0.26	190	1.9	0.13
Si	2.6	131	163	0.28	120	41	0.25
SiC[25]	2.2	466	214	0.21	500?	82	0.39
Si_3N_4	~1.4	166[27]	33		230[26]	~24	~0.74‡
AlN[7]	~4.9§	294	320[28]	0.26	225	37	0.12
2-color (long wave + midwave)							
NaCl (hot forged)	40.0	40	6.5	0.25	10	0.03	0.005
CaF_2 (hot forged)	18.9	76	10	0.28	55	0.3	0.03
ZnS (standard)	7.0	74	19	0.29	100	2.6	0.14
ZnS (multispectral)	7.0	88	27	0.32	70	2.1	0.08
ZnSe	7.6	70	16	0.28	50	1.1	0.03
GaAs	5.7	83	55	0.31	100	8.0	0.15
Ge	6.1	103	59	0.28	90	6.1	0.10
Long wave							
Diamond (CVD)	0.8	1143	2000	0.07	200	410	0.20

*Units of R' are (10^6 Pa)(W/m·K)/(10^{-6} K^{-1})(10^9 Pa) = 10^3 W/m.
Units of R are 10^3 K.

†GaP has some long wave capability from 8-9 μm.

‡Assuming $\nu \approx 0.25$.

§Estimated as $(^1/_3)(2\alpha_a + \alpha_c)$, where α_a is the expansion coefficient parallel to the a-axis of the hexagonal crystal and α_c is the expansion coefficient parallel to the c-axis.

Among the potential midwave (3-5 μm) transmitting materials in Table 4.2, sapphire has a high thermal shock figure of merit, R', and gallium phosphide is outstanding. Silicon is excellent, but its upper temperature is limited to ~260°C by free carrier absorption (Fig. 1.30). Silicon carbide (Fig. 5.24), aluminum nitride and silicon nitride would be strong candidates for thermal shock applications, but none of these materials is commercially available in transparent form. Magnesium oxide might be similar to sapphire, but it is not commercially available in transparent form.

Among 2-color (3-5 and 8-14 μm) materials, zinc sulfide and gallium arsenide are very good. However , gallium arsenide has an upper operating temperature limit of ~460°C, which restricts its applicability. Germanium is excellent with respect to thermal shock, but cannot be used much above 100°C (Fig. 1.30).

Diamond is in a class by itself, with 1 or 2 orders of magnitude of superior thermal shock resistance over available rivals. Inspection of Table 4.2 shows that the main reason for this is that the thermal conductivity of diamond is 2 orders of magnitude greater than that of common ceramics. The low expansion coefficient of diamond also contributes to its thermal shock resistance. Diamond requires more development before it is available in required sizes and shapes and at affordable prices. Diamond is oxidized (burned) in the air above 700°C, but can probably be protected up to 1000°C by oxidation-resistant coatings. Unlike other materials in Table 4.2, diamond has no midwave capability at elevated temperature because of emittance from the 2- and 3-phonon bands in Fig. 1.43. However, diamond would be suitable for shorter wavelength (e.g., 2-2.5 μm near infrared) atmospheric transmission windows in Fig. 0.2.

The thermal shock figure of merit decreases for most materials at elevated temperature.[15,29] R' changes because strength, thermal conductivity, expansion coefficient and Young's modulus all vary with temperature. Since missile flights attain high temperature rapidly, what is the appropriate temperature at which to evaluate the thermal shock figure of merit?

Figure 4.11 shows the calculated temperature on the outside surface of a dome in a missile launch under a particular set of conditions. The stress on the dome reaches a

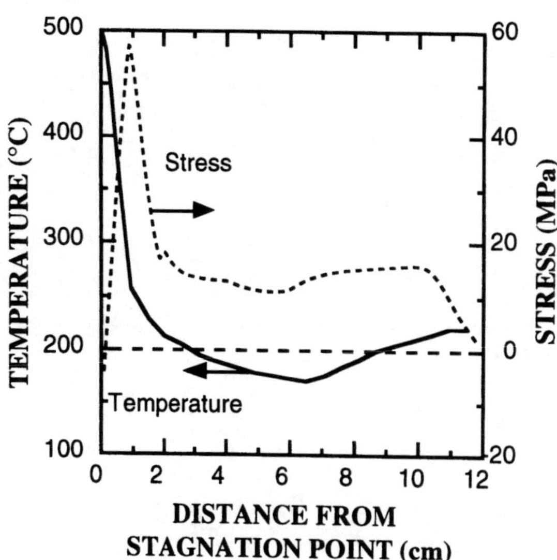

Fig. 4.11. Calculated stress and outside surface temperature as a function of position on a dome during a tactical missile launch. (Courtesy W. R. Compton, Naval Air Warfare Center.) An informative example of the measured and calculated thermal and mechanical response of a dome during a simulated high speed launch can be found in Ref. 28.

maximum approximately 1 cm from the nose of the dome, where the temperature is approximately 250°C. In a relatively high speed launch of a tactical missile, it is typical to have a maximum gradient of 100-200°C across a 2-3 mm dome thickness. In lower speed launches, the temperature gradient is smaller. Examination of numerous flight profiles shows that it is common for the temperature at maximum stress to come between 25° and 500°C. The thermal shock figure of merit at 25°C in Table 4.2 probably provides an appropriate qualitative comparison of materials.

Figure 4.12 shows the value of the thermal shock figure of merit, R', required to survive instantaneous exposure to the atmosphere at a given speed and altitude. For example, the upper curve tells us that $R' = 2.3$ is required to survive sudden exposure at Mach 4.5 at sea level. A value of R' near 0.5 is all that is required to survive the same exposure at an altitude of 15 km (50,000 feet). No safety factor is built into the graph. Fig. 4.12 is only a first order prediction. *There is no substitute for a careful finite element aerothermal stress analysis using the best available temperature-dependent data for the material to be employed.*

The wind tunnel in Fig. 4.13 allows the performance of a missile dome to be measured in a flight-like environment. Hot air is created by burning hydrogen in a high-speed stream of air. The temperature and air pressure at the surface of the dome are independent variables. After establishing the desired flow, a mounted dome can be rapidly inserted into the flow. Alternatively, a covered dome can be in the flow the whole time and the cover quickly removed when required flow conditions are attained. The inside

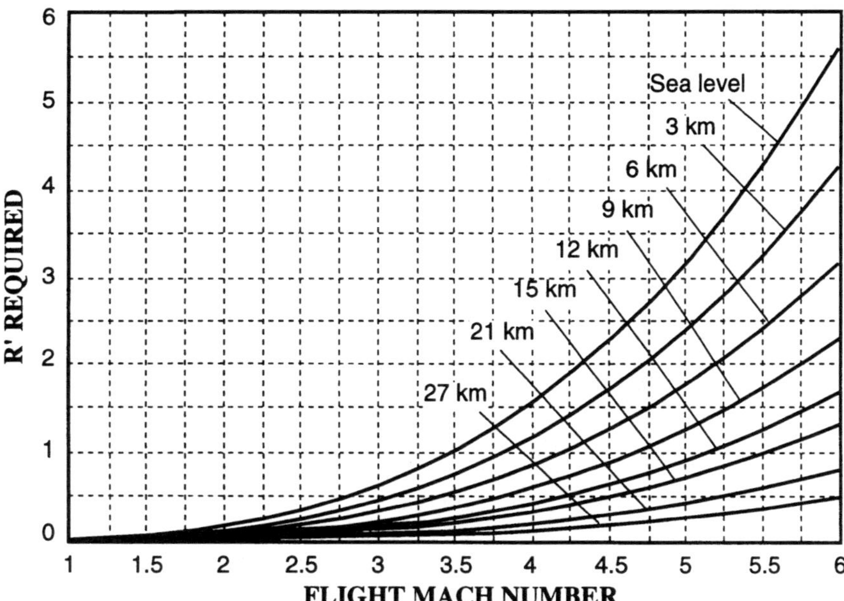

Fig. 4.12. Value of R' (units = 10^3 W/m) required for a nose-mounted dome (outer radius = 3.6 cm, thickness = 0.35 cm) to survive instantaneous exposure to an air stream at a given speed and altitude in a standard day atmosphere, assuming that the dome is initially at 15°C. There is no safety factor built into the calculation. (Courtesy W. R. Compton, Naval Air Warfare Center.)

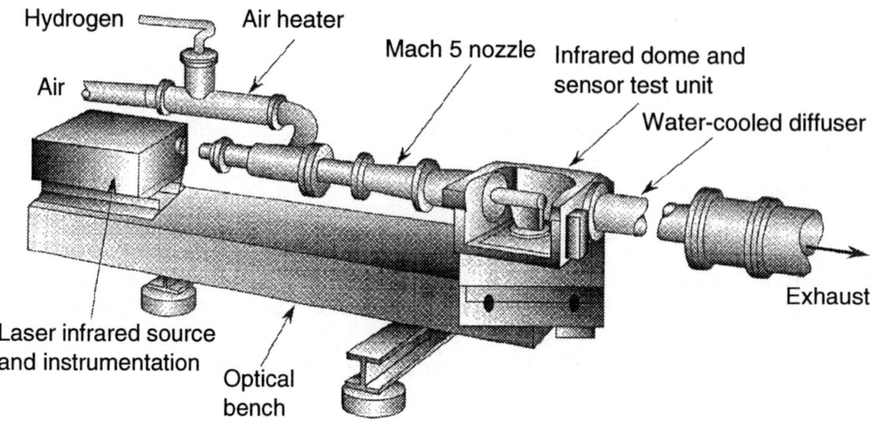

Fig. 4.13. Aerothermal/Infrared Test Facility at the Johns Hopkins University Applied Physics Laboratory, Laurel, MD.[31]

surface of a dome can be instrumented with strain gauges and thermocouples to measure strain and temperature as the dome is heated. Figure 4.14 shows the fracture pattern in a spinel dome that shattered from thermal shock in the wind tunnel.

Fig. 4.14. Fracture pattern in spinel dome after exposure to a Mach 4.6 airstream with a total pressure of 6.2 MPa and a total temperature of 1390 K.

Can we actually predict the behavior of a missile dome under aerothermal heating? Figure 4.15 is a carefully selected case in which the answer is fairly close to "yes!" The graph shows the stress in a lanthana-doped yttria dome, as measured by strain gauges and calculated by finite element methods. The dome shattered from thermal shock at 1.09 s.

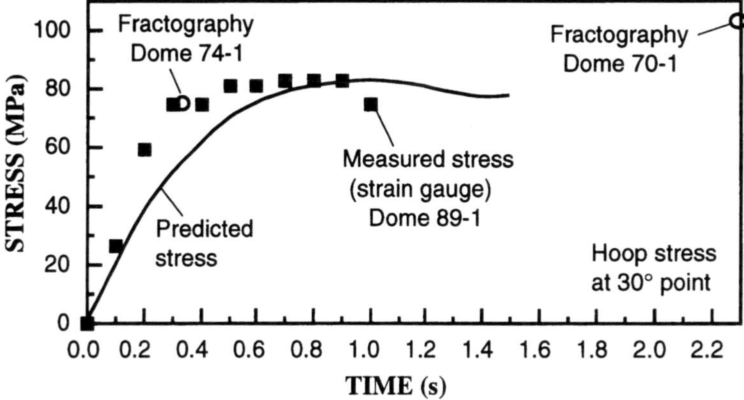

Fig. 4.15. Comparison of measured (squares) and predicted (solid line) thermal stress in lanthana-doped yttria dome in hypersonic wind tunnel test. Measurements were made by strain gauges located 30° off axis inside of the dome. Circles show stresses estimated by fractography using Eq. (3-35) on two other domes run under similar conditions. (Courtesy W. R. Compton, Naval Air Warfare Center.)

The agreement between the solid line and square data points in Fig. 4.15 is noteworthy. In addition, the remains of two other domes tested under similar conditions were examined to locate the fracture origins. The stress at failure was calculated with Eq. (3-35) from the size of the critical flaw and the known fracture toughness of the material. These experiments gave the two circle data points in Fig. 4.15, which are in remarkable agreement with the other results in the figure. Table 4.3 lists the heat flux required to break nose-mounted domes in wind tunnel tests at the facility in Fig. 4.13.

Table 4.3. Wind tunnel heat flux capabilities of infrared dome materials[31]

Material	Approximate survival limit for stagnation heat flux (W/cm^2)
Spinel	90 - 120
Yttria	90
Lanthana-doped yttria	90
Aluminum oxynitride (ALON)	100
Germania glass	45
Zinc sulfide (standard grade)	>200
	225*
Sapphire	>200
	400*

*Calculated limit for 3.56-cm-radius dome with a base diameter of 6.56 cm instantaneously exposed to a Mach 4.6 flow with a total pressure of 6.3 MPa and a total temperature of 1194 K.

4.3.2 Klein figure of merit for minimum thickness dome

If a dome is made thinner, less temperature difference builds up between the two surfaces during rapid heating and there is less thermal stress. Klein[18,32,33] has argued that domes should be made as thin as necessary to withstand aerodynamic pressures to obtain the maximum thermal shock resistance and, incidentally, the minimum infrared emittance. The higher the strength, S, of a material, the thinner it needs to be to withstand a given pressure. This notion leads to an alternative thermal shock figure of merit that applies to minimum-thickness domes:[32,33]

Klein mild thermal shock figure of merit for $\beta < 1$ in Eq. (4-5):

$$\text{Minimum thickness domes:} \quad K' = \frac{S^{5/3}(1-\nu)\,k}{\alpha E} \tag{4-8}$$

$$\text{Minimum thickness windows:} \quad K' = \frac{S^{3/2}(1-\nu)\,k}{\alpha E} \tag{4-9}$$

in which ν is Poisson's ratio, k is thermal conductivity, α is the expansion coefficient, and E is Young's modulus. The difference between Klein's and Hasselman's mild thermal shock figure of merit is the exponent on the strength term. Klein gives more weight to the strength because a stronger dome can be made thinner. For severe thermal shock, there is no dependence on the strength of the material and Klein's figure of merit is the same as Hasselman's Eq. (4-6).

Klein's figure of merit shown in Fig. 4.16 ranks dome materials in a generally similar manner to Hasselman's figure of merit. Three of the best materials, silicon nitride, aluminum nitride and silicon carbide, are not available in infrared-transparent form at this time, and silicon with an upper limit of only ~260°C is not a serious candidate for high-temperature operation. Both figures of merit rank diamond at the head of the list, followed by gallium phosphide, gallium arsenide and sapphire. Gallium arsenide is restricted by its upper operating limit of ~460°C.

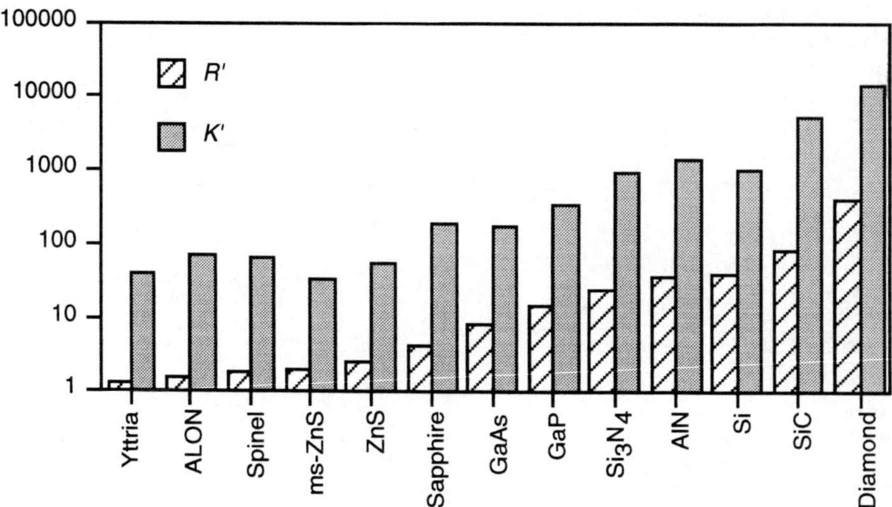

Fig. 4.16. Hasselman (R') and Klein (K') figures of merit for candidate infrared dome materials based on properties in Table 4.2. Note the logarithmic ordinate.

How thin are "minimum thickness" domes that can withstand the aerodynamic pressure of a missile flight? For the dome in Fig. 3.26 flying the medium altitude scenario in Table 3.5, which is quite stressing, Fig. 3.27 shows the maximum aerodynamic tensile stress as a function of dome thickness. Applying a safety factor of 4 to the mean strengths of dome materials in Table 4.2, these are the required thicknesses for various materials:

zinc sulfide - 5.1 mm	spinel - 3.3 mm	sapphire - 2.4 mm
gallium phosphide - 4.8 mm	diamond - 3.2 mm	ALON - 2.4 mm

To calculate the thickness of diamond, for example, we look up the strength in Table 4.2 and find 200 MPa. With a safety factor of 4, the maximum stress should not exceed 50 MPa. In Fig. 3.27, the thickness at 50 MPa is 3.2 mm. This is a crude calculation that makes no attempt to scale the strength for the area or volume under load and takes no account of any temperature dependence of the strength.

4.3.3 Mach-altitude limits for a dome

Suppose that the dome in Fig. 3.26 is initially at ambient temperature (the free-stream temperature) at a selected altitude. Then it instantaneously accelerates to a chosen Mach number at the same altitude. How fast can the dome fly without exceeding its thermal shock limits? For the remainder of this section, we assume that the thermal shock condition is "mild," which means that $\beta < 1$ in Fig. 4.10.

Following the approach of Klein,[18] we first find the heat transfer coefficient, h_{st} ($W/m^2 \cdot K$), at the stagnation point, which is the nose of the dome where the greatest heating occurs under laminar flow conditions:

$$h_{st} \approx 1.2 C_p \left(\frac{\rho_\infty \mu_\infty \alpha_\infty}{r} \right)^{1/2} M_\infty \left[1 + \frac{\gamma - 1}{2} M_\infty^2 \right]^{0.1} \tag{4-10}$$

where C_p is the heat capacity of air, ρ_∞ is the density of air, μ_∞ is the viscosity of air, α_∞ is the speed of sound, r is the radius of the dome in Fig. 3.26, M_∞ is the Mach number of the missile, and γ is the heat capacity ratio of air (heat capacity at constant pressure / heat capacity at constant volume). Over the range of conditions that we will consider, C_p = 1004 J/(kg·K) and γ = 1.4. Air density, viscosity and the speed of sound are functions of altitude listed in Table 4.4.

The heat flux into the dome at the stagnation point, Q_{st} (W/m^2), is

$$Q_{st} = h_{st} (T_{st} - T_{iw}) = h_{st} (T_{st} - T_\infty) \tag{4-11}$$

where T_{st} is the stagnation temperature and T_{iw} is the temperature of the inner wall of the dome, which is the initial temperature of the entire dome. We assume that T_{iw} is equal to the ambient temperature at the flight altitude, which is the free-stream temperature, T_∞. The stagnation temperature is related to the free-stream temperature by

$$T_{st} = T_\infty \left[1 + \left(\frac{\gamma - 1}{2} \right) M_\infty^2 \right] = T_\infty \left[1 + 0.20 M_\infty^2 \right] . \tag{4-12}$$

With T_{st} from Eq. (4-12), the difference $(T_{st} - T_\infty)$ can be written to transform Eq. (4-11) in the following manner:

Table 4.4. Properties of the atmosphere*

Altitude (km)	Temperature (T_∞, K)	Pressure (P_∞, Pa)	Density (ρ_∞, kg/m^3)	Speed of sound (α_∞, m/s)	Viscosity (μ_∞, kg/[m·s])
-1	294.65	113.931×10^3	1.3470	344.11	1.8206×10^{-5}
0	288.15	101.325×10^3	1.2250	340.29	1.7894×10^{-5}
3	268.66	70.121×10^3	0.90925	328.58	1.6938×10^{-5}
5	255.68	54.048×10^3	0.73643	320.54	1.6282×10^{-5}
10	223.25	26.500×10^3	0.41351	299.53	1.4577×10^{-5}
15	216.65	12.112×10^3	0.19475	295.07	1.4216×10^{-5}
20	216.65	5.529×10^3	0.08891	295.07	1.4216×10^{-5}
25	221.55	2.549×10^3	0.04008	298.39	1.4484×10^{-5}
30	226.51	1.197×10^3	0.01841	301.71	1.4753×10^{-5}
40	250.35	0.2871×10^3	0.00400	317.19	1.6009×10^{-5}
50	270.65	0.0798×10^3	0.00103	329.80	1.7037×10^{-5}
60	255.77	0.0225×10^3	0.000306	320.61	1.6287×10^{-5}
70	219.70	0.00552×10^3	0.0000875	297.14	1.4389×10^{-5}
80	2180.65	0.00104×10^3	0.0000200	269.44	1.216×10^{-5}

*U.S. Standard Atmosphere, 1962 [American Institute of Physics Handbook, 3rd ed, McGraw-Hill, New York (1972)].

$$Q_{st} = h_{st} T_\infty \left(\frac{\gamma-1}{2}\right) M_\infty^2 = 0.20 h_{st} T_\infty M_\infty^2 . \tag{4-13}$$

Now the key postulate introduced by Klein,[18] based on experimental observations, is that the limiting allowed heat flux is

$$Q_{lim} \approx 2\frac{R'}{d} \tag{4-14}$$

where R' is the Hasselman thermal shock figure of merit in Eq. (4-7) and d is the thickness of the dome in Fig. 3.26. This condition is independent of the dome radius, the dome material, and the Mach number. Very roughly, we can think of Q_{lim} as the heat flux that will induce failure at the 50% probability level. For our purposes, we take it that Q_{lim} represents an approximate ultimate limit for the allowable heat flux.

Equating the stagnation point heat flux in Eq. (4-13) with Q_{lim} and using Eq. (4-10) for the heat transfer coefficient, we can write an equation to find the maximum allowed Mach number for a given altitude:

$$Q_{lim} = 0.20 h_{st} T_\infty M_\infty^2$$

$$\frac{2R'}{d} = (0.20)(1.2)C_p \left(\frac{\rho_\infty \mu_\infty \alpha_\infty}{r}\right)^{1/2} M_\infty \left[1 + 0.20 M_\infty^2\right]^{0.1} T_\infty M_\infty^2$$

$$M_\infty{}^3\left[1 + 0.20M_\infty{}^2\right]^{0.1} = \frac{8.33R'}{dC_pT_\infty}\left(\frac{r}{\rho_\infty\,\mu_\infty\,\alpha_\infty}\right)^{1/2} . \qquad (4\text{-}15)$$

With C_p = 1004 J/(kg·K) and values of T_∞, ρ_∞, μ_∞ and α_∞ from Table 4.4, the right side of Eq. (4-15) is known, so we can solve for the allowed Mach number on the left side by varying M_∞ until the two sides of the equation are equal.

Example: Maximum allowed Mach number. Let's find the maximum allowed Mach number for a zinc sulfide dome with a radius r = 50 mm = 0.050 m and thickness d = 5.1 mm = 0.0051 m in Fig. 3.26 at an altitude of 3 km. We need to solve Eq. (4-15) with the following values from Table 4.4: T_∞ = 269 K, ρ_∞ = 0.909 kg/m^3, μ_∞ = 1.69 × 10^{-5} kg/(m·s), α_∞ = 329 m/s. You can verify that if all quantities are expressed in SI units, the right side of Eq. (4-15) is dimensionless. Noting that R' in Table 4.2 has the units 10^3 W/m, the Hasselman figure of merit for ZnS is R' = 2.6 × 10^3 W/m. With all these numerical values, Eq. (4-15) becomes

$$M_\infty{}^3\left[1 + 0.20M_\infty{}^2\right]^{0.1} = \frac{8.33R'}{dC_pT_\infty}\left(\frac{r}{\rho_\infty\,\mu_\infty\,\alpha_\infty}\right)^{1/2}$$

$$= \frac{8.33(2.6\times 10^3)}{(0.0051)(1004)(269)}\left(\frac{0.050}{(0.909)(1.69\times 10^{-5})(329)}\right)^{1/2} = 12.36 .$$

We retain extra digits for the right side of the equation to avoid roundoff error in solving for the Mach number. Now we can vary M_∞ on the left side of the equation (e.g., by using a spreadsheet) to find M_∞ = 3.52.

The maximum allowed Mach number of 3.52 in the preceding example is less than M_∞ = 4 that was used to find the dome thickness of 5.1 mm from Fig. 3.27. That is, it would be safe to use a thinner dome, which would allow a greater speed. If we guess a thickness of d = 3.5 mm in Eq. 4-15, we find an allowed Mach number of 3.97. A Mach number of 3.97 in Eq. (3-30) gives a pressure of 1.38 MPa across the dome, which requires a dome thickness of 5.1 mm in Eq. (3-29) if the maximum allowed tensile stress for ZnS is 1/4 of the mean strength (= 25 MPa). That is, d = 3.5 mm is too thin to withstand the pressure load.

Continuing with successive trials, if we insert d = 4.0 mm in Eq. (4-15), we find an allowed Mach number of 3.81, which gives a pressure difference of 1.27 MPa, which requires a 4.8 mm dome thickness. Trying a thickness of 4.4 mm gives a Mach number of 3.69, which produces a pressure difference of 1.19 MPa, which requires a thickness of 4.6 mm. Trying a thickness of 4.6 mm gives a Mach number of 3.65, which gives a pressure difference of 1.16 MPa, which requires a thickness of 4.5 mm. This iterative approach suggests that a dome thickness of 4.6 mm will allow a speed near Mach 3.65 at 3 km altitude and be able to survive the pressure load.

Now we must verify that the mild thermal shock condition still holds because Eq. (4-14) was predicated on being in the mild thermal shock regime. With a Mach number of 3.65, Eq. (4-10) gives a heat transfer coefficient of h_{st} = 1590 W/(m^2·K). The Biot number in Eq. (4-5) is therefore

$$\beta = \frac{th}{k} = \frac{0.0046 \text{ m} \times 1590 \text{ W/(m}^2\text{·K)}}{19 \text{ W/(m·K)}} = 0.38 \tag{4-16}$$

which is less than 1. We have verified that the thermal shock regime is mild. (Note that Eq. (4-5) uses *t* for thickness and Fig. 3.26 uses *d*.)

Example: Temperature difference at maximum stress. What is the temperature difference across the thickness of the nose of the dome at the time of maximum stress? In Fig. 4.10 we see that the temperature difference will be $(T_{st} - T_{iw}) \times \beta$. From Eq. (4-11) we can say that $(T_{st} - T_{iw}) = Q_{st}/h_{st}$. The quotient Q_{st}/h_{st} is available from Eq. (4-13): $Q_{st}/h_{st} = 0.20 T_\infty M_\infty^2$. With T_∞ = 269 K and M_∞ = 3.65, Q_{st}/h_{st} = 717 K. With a Biot number of 0.38, the temperature difference across the thickness of the dome is (0.38)(717) = 272 K.

The equations given in this section are approximations that apply to mild thermal shock under laminar flow conditions. If the flow becomes turbulent as the air moves around the dome, heating at the skirt can be higher than heating at the nose. For actual design, there is no substitute for a full finite element analysis with temperature-dependent material properties for the dome and experimental measurements of strain and temperature to verify that the calculations are accurate.

4.4 Aerodynamic domes

The ability of a rocket engine to fly fast readily exceeds the ability of most infrared windows to survive thermal shock. Diamond is a notable exception which can probably withstand the thermal shock of any tactical missile flight because of the high thermal conductivity of diamond. Several approaches can be taken to promote window survival. The window could be actively cooled by blowing cold gas over the exposed surface. The mixing of cold gas with hot air creates density and refractive index fluctuations that might blur the scene being viewed. This effect is called *aero-optic distortion.* Another idea is to pass coolant through channels in the window, as in Fig. 2.15. Disadvantages of this scheme include blockage of the scene by the channels, wavefront distortion by temperature gradients in the window, reflections and diffraction from the channel walls, and emittance from the coolant. Any active cooling system adds weight, volume, and complexity to the system.

We saw in Fig. 4.11 that the temperature on a dome falls rapidly as we move off the nose (if the flow remains laminar). One way to take advantage of this effect is to mount a flat infrared window on the rear portion of a structural dome, as in Fig. 4.17a. Such a window does not get very hot, but the field of view is quite limited compared to that of a nose-mounted hemispheric dome. An additional advantage of the slanted window is that its angle of attack is always far from perpendicular to the airstream, so it can survive impacts with raindrops and solid particles.

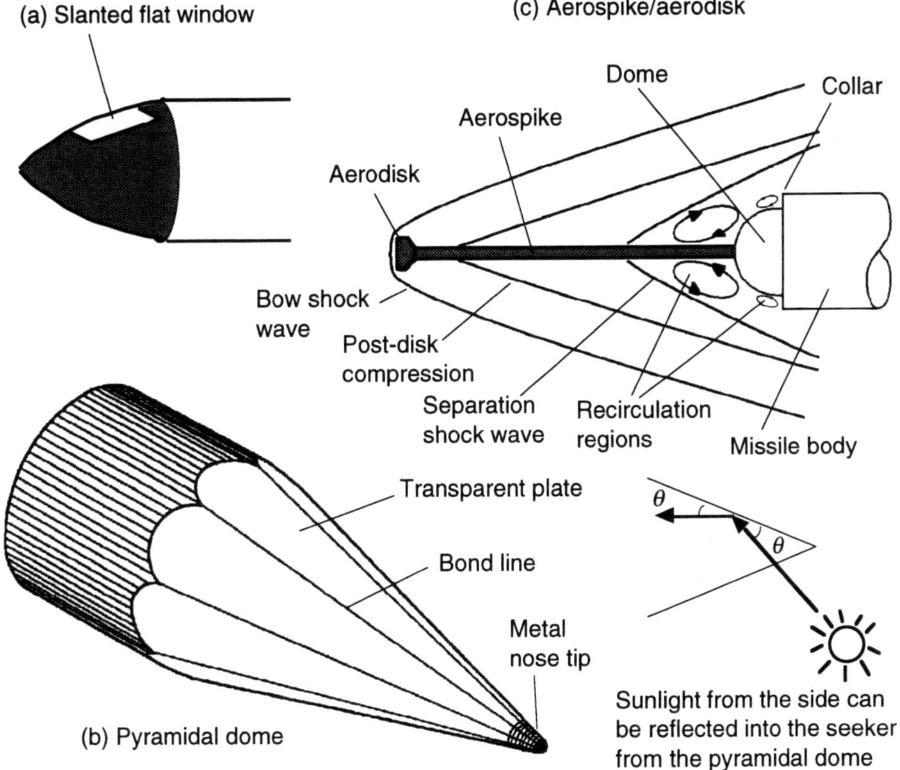

Fig. 4.17. Aerodynamic domes.

The faceted dome in Fig. 4.17b is an 8-sided pyramid with spinel panes and a heat-resistant metal nose tip.[34] The French Mistral missile has a similar dome with magnesium fluoride panes. The faces of the pyramid remain much cooler than the nose of a hemisphere. The aerodynamic design reduces drag and thereby increases range, speed, or payload. The biggest disadvantage of the design is multiple internal reflections of sunlight that lead to glint whenever the sun is in the forward hemisphere of view. Joining of the panes and overall cost are also significant considerations in this design.

The aerospike/aerodisk in Fig. 4.17c reduces drag and deflects the bow shock from the hemispheric dome behind the spike.[35] The collar created by the missile body that is wider than the dome prevents reattachment of the shock wave onto the dome at low angles of attack. Wind tunnel tests of the aerospike in Fig. 4.17c show that the shock wave reattaches to the dome at angles of attack greater than 5°, as seen in Fig. 4.18. Reattachment generates higher temperatures and pressures on the side of the dome than would be found at the nose of the dome in the absence of a spike.

4.5 Thermal stability of window materials

Even if a window or dome is heated gradually so that thermal shock is not a concern, every material has an upper operating temperature above which it cannot be used. The

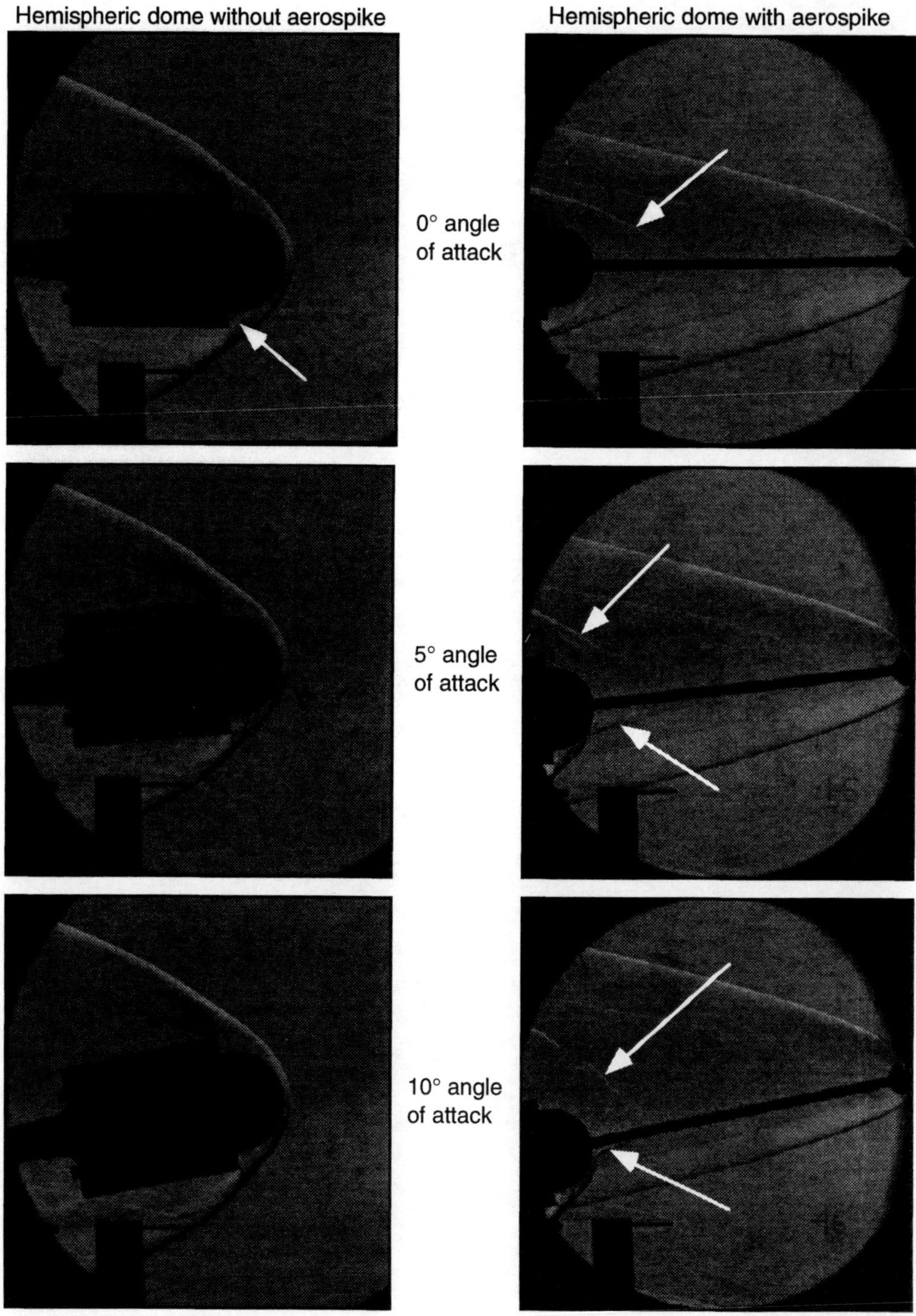

Fig. 4.18. Schlieren patterns from Mach 6 wind tunnel tests of a hemispheric dome with or without an aerospike/aerodisk (shown in Fig. 4.17c).[35] Arrows denote separation shock waves In the upper left photo, the shock wave hits the collar around the dome. In the upper right photo, the shock waves miss the dome. Some shock waves in the center right and lower right photos hit the dome directly. Schlieren patterns highlight regions of different density. [Courtesy L. D. Huebner, NASA.]

temperature might be dictated by loss of strength or loss of optical transmission (giving increased emission) resulting from reversible physical changes (such as free carrier absorption in Section 1.10) or irreversible chemical changes. A material might be usable at a particular temperature for the short duration of a missile flight, but not usable for a longer period at the same temperature. Little data exist in the literature to establish upper operating temperatures for short duration exposure.

We have already discussed a simple reversible physical change in which thermal population of the conduction band limits the transmission of semiconductors such as germanium (Fig. 1.28) or gallium arsenide at elevated temperature. An example of an irreversible chemical change is illustrated in Fig. 4.19 in which zinc sulfide windows were placed in a preheated oven with an atmosphere of air at each temperature for 5 minutes.[29] It was estimated that the samples attained the oven temperature for at least 3 minutes in each experiment. After the heating, the samples were cooled to room temperature and their mass and transmission were measured.

Expected chemical changes include oxidation and decomposition:

Oxidation: $ZnS(s) + 2O_2(g) \rightarrow ZnSO_4(s)$

Decomposition: $ZnS(s) \rightarrow Zn(s) + S(g)$.

Oxidation (which requires air) ought to occur at lower temperature than decomposition and results in a mass *gain* by the ZnS. Decomposition is expected to result in mass *loss*. While decomposition should reduce the transmission of the window, mild surface oxidation could increase or decrease the transmission. An increase will result from formation of a thin coating of lower refractive index $ZnSO_4$ on the higher refractive index ZnS window. A decrease will occur if a thick enough layer of $ZnSO_4$ forms, since $ZnSO_4$ absorbs at 10 μm.

Figure 4.19 gives no evidence for oxidation in 5 minutes at temperatures below 800°C. Note, however, that the transmission was measured after the sample was cooled

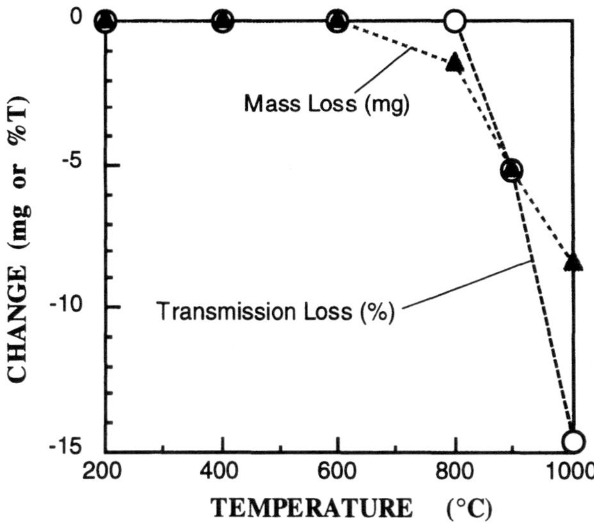

Fig. 4.19. Effect of temperature on transmittance at 10 μm wavelength and mass loss by zinc sulfide after 5 minutes of exposure to each temperature in the air.[29]

back to 25°C. We are certain that while the window is at 800°C, its transmission will be less than the transmission at 25°C (Fig. 1.24). The only way to know if zinc sulfide will be useful at 800°C is to measure its transmission at 800°C in an atmosphere representative of an operational atmosphere for a time representative of an operational time. Such experiments have rarely been done. In the case of zinc sulfide, oxidation-resistant coatings can protect the material from reaction with air at 1000°C for short periods of time.

Oxides such as sapphire, spinel, ALON and yttria are stable at temperatures above 1000°C. Substances whose anions are not oxide (such as zinc sulfide, gallium arsenide and gallium phosphide) are less stable, with upper operating limits below 1000°C. For long term use (days or years instead of seconds or minutes), much lower operating temperature limits are recommended. For example, zinc sulfide is considered safe for indefinite use at 450°C and zinc selenide is considered safe at 350°C.

References

1. C. A. Perottoni and J. A. H. da Jornada, "Pressure-Induced Amorphization and Negative Thermal Expansion in ZrW_2O_8," *Science*, **280**, 886-889 (1998).
2. J. S. O. Evans, T. A. Mary, T. Vogt, M. A. Subramanian and A. W. Sleight, "Negative Thermal Expansion in ZrW_2O_8 and HfW_2O_8," *Chem. Mater.*, **8**, 2809-2823 (1996).
3. W. J. Tropf and D. C. Harris, "Mechanical, Thermal, and Optical Properties of Yttria and Lanthana-Doped Yttria," *Proc. SPIE*, **1112**, 9-19 (1989).
4. P. Klocek, ed., *Handbook of Infrared Optical Materials*, Marcel Dekker, New York (1991).
5. R. G. Munro, "Evaluated Material Properties for a Sintered α-Alumina," *J. Am. Ceram. Soc.*, **80**, 1919-1928 (1997).
6. Y. Kumashiro, T. Mitsuhashi, S. Okaya, F. Muta, T. Koshiro, Y. Takahashi and M. Mirabayashi, "Thermal Conductivity of a Boron Phosphide Single-Crystal Wafer up to High Temperature," *J. Appl. Phys.*, 65, 2147-2148 (1989).
7. W. J. Tropf, M. E. Thomas and T. J. Harris, "Properties of Crystals and Glasses," in *Handbook of Optics* (M. Bass, E. W. van Stryland, D. R. Williams and W. L. Wolfe, eds.), Vol. II, Chap. 33, McGraw-Hill, New York (1995).
8. R. Berman, *Thermal Conduction in Solids*, Clarendon Press, Oxford (1976).
9. Y. S. Touloukian, R. W. Powell, C. Y. Ho and P. G. Klemens, *Thermophysical Properties of Matter*, Vols. 1 and 2, IFI Plenum Press, New York (1970).
10. R. C. Zeller and R. O. Pohl, "Thermal Conductivity and Specific Heat of Noncrystalline Solids," *Phys. Rev. B*, **4**, 2029-2041 (1971).
11. G. S. Sheffield and J. R. Schorr, "Comparison of Thermal Diffusivity and Thermal Conductivity Methods," *Ceram. Bull.*, **70**, 102-106 (1991).
12. M. W. Price and T. E. Hubbert, *Mechanical and Thermal Properties of Four IR Dome Materials*, Southern Research Institute Report SRI-EAS-87-1272-6225, February 1988.
13. G. G. Gadzhiev and G. N. Dronova, "Thermal Conductivity of Polycrystalline Zinc Sulfide," *Izv. Akad. Nauk SSSR, Neorg. Mater.*, **19**, 1087-1089 (1983).
14. G. E. Potter, A. K. Knudsen, J. C. Tou and A. Choudhury, "Measurement of the Oxygen and Impurity Distribution in Polycrystalline Aluminum Nitride with Secondary Ion Mass Spectrometry," *J. Am. Ceram. Soc*, **75**, 3215-3224 (1992).
15. R. L. Gentilman, "Current and Emerging Materials for 3-5 Micron IR Transmission," *Proc. SPIE*, **683**, 2-11 (1986).

16. D. P. H. Hasselman, "Thermal Stress Resistance Parameters for Brittle Refractory Ceramics: A Compendium," *Ceram. Bull.*, **49**, 1033-1037 (1970).
17. D. P. H. Hasselman, "Thermal Stress Resistance in Engineering Ceramics," *Mater. Sci. Eng.*, **71**, 251-264 (1985).
18. C. A. Klein, "Thermal Shock Resistance of Infrared Transmitting Windows and Domes," *Opt. Eng.*, **37**, 2826-2836 (1998).
19. D. Lewis III, "Thermal Shock and Thermal Shock Fatigue Testing of Ceramics with the Water Quench Test," in *Fracture Mechanics of Ceramics* (R. C. Bradt, A. G. Evans, D. P. H. Hasselman, and F. F. Lange, eds), Vol. 6, pp. 487-495, Plenum Press, New York (1983).
20. T. Andersson and D. J. Rowcliffe, "Indentation Thermal Shock Test for Ceramics," *J. Am. Ceram. Soc.*, **79**, 1509-1514 (1996).
21. G. A. Schneider and G. Petzow, "Thermal Shock Testing of Ceramics -- A New Testing Method," *J. Am. Ceram. Soc.*, **74**, 98-102 (1991).
22. C. S. J. Pickles and J. E. Field, "The Laboratory Simulation of Thermal Shock Failure," *J. Phys. D*, **29**, 436-441 (1996).
23. D. H. Platus, O. Esquivel, J. D. Barrie and P. D. Chaffee, "Sapphire Window Statistical Thermal Fracture Characterization Using a CO_2 Laser," *Proc. SPIE*, **3060**, 236-245 (1997).
24. G. C. Wei and J. Walsh, "Hot-Gas-Jet Method and Apparatus for Thermal-Shock Testing," *J. Am. Ceram. Soc.*, **72**, 1286-1289 (1989).
25. Y. Kim, A. Zangvil, J. S. Goela and R. L. Taylor, "Microstructure Comparison of Transparent and Opaque CVD SiC," *J. Am. Ceram. Soc.*, **78**, 1571-1579 (1995); J. S. Goela and R. L. Taylor, "Transparent SiC for Mid-IR Windows and Domes," *Proc. SPIE*, **2286**, 46-59 (1994).
26. R. A. Tanzilli, J. J. Gebhardt and J. D'Andrea, *Processing Research on Chemically Vapor Deposited Silicon Nitride — Phase III*, General Electric Co. Report 81SDR2111, October 1981, p. 3-5.
27. S. C. Danforth and M. C. Richman, "Strength and Fracture Toughness of Reaction-Bonded Si_3N_4," *Ceram. Bull.*, **62**, 501-504, 516 (1983).
28. G. A. Slack, R. A. Tanzilli, R. O. Pohl and J. W. Vandersande, "The Intrinsic Thermal Conductivity of AlN," *J. Phys. Chem. Solids*, **48**, 641-647 (1987).
29. R. W. Tustison and R. L. Gentilman, "Current and Emerging Materials for LWIR External Windows," *Proc. SPIE*, **968**, 25-34 (1988).
30. F. A. Strobel, "Thermostructural Evaluation of Spinel Infrared (IR) Dome," *Proc. SPIE*, **297**, 125-136 (1981).
31. J. S. Lin and W. B. Weckesser, "Thermal Shock Capabilities of Infrared Dome Materials," *Johns Hopkins APL Technical Digest*, **13** [3], 379-385 (1992).
32. C. A. Klein, "Infrared Missile Domes: Is There a Figure of Merit for Thermal Shock?" *Proc. SPIE*, **1760**, 338-347 (1994).
33. C. A. Klein, "Infrared Missile Domes: Heat Flux and Thermal Shock," *Proc. SPIE*, **1997**, 150-169 (1993).
34. B. S. Fraser and A. Hemingway, "High Performance Faceted Domes for Tactical and Strategic Missiles," *Proc. SPIE*, **2286**, 485-492 (1994).
35. L. D. Huebner, A. M. Mitchell and E. J. Boudreaux, "Experimental Results on the Feasibility of an Aerospike for Hypersonic Missiles," AIAA Paper 95-0737, 33rd Aerospace Sciences Meeting, January 1995.

Chapter 5

FABRICATION OF INFRARED MATERIALS

Although some infrared-transmitting materials are found in nature, all practical bulk materials for windows and domes are synthetic. This chapter outlines the principal methods of fabrication and should familiarize you with some terminology of ceramics. Our discussion should give you an idea of how some of the properties of infrared materials can be modified through the fabrication process.

5.1 Classes of infrared materials

Bulk infrared window materials are either single crystals, polycrystalline, or glasses. Some examples of each are given in Table 5.1. In a perfect *crystal*, all of the atoms are

Table 5.1. Classes of bulk infrared materials

Class of material	Examples
Single-crystal	Sapphire (Al_2O_3), Si, Ge, GaAs, CaF_2, LiF, KBr, NaCl
Polycrystalline	Hot pressed: MgF_2 and other Irtran®* materials Hot isostatically pressed: Y_2O_3 Hot pressed and hot isostatically pressed: Spinel ($MgAl_2O_4$) Sintered: ALON ($9Al_2O_3{\cdot}5AlN$), lanthana-doped yttria Chemical vapor deposited: ZnS, ZnSe, diamond, Si, SiC Melt growth with cm-size grains: Si, Ge, GaAs, GaP
Glass	Midwave materials (3-5 μm) Calcium aluminate (43%CaO-47%Al_2O_3-10%BaO) Germanate (33%GeO_2-37%Al_2O_3-20%CaO-5%BaO-5% ZnO) Fluoride[1] ($ZrF_4/HfF_4/BaF_2$) Fused silica (SiO_2) Long wave materials[2] (8-12 μm) Chalcogenides (S, Se, Te + other elements) AMTIR-1®† ($Ge_{33}As_{12}Se_{55}$) Arsenic trisulfide[3] (As_2S_3) Chalcogenide halide[4] ($Te_2Se_3IAs_4$)

*Irtran is a trademark for materials formerly manufactured by Eastman Kodak. The compositions are Irtran 1 = MgF_2, Irtran 2, = ZnS, Irtran 3 = CaF_2, Irtran 4 = ZnSe, Irtran 5 = MgO and Irtran 6 = CdTe.

†AMTIR is a trademark of Amorphous Materials, Inc.

Fig. 5.1. Structure of α-quartz, SiO_2. Each silicon atom (dark sphere) is bound to four oxygen atoms (light spheres), and each oxygen atom is attached to two silicon atoms.

aligned in a regular manner. Figure 5.1 shows the structure of quartz (SiO_2), which is a repeating arrangement of SiO_4 tetrahedra. Every silicon atom is attached to four oxygen atoms, and every oxygen atom is attached to two silicon atoms. The black lines outline two unit cells. The arrangement of atoms within each unit cell is identical throughout the crystal.

In a *glass* such as fused silica, which also has the formula SiO_2, there is no regular arrangement of atoms. Every silicon atom still has four tetrahedrally arranged oxygen neighbors, and each oxygen atom is attached to two silicon atoms. However, the attachment of one tetrahedron to another twists and turns so that the structure does not repeat itself from one region to another. We call such a material *amorphous*, because it has no crystalline form. We know that the structure of fused silica is more open than that of quartz (i.e., it has more empty space), because the density of fused silica is approximately 2.2 g/mL, while the density of α-quartz is 2.655 g/mL.

A *polycrystalline* material is made up of randomly oriented crystals, called *grains*. The box in Section 1.2.2 shows a network of grains in polycrystalline yttria. Figure 5.2 is a greatly magnified view of *grain boundaries* in silicon nitride (β-Si_3N_4). At the left, two grains with different crystalline orientation touch each other directly. At the right, there is a 1.1-nm-thick layer of amorphous (noncrystalline) material at the grain boundary. The amorphous material could be Si_3N_4 or it might be another material. Crystals tend to exclude foreign atoms that do not belong in the crystal, so impurities tend to aggregate at grain boundaries. The grain boundaries in yttria in the box in Section 1.2.2 are rich in silica, which is an impurity in the Y_2O_3 powder from which the ceramic was made. The thickness and composition of the grain boundary may affect strength, since grain boundaries are the "glue" holding the material together. Grain boundaries can also affect thermal conductivity, since heat does not flow across the boundaries at the same rate as it flows within each crystallite. Grain sizes in polycrystalline materials typically range from tenths to hundreds of micrometers.

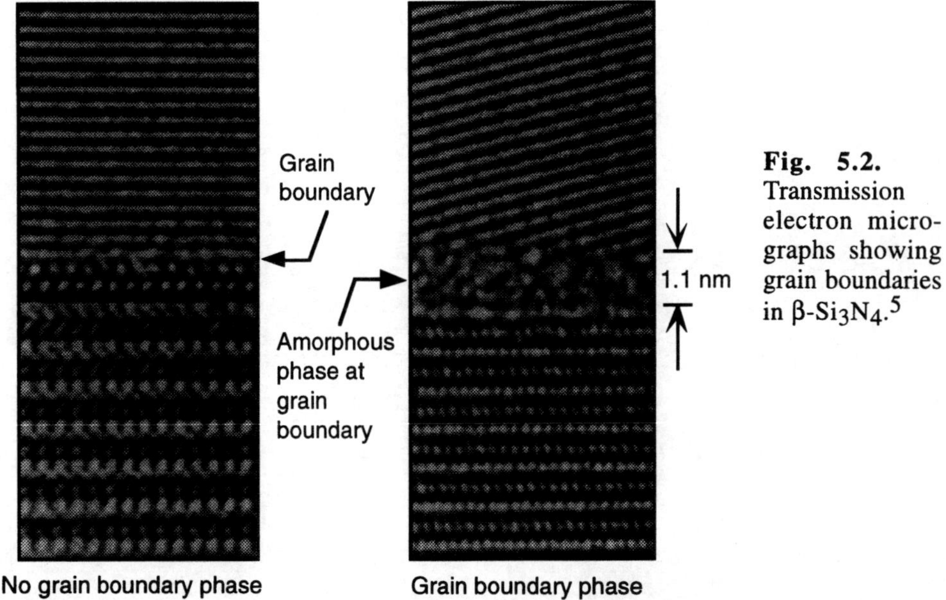

Fig. 5.2. Transmission electron micrographs showing grain boundaries in β-Si_3N_4.[5]

Single-crystal infrared optical materials in Table 5.1 include sapphire, germanium and gallium arsenide. While these can be produced as true single crystals, they sometimes contain centimeter-sized grains. The properties of materials with such a large grain size tend to be the same as those of single crystals, which are different from those of polycrystalline materials. Single crystals tend to have lower optical scatter (Table 5.2) and lower absorption than polycrystalline materials.

Polycrystalline materials in Table 5.1 include such common windows as magnesium fluoride and zinc sulfide. Polycrystalline materials are usually cheaper than single crystals of the same material and tend to be tougher and stronger. Foreign material at the grain boundaries with a refractive index different from that of the grains scatters light (diverting the light from its original path). Other sources of scatter in polycrystalline materials are the presence of multiple phases (such as cubic and hexagonal ZnS) and birefringent grains (MgF_2, Section 1.2.1).

Glasses are easiest to fabricate because they are made simply by melting elements or compounds together. They can be made in large shapes with good optical quality and uniformity. Unfortunately, glasses have inferior mechanical properties compared to many polycrystalline and single-crystal materials, and are of little use in external domes and

Table 5.2. Optical scattering

Class of material	Typical infrared scatter
Single-crystal	10^{-2}%
Glass	10^{-1}%
Polycrystalline	10^{-1} - 10^{1}%

windows for which durability is a criterion. Glasses tend to be susceptible to erosion by rain and particles and have low strength, poor resistance to thermal shock and low melting temperatures. Calcium aluminate is one of the few glasses used as an external window.

5.1.1 Glass-ceramics

Before leaving the subject of glass, mention should be made of *glass-ceramics*,[6] which are made by heating a glass to a temperature at which it begins to crystallize. A glass-ceramic contains crystalline grains embedded in the residual glass matrix. Depending on the time and temperature, the material could be nearly all crystalline.

One notable glass-ceramic is Zerodur®, a commercial material from Schott Glaswerke (Mainz, Germany) with near-zero expansion coefficient near 300 K (Fig. 5.3). The mean expansion coefficient from 20 to 300°C is 0.05 (±0.10) × 10^{-6} K^{-1} and from -180 to 20°C it is -0.16 (±0.10) × 10^{-6} K^{-1}. Zerodur contains 70 - 78 wt% crystalline SiO_2 with a negative expansion coefficient embedded in a glass phase with a positive expansion coefficient. The mean crystallite size is ~50 nm. Visible transmission is ~90%, but the mid-infrared region is obscured by a strong OH absorption near 3 μm and a cutoff at 4 μm (Fig. 5.4). The infrared transmission can be substantially improved by a proprietary dehydration process that removes OH absorption.[7]

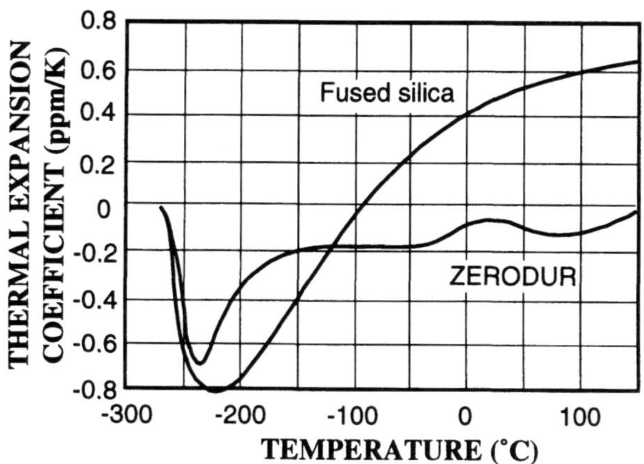

Fig. 5.3. Thermal expansion coefficient of Zerodur glass-ceramic. (Data from Schott Glaswerke.)

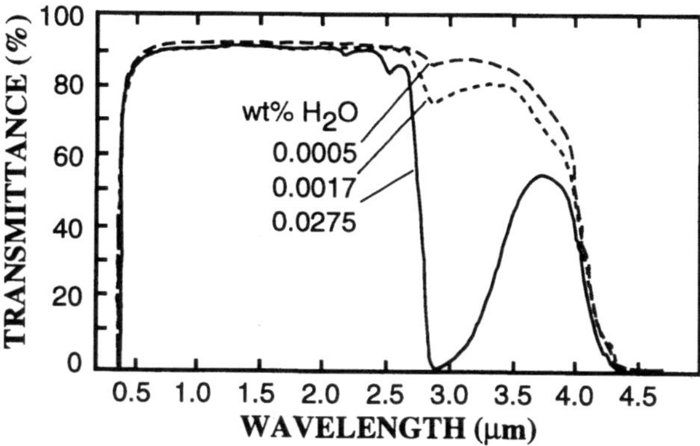

Fig. 5.4. Optical transmission of 3.0-mm-thick Zerodur 8563 glass-ceramic.[7] Mid-infrared transmittance can be improved by a proprietary dehydration process.

The infrared transmission of a germanate glass-ceramic derived from barium gallo-germanate glass ($20BaO{\cdot}10Ga_2O_3{\cdot}70GeO_2$) is shown in Fig. 5.5.[8] The glass has excellent visible transmission, but the visible transmission in the glass-ceramic is limited by optical scatter from the 0.2- to 0.5-μm-diameter crystallites. Table 5.3 shows the improvement in physical properties upon transforming the glass into a glass-ceramic. Fig. 5.6 shows that replacing oxide by sulfide or selenide in the glass moves the infrared cutoff to a longer wavelength.

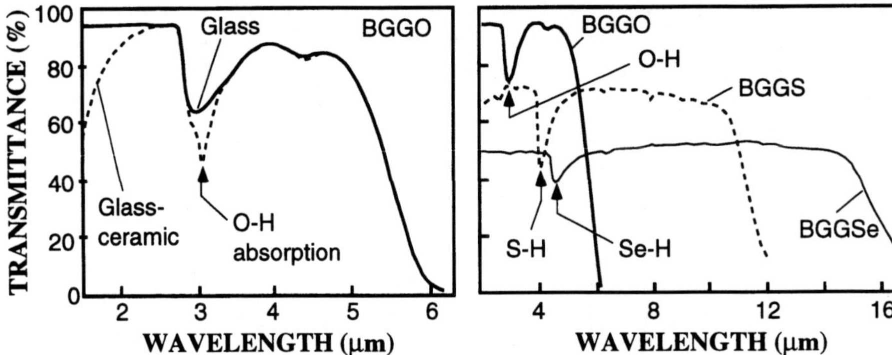

Fig. 5.5. Comparison of optical transmittance of barium gallo-germanate (BGGO) glass and BGGO glass-ceramic.[8] OH absorption near 3 μm can be decreased by physical and chemical treatments to reduce the water content.

Fig. 5.6. Substitution of sulfur (S) or selenium (Se) for oxygen (O) in barium gallo-germanate (BGGO) glass pushes the infrared cutoff into the long wave region.[8] Section 1.11 explains why heavier atoms move the infrared cutoff to a longer wavelength.

Table 5.3. Properties of barium gallo-germanate glass and glass-ceramic[8]

Property	BGGO glass	BGGO glass-ceramic
Hardness (kg/mm^2)	400	560
Young's modulus (GPa)	70	120
Strength (MPa)	60	130
Fracture toughness ($MPa\sqrt{m}$)	0.7	1.5
Rain damage threshold velocity (m/s) (single water jet impact)	~186	335
Thermal expansion coefficient (ppm/K)	7.6	5.9
Thermal conductivity (W/(m·K))	0.7	0.72

Glasses can be toughened by intentionally incorporating second phases with shapes such as fibers or platelets that divert cracks and make it harder for a fracture to propagate.[9] It is difficult to make a transparent composite material because the refractive indexes of the two phases must be matched. Calcium aluminate glass is strengthened by a factor of 3 and its rain erosion rate (measure by loss in transmission) slowed by a factor of 4 by controlled heating to crystallize the outer 50-100 μm of the bulk material.[10]

5.2 Fabrication of polycrystalline materials by powder processing

Figure 5.7 compares costs of the steps in fabricating single-crystal sapphire with the costs of fabricating three less expensive polycrystalline materials.[11] Magnesium fluoride, ALON (aluminum oxynitride)[12] and yttria are made by powder consolidation processes, whereas sapphire is grown from a melt. Sapphire is superior to the other three materials in Fig. 5.7 in many respects, including greater erosion and thermal shock resistance, greater strength and lower optical scatter. High cost limited its use in the past, but sapphire is now the material of choice for the most demanding systems.

We see in Fig. 5.7 that the costs of raw materials, blank fabrication (making an unpolished bulk material) and polishing are all greater for sapphire than for the other materials. Crystal growth to make a blank is more expensive than powder consolidation for most materials. The same physical characteristics that give sapphire its exceptional durability also make it difficult and expensive to polish. An additional problem in polishing a sapphire dome is that it is mechanically anisotropic — some crystal directions polish more rapidly than others, making it difficult to achieve a perfect hemispheric shape. In the remainder of this section we explore the various steps in fabricating a window or dome by powder processing technology.

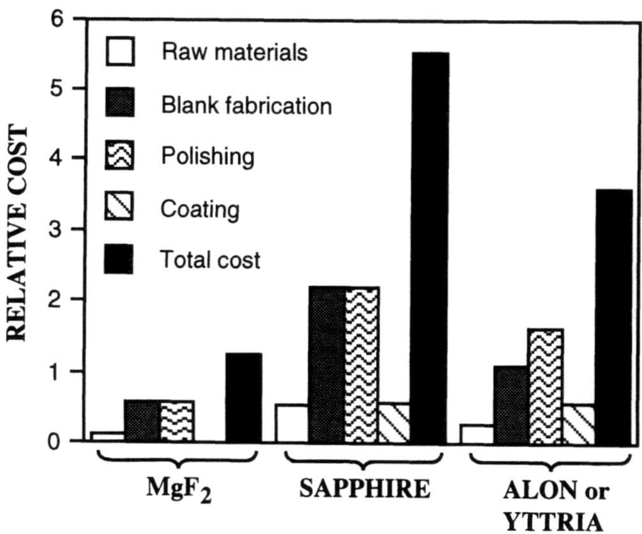

Fig. 5.7. Estimated relative costs of fabrication of some mid-wave infrared window materials.[11]

5.2.1 Yttria: an example of dome fabrication from a powder

The fabrication of yttria domes[13] begins with the purest commercially available grade of yttrium oxide powder, containing 99.99% Y_2O_3. Variation in the impurities in commercial powder is a vexing and costly problem. The grain boundaries of the material shown in the box in Section 1.2.2 are highlighted by a silica impurity from the starting powder that contained just 130 ppm silicon. Previous and subsequent shipments of powder from the same vendor did not contain so much silica and did not give rise to the distinct grain boundary phase. Needless to say, the appearance of the grain boundary phase was a costly problem for the ceramists using the powder.

The starting yttria powder typically has a particle size of 1-3 μm, but contains some clumps of particles (*agglomerates*) that prevent uniform filling of a mold. *Deagglomeration* is accomplished by preparing a slurry (a mixture of powder and liquid) in methanol and *milling* in a polyethylene or rubber-lined jar for 5-24 h using pellets of Y_2O_3 as the grinding medium. In the milling process, the jar is slowly rotated so that the grinding medium rolls over the powder and breaks it up into finer particles. Other ceramics such as ZrO_2 can be used as the grinding medium, but this could introduce ZrO_2 impurity into the Y_2O_3 powder. The slurry is finally poured through a 400 mesh sieve to remove remaining agglomerates, and then contains 1-2 μm diameter Y_2O_3 particles.

To prepare the milled particles for further use, an organic binder (polyvinyl pyrrolidone) and a dispersant (acetic acid) are added, and the slurry is passed through an ultrasonic horn for further deagglomeration. The slurry then enters a spray drier in which solvent is removed as the particles are sprayed through a hot zone.

We are now ready for *cold isostatic pressing*. The powder is poured into the hemispheric mold in Fig. 5.8, which is then evacuated and sealed. A uniform (isostatic) pressure of 180 MPa is applied by gas pressure at 25°C to compact the powder into a very porous *green body*. (Ceramists refer to a compacted body that has not been heated as a green body.) *Binder burnout* is then accomplished by heating the green body in the air at 1400°C for 90 minutes. At this point the ceramic is about 75% as dense as fully dense yttria and the dimensions have shrunk by 10% from those of the green body.

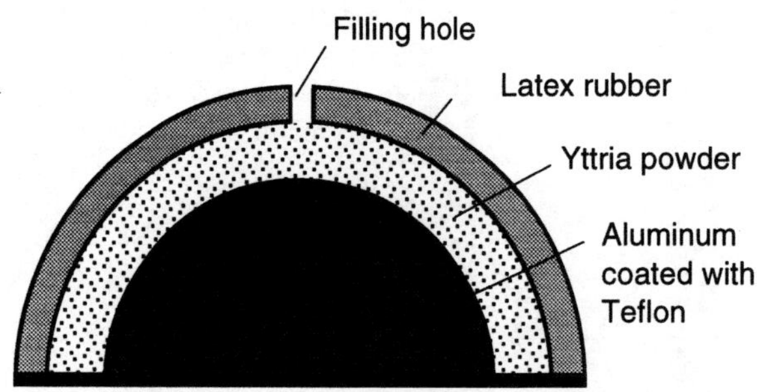

Fig. 5.8. Mold used to fabricate yttria dome.

Next comes *sintering*, which involves heating at a high enough temperature to promote growth and coalescence of the grains (Fig. 5.9). To accomplish this, the ceramic is placed in an yttria enclosure inside a tungsten furnace and heated at 1700-1900°C for an hour to reach 95% density. The resulting body has *closed porosity*, in which most of the remaining pores have no connection to the outside of the ceramic.

Sintering is carried out inside an yttria enclosure to minimize the change in *stoichiometry* (chemical composition) of the ceramic that is being sintered. When a tungsten furnace is used alone, there is some loss of oxygen (giving the formula Y_2O_{3-x}) and also some surface contamination with tungsten. A graphite furnace is worse, since graphite reacts readily with oxygen and reduces the oxygen content even further away from the composition Y_2O_3. The yttria enclosure maintains an approximately correct vapor pressure of oxygen in the atmosphere around the piece being fired to prevent gross departure from the stoichiometry Y_2O_3.

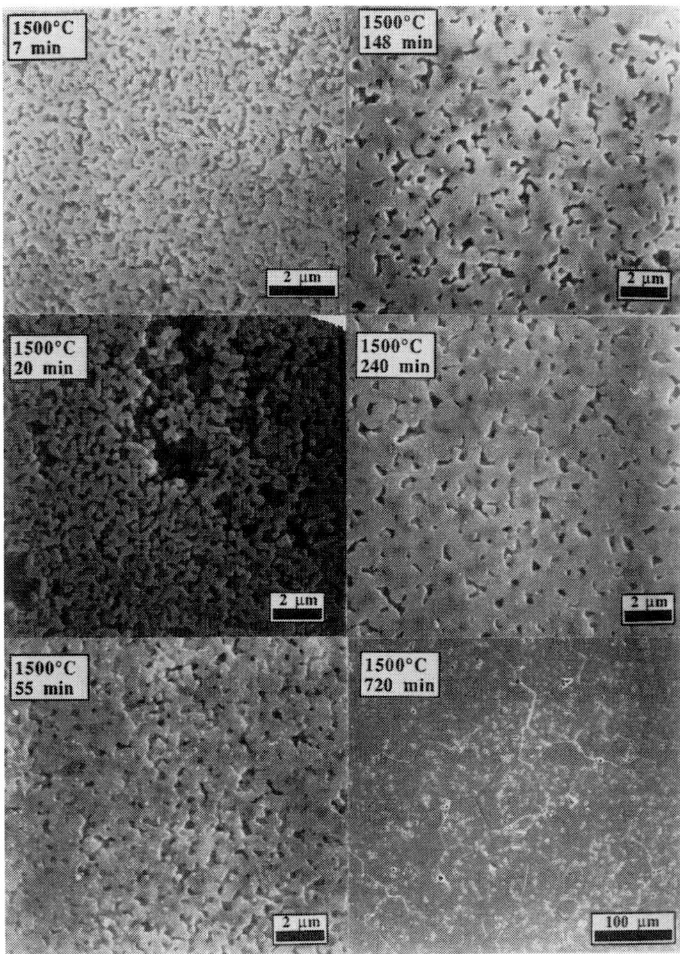

Fig. 5.9. Grain growth during sintering of Y_2O_3.[14] Scanning electron micrographs show surfaces that have been polished and etched with concentrated sulfuric acid to highlight the grain structure. (Photographs courtesy M. Akinc, Iowa State University.)

Once closed porosity is reached, final densification to 99.9+% of theoretical density can be achieved by *hot isostatic pressing* (HIPing) inside an yttria enclosure at 1700-1900°C for 5-10 h using argon gas to apply a uniform pressure of 180 MPa. Pressure applied to the ceramic body causes collapse to full density. Closed porosity is a prerequisite condition, or else the argon gas would penetrate the network of open pores and exert pressure from inside the body as well as from outside. The pores must migrate to the surface and release their gaseous content during hot isostatic pressing. The voids left behind are filled by the same atomic migration that occurs in the sintering step.

The fully dense, hot isostatically pressed ceramic is generally dark in appearance and is slightly deficient in oxygen. The correct stoichiometry is restored by annealing in air at 1450-1800°C for 30-60 minutes to produce a milky white ceramic body. This is the *dome blank* that must be ground and polished to produce a finished product.

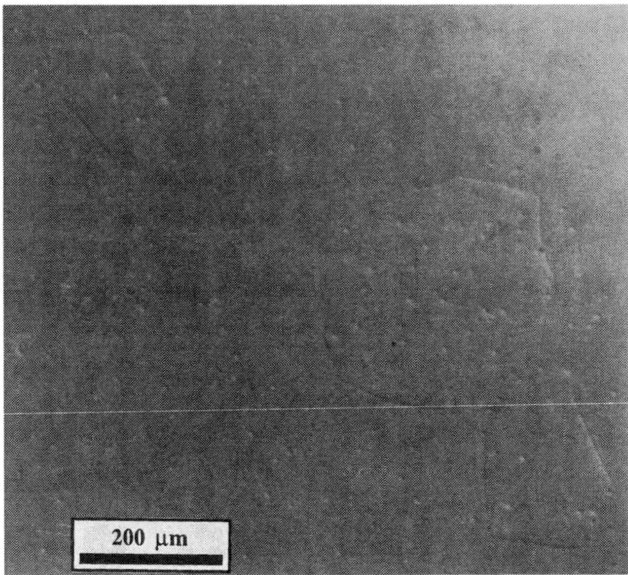

Fig. 5.10. Optically polished surface of yttria, showing the boundary of a large grain. Tiny pinholes occur where pores left from incomplete densification intersect the surface. (Photo courtesy Marian Hills, Naval Air Warfare Center.)

Figure 5.10 shows the polished surface of an optical quality yttria disk. One large grain and many tiny pinholes are visible. The pinholes exist where pores that are left from incomplete densification (sintering and hot isostatic pressing) intersect the surface of the disk.

Figure 5.11 shows the appearance of material after each step in the fabrication of an ALON dome, which is also produced by powder processing. One of the potential applications of ALON is in bulletproof windows, as shown in Fig. 5.12.

Fig. 5.11. Stages in ALON dome fabrication. Aluminum oxynitride powder is pressed into a green body shown at the left. Sintering produces the dome blank at top center, which is ground and polished into a finished dome at the right. (Courtesy Raytheon Co.)

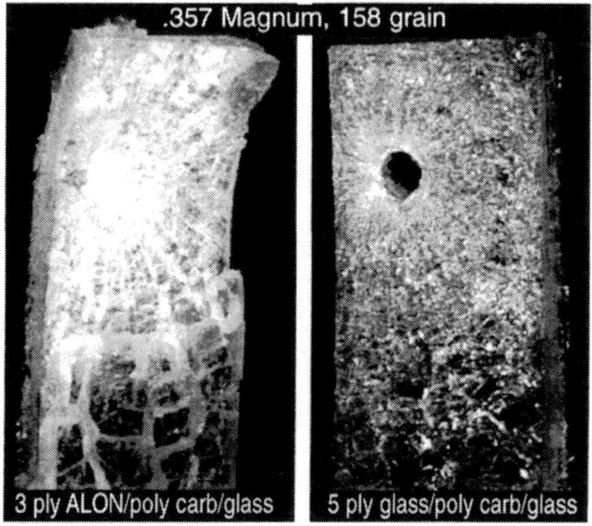

Fig. 5.12. Bullet impact damage in 3-ply ALON laminate (*left*) and commercial 5-ply glass laminate (*right*). (Courtesy Raytheon Co.)

5.2.2 Methods of densifying ceramics: sintering, hot pressing and hot isostatic pressing

Sintering involves the application of heat to cause the coalescence and growth of ceramic particles (Fig. 5.9). Large particles grow at the expense of small particles because the energy of atoms on the surface of a particle is greater than the energy in the interior. Since a large particle has a greater volume-to-surface ratio than does a small particle, the large particle is favored. Given sufficient thermal energy, atoms migrate along surfaces and through the vapor so that large particles grow at the expense of small particles. The thermodynamic driving force for elimination of pores is the higher energy of an atom at a solid/gas interface than at a solid/solid interface.

Optical ceramics must be fully dense, because pores scatter light. Figure 5.13 shows the rate of densification of yttria powder at three different sintering temperatures.[14] The greater the temperature, the greater the rate of increase in density. Figure 5.14 shows that the grain size increases at all three temperatures as full density is approached.[14] Sintering toward full density usually produces very large grain sizes. Since strength and fracture toughness may decrease with increasing grain size (Section 3.8.1), sintering to full density may not be desirable from the standpoint of mechanical properties.

Sintering aids are substances added to ceramic powders to facilitate the sintering process. Lanthana (La_2O_3) is a very special example of a sintering aid that allows yttria (Y_2O_3) to be sintered to full density without excessive grain growth.[15] Yttria containing 9 mole percent lanthana forms a solid solution with a cubic crystal structure. Above approximately 2050°C, cubic and hexagonal solid solutions coexist in equilibrium. When particles of the cubic solid solution are sintered at 2170°C, small particles of hexagonal phase form on the surface of particles of the cubic phase. The ceramic slowly sinters to full density as pores diffuse out, but the grains do not grow to large sizes. Apparently the long diffusion path for atoms moving from cubic grain to cubic grain around the hexagonal grains prevents grain growth.

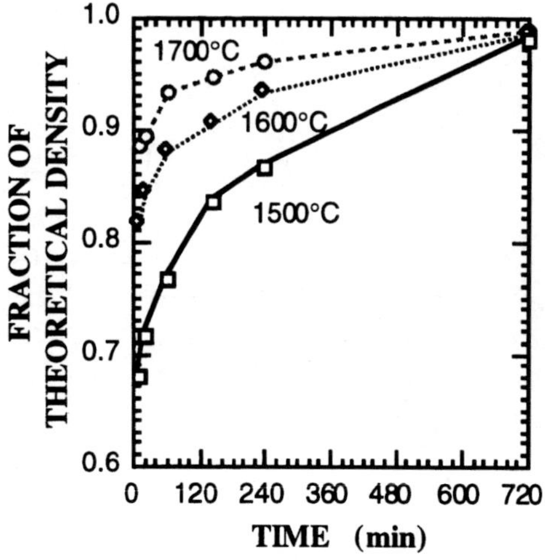

Fig. 5.13. Effect of sintering temperature on the rate of increase of density of yttria.[14]

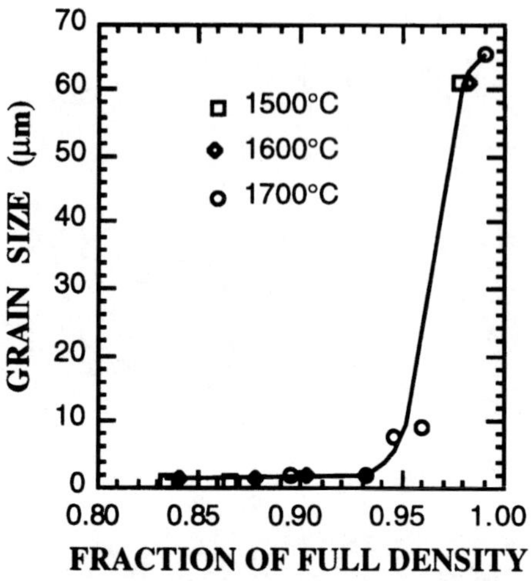

Fig. 5.14. Grain growth as a function of density during sintering of yttria.[14]

After sintering to full density at 2170°C, the lanthana-doped yttria is a poor optical material that scatters a great deal of light because it is a mixture of cubic and hexagonal grains with different refractive indexes. By *annealing* (holding at a constant temperature) for several hours at 1920°C, at which temperature only the cubic phase is thermodynamically stable, the hexagonal particles grow back into the cubic particles. Grain growth of the cubic particles at 1920°C is substantial, but not nearly as great as it would have been during sintering to full density in the absence of hexagonal phase. Figure 5.15 shows these effects. By varying the annealing time, it is possible to control the extent of conversion of hexagonal phase back to cubic phase. By allowing a small amount of hexagonal phase to remain, it is possible to trade off optical transmission and fracture toughness, since the presence of second phase makes the material tougher.[16]

The effect of lanthana on yttria is to allow sintering to a fully dense optical material with moderate grain size. In the absence of lanthana, hot isostatic pressing is required to fully densify the material, and grains grow to a larger size. Sintering is more economical and easier to scale up than hot isostatic pressing. The strength of lanthana-doped yttria is slightly greater than that of undoped yttria, and, as noted above, fracture toughness can be increased in the lanthana-doped material at some expense in transmission. Changing the chemical composition from pure yttria to 9 mole percent lanthana in yttria significantly decreases thermal conductivity (Fig. 4.7).

In *hot pressing,* both pressure and heat are applied to densify a ceramic. Pressure increases the rate of densification and allows a particular density to be reached without as much grain growth as would occur in reaching the same density by sintering. Figure 5.16 shows a typical example of *uniaxial* hot pressing, in which pressure is applied along one direction to a powder contained in a graphite or refractory metal die inside a furnace. A pressure of 300 MPa can be applied in a graphite die. Magnesium fluoride and all of the formerly manufactured Kodak Irtran® materials are hot pressed.

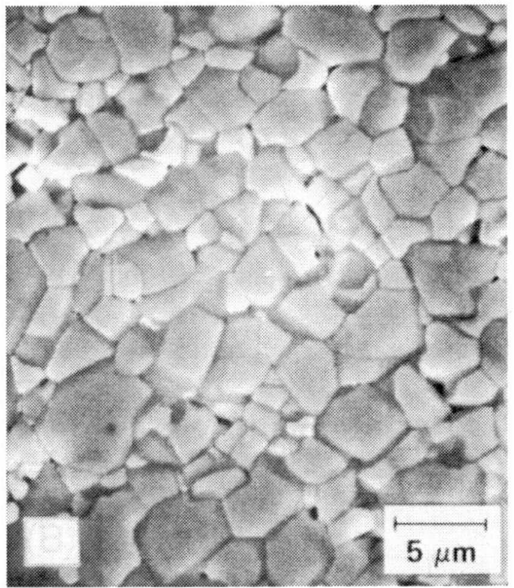

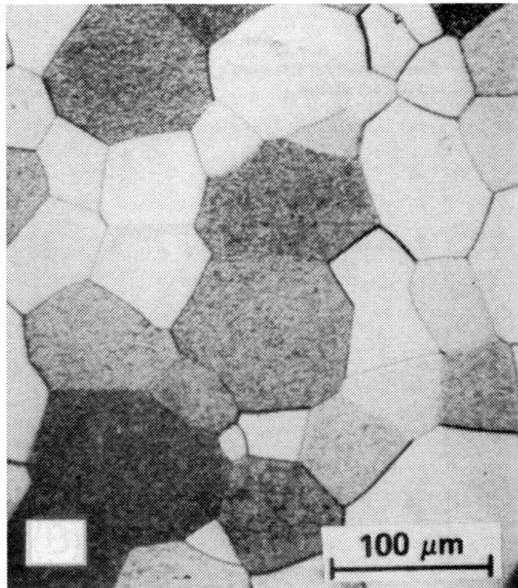

Fig. 5.15. Microstructure of lanthana-doped yttria after (*left*) sintering for 6 h at 2170°C and (*right*) subsequent annealing for 5 h at 1920°C. Surfaces were polished and then etched with boiling 20% HCl to highlight the grains.[15]

In *hot isostatic pressing*, a pressure of 10-1000 MPa is applied by argon gas equally in all directions to the sample that may be as hot as 3000°C. As in hot pressing, this permits densification with less grain growth than would occur in simple sintering. Figure 5.16 shows one type of hot isostatic pressing in which an unconsolidated powder is sealed in a glass or metal container. The glass melts and the metal is crushed during hot isostatic pressing, but both retain their contents. A significant problem with glass is removing it from the ceramic after consolidation. Typically, the molten glass mixes

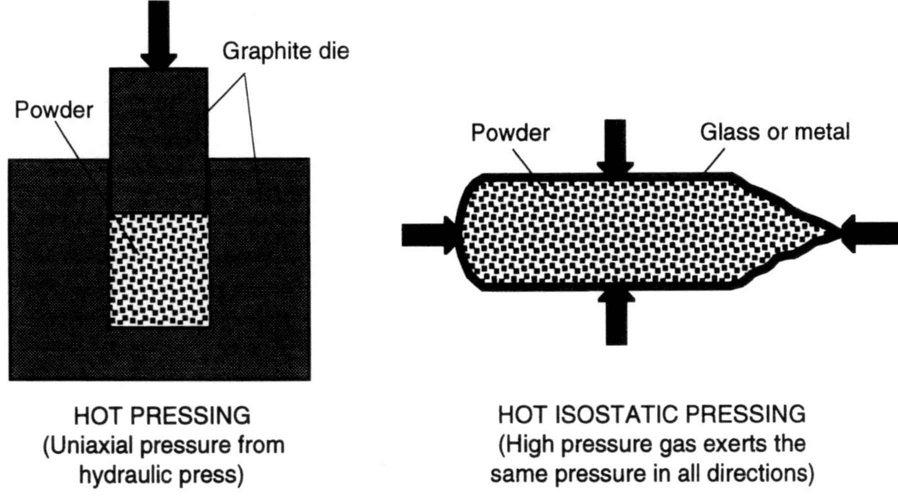

Fig. 5.16. Schematic diagrams illustrating hot pressing and hot isostatic pressing.

intimately with the outer part of the ceramic before densification is complete. Another serious problem is the mismatch of thermal expansion between the glass and ceramic, which may lead to cracking of the ceramic during cool down. Still another problem is chemical reaction between the container and the ceramic powder at elevated temperature.

In the fabrication of undoped yttria described in Section 5.2.1, the powder was first sintered to closed porosity. It could then be hot isostatically pressed without encapsulation, because the high pressure argon could not penetrate to the interior of the ceramic through a network of connected pores. Hot isostatic pressing of a body with closed porosity is a convenient and clean procedure that eliminates problems associated with the container.

5.2.3 Annealing

Annealing involves holding a material at a particular temperature in a particular atmosphere to effect a change in physical or chemical properties. In Section 5.2.1 we saw that undoped yttria is annealed in air at 1450-1800°C to restore the oxygen content that is depleted during hot isostatic pressing. Figure 5.17 shows an example in which lanthana-doped yttria was annealed at 1400°C under argon containing 2×10^{-6} atm O_2 to remove the hydroxyl (OH) impurity in the ceramic.[17] This treatment nearly eliminates absorption at 3000 cm^{-1} (3.3 μm) that would otherwise seriously interfere with the performance of the material as a midwave infrared window.

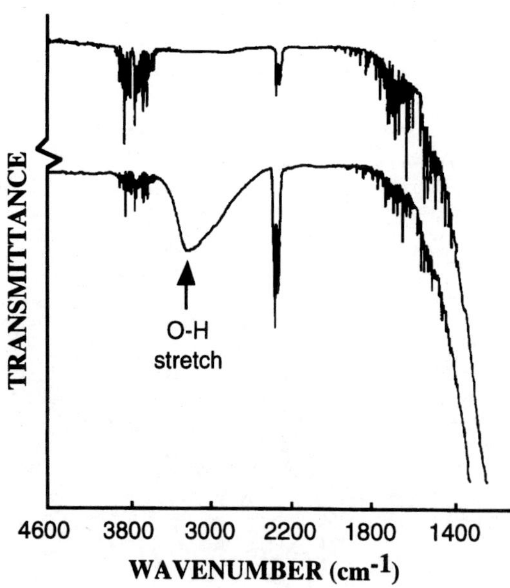

Fig. 5.17. Effect of annealing lanthana-doped yttria at 1400°C under argon containing 2×10^{-6} atm O_2.[17] The band near 3000 cm^{-1} that arises from OH stretching is removed by this process. The multitude of sharp lines that look like noise near 3600-3800 and 1400-1800 cm^{-1} are due to water vapor, not to the ceramic. The sharp doublet near 2300 cm^{-1} is from carbon dioxide.

Annealing can reduce stress left from polishing ceramics to an optical quality finish and can heal the microscopic cracks left from polishing (Section 5.5). Microscopic scratches left in the surface can serve as fracture origins (Section 3.3) and thereby reduce the strength of the material. Annealing at a sufficiently high temperature allows recrystallization of the damaged surface and smoothing of the microscopic flaws by mobile atoms that migrate to reduce the exposed surface.

In one study,[18] ten disks of lanthana-doped yttria were polished and then annealed at 1000°C for 1 h. Their average strength was 123 ± 19 MPa. Another set of ten identically prepared disks was annealed at 1200°C for 1 h and had an average strength of 192 ± 32 MPa. Apparently, 1000°C is not sufficient for healing of the surface damage, but 1200°C is sufficient. The strength increased by more than 50% from annealing.

5.3 Chemical vapor deposition

Chemical vapor deposition (CVD) is a process in which gaseous reactants are passed over a hot substrate on which they decompose to produce a solid product. Figure 5.18 shows a rough gallium phosphide deposit and a transparent, polished plate produced from it.[19] While CVD optical materials are polycrystalline, single-crystal electronic materials can be grown by chemical vapor deposition on single-crystal substrates.

5.3.1 Zinc sulfide and zinc selenide

Zinc sulfide and zinc selenide are, by far, the most important infrared optical materials produced by chemical vapor deposition. Figure 5.19 shows an industrial reactor in which ZnS or ZnSe is grown over areas up to 1 m^2 and thicknesses up to 3 cm at rates of 60 μm/h from zinc vapor and either hydrogen sulfide or hydrogen selenide gas[20]:

$$Zn(g) + H_2S(g) \rightarrow ZnS(s) + H_2(g) \qquad \text{(at 630-730°C)} \qquad (5\text{-}1)$$

$$Zn(g) + H_2Se(g) \rightarrow ZnSe(s) + H_2(g) \qquad \text{(at 730-825°C).} \qquad (5\text{-}2)$$

Gases are injected as 10-25% mixtures with argon at pressures of 20-60 torr. Control of flow patterns is critical to uniform growth. Deposition takes place on large, flat graphite surfaces or in as many as 200 graphite wells that give dome-shaped blanks.

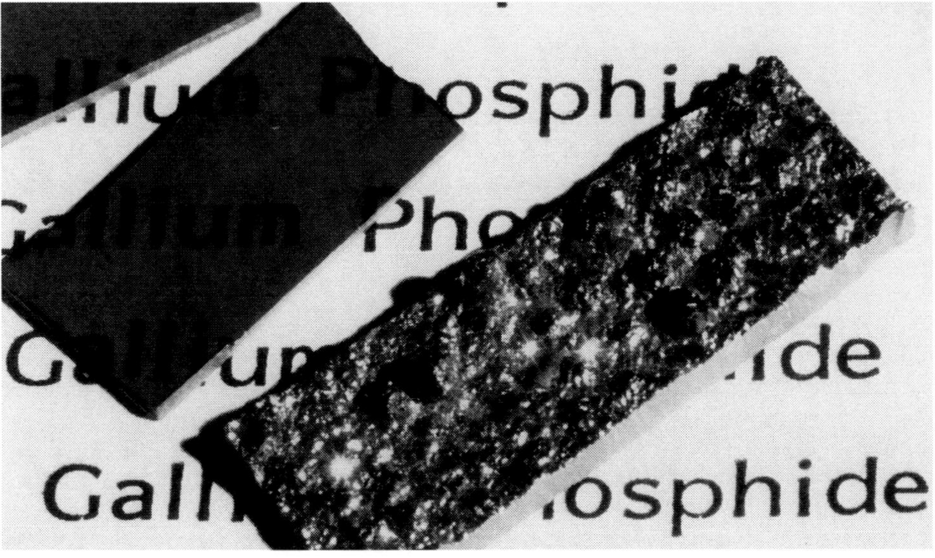

Fig. 5.18. Gallium phosphide produced by chemical vapor deposition. The 1-mm-thick piece at the bottom shows the rough appearance of as-grown material, while the upper piece has been polished to transparency. (Courtesy Raytheon Co.)

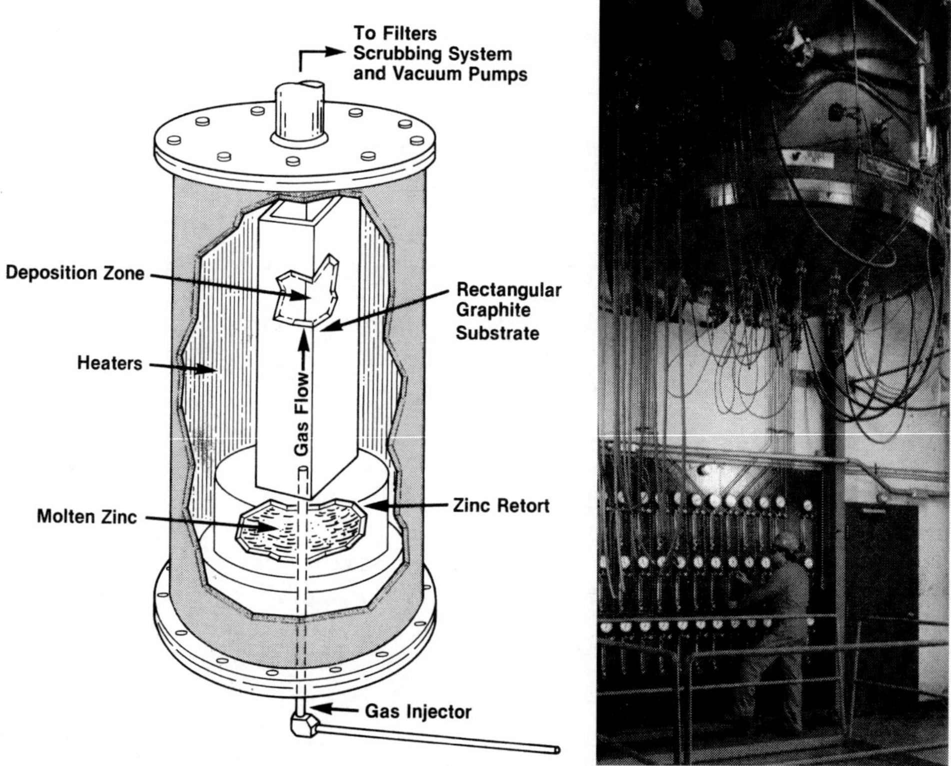

Fig. 5.19. *Left*: Diagram of chemical vapor deposition reactor for producing ZnS and ZnSe. *Right*: Industrial reactor suspended above its operator. The center portion is lowered from the furnace to unload product. (Courtesy Morton International.)

The grain size of CVD ZnS depends mainly on deposition temperature and varies from 0.6 μm at 500°C to 850 μm at 1000°C.[21] Fracture toughness increases slightly as the grain size increases from 0.6 to approximately 20 μm, and then decreases at larger grain size.[21,22] Hardness decreases somewhat as grain size increases up to 20 μm, and thereafter remains fairly constant. "Standard grade" commercial ZnS has a grain size in the 2-8 μm range.

The standard grade CVD ZnS (also called "FLIR grade" for *f*orward *l*ooking *i*nf*r*ared military sensors) has a dark yellow to amber to red color, is a poor visible transmitter, and has a medium intensity infrared absorption near 6 μm (Fig. 5.20). It has significant short wavelength infrared and visible scatter because it is a mixture of cubic and hexagonal crystal phases[23] with slightly different refractive indexes.[24] The 6 μm absorption band is thought to arise from a zinc hydride species associated with a sulfur vacancy, which also gives rise to the visible color and is correlated with the degree of optical scatter.[25] Post-deposition hot isostatic pressing converts standard grade ZnS into transparent, colorless multispectral ZnS (also called Cleartran® or water-clear grade, Fig. 5.20). Hydrogen is lost from the crystal, the Zn:S stoichiometry is restored, and hexagonal grains are converted to the cubic crystal structure during hot isostatic pressing. Grain size increases from 2-8 μm up to 20-200 μm.

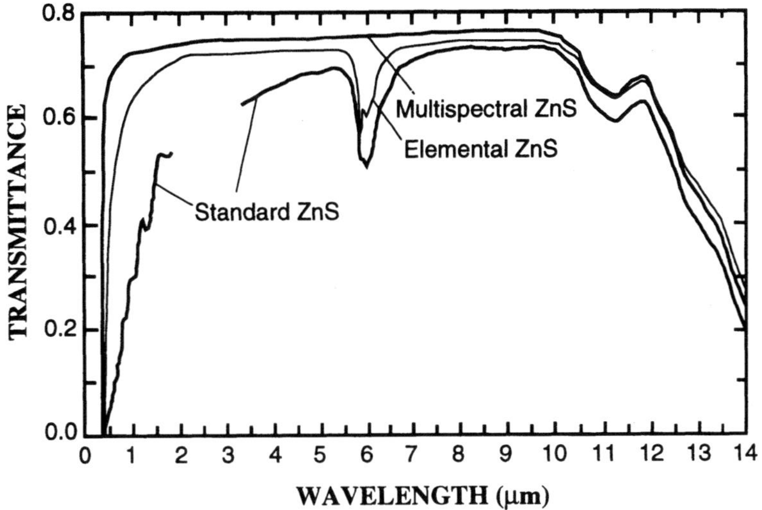

Fig. 5.20. Infrared and visible transmission of standard (6.0 mm thick), multispectral (5.2 mm thick) and elemental (4.5 mm thick) ZnS. Total integrated scatter collected between 2.5° and 70° from normal is listed below:

ZnS grade	3.39 μm Forward scatter	3.39 μm Backward scatter	10.59 μm Forward scatter	10.59 μm Backward scatter
Standard	8.3%	2.0%	0.4%	0.2%
Multispectral	0.3%	0.07%	0.1%	0.03%
Elemental	1.2%	---	0.09%	---

Figure 5.20 also shows the transmission of "elemental" zinc sulfide, which is grown at Raytheon by a CVD process that begins with elemental sulfur, rather than gaseous H_2S. Elemental ZnS is a transparent, light yellow material with strength and hardness similar to that of standard ZnS, but optical transmission similar to that of multispectral ZnS in the midwave infrared region. The rain erosion performance of elemental ZnS is also similar to that of standard ZnS. Table 5.3 compares the properties of all three varieties of ZnS.

Zinc selenide is the other major infrared window material produced by chemical vapor deposition. Its infrared optical quality is superior to that of any grade of ZnS (Fig. 1.18d), but ZnSe is weaker and softer than ZnS (Table 5.3). ZnSe is an excellent infrared laser window material (for which very low absorption is critical), but it is not durable enough for external use. Perhaps a very durable coating would render ZnSe usable for low speed applications. A hybrid material which goes by such trade names as Tuftran® consists of bulk ZnSe with a 1-mm-thick layer of ZnS deposited on the outer surface for increased erosion resistance. Spectra of ZnSe and Tuftran are displayed in Figs. 5.21 and 5.22. ZnSe can be grown to sizes of more than 1 m, as illustrated in Fig. 5.23.

A hot pressed composite material made from 0.5 μm ZnS powder and 0.01-1 μm diamond powder was fabricated in an attempt to obtain enhanced mechanical properties.[27] The long wave transmission was within ~5% of ZnS. With 10 wt % diamond, the

Table 5.3. Comparison of ZnS and ZnSe properties*

Property	Standard ZnS	Multispectral ZnS	Elemental ZnS	ZnSe
Absorption coefficient at 10.6 μm (cm^{-1})	0.2	0.2	–	0.003
Knoop hardness (kg/mm^2)	250	160	215	105
Strength at 20°C (MPa)	103	69	~100	52†
Strength at 260°C (MPa)	124	–	–	72
Young's modulus (GPa)	74	88	–	70
Poisson's ratio	0.29	0.28	–	0.28
Density (g/mL)	4.09	4.09	4.09	5.27
Thermal conductivity [W/(m·K)]	17	27	19	18
Heat capacity [(J/(g·K)]	0.49	0.46	0.52	0.34
Expansion coefficient (10^6 K^{-1})	6.6	6.3	–	7.1
Refractive index (n) at 10.6 μm	2.19	2.19‡	–	2.40‡
Refractive index (n) at 0.6 μm	2.36	2.36‡	–	2.61‡
dn/dT at 10.6 μm (10^5 K^{-1})	4.1	–	–	6.1
dn/dT at 0.6 μm (10^5 K^{-1})	–	5.4	–	10.7
Dielectric constant (microwave)	8.35	8.39	–	8.98
Loss tangent (microwave)	0.0024	–	–	0.0017

*Extracted from manufacturers' catalogs, which do not give identical values for all properties. Data for elemental ZnS from Raytheon are considered "preliminary." This material is still under development.

†The dependence of ZnSe strength on grain size is shown in Fig. 3.34.

‡The inhomogeneity in refractive index of multispectral ZnS and ZnSe is ~2×10^{-6}.[26]

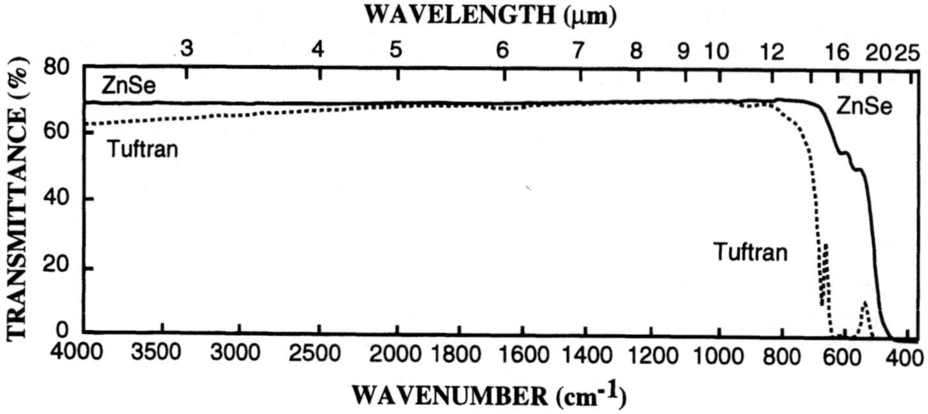

Fig. 5.21. Infrared spectra of ZnSe and Tuftran,® which is ZnSe with a thin outer cladding of ZnS. (Data from Morton International.)

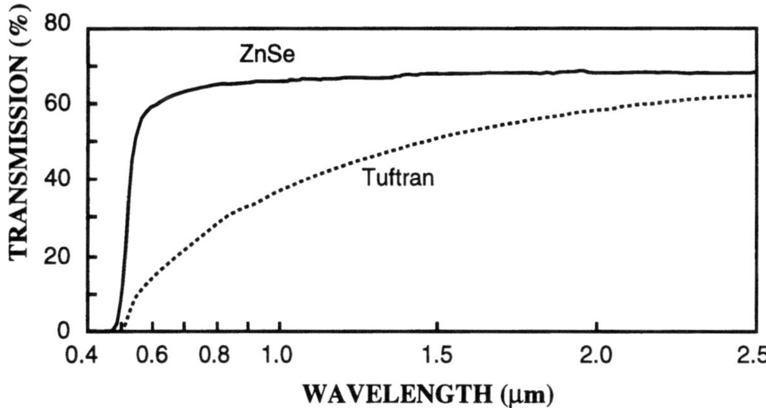

Fig. 5.22. Visible and near-infrared spectra of ZnSe and Tuftran.® (Data from Morton International.)

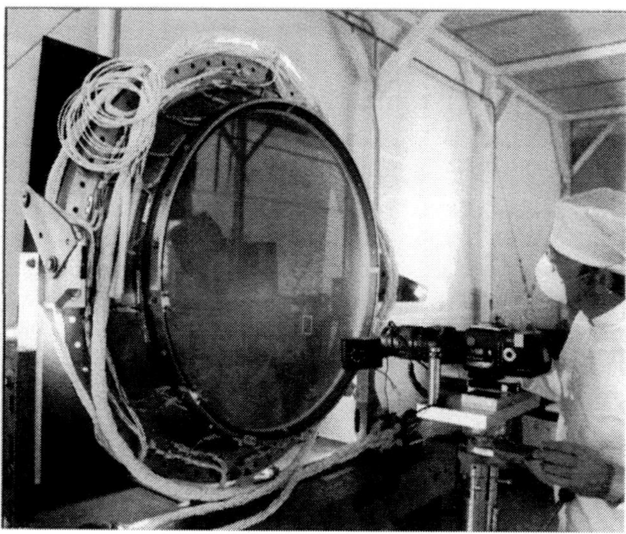

Fig. 5.23. Largest known ZnSe window manufactured for a cryo-vacuum chamber. (Courtesy Raytheon Danbury Optical Systems.)

fracture toughness of ZnS was doubled, but the hardness and yield strength were unchanged. The single-impact waterjet damage threshold velocity (Chapter 7) increased from 170 to 280 m/s, but the multiple impact (400 shots) damage threshold velocity was within 10% of the 130 m/s observed for pure ZnS.[28]

In another attempt to improve the mechanical properties of zinc sulfide, a powder containing a solid solution of ZnS and GaP was hot pressed to 99% of theoretical density.[29] With 21 wt% GaP, the hardness increased from 1.4 to 2.7 GPa and the fracture toughness increased from 0.57 to 0.89 $MPa\sqrt{m}$.

5.3.2 Silicon carbide and silicon nitride

Silicon carbide is a potentially highly thermal-shock-resistant material (Table 4.2, Fig. 4.16) that has been made in transparent form on an experimental basis by chemical vapor deposition from methyltrichlorosilane (CH_3CCl_3) in the presence of excess H_2 and

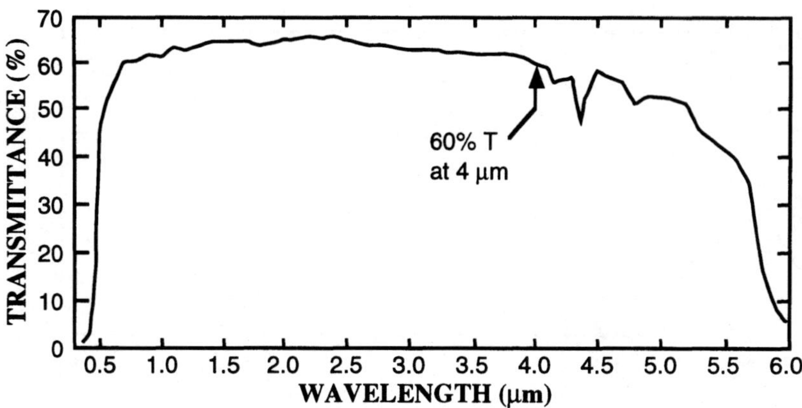

Fig. 5.24. Transmission spectrum of transparent CVD cubic β-silicon carbide (0.25 mm thick).[31]

HCl at ~1400°C.[30,31] The transmission spectrum in Fig. 5.24 shows weak absorption bands in the 4-5 μm region and a peak transmittance of about 66% at 2.3 μm in this very thin specimen.

Example: Estimating the absorption coefficient from the transmission spectrum. Let's estimate the absorption coefficient of silicon carbide at 4 μm from the measured transmittance of 60% in Fig. 5.24. We need to know the theoretical transmittance for nonabsorbing SiC at this wavelength, which we can calculate from the refractive index of $n = 2.508$ listed for cubic SiC at 4 μm in Table C.1 in Appendix C. We calculate the theoretical transmittance from Eq. (1-13):

$$\text{Theoretical transmittance} = \frac{2n}{n^2+1} = \frac{(2)(2.508)}{2.508^2+1} = 0.691 .$$

The remaining 30.9% of the light is reflected. To *estimate* the absorption coefficient, we consider the internal transmittance of the specimen to be the measured transmittance divided by the theoretical transmittance:

$$\text{Internal transmittance} \approx \frac{\text{measured transmittance}}{\text{theoretical transmittance}} = \frac{0.60}{0.691} = 0.868 .$$

The absorption coefficient (α) is then found from Eq. (1-1) with a thickness of $b = 0.025$ cm for the specimen in Fig. 5.24:

$$\text{Internal transmittance} = e^{-\alpha b}$$

$$0.868 = e^{-\alpha(0.025)} \Rightarrow \ln(0.868) = -\alpha(0.025) \Rightarrow \alpha \approx 6\ \text{cm}^{-1} .$$

The "absorption" coefficient that we just calculated includes contributions from scatter and is sometimes called the "extinction coefficient." Because the uncertainty in measured transmittance is rarely less than ±1%, there is significant uncertainty in our estimate of the internal transmittance when the observed transmittance is close to the theoretical transmittance.

Fig. 5.25 shows the absorption coefficient measured in a careful experiment with a thicker (0.556 mm) specimen.[32] The optical scatter was measured and subtracted from the apparent absorption to obtain the true absorption. The measured absorption coefficient at 4 μm (2500 cm^{-1}) is 2.1 cm^{-1} at 291 K and 3.3 cm^{-1} at 912 K. A 2-mm-thick window would have an emittance of $\varepsilon = 0.44$ at 4 μm at 912 K (calculated with Eq. (1-27) using $n = 2.489$ from Appendix C). This emittance is too great for an infrared window at elevated temperature. At best, silicon carbide would have a useful window in the 3-3.5 μm region, or in shorter wavelength atmospheric windows, such as 2-2.5 μm.

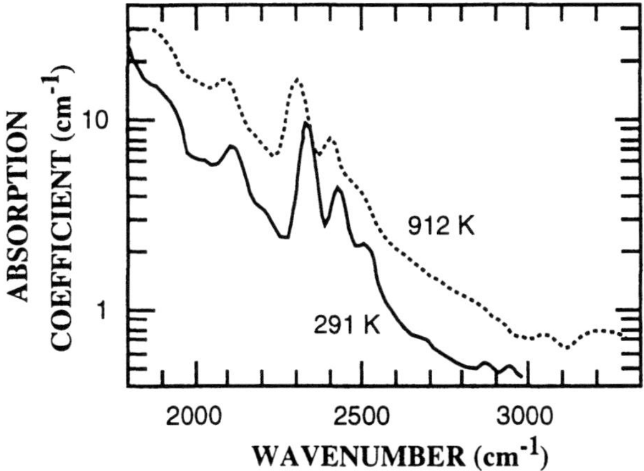

Fig. 5.25. Absorption coefficient of CVD β-silicon carbide obtained after subtracting the measured optical scatter from the measured transmittance of a 0.556-mm-thick specimen.[32]

Another material that ought to have superior thermal shock resistance is silicon nitride, Si_3N_4, which is a candidate for such demanding applications as combustion engine structural components. An effort was made around 1980 to deposit transparent silicon nitride by chemical vapor deposition from precursors such as SiF_4 or $SiCl_4$ and NH_3.[33] The best material that resulted was white in color, contained 1.1 wt% oxygen, and had a maximum specular transmittance of ~65% in the 3-4 μm region (Fig. 5.26).

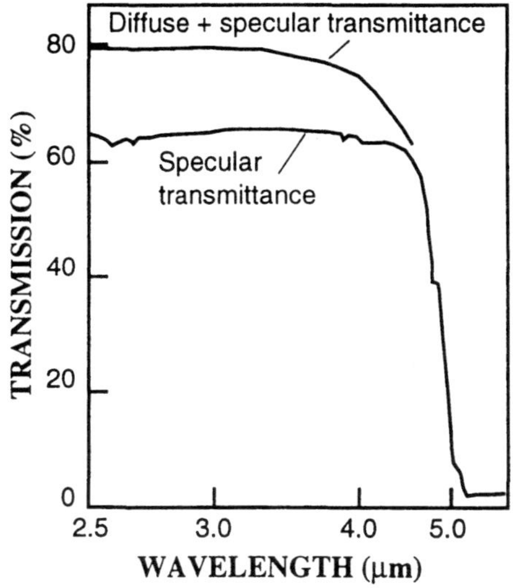

Fig. 5.26. Transmission spectrum of white CVD silicon nitride (0.254 mm thick).[33]

5.4 Single-crystal materials

Large single crystals of optical materials from which windows or domes can be cut are usually grown by slow cooling of a melt. In the *Stockbarger-Bridgman* method (Fig. 5.27), a small single-crystal seed is placed in the bottom of a graphite or quartz vessel filled with powder of the same material as the seed. The vessel is placed in a furnace with a carefully controlled temperature gradient that melts the powder but not the seed. Upon slow withdrawal of the vessel from the hot zone, the liquid solidifies on the surface of the seed and a large single crystal results. A variant of this technique uses a powder charge in a horizontal boat that is slowly withdrawn from the heated region.

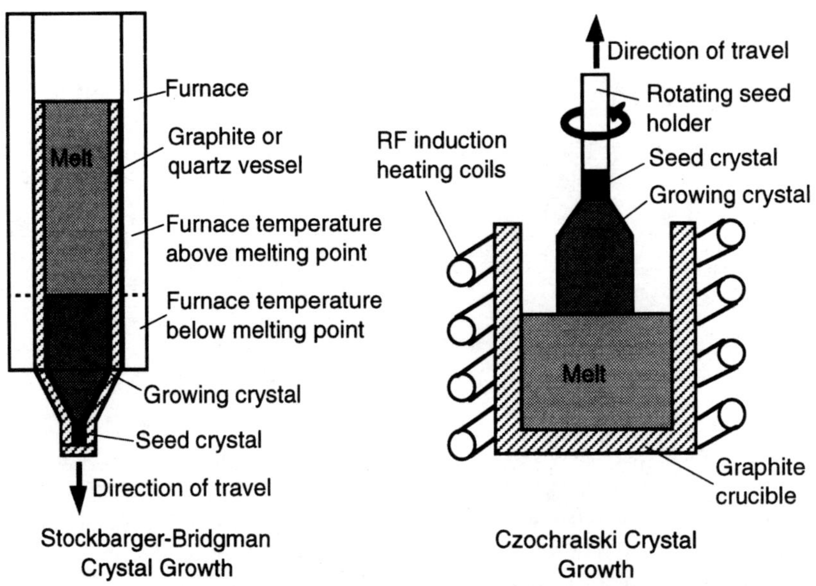

Fig. 5.27. Growth of single crystals by vertical Stockbarger-Bridgman and Czochralski methods.

Czochralski (pronounced cho-**kral**-ski) crystal growth (Fig. 5.27) begins by dipping a rotating single-crystal seed onto the surface of molten material. As the seed is slowly withdrawn, a single-crystal cylindrical boule grows out of the melt. Figure 5.28 shows the seed end of a sodium chloride boule grown by this method.

5.4.1 Gallium arsenide, gallium phosphide, germanium and silicon

Gallium arsenide is an example of an infrared material grown by a variation of the horizontal Stockbarger-Bridgman method.[34] A graphite boat containing GaAs powder is placed in one end of a sealed quartz ampule with solid As at the other end. The As is kept at an elevated temperature to maintain sufficient vapor pressure to prevent loss of As from molten GaAs. The furnace is programmed to melt the GaAs powder and then very slowly freeze the contents of the boat from one end to the other. Large flat pieces of crystalline GaAs containing centimeter-sized single-crystal regions result from this procedure (Fig. 5.29).[35]

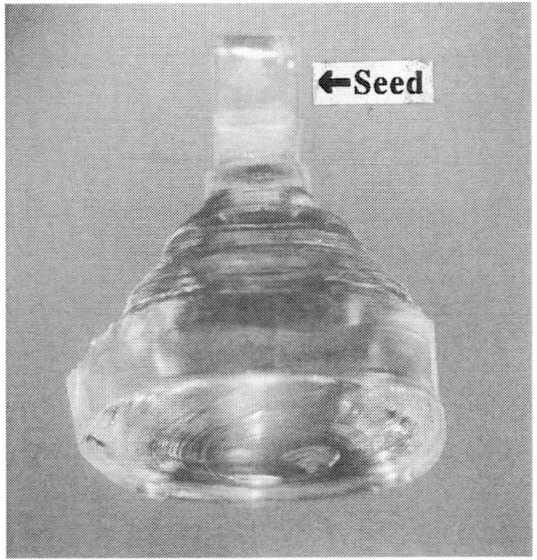

Fig. 5.28. Seed end of 2.5-cm-diameter boule of single-crystal NaCl grown by Czochralski method. The bulk of the boule has been cut off. (Courtesy Marian Hills, Naval Air Warfare Center).

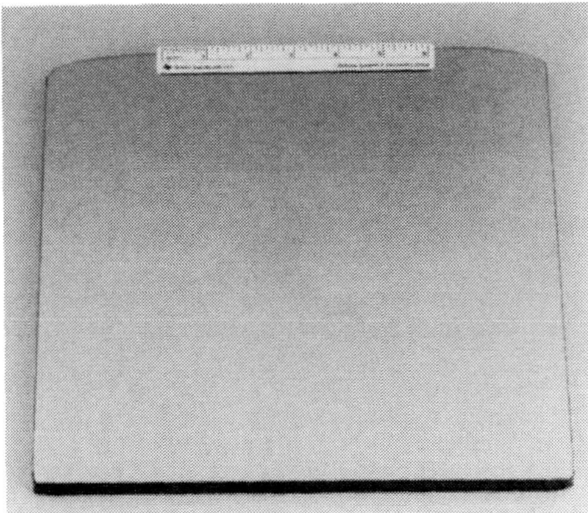

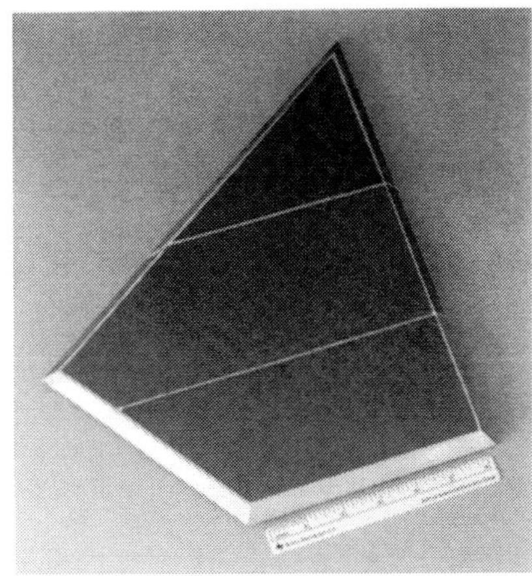

Fig. 5.29. *Left:* Monolithic GaAs window (29 × 31 × 1.0 cm). *Right:* GaAs window made from three segments bonded together. (Courtesy Raytheon Systems Co.[35])

Single-crystal gallium phosphide is grown by the Czochralski method inside a chamber filled with 40 atm of Ar to help retard evaporation of phosphorus from the melt.[34] An additional means used to help prevent decomposition is to coat the melt with a layer of molten boric oxide (B_2O_3) that floats on and encapsulates the GaP liquid.

High quality germanium windows are fabricated from 45 kg boules of single-crystal material with a 24-cm diameter grown by the Czochralski method.[36] "Polycrystalline" Ge windows consist of multicentimeter-sized grains.

Float zone crystal growth in Fig. 5.30 can produce single-crystal rods of relatively conductive materials such as silicon.[34] The lower end of a polycrystalline silicon rod is positioned inside induction heating coils that melt a narrow zone which is held in place

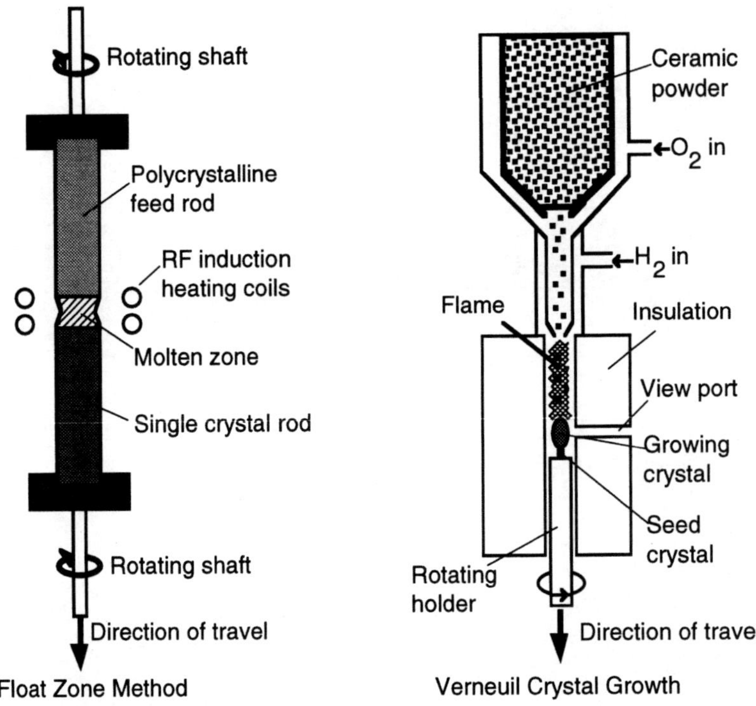

Fig. 5.30. Float zone and Verneuil crystal growth methods.

by surface tension. Rotation of the rod maintains a uniform radial temperature distribution as the rod is slowly translated downward. The freshly cooled material below the molten zone solidifies as a single crystal and impurities are carried by the melt to the top of the rod. Repeated passes further purify the material.

High quality, single-crystal silicon is usually made by the Czochralski method with quartz (SiO_2) crucibles which introduce oxygen contamination at a level of ~2×10^{18} atoms/cm^3. Alternatively, float zone crystal growth does not require a crucible and produces "oxygen-free" silicon. Either kind of silicon is an excellent midwave infrared material, as shown in Fig. 1.18e. The oxygen impurity in Czochralski silicon gives rise to significant absorption near 9 µm which is almost absent in float zone silicon (Fig. 5.31).[37,38] The remaining absorptions in "oxygen-free" silicon are low enough so that the material could have some long wave applications. Since the upper operating temperature of silicon is only ~260°C, emittance from the weak absorptions is not too serious. Polycrystalline silicon made by chemical vapor deposition is discussed in Section 7.8.5. Its absorption spectrum is close to the "oxygen-free" trace in Fig. 5.31.

Verneuil (pronounced ver-**noy**) crystal growth, also called *flame fusion*, is shown in Fig. 5.30.[39] It employs a hydrogen/oxygen flame to melt fine particles of oxide materials such as sapphire (Al_2O_3) and spinel ($MgAl_2O_4$), which then solidify onto the surface of a seed crystal at the bottom of the growth chamber. The crystal grows as the seed is slowly withdrawn from the apparatus. A variation of this technique uses a radio frequency plasma with an inert gas instead of a combustion flame. The Verneuil process is not commonly used for infrared optical materials at present.

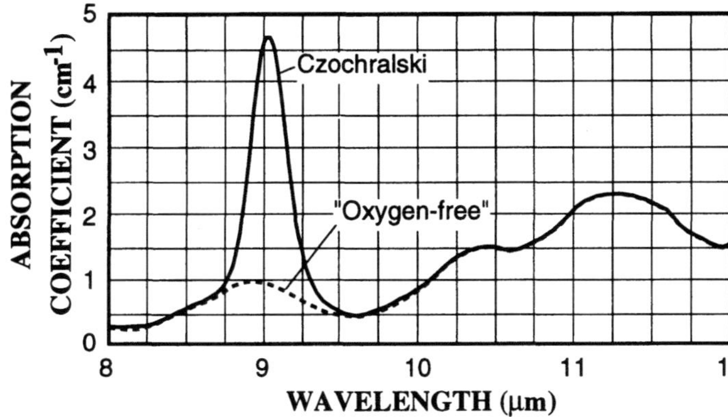

Fig. 5.31. Long wave infrared absorption of ordinary Czochralski silicon and "oxygen-free" silicon.[37]

5.4.2 Sapphire

Single-crystal aluminum oxide (α-Al_2O_3), better known as sapphire, is currently the material of choice for infrared windows and domes that must survive in the most demanding environments. The *heat exchanger method*[40-44] (Fig. 5.32) is one way in which sapphire is grown. In this technique, aluminum oxide is heated above its melting point of 2040°C in a molybdenum crucible containing a single-crystal sapphire seed at the bottom. The seed is mounted on a cold finger that is cooled by helium gas. Evacuation of the furnace permits volatile impurities to escape prior to crystallization, which is begun at the seed by lowering the furnace temperature and increasing the coolant supply to the cold finger. The entire charge solidifies over several days, after which it is annealed just below the melting point to remove residual stress. A very slow cooldown results in a huge, stress-free single crystal with a mass up to 65 kg and a diameter up to 34 cm (Fig. 5.33). Efforts are in progress to scale the HEM process to produce boules of 50- or 75-cm diameter with masses upwards of 100 kg.

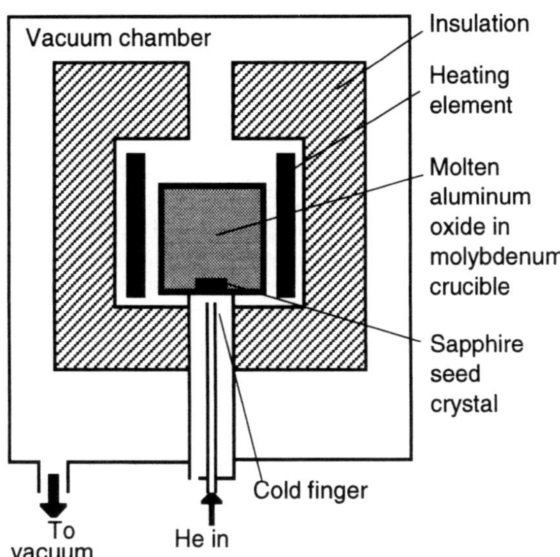

Fig. 5.32. Heat exchanger method for single-crystal sapphire growth.[17-19]

Fig. 5.33. A 34-cm-diameter, 65-kg crystal of sapphire grown by the heat exchanger method. (Photo courtesy Crystal Systems, Inc.)

HEM sapphire is classified into five optical grades and three purity grades. The optical grades, in order of decreasing quality, are designated "Hemex" (best grade), "Hemlux," "Hemlite," "Hemcore," and "Hemverneuil" (lowest grade).[44] Hemex material has no visible scatter when observed with strong white light by the naked eye and no evidence of lattice distortion when viewed through crossed polarizers. Hemlux grade sapphire has barely detectable scatter or distortion and Hemlite grade has clearly detectable scatter or distortion. Grading is done by a skilled technician using standard specimens for comparison. Hemlite grade is the most common commercial form of HEM sapphire. The root-mean-square inhomogeneity in refractive index of 1-cm-thick × 5-cm-diameter disks of all five grades of HEM sapphire is $2\text{-}7 \times 10^{-7}$.

The left side of Fig. 5.34 compares the infrared emittance of Hemex and Hemlite HEM sapphire. The weak emission band near 3000 cm^{-1} (3.3 μm) in Hemlite is potentially significant because this grade of sapphire could be used for missile domes and windows that must operate at high temperature.

The three purity grades of HEM sapphire, in order of decreasing chemical purity, are called "Ultra," "White," and "Standard." The right side of Fig. 5.34 compares the ultraviolet transmission of two samples of "Ultra" purity HEM sapphire. The "vacuum ultraviolet grade" specimen has received a proprietary treatment to remove the absorption band near 200 nm.*

The least efficient way to carve a dome from a crystal of sapphire is to start with a boule and grind away all of the excess material. More than twice as many domes can be fabricated from the same volume if consecutive hemispheres are "scooped" out of the boule (Fig. 5.35) with a hemispheric diamond-plated cutting tool.[42]

*The portion of the ultraviolet region below 200 nm is called the "vacuum" ultraviolet region because oxygen in the air absorbs ultraviolet radiation in this region. Spectrometers are usually evacuated to record spectra in this region.

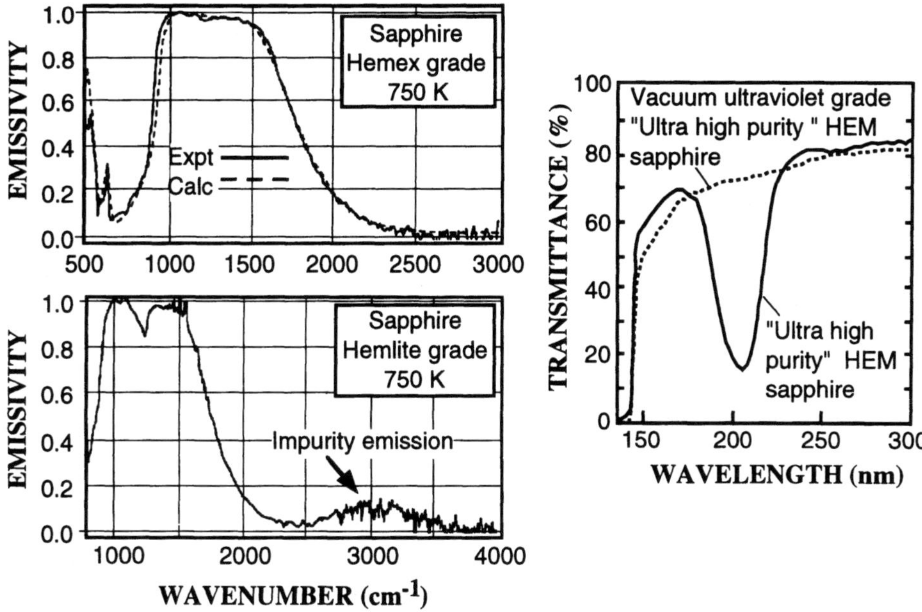

Fig. 5.34. Spectroscopic properties of different grades of HEM sapphire. *Left:* Emission spectra show an impurity (probably OH) in Hemlite grade that is absent in Hemex grade material.[45] *Right:* An ultraviolet absorption band observed in "Ultra" purity material is eliminated by a proprietary treatment.[46,47] The absorption band near 200 nm is attributed to an oxygen vacancy occupied by 2 electrons (an F center).[48]

Since it is a very hard material, sapphire is difficult and expensive to cut and polish. In Fig. 5.7 it was estimated that polishing a sapphire dome costs as much as growing the single-crystal material. It is therefore desirable to grow sapphire in the shape of a hemisphere to minimize the amount of grinding required to attain the final dimensions. In Israel, a vertical Stockbarger-Bridgman method (also called gradient solidification[81]) is employed to grow near-net-shape sapphire dome blanks in hemispheric molybdenum crucibles.[49]

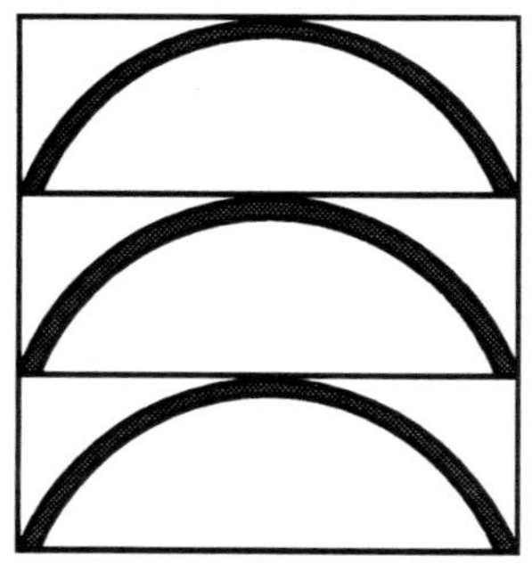

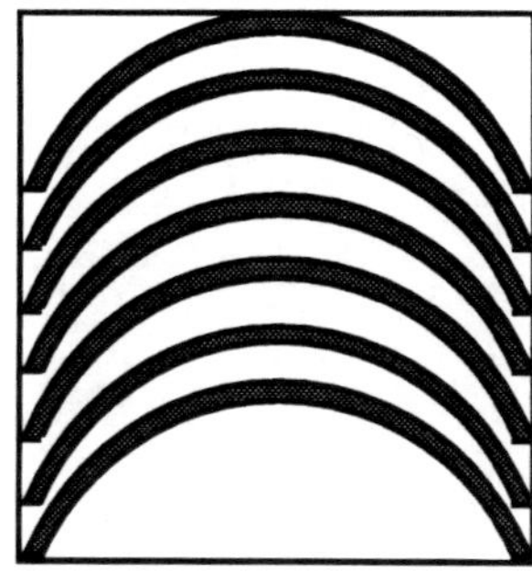

Fig 5.35. Seven domes are scooped from the same volume needed for three domes. (Courtesy Precision Lapping and Optical Co.)

An important near-net-shape process for sapphire is *e*dge-defined, *f*ilm-fed *g*rowth, called the EFG method.[50-54] In this variation of the Czochralski technique, a seed crystal is dipped into the top of a capillary tube containing molten aluminum oxide. The shape of the capillary determines the shape of the resulting crystal. A round capillary gives a continuous sapphire filament. A thin rectangular capillary gives a flat sapphire ribbon or sheet (Fig. 5.36), while a tubular capillary gives a tube-shaped rod. A major application for flat sheets of EFG sapphire is in grocery store scanners which contain a thin sheet of sapphire laminated to a glass plate. The sapphire is extremely resistant to scratching and eliminates the frequent need to replace scratched glass in the scanner.

EFG crystal growth with a semicircular opening and a semicircular seed crystal produced the hemispheric shape in Fig. 5.37.[53] With proper orientation of the seed, the optical axis of the sapphire crystal (Fig. 1.6) is aligned with the axis of the hemisphere. This method of near-net-shape dome growth has not been put into production.

Although the crystal quality of EFG sapphire is good, it is not as good as that of HEM sapphire, which is grown much more slowly and annealed *in situ* prior to removing the boule from the furnace. EFG sapphire tends to have gas bubbles near the surface and

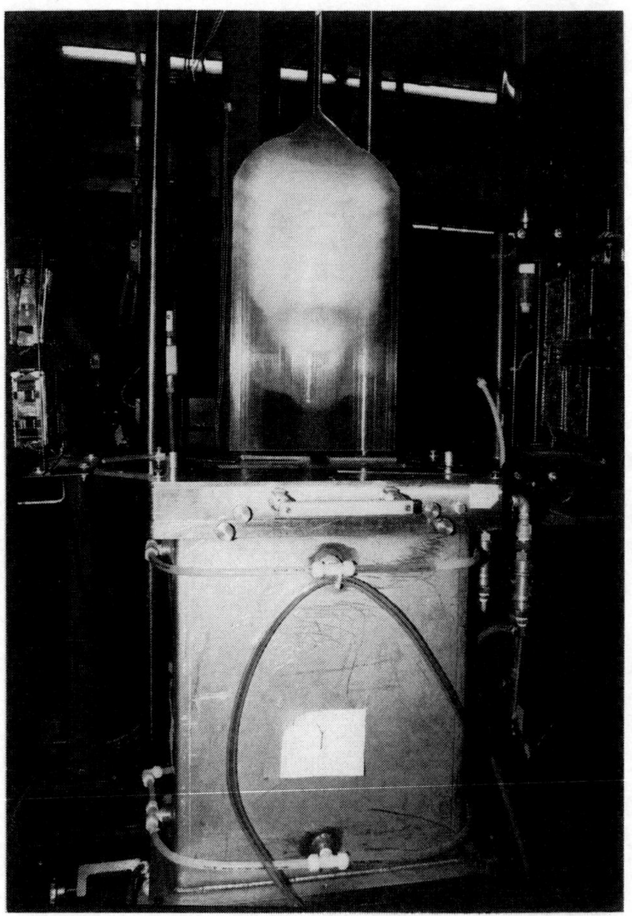

Fig. 5.36. Large flat sheet of sapphire being drawn from a crucible of molten alumina by the EFG method. The single-crystal sapphire seed (15 cm long × 0.75 cm wide) widened to 30 cm during the initial growth process. The final sheet was 30 cm wide for a length of 46 cm. (Courtesy Saphikon, Inc.)

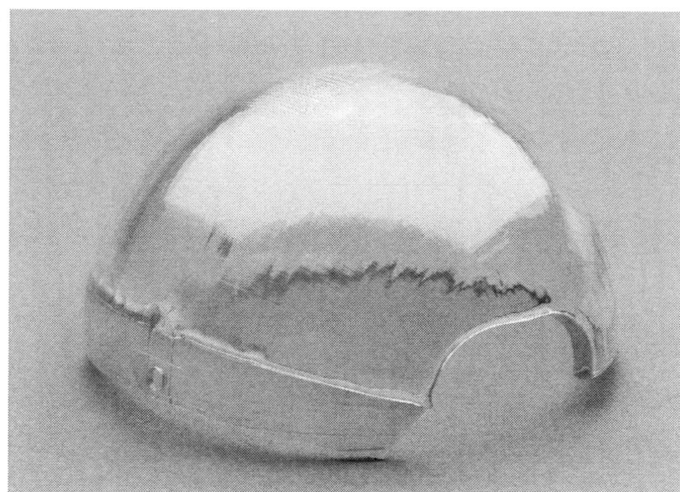

Fig 5.37. Near-net-shape EFG sapphire dome with 7-cm base diameter from Saphikon. Cutouts near the base are from the mechanism by which the seed was rotated through an arc during crystal growth. The process was later modified to eliminate the cutouts.

subgrain boundaries, which are adjoining crystals that are slightly misoriented from each other.* It is reported that an automated system that uses the weight of the growing crystal to control the growth process improves the crystal quality of EFG sapphire.[54]

Large sapphire windows are desired for applications requiring excellent wear resistance. A monolithic, polished window with dimensions 25 × 30 × 0.025 cm has been fabricated from a sheet of EFG sapphire.[56] Four pieces of HEM sapphire were bonded at their edges with a glass frit (a ceramic braze) to produce a polished circular window with a diameter of 20 cm and a thickness of 1.1 cm.[56] A thick, monolithic rectangular window with a diagonal length of ~27 cm has also been fabricated from HEM sapphire.[57]

5.4.3 Hot forging

We can strengthen some single-crystal materials by *hot forging*, in which uniaxial pressure (possibly with heat) severely compresses and distorts the solid (Fig. 5.38). In favorable cases, slipping of one atomic plane past another leads to an effectively polycrystalline material with very small grain size. Small grains increase the fracture toughness and strength, as described in Section 3.8.1. Figure 5.38 shows the effect of hot forging on the strength of potassium chloride. The strength increases by a factor of 8 from the single-crystal value at the lower left of the graph to the polycrystalline value at the upper right.

5.5 Optical finishing

A smooth finish reduces optical scatter that occurs on rough surfaces (Fig. 1.10). Figure 5.39 shows surface profiles of two typical optically polished domes with root-

---

*Native strength-determining defects in EFG fibers have been carefully characterized.[55] Fibers with a diameter of 0.14 mm whose tensile strength is controlled by internal flaws have a Weibull modulus of m = 12.1 and a Weibull scaling factor of S_o = 3.5 GPa. Fibers with strength-controlling surface flaws have m = 6.1 and S_o = 3.4 GPa. Thus the mean strength of sapphire fibers is extremely high, but surface flaws are more variable than internal flaws.

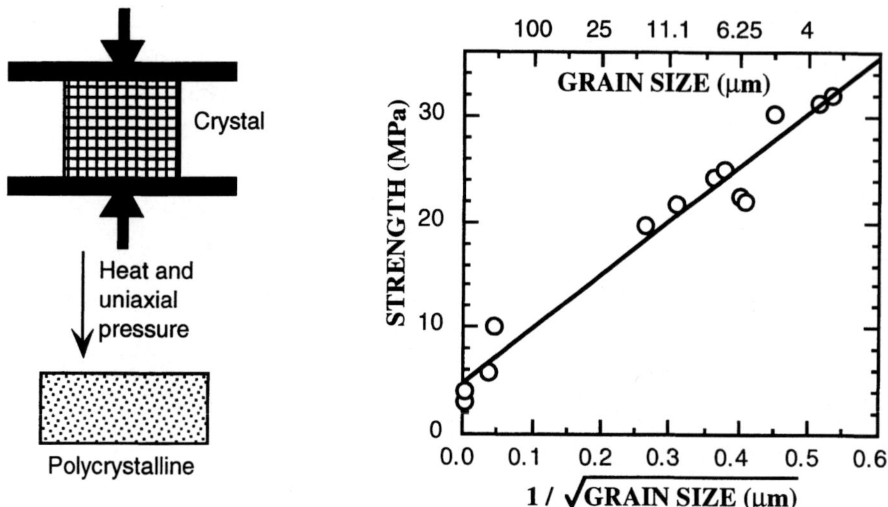

Fig. 5.38. *Left*: Hot forging compresses a single crystal until planes of atoms slip past one another, creating a polycrystalline material. *Right*: Grain size dependence of strength of hot forged KCl[58] follows the Petch equation (3-36). The smaller the grain size, the greater the strength.

mean-square roughness* near 2 nm. Both domes in Fig. 5.39 are acceptably smooth, but notice the distinct scratch marks in Dome B that are absent in Dome A.

Roughness refers to irregular surface features separated by distances in the range 10^{-10} - 10^{-4} m. Features of an optical surface separated by 10^{-4} - 10^{-3} m are called *waviness*, while features with an even larger separation define the *optical figure*, or deviation from the ideal shape of an optical element.[59-61]

The total integrated scatter (Section 2.2) from a single rough surface into one hemisphere is related to the rms roughness, δ, and wavelength of light, λ, by the approximate equation[59]

$$\text{Total integrated scatter} \approx \left(\frac{4\pi\delta}{\lambda}\right)^2 . \qquad (5\text{-}3)$$

Table 5.4 tells us that the dome surface roughness (~2 nm) in Fig. 5.39 gives <1% visible scatter, and a roughness 10 times greater could be tolerated for infrared operation.

*Suppose we have N equally spaced measurements of surface elevation, z_i, with the average elevation designated $\bar{z}$. Root-mean-square (rms) roughness is defined as

$$\text{rms roughness} = \delta = \sqrt{\frac{1}{N}\sum_{i=1}^{N}(z_i-\bar{z})^2} .$$

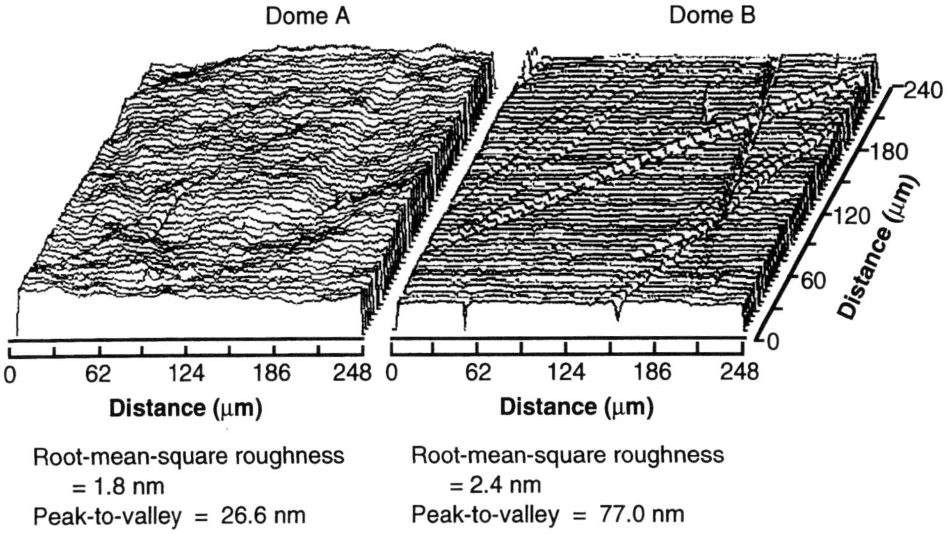

Fig. 5.39. Three-dimensional surface profiles of yttria domes.

Surface quality is measured by several quantitative and qualitative methods. The measurement in Fig. 5.39 was made with an optical surface profiler, which gives a picture of the surface topography, a root-mean-square roughness for the entire surface and a peak-to-valley elevation difference. Vertical sensitivity in the range 0.02 - 0.3 nm is available on such instruments, whose lateral resolution can be 0.5-30 μm. In a mechanical profiler a diamond stylus is moved in a line along a surface and records the change in elevation at each step in its path. Vertical resolution is approximately 0.05 nm, with a lateral resolution of 0.2 μm. Another quantitative measure of surface roughness is based on measuring total integrated scatter.[62] The modulation transfer function (Section 2.3) of a finished optical flat is a good indicator of the overall quality of both the bulk material and its surface finish.

Different methods of measuring surface roughness do not measure exactly the same features. Therefore an identification of the method and the range of spatial wavelengths to which it is sensitive should accompany a reported value for surface roughness.[60] For example, the resolution of a mechanical profiling system is limited by the size of the stylus relative to the sharpness of the surface relief. A "fat" stylus" cannot get into a "thin" scratch and therefore does not sense the scratch.

Table 5.4. Surface roughness and optical scatter from Eq. (5-3)

Wavelength	Surface roughness to give indicated scatter	
	Total integrated scatter = 10^{-4}	Total integrated scatter = 10^{-2}
0.5 μm	0.4 nm	4 nm
3 μm	2.4 nm	24 nm
10 μm	8.0 nm	80 nm

5.5.1 Scratch/dig specifications

A qualitative measure of surface finish is the *scratch/dig specification*.[63,64] U.S. scratch numbers in Table 5.5 give the width of a scratch in 0.1-µm units. Dig number refers to the diameter of a pit, bubble, pinhole or inclusion in 10-µm units. An optical surface that meets the 60/40 scratch/dig specification has no scratch wider than 6 µm and no dig diameter larger than 400 µm. We consider the scratch/dig specification to be qualitative because inspection is done visually by an experienced observer. Standard specimens for visual comparison of scratches and digs are available.

Table 5.5. Scratch/dig specifications

MIL-0-13830A[a]			MIL-F-48616[b]				
Number	Maximum scratch width	Maximum dig diameter	Letter	Scratch width	Dig diameter	Disregard scratch width	Disregard dig diameter
10	1 µm	100 µm	A	5 µm	50 µm	<1 µm	<10 µm
20	2 µm	200 µm	B	10 µm	100 µm	<2.5 µm	<25 µm
40	4 µm	400 µm	C	20 µm	200 µm	<5 µm	<50 µm
60	6 µm	600 µm	D	40 µm	300 µm	<10 µm	<50 µm
80	8 µm	800 µm	E	60 µm	400 µm	<10 µm	<100 µm
			F	80 µm	500 µm	<20 µm	<100 µm
			G	120 µm	700 µm	<20 µm	<200 µm
			H		1000 µm		<250 µm

[a]MIL-0-13830A states that for a circular element "the combined length of maximum size scratches ... shall not exceed one quarter the diameter of that element. When a maximum size scratch is present, the sum of the products of the scratch numbers times the ratio of their length to the diameter of the element ... shall not exceed one half the maximum scratch number. When a maximum size scratch is not present, the sum of the products of the scratch numbers times the ratio of their length to the diameter of the element ... shall not exceed the maximum scratch number." For irregular shaped digs, the diameter is taken as the average of the maximum length and maximum width. "The permissible number of maximum size digs shall be one per each 20 mm of diameter ... on any single optical surface. The sum of the diameters of all digs ... shall not exceed twice the diameter of the maximum size specified per 20 mm diameter. Digs less than 2.5 microns shall be ignored. All digs ... whose quality is number 10 or smaller shall be separated edge to edge by at least 1 mm."

[b]MIL-F-48616 states that "scratches are permissible provided the width does not exceed that specified by the scratch letter. The accumulated length of all maximum scratches shall not exceed 1/4 of the average diameter of the element." Integrating scratches: If no maximum width scratch is present, "all scratches of widths less than or equal to the maximum allowable scratch width, and greater than or equal to the minimum scratch width to be considered ... shall be included in the integration. The length of each scratch shall be multiplied by the scratch width. These products are to be added and the sum divided by the average diameter of the element. If a maximum scratch is present, this resulting value shall not exceed 1/2 the maximum allowed scratch width. If no maximum scratch is present, this value shall not exceed the maximum allowed scratch width." No more than one maximum size dig shall occur in any 20 mm diameter circle. Integrating digs: "All digs of diameters less than or equal to the maximum allowable dig diameter and greater than or equal to the minimum dig diameter to be considered shall be included in the integration. All digs shall be accumulated such that the sum of the diameters does not exceed twice the diameter of the maximum allowed dig for any 20 mm diameter circle.... All digs of size B or smaller shall be separated by 1.0 mm minimum."

A second scratch/dig specification in Table 5.5 uses letters to designate allowable sizes. For example, the specification F/E means that no scratch width can exceed 80 μm and no dig diameter can exceed 400 μm. Scratch widths in MIL-F-48616 are an order of magnitude larger than those of MIL-0-13830A.

5.5.2 Optical polishing

Optical polishing usually employs a series of progressively finer abrasive media to grind away the high points of a surface until it is smooth. Common abrasive powders include diamond, silicon carbide, alumina and silica. Opticians and machinists have different terminologies to describe the coarse, medium and fine stages in optical finishing:

	Optician's terminology	Machinist's terminology
Coarse:	machining	grinding
Medium:	grinding	lapping
Fine:	polishing	polishing

Typically, each step is carried out until the depth of material removed is at least four times as great as the diameter of the previous abrasive. The objective is to remove the damage layer produced by the previous step. Using the machinist's terminology, for example, grinding might be carried out with 45-μm-diameter diamond particles until the desired shape is achieved. Lapping with 9-μm-diameter alumina might then be used to remove at least $4 \times 45 = 180$ μm of thickness. Final polishing with colloidal silica (<1 μm diameter) would then be carried out until at least $4 \times 9 = 36$ μm of additional thickness is gone. For a higher quality finish, more intermediate sizes of abrasives might be used and more material might be removed in each step. Grinding abrasives are designated "fixed" or "loose" depending on whether they are embedded in a soft matrix (such as a metal or resin) or are used as a liquid slurry.

Mild techniques have been developed to minimize subsurface damage from polishing. In *float polishing*, relatively little pressure is placed on the rotating workpiece which is submerged in a slurry of polishing powder in deionized water above a compliant, rotating lap made of tin, copper or nylon. The workpiece floats on a ~1 μm layer of fluid in which solid particles flow past the surface being polished. The solid particles can range from 1 nm in size (colloidal silica) to 100 nm (titania, alumina, ceria or magnesia). If the material removal mechanism is largely chemical in nature, the process is called *chemo-mechanical polishing*. Chemo-mechanical polishing of single-crystal calcium fluoride laser windows with colloidal silica produces a sufficiently damage-free surface to increase the laser damage threshold fluence by an order of magnitude compared to windows polished by conventional means.[65]

Magnetorheological (MRF) finishing is a still-experimental polishing technique invented in Belarus and under development at the Center for Optics Manufacturing at the University of Rochester. The magnetorheological fluid is an aqueous suspension of microscopic iron particles and ceria abrasive that flows past the workpiece. Application of a magnetic field stiffens the fluid, which replaces the lap of conventional polishing. Computer control of the fluid's shape and stiffness by adjusting the magnetic field allows the deterministic production of various shapes for the polished part.

Another common method of finishing optical elements, especially those with nonspherical surfaces, is *single-point diamond turning*. In this technique, a computer-operated diamond cutting tool under exquisite control mills the desired shape into the part,

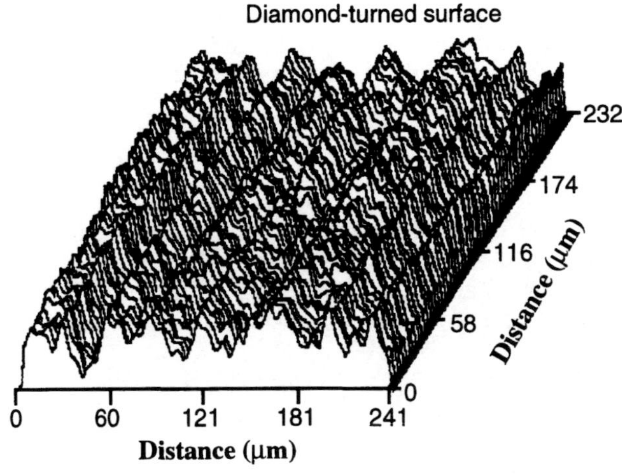

Fig. 5.40. Topographic map of aluminum surface with an optical finish produced by single-point diamond turning.

which need not be polished prior to turning. A diamond-turned surface has almost as fine a finish as a highly polished surface. For some materials in some crystallographic directions the diamond tool takes tiny chips out of the surface, which degrades the quality of the finish. The diamond-turned surface in Fig. 5.40 exhibits characteristic grooves left by the cutting process. Diamond turning is effective with such materials as ZnS, ZnSe, Si, Ge, GaAs and MgF_2,[66] but oxides such as sapphire, spinel and yttria wear down the diamond point too rapidly to be machined.

The *flatness* of a disk or window is measured with an optical interferometer, as shown in Fig. 5.41. A work piece of unknown flatness is placed on a standard optical flat made of fused silica and monochromatic light is passed upward through the optical flat. Reflection (shown at the right in Fig. 5.41) from the lower surface of the work piece interferes with reflection (not shown) from the upper surface of the optical flat. Whenever the separation between the work piece and the optical flat is a half multiple of the wavelength of light, destructive interference occurs.

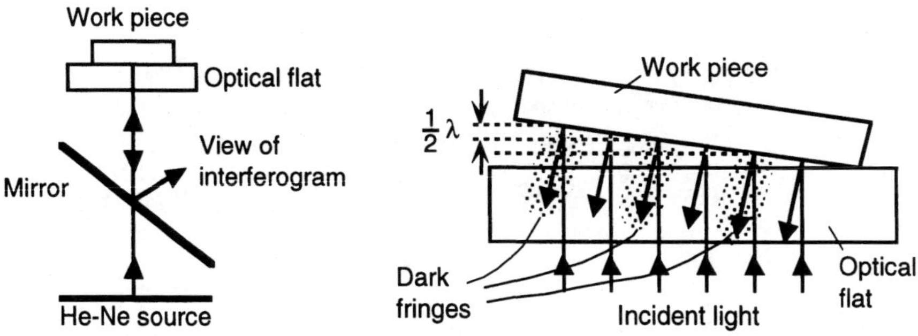

Fig. 5.41. Interferometric measurement of flatness.

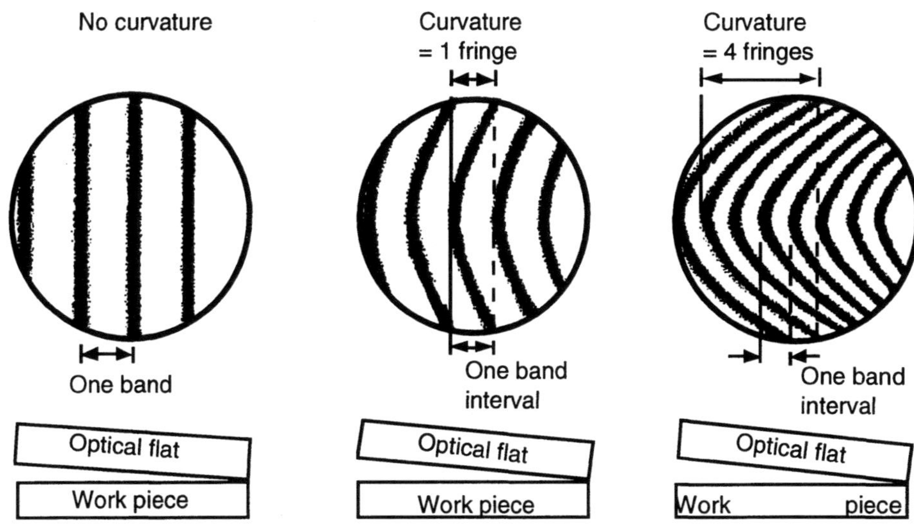

Fig. 5.42. Curvature in the interference patterns from the experiment in Fig. 4.3 provides a measure of the curvature of the surface of the work piece being polished.

The experiment in Fig. 5.41 might produce one of the patterns in Fig. 5.42. If the work piece is flat, the dark fringes form straight lines. When the surface of the work piece is curved, the interference pattern is also curved. For every two fringes of curvature at the right of Fig. 5.42, the irregularity in the work piece is equal to one wavelength of light. If helium-neon light of wavelength 0.63 μm gives 4 fringes of curvature, the irregularity of the surface is 2×0.63 μm ≈ 1.3 μm.

Different crystal planes of single-crystal materials polish at different rates. In the case of sapphire, the *a*- and *m*-planes (Fig. 1.6) lose material at 4 times the rate of the *r*-plane and 2 times the rate of the *c*-plane.[43] Since all of these planes are exposed in different locations on the surface of a dome, polishing naturally leads to hills and valleys on the dome unless the process is carefully monitored.

5.6 The effect of surface finish on mechanical strength

The most important factor governing the mechanical strength of an optical ceramic is the quality of the surface finish.

In Section 3.3 we noted that ceramics usually fail at pre-existing flaws. Internal flaws, such as large voids or inclusions, can be left from the ceramic fabrication process. More commonly, damage from grinding and polishing provides the largest, strength-limiting flaws. Fig. 5.43 shows subsurface damage in sapphire after grinding and polishing.[67] The large crack in the ground specimen extends ~14 μm below the original surface. In the polished sample, no subsurface cracks are evident and the damaged region of the crystal extends only 100 nm (0.1 μm) below the surface. Sometimes the polishing step does not remove deep cracks created by grinding and the residual subsurface damage weakens the material.

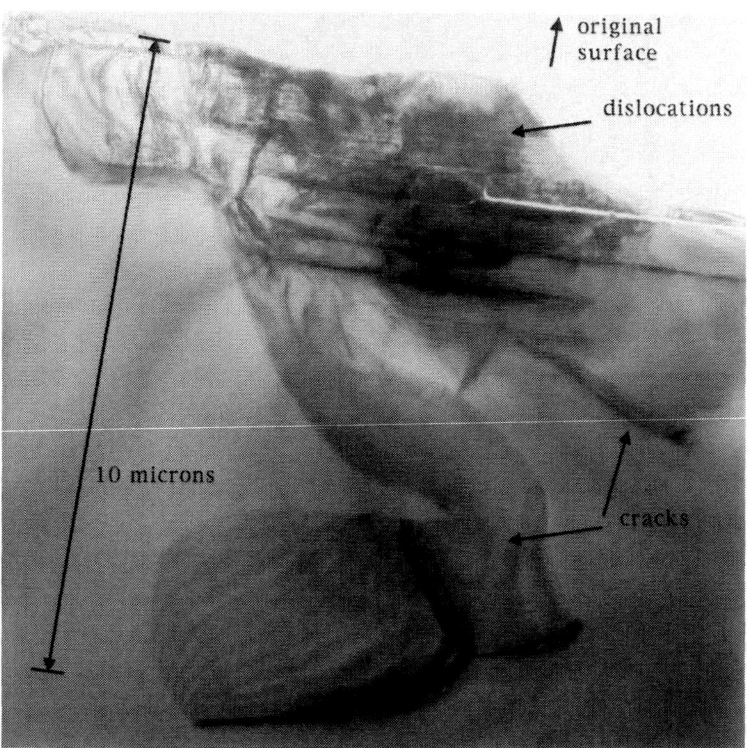

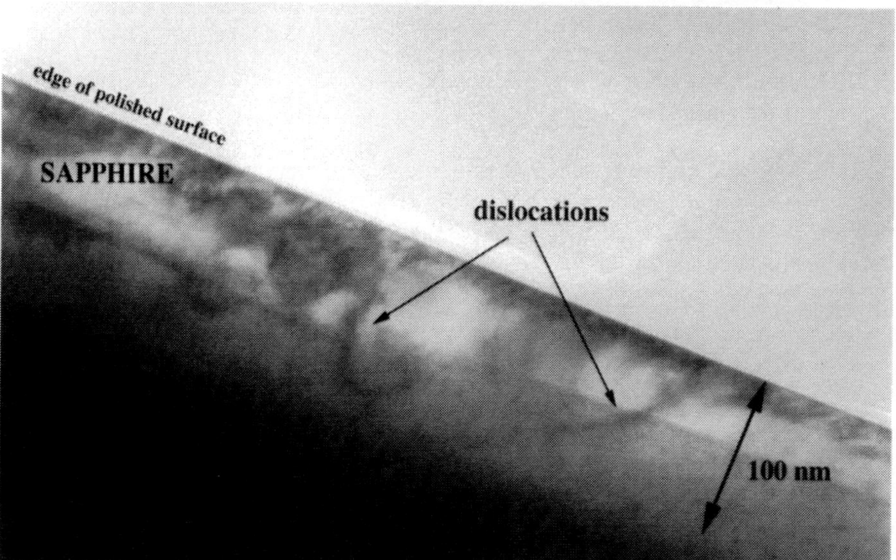

Fig. 5.43. Grinding damage (*top*) and polishing damage (*bottom*) observed by transmission electron microscopy beneath the *c*-plane surface (Fig. 1.6) of sapphire.[67] In the upper photo, the original surface was lost when the sample was thinned for microscopy by ion milling. The upper surface in the figure is ~3-4 μm beneath the original surface. Dislocations labeled in the figures are misalignments of the atoms in the crystal. [Courtesy B. J. Hockey and D. Black, National Institute of Standards and Technology.]

To illustrate the effect of surface finish, consider the strengths of calcium fluoride and strontium fluoride polished to an optical finish by two different procedures:[68]

	Strength (MPa)	
	Optical polish A	Optical polish B
Single-crystal CaF_2	172 ± 23	104 ± 35
Fusion cast SrF_2	83 ± 39	163 ± 30

One polish gave a strength that is around twice as great as the other , but the effects were reversed for the two materials. The variable effect of different polishing procedures on mechanical strength is one of the most vexing problems in engineering with optical materials.

The data for CaF_2 and SrF_2 make a case for this author's opinion that extensive, expensive mechanical testing of optical ceramics for the purpose of an engineering database is rarely warranted. No matter how well one lot of material is characterized, there is no guarantee that the next lot polished by the next person in the next shop will have the same characteristics. *The most variable factor governing the mechanical strength of an optical ceramic is the surface finish.*

Subsurface cracks are not usually evident when the polished surface is inspected. However, a method called *dimpling* can be used to measure the depth of subsurface damage.[43] In this procedure, a small circular pit is ground into the polished surface by lapping with 0.5-μm diamond abrasive and a steel ball of radius r, as shown in Fig. 5.44. When the resulting pit (the dimple) is observed with a microscope, as shown at the right of Fig. 5.44, there is a rough annulus where the original subsurface damage intersects the dimple. From the geometry of the dimple, the depth of subsurface damage, Δ, is

$$\Delta = \frac{(d_1/2)^2}{r} - \frac{(d_2/2)^2}{r} \tag{5-4}$$

where d_1 and d_2 are dimensions defined in Fig. 5.44. Dimpling can discern gross subsurface damage from grinding, but may not detect subtle damage left from polishing.

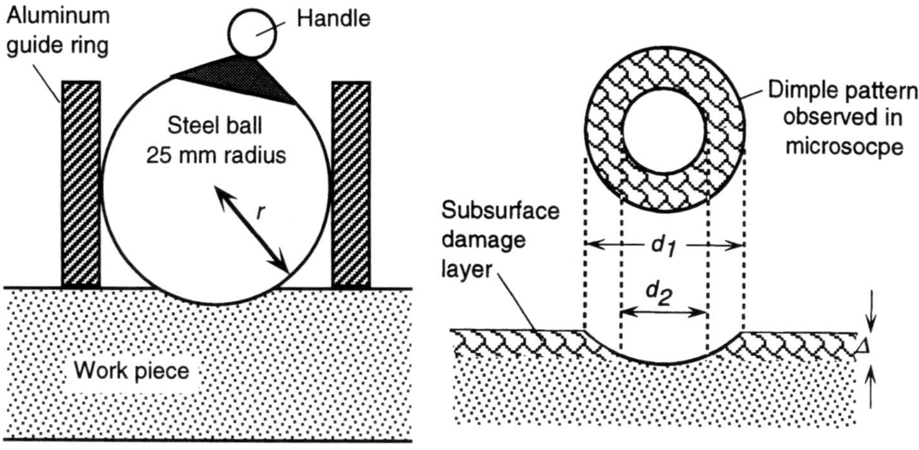

Fig. 5.44. In dimpling, a small pit is lapped into the surface of a work piece. The diameter of the damage ring seen in the pit is related to the depth of subsurface damage.

Optical ceramics have a wide distribution of strengths (a small Weibull modulus) because their strength is determined by a small number of subsurface cracks that were not removed by polishing. A ground surface has a lower strength than a polished surface, but the Weibull modulus for the ground surface could be higher because there are many more flaws of similar depth from which fracture can originate.

Usually, the finer the grit used to grind a surface, the stronger is the resulting ceramic. For the experiment in Fig. 5.45, polycrystalline silicon nitride was cut into 3 × 4 × 40 mm bars and all surfaces were ground parallel to the long axis of the bar with an 800-grit diamond abrasive wheel.[69] Grit size is inversely related to the diameter of the abrasive particles: the higher the grit size, the finer the particles. The long edges were then *chamfered*, which means grinding or polishing to remove the sharp edge where stress reaches a maximum and from which failure often occurs. One 4 × 40 mm face of each bar was then reground parallel to the long axis with a coarser diamond abrasive of either 80, 200 or 400 grit size. This face was then placed in tension in a 4-point bending test (Fig. 3.8) to measure the strength, which is shown in Fig. 5.45. The coarse-ground surface has less than half the strength of the fine-ground surface.

A second experiment demonstrates that removing the abrasion-damaged surface layer increases the strength of the silicon nitride bars.[69] The fine-ground bars were roughened on their tensile surface with an 80-grit diamond wheel. Then various depths of the rough surface were removed by lapping with 6-μm loose diamond abrasive. Figure 5.46 shows that lapping 40 μm off the coarse surface restores the ceramic to its original strength and that lesser depths of material removal have smaller effects.

An example of the effect of abrasive grit size on the strength of an optical ceramic is provided by an experiment with single-crystal magnesium fluoride.[70] Lapping with 12-μm diamond abrasive produced a strength of 137 ± 29 MPa. Lapping with 5-μm diamond increased the strength to 167 ± 32 MPa.

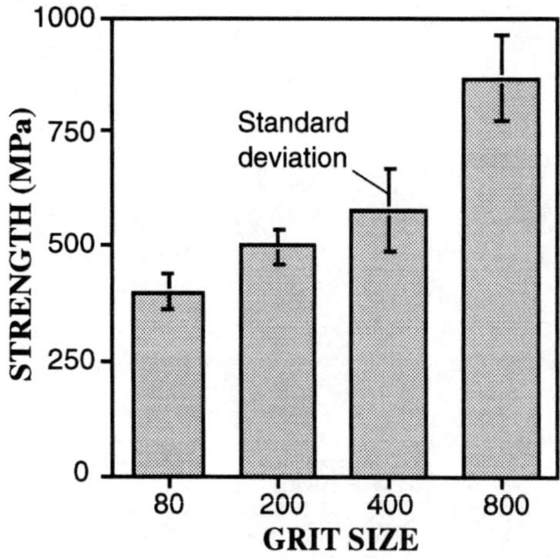

Fig. 5.45. Effect of grinding grit size on the strength of silicon nitride flexure bars.[69] Standard deviations are for 8 replicate tests.

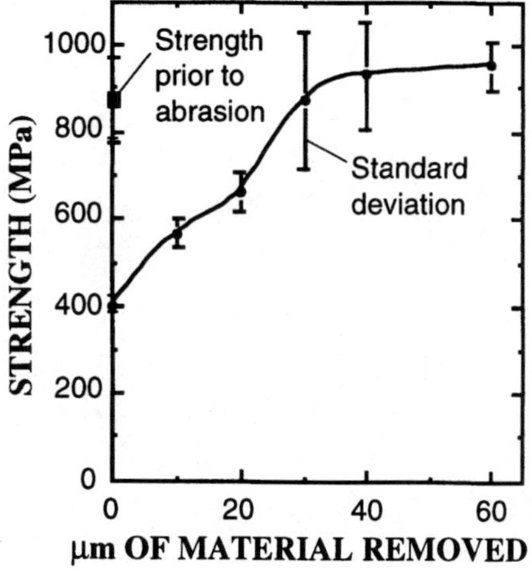

Fig. 5.46. Effect of depth of removal of grinding damage on the strength of silicon nitride flexure bars.[69]

The orientation of grinding damage relative to the tensile field is an important factor in determining mechanical strength.[71] Fig. 5.47 shows 3-point flexure bars that were ground either parallel or perpendicular to the tensile axis with a 325 grit diamond abrasive wheel. For polycrystalline magnesium fluoride (grain size <1 μm), the strength of bars ground perpendicular to the tensile axis was 61% of the strength of bars ground parallel to the tensile axis (53 ± 2 versus 87 ± 2 MPa). For single-crystal spinel, material ground perpendicular to the tensile axis was 74% as strong as material ground parallel to the axis (200 ± 10 versus 269 ± 7 MPa). Subsequent studies showed a grain size dependence to the effect of grinding direction.[72,73] The effect was largest for fine-grain material, then nearly zero at intermediate grain size, and grew large again for single crystals.

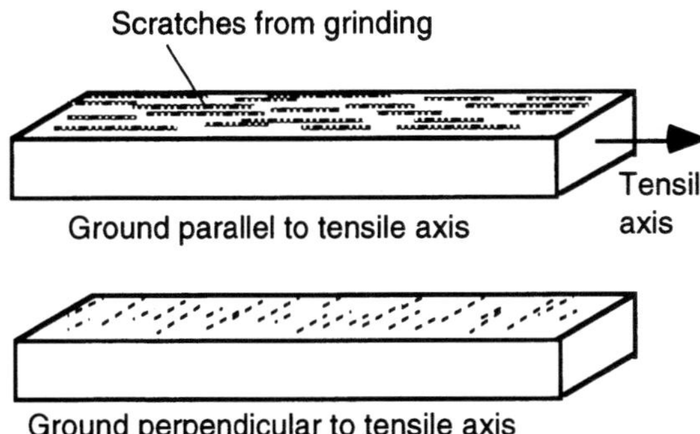

Fig. 5.47. Residual scratches from grinding perpendicular to the tensile axis lower the strength more than do scratches oriented parallel to the tensile axis.

Chemical etching is one way to remove grinding/polishing damage and strengthen ceramics. However, optical figure is difficult to maintain during chemical etching. In a study of garnet laser rods (such as yttrium aluminum garnet, YAG), conventional high-quality optical polishes left subsurface damage as deep as 50 μm below the surface.[74] Deep chemical etching with 85 wt% phosphoric acid at 200°C for 20 min increased the flexure strength of YAG from 175 to 2360 MPa. However, the sample was no longer optically flat. The optical figure could be restored by a fine polish after the etch, but two thirds of the strength of the etched surface was lost. In another study, 1-mm-diameter sapphire rods were chemically etched in molten borax at 965° or 1100°C for periods ranging from 10 to 30 min.[75] The bending strength increased from the as-received ground-finish value of 500 MPa to numbers in the range 1000-7000 MPa. The results were highly variable, with no correlation between strength and depth of material removed. Strength was probably limited by flaws inside the original material.

Flame polishing is another way to increase the strength of a ceramic that can be heated near its melting point without decomposition. In the case of 1-mm-diameter sapphire rods, flame polishing could increase the strength from 500 MPa up to as high as 7000 MPa.[75,76] Flame or laser polishing of large windows would be difficult because of thermal shock induced by localized heating.

Annealing has proven value in strengthening ceramics. In annealing, the material is slowly heated to a temperature at which atoms gain enough mobility to recrystallize and heal subsurface damage. The object is left at the annealing temperature for a specified period and then slowly cooled. Fig. 5.48 shows that annealing polished ruby (Cr^{3+}-doped sapphire) at 1200°C for 1 h increases the strength by more than a factor of 2 and that

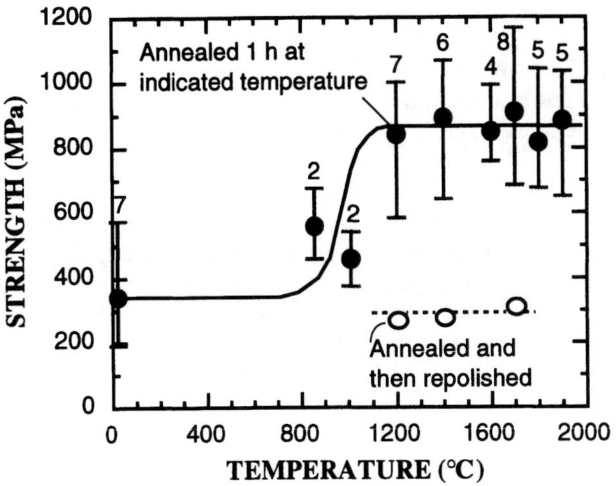

Fig. 5.48. Effect of annealing on the strength of polished ruby (Cr^{3+}-doped sapphire).[77] Numbers above error bars tell how many specimens were annealed at each temperature.

heating up to 1700°C gives no further improvement.[77] The open circles demonstrate that if the annealed ruby is repolished, the strength falls back to the initial value. In two other reports on sapphire, annealing at unspecified temperatures increased the strength from 528 ± 93 to 692 ± 97 MPa[78] and from 742 ± 158 to 1127 ± 407 MPa.[43] It is worthwhile to note that the stronger we make a material by surface treatments, the more susceptible it becomes to loss of strength by handling damage.

Because objects with different shapes are necessarily ground and polished in different manners, we cannot reliably predict the strength of curved parts from the strengths of flat coupons. Fig. 5.49 shows an example in which 25-mm-diameter lens-shaped specimens were core drilled from a yttria dome. The edges of the cores were polished to remove chips and then the ring-on-ring flexure strength of the cores was measured. Despite the care taken in edge preparation, three fourths of the cores failed at the edge, rather than inside the load ring. The mean strength of the cores that did not fail at the edge was 60% of the strength predicted from flat test coupons of the same material with an "equivalent" finish. The tentative conclusion is that yttria domes are weaker than yttria disks, which might be the result of the necessary differences in polishing domes and disks.

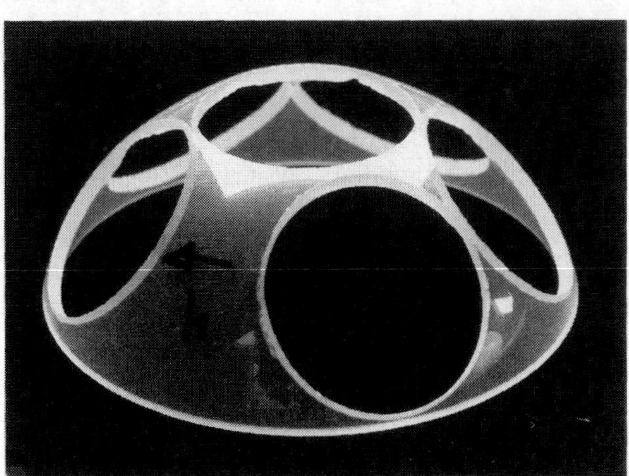

Fig. 5.49. Yttria dome (71-mm-diameter) from which six 25-mm-diameter disks were removed with a core drill.[79]

5.7 Polymer infrared windows

Raytheon manufactures a strong, durable, inexpensive, long wave infrared polyethylene window suitable for consumer products, such as night vision systems on cars and security devices for homes.[80] It is estimated that, in production, polyethylene windows could cost as little as 2% of the price of conventional windows such as silicon. The key to high strength is to begin with a woven yarn cloth of ultra-high-molecular-weight, highly crystalline polyethylene, as shown in Fig. 5.50. The cloth is then impregnated with a lower melting, amorphous polyethylene and hot pressed into a 0.1-mm-thick composite window that retains a skeleton of the strong cloth. The high tensile strength of crystalline polyethylene gives the window substantial in-plane strength. Some of the resulting material properties are listed in Table 5.6.

A critical requirement for outdoor use of a plastic material is resistance to degradation by ultraviolet sunlight. Without protection, most plastics are rapidly degraded. Figure 5.51 shows the loss of infrared transmission after an ultraviolet radiation exposure equivalent to 2 years in Miami, Florida. The figure also shows that by blending a

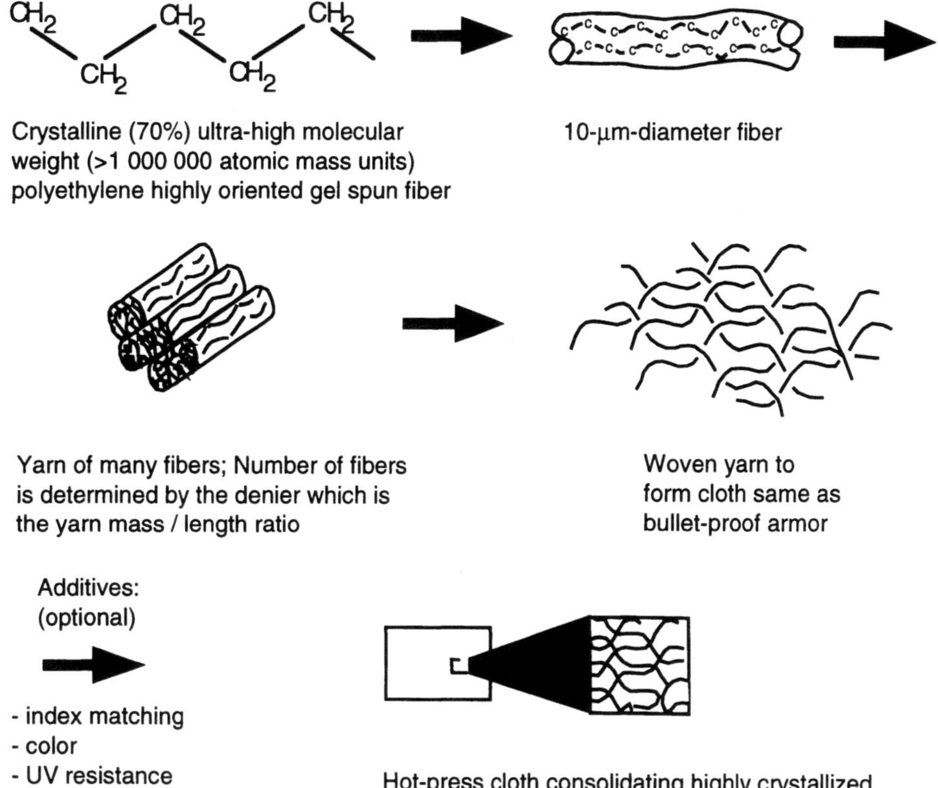
CH2
CH2
CH2
CH2
CH2
Crystalline (70%) ultra-high molecular weight (>1 000 000 atomic mass units) polyethylene highly oriented gel spun fiber
10-μm-diameter fiber
Yarn of many fibers; Number of fibers is determined by the denier which is the yarn mass / length ratio
Woven yarn to form cloth same as bullet-proof armor
Additives: (optional)
- index matching
- color
- UV resistance
Hot-press cloth consolidating highly crystallized and lower-melting-point amorphous polyethylene into a composite optical material while embossing moth eye antireflection layer onto surfaces

Fig. 5.50. Fabrication of Raytheon polyethylene for long wave infrared windows.[80]

Table 5.6. Propeties of Raytheon polyethylene window

Property	Value
Useful temperature range	-40 to +120°C
Refractive index	1.53
7.5-13 μm transmittance (with moth-eye antireflection surface)	84-90%
Optical scatter	~1%
Modulation transfer function @ 10 line pairs/mm	>95%
In-plane tensile strength	280-630 MPa
Young's modulus	>15 GPa

proprietary, anti-oxidation additive into the polyethylene, there is almost no loss of transmission in the ultraviolet exposure test.

Reflection at each surface of the low-refractive-index polyethylene window is ~4%. The loss can be reduced below 1% per surface by antireflection layers. This is conveniently achieved by warm pressing a "moth-eye" structure (described in Section 6.1.1) into each surface, for a net gain of ~7% in transmission in Fig. 5.52.

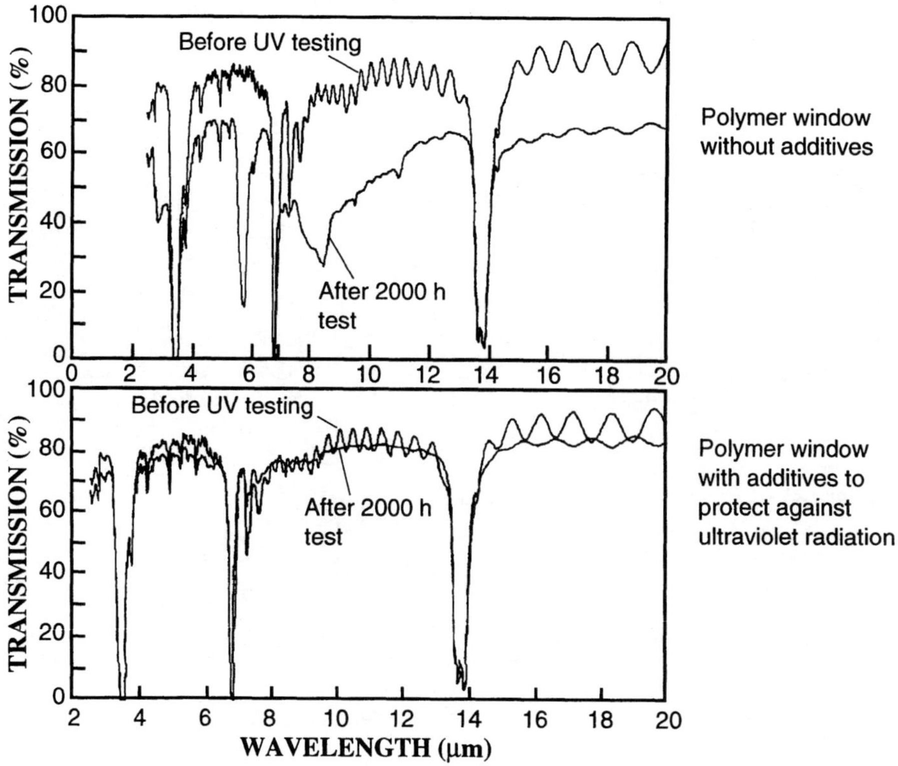

Fig. 5.51. Effect of additives on stability of polymer window in ultraviolet light.[80]

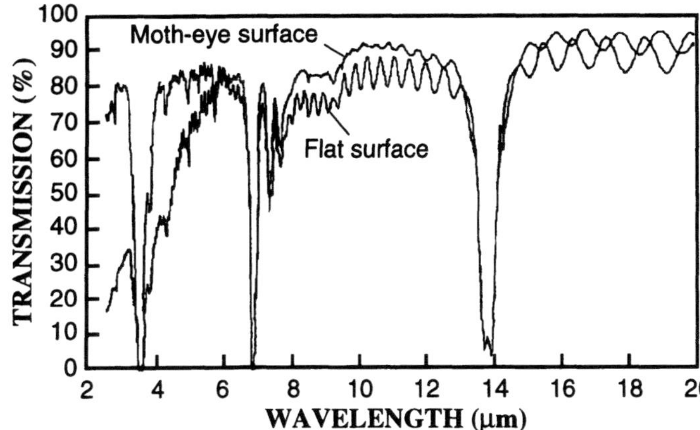

Fig. 5.52. Imprinting a moth-eye pattern in the surface of a polymer window increases transmission by 7%.[80]

References

1. M. Dejneka, R. E. Riman and E. Snitzer, "Sol-Gel Synthesis of High-Quality Heavy-Metal Fluoride Glasses," *J. Am. Ceram. Soc.*, **76**, 3147-3150 (1993).
2. J. Wasylak, "New Glasses of Shifted Absorption Edge in Infrared and Materials for Optics and Light Fiber Technique," *Opt. Eng.*, **36**, 1652-1656 (1997).
3. A. R. Hilton, Sr., A. R. Hilton, Jr., J. McCord and G. Whaley, "Production of Arsenic Trisulfide Glass," *Proc. SPIE*, **3060**, 335-343 (1997).
4. J. Lucas, X. H. Zhang, G. Fonteneau, H. L. Ma, C. Blanchetiere and J. L'Helgoualch, "Chalcogen Halide IR Glasses," *Proc. SPIE*, **1760**, 105-111 (1992).
5. M. K. Cinibulk, H.-J. Kleebe, and M. Rühle, "Quantitative Comparison of TEM Techniques for Determining Amorphous Intergranular Film Thickness," *J. Am. Ceram. Soc.*, **76**, 426-432 (1993).
6. P. W. McMillan, *Glass-Ceramics*, Academic Press, New York (1979).
7. H. G. Krolla and W. Semar, "Glass Ceramics for Optical Applications in the Visible and Infrared Regions with Low Coefficient of Thermal Expansion," *Proc. SPIE*, **2286**, 60-67 (1994).
8. S. S. Bayya, B. B. Harbison, J. S. Sanghera and I. D. Aggarwal, "Glass-Ceramization Technology for IR Window and Dome Applications," *Proc. 7th DoD Electromagnetic Windows Symp.*, pp. 53-60, Laurel, Maryland (1998).
9. H. Iba, T. Chang, Y. Kagawa, H. Minakuchi and K. Kanamaru, "Fabrication of Optically Transparent Short-Fiber-Reinforced Glass Matrix Composites," *J. Am. Ceram. Soc.*, **79**, 881-884 (1996); A. R. Boccaccini and P. A. Trusty, "Toughening and Strengthening of Glass by Al_2O_3 Platelets," *J. Mater. Sci. Lett.*, **15**, 60-63 (1996).
10. S. V. Phillips and G. Partridge, "Improved Rain-Erosion-Resistant Materials for the 3-5 μm Waveband," *Am. Ceram. Soc. Bull.*, **65**, 1490-1492 (1986).
11. R. L. Gentilman, "Current and Emerging Materials for 3-5 Micron IR Transmission," *Proc. SPIE*, **683**, 2-11 (1986).
12. T. M. Hartnett, S. D. Bernstein, E. A. Maguire and R. W. Tustison, "Optical Properties of ALON (Aluminum Oxynitride), *Proc. SPIE*, **3060**, 284-295 (1997).
13. T. Hartnett, M. Greenberg and R. L. Gentilman, "Optically Transparent Yttrium Oxide," U. S. Patent 4,761,390 (2 August 1988).
14. D. J. Sordelet, "Preparation and Sintering of Spherical, Monosized Y_2O_3 Particles," M.S. Thesis, Iowa State University, Ames, Iowa (1987).

15. W. H. Rhodes, "Controlled Transient Solid Second-Phase Sintering of Yttria," *J. Am. Ceram. Soc.*, **64**, 13-19 (1981).
16. W. H. Rhodes, G. C. Wei and E. A. Trickett, "Lanthana-Doped Yttria: A New Infrared Window Material," *Proc. SPIE*, **683**, 12-18 (1986).
17. W. H. Rhodes, E. A. Trickett and G. C. Wei, "Transparent Polycrystalline Lanthana-Doped Yttria," *Proc. SPIE*, **505**, 9-14 (1984).
18. G. C. Wei, M. R. Pascucci, E. A. Trickett, S. Natansohn and W. H. Rhodes "Enhancement in Aerothermal Shock Survivability of Lanthana-Strengthened Yttria Windows and Domes," *Proc. SPIE*, **1326**, 33-47 (1990).
19. J. M. Wahl and R. W. Tustison, "Optical, Mechanical and Water Drop Impact Characteristics of Polycrystalline GaP," *J. Mater. Sci.*, **29**, 5765-5772 (1994).
20. J. S. Goela and R. L. Taylor, "Monolithic Material Fabrication by Chemical Vapour Deposition," *J. Mater. Sci.*, **23**, 4331-4339 (1988).
21. J. A. Savage, K. L. Lewis, A. M. Pitt and R. H. L. Whitehorse, "The Role of a CVD Research Reactor in Studies of the Growth and Physical Properties of ZnS Infrared Optical Material," *Proc. SPIE*, **505**, 47-51 (1984).
22. D. Townsend and J. E. Field, "Fracture Toughness and Hardness of Zinc Sulfide as a Function of Grain Size," *J. Mater. Sci.*, **25**, 1347-1352 (1990).
23. J. A. Savage, *Infrared Optical Materials and Their Antireflection Coatings*, p. 124, Adam Hilger, Bristol (1985).
24. H. H. Li, "Refractive Index of ZnS, ZnSe and ZnTe and Its Wavelength and Temperature Derivatives," *J. Phys. Chem. Ref. Data*, **13**, 103-150 (1984).
25. K. L. Lewis, G. S. Arthur and S. A. Banyard, "Hydrogen-Related Defects in Vapour-Deposited Zinc Sulfide," *J. Crystal Growth*, **66**, 125-136 (1984).
26. S. P. Rummel, H. E. Reedy and G. L. Herrit, "Residual Stress Birefringence in ZnSe and Multispectral ZnS," *Proc. SPIE*, **2286**, 132-141 (1994).
27. L. A. Xue, D. S. Farquhar, T. W. Noh, A. J. Sievers and R. Raj, "Optical and Mechanical Properties of Zinc Sulfide Diamond Composites," *Acta Metall. Mater.*, **38**, 1743-1752 (1990).
28. C. R. Seward, C. S. J. Pickles, R. Marrah and J. E. Field, "Rain Erosion Data on Window and Dome Materials," *Proc. SPIE*, **1760**, 280-290 (1992).
29. Y. Han and M. Akinc, "Zinc Sulfide/Gallium Phosphide Composites by Chemical Vapor Transport," *J. Am. Ceram. Soc.*, **78**, 1834-1840 (1995).
30. Y. Kim, A. Zangvil, J. S. Goela and R. L. Taylor, "Microstructure Comparison of Transparent and Opaque CVD SiC," *J. Am. Ceram. Soc.*, **78**, 1571-1579 (1995).
31. J. S. Goela and R. L. Taylor, "Transparent SiC for Mid-IR Windows and Domes," *Proc. SPIE*, **2286**, 46-59 (1994).
32. S. K. Andersson and M. E. Thomas, "Infrared Properties of CVD β-SiC," *Infrared Phys. Technol.*, **39**, 223-234 (1998); *Proc. SPIE*, **3060**, 306-319 (1994).
33. R. A. Tanzilli and J. J. Gebhardt, "Optical Properties of Chemical Vapor Deposition (CVD) Silicon Nitride," *Proc. SPIE*, **297**, 59-64 (1981).
34. P. Klocek, L. E. Stone, M. W. Boucher and C. DeMilo "Semiconductor Infrared Optical Materials," *Proc. SPIE*, **929**, 65-78 (1988).
35. P. Klocek, M. W. Boucher, J. M. Trombetta and P. A. Trotta, "High Resistivity and Conductive Gallium Arsenide for IR Optical Applications," *Proc. SPIE*, **1760**, 74-85 (1992).
36. M. Azoulay, G. Gafni and M. Roth, "Recent Progress in the Growth and Characterization of Large Ge Single Crystals for IR Optics and Microelectronics," *Proc. SPIE*, **1535**, 35-45 (1991).
37. C. R. Poznich and J. C. Richter, "Silicon for Use as a Transmissive Material in the Far IR?" *Proc. SPIE*, **1760**, 112-120 (1992).
38. K. V. Ravi, "Diamond Technology for Infrared Seeker Windows," *J. Spacecraft and Rockets*, **30**, 79-86 (1993).

39. R. Facklenberg, "The Verneuil Process," in *Crystal Growth: Theory and Techniques*, (C. H. L. Goodman, ed.) Vol. 2, Plenum Press, New York (1978).
40. C. P. Khattak and F. Schmid, "Growth of Large-Diameter Crystals by HEM for Optical and Laser Applications," *Proc. SPIE*, **505**, 4-8 (1984).
41. C. P. Khattak, A. N. Scoville and F. Schmid, "Recent Developments in Sapphire Growth by Heat Exchanger Method (HEM)," *Proc. SPIE*, **683**, 32-35 (1986).
42. F. Schmid and C. P. Khattak, "Current Status of Sapphire Technology for Window and Dome Applications," *Proc. SPIE*, **1112**, 25-30 (1989).
43. F. Schmid, M. B. Smith and C. P. Khattak, "Current Status of Sapphire Dome Production," *Proc. SPIE*, **2286**, 2-15 (1994).
44. C. P. Khattak, F. Schmid and M. B. Smith, "Correlation of Sapphire Quality with Uniformity and Optical Properties," *Proc. SPIE*, **3060**, 250-257 (1997).
45. R. M. Sova, M. J. Linevsky, M. E. Thomas and F. F. Mark, "High Temperature Properties of Oxide Dome Materials," *Proc. SPIE*, **1760**, 27-40 (1992).
46. F. Schmid, C. P. Khattak and D. M. Felt, "Producing Large Sapphire for Optical Applications," *Am. Ceram. Soc. Bull.*, **73**, 39-44 (1994).
47. M. E. Thomas, W. J. Tropf and S. L. Gilbert, "Vacuum-Ultraviolet Characterization of Sapphire, ALON, and Spinel near the Band Gap," *Opt. Eng.*, 32, 1340-1343 (1993).
48. K. H. Lee and J. H. Crawford, Jr., "Additive Coloration of Sapphire," *Appl. Phys. Lett.*, **33**, 273-275 (1978); B. G. Draeger and G. P. Summers, "Defects in Unirradiated α-Al_2O_3," *Phys. Rev. B*, **19**, 1172-1177 (1979).
49. S. Biderman, A. Horowitz, Y. Einav, G. Ben-Amar, D. Gazit, A. Stern and M. Weiss, "Production of Sapphire Domes by the Growth of Near Net Shape Single Crystals," *Proc. SPIE*, **1535**, 27-34 (1991).
50. H. E. LaBelle, "EFG the Invention and Application to Sapphire Growth," *J. Crystal Growth*, **50**, 8-17 (1980).
51. H. E. LaBelle, J. Serafina and J. J. Fitzgibbon, "Recent Developments in Growth of Shaped Sapphire Crystals," *Proc. SPIE*, **683**, 36-40 (1986).
52. J. W. Locher, H. E. Bennett, C. P. Archibald and C. T. Newmyer, "Large Diameter Sapphire Dome: Fabrication and Characterization," *Proc. SPIE*, **1326**, 2-10 (1990).
53. J. W. Locher, H. E. Bates, W. C. Severn, B. G. Pazol and A. C. DeFranzo, "80-mm-EFG Sapphire Dome Blanks Yield High Quality, Low Cost Single Crystal Domes," *Proc. SPIE*, **1760**, 48-54 (1992).
54. V. N. Kurlov and S. N. Rossolenko, "Growth of Shaped Sapphire Crystals Using Automated Weight Control," *J. Crystal Growth*, **173**, 417-426 (1997).
55. E. R. Trumbauer, J. R. Hellmann, D. L. Shelleman and D. A. Koss, "Effect of Cleaning and Abrasion-Induced Damage on the Weibull Strength Distribution of Sapphire Fiber," *J. Am. Ceram. Soc.*, **77**, 2017-2024 (1994).
56. B. G. Pazol, R. DeVito, P. J. Giguere and P. S. Kiefner, "Development of Sapphire Windows for Use in High Quality Imaging Systems," *Proc. SPIE*, **1760**, 55-65 (1992).
57. J. Askinazi, "Large Aperture, Broadband Sapphire Windows for Common Aperture, Target Acquisition, Tracking and Surveillance Systems," *Proc. SPIE*, **3060**, 214-225 (1997).
58. P. F. Becher and R. W. Rice, "Strengthening Effects in Press Forged KCl," *J. Appl. Phys.*, **44**, 2915-2916 (1973).
59. J. M. Bennett and L. Mattson, *Introduction to Surface Roughness and Scattering*, Optical Society of America, Washington, D.C. (1989).
60. J. Bennett, V. Elings and K. Kjoller, "Recent Developments in Profiling Optical Surfaces," *Appl. Opt.*, **32**, 3442-3447 (1993).
61. J. M. Bennett, "Characterization of Surface Roughness" in *Characterization of Optical Materials* (G. J. Exarhos and L. E. Fitzpatrick, eds), pp. 9-26, Butterworth-

Heinemann, Boston (1993).

62. "Standard Test Method for Measuring the Effective Surface Roughness of Optical Components by Total Integrated Scattering," American Society for Testing and Materials (ASTM) Document F1048-87, Philadelphia (1987).
63. Military Specification MIL-0-13830A; American National Standard PH3-617, American National Standards Institute, New York (1980).
64. L. R. Baker and J. Singh, "Comparison of Visibility of Standard Scratches," *Proc. SPIE*, **525**, 64-68 (1985).
65. H. Johansen and G. Kästner, "Surface Quality and Laser-Damage Behavior of Chemo-Mechanically Polished CaF_2 Single Crystals Characterized by Scanning Electron Microscopy," *J. Mater. Sci.*, **33**, 3839-3848 (1998).
66. M. C. Gerchman and B. E. McLain, "An Investigation of the Effects of Diamond Machining on Germanium for Optical Applications," *Proc. SPIE*, **929**, 94-98 (1988); R. L. Rhorer and C. J. Evans, "Fabrication of Optics by Diamond Turning," in *Handbook of Optics* (M. Bass, E. W. van Stryland, D. R. Williams and W. L. Wolfe, eds.), Vol. I, Chap. 41, McGraw-Hill, New York (1995).
67. D. Black, R. Polvani, L. Braun, B. Hockey and G. White, "Detection of Sub-Surface Damage: Studies in Sapphire," *Proc. SPIE*, **3060**, 102-114 (1997).
68. P. Miles, "High Transparency Infrared Materials — A Technology Update," *Opt. Eng.*, **15**, 451-459 (1976).
69. S.-J. Cho, Y.-H. Huh, N. Kawashima, T. Kuroyama and T. Ogawa, "Effects of Grinding Conditions on Flexural Strength of Si_3N_4," *J. Mater. Sci. Lett.*, **14**, 1141 (1995).
70. R. W. Sparrow, H. Herzig, W. V. Medenica and M. J. Viens, "Influence of Processing Techniques on the VUV Transmittance and Mechanical Properties of Magnesium Fluoride Crystal," *Proc. SPIE*, **2286**, 33-45 (1994).
71. R. W. Rice, J. J. Mecholsky, Jr., and P. F. Becher, "The Effect of Grinding Direction on Flaw Character and Strength of Single Crystral and Polycrystalline Ceramics," *J. Mater. Sci.*, **16**, 853-862 (1981).
72. R. W. Rice, "Correlation of Machining-Grain Size Effects on Tensile Strength with Tensile Strength-Grain Size Behavior," *J. Am. Ceram. Soc.*, **76**, 1068-1070 (1993).
73. R. W. Rice, "Porosity Effects on Machining Direction-Strength Anisotropy and Failure Mechanisms," *J. Am. Ceram. Soc.*, **77**, 2232-2236 (1994).
74. J. Marion, "Strengthened Solid-State Laser Materials," *Appl. Phys. Lett.*, **47**, 694-696 (1985).
75. F. P. Mallinder and B. A. Proctor, "Preparation of High-Strength Sapphire Crystals," *Proc. Brit. Ceram. Soc.*, 9-16 (1966).
76. F. P. Mallinder and B. A. Proctor, "The Strengths of Flame-Polished Sapphire Crystals," *Phil. Mag.*, **B13**, 197-207 (1966).
77. L. M. Belyayev, ed., *Ruby and Sapphire*, p. 321, Nauk Publishers, Moscow (1974) in English translation by P. M. Rao, Amerind Publishing Co, New Delhi. Available from U.S. National Technical Information Service, Springfield, VA.
78. M. R. Borden and J. Askinazi, "Improving Sapphire Window Strength," *Proc. SPIE*, **3060**, 246-249 (1997).
79. D. C. Harris, G. A. Hayes, N. A. Jaeger, L. D. Sawyer, R. C. Scheri, M. E. Hills, K. R. Hayes, S. E. Homer, Y. L. Tsai and J. J. Mecholsky, Jr, "Mechanical Strength of Hemispheric Domes of Yttria and Lanthana-Doped Yttria," *J. Am. Ceram. Soc.*, **75**, 1247-1253 (1992).
80. W. A. Weimer and P. Klocek, "Advances in Low-Cost Long-Wave Infrared Polymer Windows," *Proc. SPIE*, **3705** (1999).
81. A. Horowitz, S. Biderman, G. Ben Amar, U. Laor, M. Weiss and A. Stern, "The Growth of Single Crystals of Optical Materials via the Gradient Solidification Method," *J. Crystal Growth*, **85**, 215-222 (1987).

Chapter 6

OPTICAL COATINGS

Thin coatings are widely used to improve the transmittance of a window by reducing reflection and to reject unwanted optical or radio frequency wavelengths.[1-3] Protective coatings endure rain and particle impacts and provide scratch resistance. Occasionally a thin coating can enhance mechanical strength. This chapter describes the use of coatings to reduce reflection and to reject radio and microwave frequencies. Coatings to protect against rain and sand erosion are discussed in the next chapter.

6.1 Antireflection coatings

Whenever light encounters an interface at which the refractive index changes, partial reflection occurs. In Table 1.3 we saw that approximately 4% reflection occurs at the interface between air and glass. Reflection from a zinc sulfide surface is 14%, while that from germanium is 36%. Since light entering almost any optical system encounters a series of interfaces, it is imperative that reflection be reduced at each interface; otherwise, the light will be attenuated to almost nothing.

Figure 6.1 illustrates the principle of operation of a quarter-wave antireflection coating. For simplicity, we describe the case of plane-polarized light whose electric field oscillates in the plane of the page. Consider light with wavelength λ_o traveling from air with index of refraction n_o (= 1) into a nonabsorbing antireflection coating of refractive index n_1, and then into a bulk optical material of refractive index n_2. The wavelength of light inside the coating is λ_o/n_1.

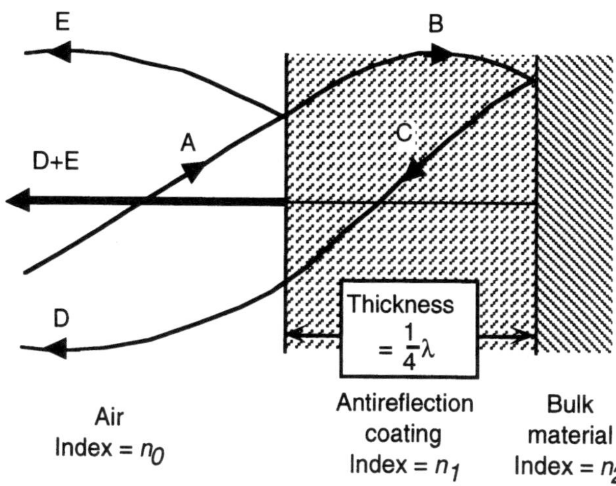

Fig. 6.1. Quarter-wave antireflection coating. It is assumed that the coating absorbs no light and that there is negligible reflection as light C exits the coating to become D.

Curve A in Fig. 6.1 represents the electric field of the incident light. When light strikes the first surface of the antireflection coating, some is reflected (E) and some enters the coating (B). When light wave B strikes the bulk material, some is reflected (C) and some is transmitted (not shown). Finally, light wave D exits the antireflection coating. (We will ignore the small part of C that is reflected back toward the right as C exits the coating.) The two waves that exit the coating back toward the source of the incident light are D and E. Their sum represents the total reflected light. *The antireflection coating is designed so that waves D and E are equal in magnitude but opposite in sign. Therefore they cancel and the net reflection is zero.*

For the coating to function, the intensity of light reflected from the air/coating interface must equal the intensity reflected from the coating/bulk interface. Equation (1-10) told us that the reflected power of normal-incidence light at each interface is

$$\text{Reflectance from outer surface} = \left(\frac{n_1 - n_o}{n_1 + n_o}\right)^2 \tag{6-1}$$

$$\text{Reflectance from inner surface} = \left(\frac{n_2 - n_1}{n_2 + n_1}\right)^2 . \tag{6-2}$$

Setting the two reflectances equal to each other allows us to solve for the refractive index of the antireflection coating*:

$$n_1 = \sqrt{n_o n_2} . \tag{6-3}$$

If the condition in Eq. (6-3) is true, then the intensity of light reflected at each surface will be the same. To achieve a 180° phase difference between the two reflected waves, the thickness of the antireflection coating should be one-quarter of the wavelength of light in the coating ($= {}^1/_4\, \lambda_o/n_1$). When these two conditions are met, the reflection of light of wavelength λ_o will be nearly zero. Such a simple coating is tuned to operate only near λ_o. Other wavelengths will not have zero reflection. Table 6.1 lists some candidate antireflection coating materials.

Example: Quarter-wave antireflection coating for ZnS. What should be the thickness and refractive index of an antireflection coating for ZnS with maximum transmission at 10 μm? The refractive index is given by Eq. (6-3), where n_2 is the refractive index of zinc sulfide (2.2) and n_o is the refractive index of air (1):

$$n_1 = \sqrt{(1)(2.2)} = 1.48 .$$

The thickness should be one-fourth of the wavelength of light in the coating:

$$\text{Thickness} = {}^1/_4\, \lambda_o/n_1 = {}^1/_4\, (10\ \mu\text{m}) / (1.48) = 1.69\ \mu\text{m} .$$

Table 6.1 tells us that SiO_2 or ThF_4 are candidate coating materials, because each has a refractive index near 1.48. Thorium compounds are radioactive and are being phased out.

---

*Closed-form equations exist for designing 1-, 2- and 3-layer antireflection coatings.[4-6]

Table 6.1. Potential antireflection coating materials*

Material	Infrared refractive index	Useful wavelength range (μm)
AlF_3	1.31	? - 11
Fluorocarbon polymer	1.35	? - 7
MgF_2	1.35	0.11 - 9
LaF_3	1.35	? - 12
HfF_4	1.36	? - 12
YF_4	1.37	? - 12
ZrF_4	1.40	? - 12
BaF_2	1.40	<0.2 - 14
SiO_2	1.44	0.2 - 4.5
ThF_4	1.5	0.26 - 12
NdF_3	1.58	? - ?
Al_2O_3	1.6-1.7	0.17 - 6.5
MgO	1.68	0.23 - 9
SiO	1.7	0.55 - 8
PbF_2	1.7	0.25 - 17
ThO_2	1.74	? - ?
Y_2O_3	1.9	0.3 - 12
ZrO_2	1.9	? - 13
HfO_2	1.95	? - 13
Diamondlike carbon (DLC)	1.9-2.6	? - >25
CeO_2	2.1	? - 14
AlN	2.1	? - 11
ZnS	2.2	0.35 - 14.5
Diamond	2.4	0.25 - >100
ZnSe	2.4	0.5 - 20
As_2S_3	2.41	0.6 - 13
SiC	2.67	0.5 - 10
As_2Se_3	2.79	0.8 - 18
GaP	2.90	0.6 - 12
GaAs	3.27	1 - 16
Si	3.44	1.1 - >100
Ge	4.10	1.8 - 23

*References 7 and 8 provide information on AlF_3, LaF_3, HfF_4, YF_4, ZrF_4 and fluorocarbon polymer as coating materials.

Figure 6.2 shows a midwave antireflection coating on silicon tuned for optimum transmission near 4.5 μm. Excellent broadband antireflection behavior can be achieved with multilayer designs.[1-3,9] Figure 6.3 shows a two-layer design used to reduce the reflection of a ZnSe CO_2 laser window below 0.06% at 10.6 μm.[10]

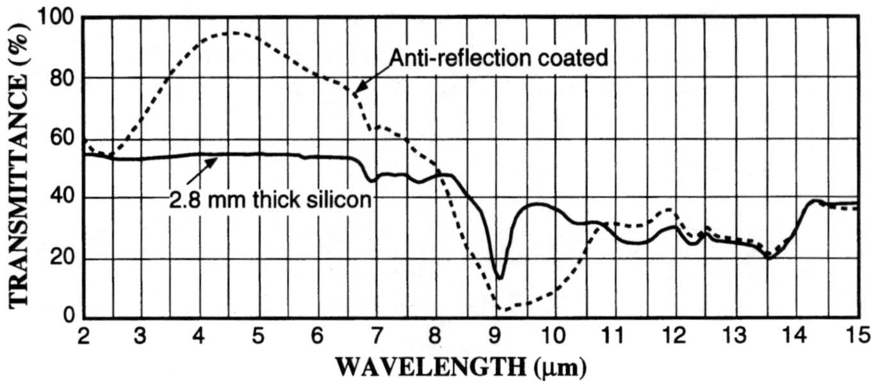

Fig. 6.2. Spectrum of silicon with and without a midwave antireflection coating.

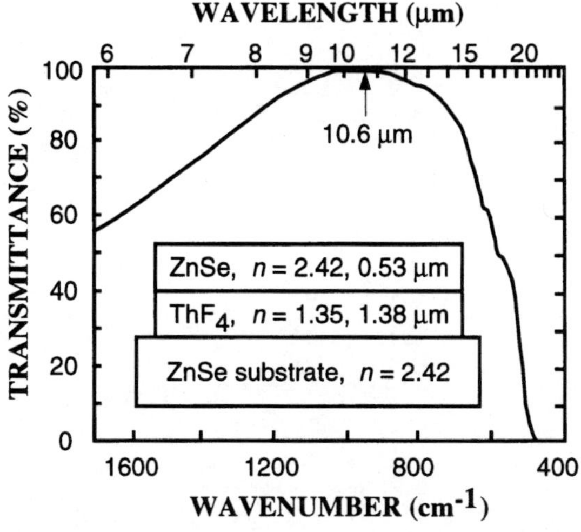

Fig. 6.3. Spectrum of ZnSe laser window with two-layer antireflection coating that reduces reflection to <0.06% at 10.6 μm.[10]

A coating with a continuously graded refractive index reduces reflection over a wide range of wavelengths and up to fairly high angles of incidence. The left side of Fig. 6.4 shows the profile of a coating with 16 layers that approximate a continuously graded refractive index.[11] Each layer was made by co-evaporating different proportions of ZnSe ($n = 2.4$) and ThF_4 ($n = 1.5$). The greater the proportion of ThF_4, the lower the refractive index of the layer. The right side of Fig. 6.4 shows the observed excellent broadband antireflection performance of the graded coating from visible through long wave infrared wavelengths. The infrared reflectance of this coating does not increase significantly until the angle of incidence exceeds 60°.

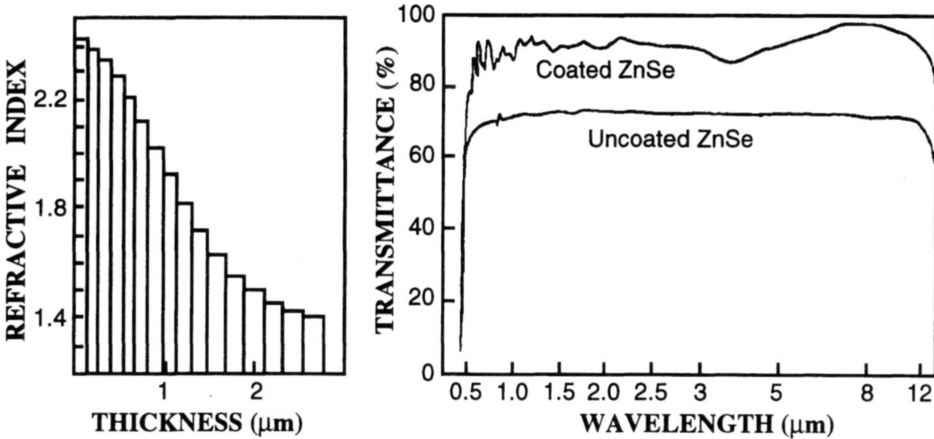

Fig. 6.4. *Left:* Profile of a multi-layer antireflection coating that approximates a continuously graded refractive index. *Right:* Measured performance of the multi-layer coating on a 2-mm-thick ZnSe window.[11]

6.1.1 Moth eye surfaces

Another way to reduce reflection from a surface is with a "moth eye" structure, which effectively grades the refractive index between that of air and the coating material. Figure 6.5 shows a moth eye surface on diamond.[12] Such surfaces on optical elements were discovered in moths[13] and have been designed into man-made structures to reduce reflection.[14] In Fig. 6.5, the multilayer structure with a flat diamond surface has a reflectance of about 18% at 10 μm, which is dominated by the single-surface reflectance from the front face of the outer diamond layer (15%). When the flat diamond outer surface is replaced by a moth eye structure, reflectance is reduced to 7% at the design wavelength of 10 μm.

The moth eye diamond surface was created by first etching a reverse moth eye structure into silicon by lithographic techniques. Then diamond was grown on the etched surface by chemical vapor deposition. When the silicon was dissolved, the remaining diamond had a moth eye structure. An antireflection surface on GaAs can be fabricated by etching a series of steps into the surface to simulate a moth eye structure.[16]

Moth eye structures are feasible for infrared wavelengths, because we can etch features with micron-size dimensions. It would be considerably more difficult to create features whose dimensions are tenths of a micron for visible wavelengths. For microwave radiation, with wavelengths of centimeters or millimeters, moth eye structures are employed on the walls of anechoic test chambers to reduce reflection. These chambers are used to measure radar reflections from test objects without complication by extraneous reflections from the walls of the chambers. It is also possible to put a conventional quarter-wave antireflection coating on microwave optical elements, as shown in Fig. 6.6. The principle of operation of this coating is the same as that of the infrared coating in Fig. 6.1.

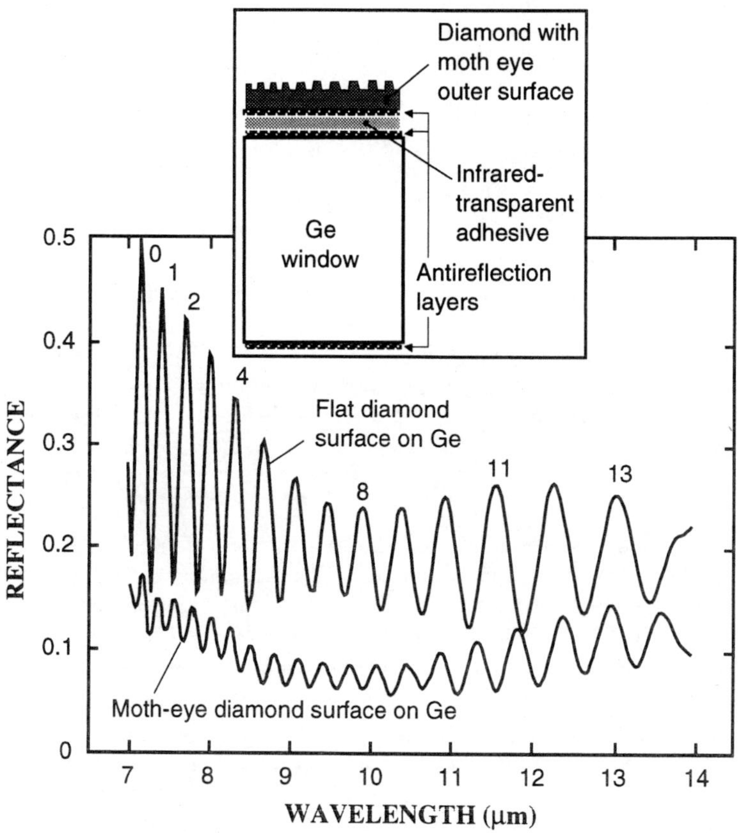

Fig. 6.5. *Upper:* Electron micrographs of moth eye surface on diamond film.[12,15] *Lower:* Reflectance spectrum of flat diamond coating on germanium and moth eye diamond coating on germanium. (Courtesy A. Harker, Rockwell Science Center.)

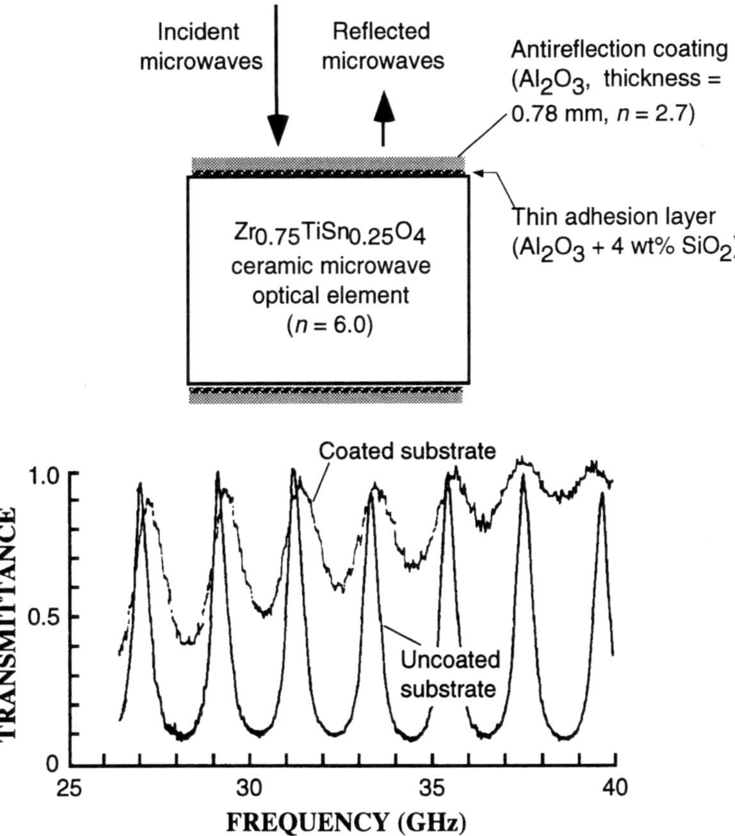

Fig. 6.6. Performance of an antireflection coating at microwave frequencies.[17] The refractive indexes (n) were measured at 72 GHz. The Al_2O_3 antireflection coating has an absorption coefficient of 0.093 cm^{-1} at 72 GHz.

6.1.2 Interference fringes for measuring coating thickness

Figures 6.5 and 6.6 both exhibit interference fringes arising from multiple reflections within the antireflection coatings, as shown in Fig. 6.7. Whenever the reflected and unreflected rays are in phase, maximum transmittance is observed. This phenomenon is the same as the etalon effect described in Section 1.3.2. Knowing the refractive index (n) of the coating, we can calculate its thickness from the period of the oscillations in the transmission or reflectance spectrum.

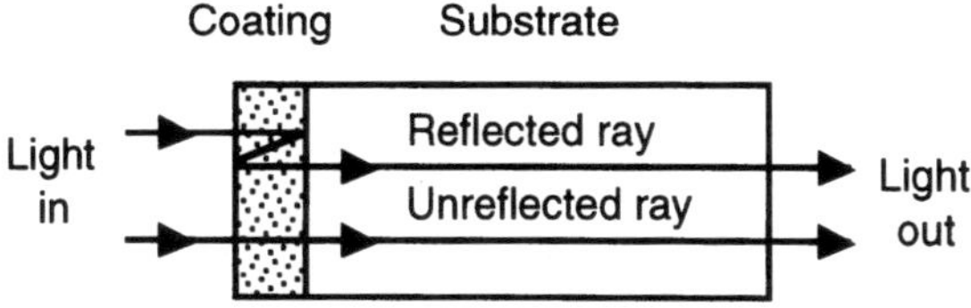

Fig. 6.7. Interference fringes: A maximum occurs when the reflected ray interferes constructively with the unreflected ray.

If N maxima occur between wavenumbers $\bar{\nu}_2$ and $\bar{\nu}_1$, then the thickness of the coating is*

$$\text{Thickness} = \frac{N}{2n}\left(\frac{1}{\bar{\nu}_2 - \bar{\nu}_1}\right) \quad \text{(using normal incidence light).} \qquad (6\text{-}4)$$

Example: Thickness of diamond coating. How thick is the flat diamond coating that gives a strong interference pattern in Fig. 6.5? The maximum labeled 0 is at 7.15 μm (1399 cm^{-1}) and maximum designated 13 is at 13.05 μm (766 cm^{-1}). Appendix C lists a refractive index of 2.38 for diamond in this region of the spectrum. Now we have enough information to use Eq. (6-4):

$$\text{Thickness} = \frac{13}{2(2.38)}\left(\frac{1}{1399\ \text{cm}^{-1} - 766\ \text{cm}^{-1}}\right) = 1.73 \times 10^{-3}\ \text{cm} = 17.3\ \mu\text{m}\ .$$

The diamond layer is calculated to be 17.3 μm thick.

Measurement of thickness by interference fringes can be done after coating an article, or in real time during deposition. Another way that coating thickness is controlled during deposition is with a quartz oscillator microbalance. This is a quartz crystal whose natural oscillation frequency changes in response to mass deposited on its surface. Such a crystal can be coated in a deposition chamber and used to calibrate the deposition rate as a function of power, pressure, *etc*., in the growth chamber.

6.1.3 Adherence of coatings

Rudimentary standards for adherence of coatings are given in such documents as U.S. Military Specification MIL-C-675C, which deals with magnesium fluoride antireflection coatings on glass. For example, the coating shall show no evidence of flaking, peeling, cracking or blistering after exposure to 95-100% humidity at 49°C (120°F) for 24 h, or after immersion in a sodium chloride solution (45 g/L) at room temperature for 24 h. It must also withstand 24 h of a salt spray fog test. An abrasion test requires that the coating not be damaged when a standard pencil eraser is rubbed across the surface for 20 cycles (40 strokes) with an applied load of 10 N. The standard test of adherence is that the coating remain attached to its substrate when a piece of cellophane tape is pressed firmly to the coating and then pulled off quickly at an angle normal to the surface. Endurance of other optical coatings for external use may be specified in terms of surviving windshield wiping with an abrasive solution, impact by raindrops or impact by sand particles without exceeding an allowed transmission loss.

6.1.4 Emittance from coatings

Antireflection coatings are usually thin enough so that they do not absorb very much light. Nonetheless, Fig. 2.14 showed that just a few percent emissivity in the midwave region could have a significant effect on the signal-to-noise ratio from an infrared sensor at elevated temperature.

---

*An analysis that is more complex than Eq. (6-4) uses the fringe pattern to find the refractive index, absorption coefficient, thickness and surface roughness of the coating.[18]

Little published data exist on emittance from antireflection coatings. Fig. 6.8 shows the absorptance of a sapphire window coated on both sides with an unspecified antireflection coating. Just to refresh your memory, *absorptance* is the fraction of incident light that is absorbed. Emittance will be equal to the absorptance. The broad band near 2.9 μm is attributed to water loosely bound to the coating. Baking the window at 300°C in vacuum removes the absorption band, but it reappears if the window is kept in humid air. The three sharp features near 3.4 μm are attributed to hydrocarbon impurity in the coating. These features are not lost upon heating in vacuum. The net result is that this particular coating adds about 1% emittance to the sapphire window in the midwave infrared region.

Antireflection coatings are usually just a few microns thick. Erosion protection coatings discussed in the next chapter are typically tens of microns thick and can therefore be a significant source of emittance.

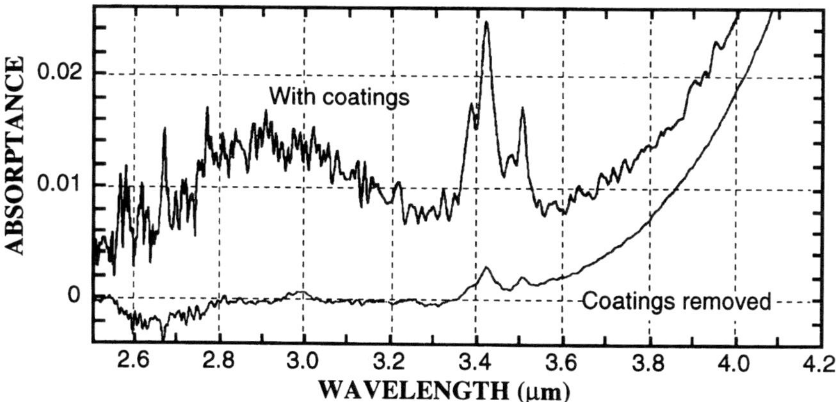

Fig. 6.8. Absorptance of coated and uncoated sapphire at 20°C. (Courtesy L. Hanssen and S. Kaplan, National Institute of Standards and Technology.)

6.1.5 Rugate filters

A *Rugate filter* is an optical filter with a sinusoidally varying refractive index designed to transmit most frequencies but to block (reflect) one or more narrow regions of the spectrum. A Rugate filter could be used to permit a broad band of wavelengths to enter a detector while filtering out a few strong laser wavelengths that could incapacitate the detector.

The sinusoidal refractive index is created by depositing a coating with varying proportions of two target materials, such as ZnS and ZnSe, by laser flash evaporation under exquisite computer control. As an example, Fig. 6.9 shows the design of a filter with 50 cycles of coating intended to reject radiation at 1.05 μm. Each cycle of the sine wave is composed of 32 layers of different composition, giving 1600 distinct layers in the coating, which is only 11.1 μm thick.

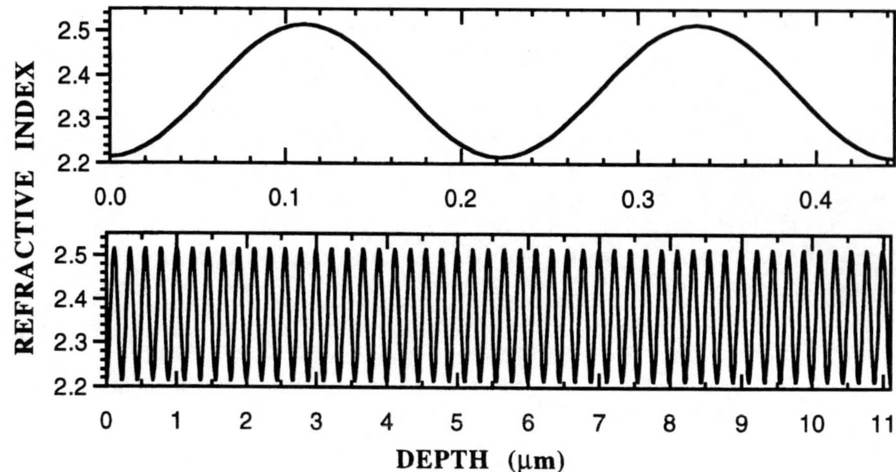

Fig. 6.9. Coating design for rejection of 1.05-μm radiation by a Rugate filter. Each of the 50 cycles of refractive index variation in this coating is composed of 32 individual layers.

To reject the wavelength λ_{rej}, each cycle of the coating should have a thickness of $t = \lambda_{rej}/(2n_{av})$, where n_{av} is the average refractive index of the coating. The refractive index of the coating will be given by $n(z) = n_{av} - \Delta n \cos(2\pi z/t)$, where z is the depth in the coating and Δn is the maximum index excursion from the mean. In Fig. 6.9 the horizontal axis is z, $n_{av} = 2.365$, $\Delta n = 0.15$, and $t = (1.05\ \mu m)/(2*2.365) = 0.222\ \mu m$. The 50-cycle structure has a thickness of $50t = 11.1\ \mu m$.

The predicted and actual performance of the Rugate filter are shown in Fig. 6.10.[20] If you want to reject several narrow wavelengths, you would need to superimpose sinusoidal oscillations of the refractive index for each of the rejection wavelengths.

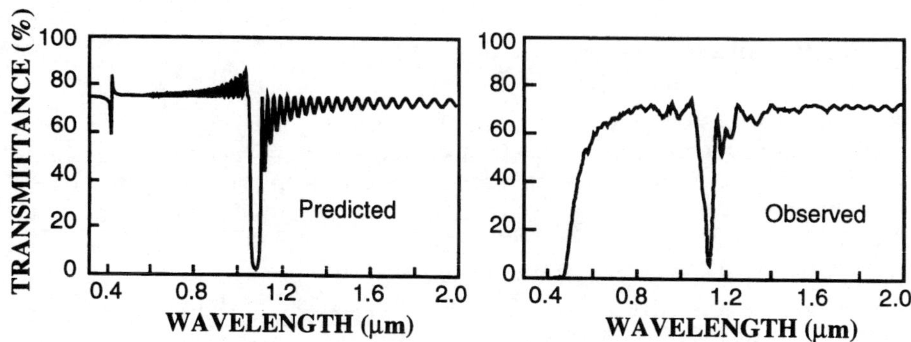

Fig. 6.10. Performance of the Rugate filter whose design is shown in Fig. 6.9.[20] The layers were deposited on a ZnS substrate. The performance was similar when the same layers were deposited on KCl or polycarbonate substrates.

6.2 Stress in coatings

Most coatings are deposited with as much as 100-1000 MPa of intrinsic stress that may be tensile or compressive. Stress arises from such factors as lattice mismatch between coating and substrate, thermal expansion mismatch, growth morphology and glass formation due to rapid cooling. Deliberate effort is required to minimize the stress in a coating by altering growth conditions. Table 6.2 shows the measured stress in some quarter-wave antireflection coatings evaporated onto fused silica.[21] Zinc sulfide is highly compressive, PbF_2 is almost unstressed, and the other coatings are in tension.

Table 6.2. Stress in antireflection coatings for 0.63 μm wavelength on fused silica[21]

Material	Thickness (nm)	Stress (MPa)	Stress type
ZnS	68	180	Compressive
PbF_2	91	5	Compressive
Na_3AlF_6 (Cryolite)	117	25	Tensile
$ThOF_2$	109	160	Tensile
MgF_2	115	340	Tensile

One way to create compression or tension is to deposit a coating with different thermal expansion from the substrate. Suppose a coating of low thermal expansion is deposited at 400°C on a substrate of high thermal expansion. As the coated sample cools, the substrate contracts more than the coating. If the coating adheres to the substrate, the contracting substrate compresses the coating, which therefore cools with compressive stress. The substrate, in turn, is in a tensile state induced by the coating. The right side of Fig. 6.11 is a greatly exaggerated view of this state of affairs. If the situation had been reversed and the coating had a higher thermal expansion than the substrate, the coating would end up in tension, as on the left of Fig. 6.11. The mean stress in the substrate in both cases is much less than the mean stress in the coating because the substrate is much thicker than the coating.

We can calculate the stress in an isotropic film due to its thermal expansion mismatch with an isotropic substrate.[22-24] Suppose that the film is deposited at

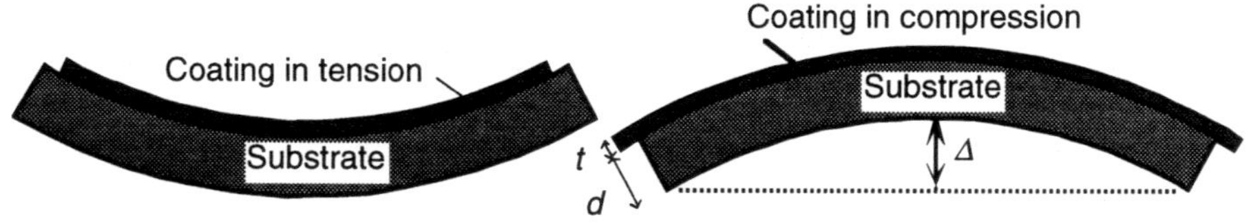

Fig. 6.11. Exaggerated view of curvature resulting from stress in thin coating.

temperature T in a stress-free condition. Suppose also that the film adheres strongly to the substrate so it is locked into position at the interface as the two cool down. When the substrate cools to temperature T_o, it shrinks by the amount $\alpha_s \Delta T$, where α_s is the expansion coefficient of the substrate and $\Delta T = T - T_o$. If it were not attached to the substrate, the film would shrink by the amount $\alpha_f \Delta T$, where α_f is the expansion coefficient of the film. In fact, because of its strong adherence to the much thicker substrate, the film shrinks by the same amount as the substrate, which is $\alpha_s \Delta T$. The strain in the film will be $\varepsilon_f = (\alpha_f - \alpha_s)\Delta T = \Delta\alpha\Delta T$. The stress in the film is equal to the strain times an appropriate modulus, which is the *biaxial modulus* given by $E_f/(1 - \nu_f)$, where E_f is Young's modulus for the film and ν_f is Poisson's ratio for the film. The net result is that the stress in the film, σ_f, due to its thermal expansion mismatch with the substrate is

$$\sigma_f = \frac{E_f \Delta\alpha\Delta T}{(1 - \nu_f)} \quad . \tag{6-5}$$

Example: Expansion mismatch stress. Suppose that an isotropic coating with a thermal expansion coefficient of 6.0×10^{-6} K^{-1} is deposited at 400°C on a substrate with an expansion coefficient of 5.0×10^{-6} K^{-1}. For the coating, Young's modulus is 100 GPa and Poisson's ratio is 0.25. If the only stress in the coating arises from the different expansion coefficients, what will be the stress at 25°C? Putting values into Eq. (6-5) gives the stress:

$$\sigma_f = \frac{(100 \text{ GPa})(1.0 \times 10^{-6} \text{ K}^{-1})(375 \text{ K})}{(1 - 0.25)} = 50 \text{ MPa} \quad .$$

The thin coating will have a stress of 50 MPa when it cools to 25°C.

Is the coating in tension or compression? The substrate has a smaller expansion coefficient than the coating, so the substrate contracts less than the coating when the part cools down. Therefore the stress in the coating is tensile, because the substrate is trying to stretch the coating to larger dimensions when they cool. If the expansion coefficient of the coating were smaller than the expansion coefficient of the substrate, the coating would be in compression when the two cool down.

In general, there can be other large stresses besides thermal expansion mismatch stress in a coating. It is not easy to predict these other stresses; their magnitude and sign are dependent upon film deposition conditions.

We can calculate the deflection of the substrate, Δ, in Fig. 6.11 from the stress in the film. If the circular substrate disk has Poisson's ratio ν_s, Young's modulus E_s, diameter D and thickness d, then the deflection is[21]

$$\Delta = \frac{3\,(1 - \nu_s)\, t\, D^2\, \sigma_f}{4\, E_s d^2} \tag{6-6}$$

where t is the thickness of the coating and σ_f is the stress in the coating.

Example: Deflection of a coated disk. Let's calculate the deflection at the center of a ZnSe disk with a thin compressive coating of ZnS. Suppose that the disk diameter is 50 mm and its thickness is 2.0 mm. Let the coating thickness be 20 μm and suppose that it has the same stress as the coating in Table 6.2 (180 MPa). With $E_s = 70$ GPa and $\nu_s = 0.28$ for ZnSe, we use Eq. (6-6) to find the deflection, Δ, in Fig. 6.11:

$$\Delta = \frac{3\,(1 - 0.28)\,(20 \times 10^{-6}\ \text{m})\,(0.0050\ \text{m})^2\,(180 \times 10^6\ \text{N/m}^2)}{4\,(70 \times 10^9\ \text{N/m}^2)\,(0.0020\ \text{m})^2} = 0.17\ \mu\text{m}\ .$$

The deflection of this window by this coating is negligible for long wave infrared applications. Thicker or more stressed coatings on windows with a smaller thickness-to-diameter ratio can cause significant optical distortion, especially at visible wavelengths.

Coatings can have significant effects on the mechanical properties of windows. Section 7.8.1 cites examples in which coatings increase flexure strength. Table 6.3 gives examples in which 1-μm-thick coatings of silicon nitride or alumina increased the indentation fracture toughness of a variety of substrates.[25,26] The compressive stress in the silicon nitride was approximately 1.5 GPa. In addition to increasing strength and toughness, hardness can also be increased by compressive stresses induced by ion implantation at the surface of a material.[27-31]

Table 6.3. Fracture toughness of windows coated with 1 μm of Si_3N_4 or Al_2O_3[27,28]

Substrate	Fracture toughness - uncoated ($\text{MPa}\sqrt{\text{m}}$)	Fracture toughness - coated Si_3N_4	Al_2O_3
Glass	0.62	1.1	0.77
Silicon	0.74	1.4	1.0
Sapphire	1.7	3.4	3.2
Germanium	0.43	1.3	–
ALON	1.43	–	1.9
Spinel	1.1	–	1.7

6.3 Conductive coatings for electromagnetic shielding

Very thin electrically conductive coatings can be transparent at visible and infrared frequencies, but opaque to microwaves and radio waves. Such coatings are used on domes and windows to shield sensitive optical and infrared detectors against harmful electromagnetic interference.[32] Common approaches to shielding include coating the optical window with an electrically conductive layer, covering the window with a metallic mesh, or doping the bulk window material to make it conductive.

Electrical *resistivity*, ρ, is a property of bulk material that relates the electrical resistance to the geometry of a specimen. A sample with cross sectional area A and length L (Fig. 6.12, *left*) will have a resistance, R, given by

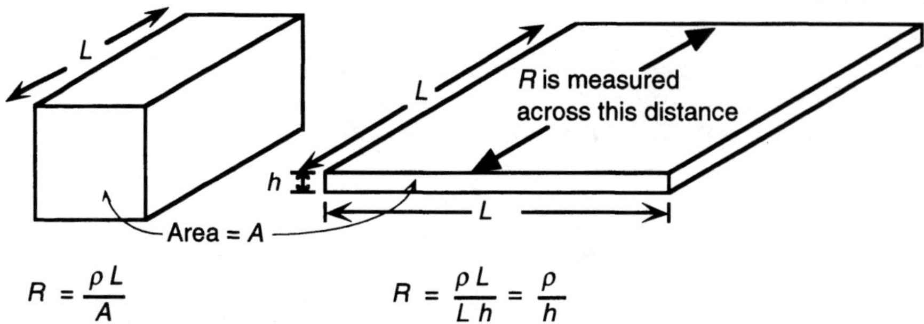

Fig. 6.12. Geometries for resistance of bulk material (*left*) and thin film (*right*).

$$\text{Resistance} = R = \frac{\rho L}{A}. \tag{6-7}$$

The greater the length and the smaller the cross section, the greater the resistance. Units of resistance are ohms (Ω) and units of resistivity are Ω·m. Now consider the thin, square sheet of conductive coating in Fig. 6.12 (*right*) with thickness h and sides of length L. Resistance measured across the length is called the *sheet resistance* or *surface resistance*:

$$\text{Sheet resistance} = R_S = \frac{\rho \times \text{length}}{\text{area}} = \frac{\rho L}{L h} = \frac{\rho}{h}. \tag{6-8}$$

The resistance across a square section of the thin layer is constant (ρ/h), regardless of the size of the square. Sheet resistance has units of ohms, but we usually call it "ohms per square" (Ω/□) to recognize it as sheet resistance.

Now consider the thin metallic film with thickness h and sheet resistance R_S on the nonconductive infrared window with thickness d and dielectric constant ε in Fig. 6.13. (The values of R_S and ε must apply at the frequency of radiation being shielded. In general, the radio frequency dielectric constant of a window is different from the infrared dielectric constant.) The *shielding effectiveness*, *SE*, is a logarithmic measure of the ability of the coating to prevent penetration of the radio frequency field. If the incident radiant power is P_o and the transmitted power is P, the transmittance is

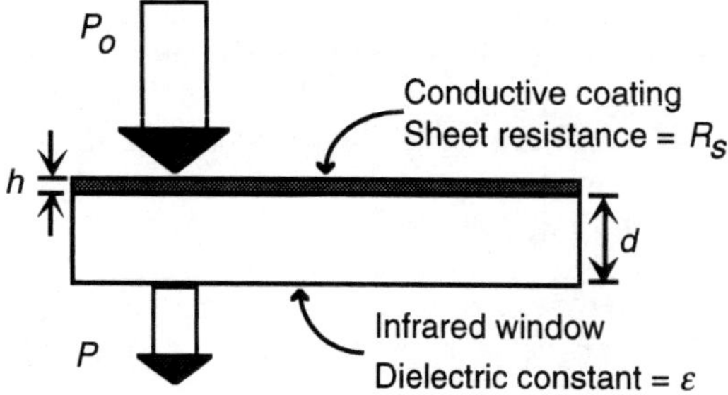

Fig. 6.13. Thin conductive coating on a transparent window.

$$\text{Transmittance} = \frac{P}{P_o} = 10^{-SE/10} . \tag{6-9}$$

Equation (6-9) says that a shielding effectiveness of 10 reduces transmittance to 10% and $SE = 30$ reduces transmittance to 0.1%. We say that the dimensionless shielding effectiveness has units of *decibels*, dB. In general, shielding arises from a combination of reflection and absorption of the incident radiation.

The shielding effectiveness for perpendicular incidence of electromagnetic radiation lies between the limits in Eqns. (6-10a) and (6-10b):[33,34]

$$SE\,(\lambda/2) = 20 \log \left(1 + \frac{188.5}{R_S}\right) \tag{6-10a}$$

$$SE\,(\lambda/4) = 20 \log \left(\frac{1+\varepsilon}{2\sqrt{\varepsilon}} + \frac{188.5}{R_S\sqrt{\varepsilon}}\right) . \tag{6-10b}$$

The limit in Eq. (6-10a) applies to a window thickness d (Fig. 6.13) of $\lambda/2$, λ, $3\lambda/2$, 2λ *etc*., where λ is the wavelength of incident radiation. The limit in Eq. (6-10b) applies to window thicknesses of $\lambda/4$, $3\lambda/4$, $5\lambda/4$, $7\lambda/4$, *etc*. A further condition for Eqns. (6-10) to be valid is that the coating thickness h must be small compared to the skin depth.

Skin depth, δ, is a measure of the attenuation of electromagnetic radiation in a conductor. The oscillating electric field decreases by the factor $e^{-z/\delta}$ when it penetrates a distance z into the conductor. You might recognize that skin depth is proportional to the reciprocal of the absorption coefficient for electromagnetic radiation. For a coating with sheet resistance R_S and thickness h, the skin depth is

$$\text{Skin depth} = \delta = \sqrt{\frac{2\,R_S\,h}{\omega\,\mu}} \tag{6-11}$$

where ω is the angular frequency of the radiation (= $2\pi \times$ frequency) and μ is the magnetic permeability of the coating. For nonmagnetic materials, μ equals the permeability of free space, which is $4\pi \times 10^{-7}$ henry/m. (The quantity $R_S\,h$ is the resistivity of the coating material, which is also the inverse of the conductivity of the coating material.) For example, for $R_S = 10\ \Omega/\square$ and $h = 10\ \mu$m, the skin depth for 100 MHz radiation is

$$\delta = \sqrt{\frac{2\,(10\ \Omega/\square)\,(10 \times 10^{-6}\ \text{m})}{(2\pi \times 10^{8}\ \text{s}^{-1})\,(4\pi \times 10^{-7}\ \text{henry/m})}} = 5.0 \times 10^{-4}\ \text{m} . \tag{6-12}$$

Henrys and ohms are compatible with SI units, so the answer comes out in meters.

Both parts of Eq. (6-10) are plotted in Fig. 6.14 for an infrared window with a dielectric constant of 4, which is representative of the dielectric constant of glass or silica. Experimental data from three sources are also superimposed on the graph.[33,35,36] We see that the radio frequency field is strongly attenuated by a thin coating with a low sheet resistance. For example, for $R_S = 10\ \Omega/\square$, the shielding effectiveness lies between 20.6 and 26.0 dB, and the radio frequency transmission lies between $10^{-20.6/10} = 0.87\%$ and $10^{-26.0/10} = 0.25\%$. *The greater the conductivity of the coating, the lower the sheet resistance and the greater the shielding effectiveness.*

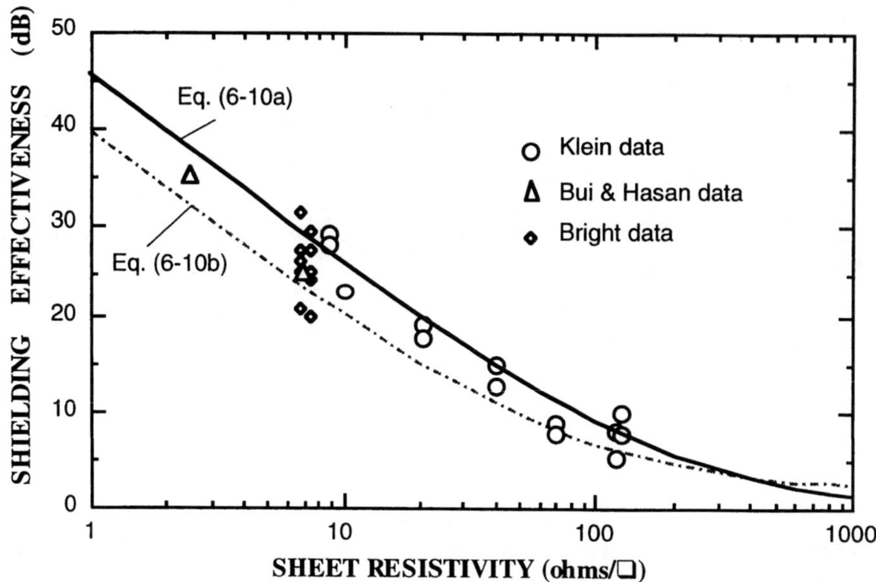

Fig. 6.14. Shielding effectiveness of a thin conductive coating on an infrared window of dielectric constant $\varepsilon = 4$, which is representative of a glass or silica substrate. Points show experimental data.[33,35,36]

The problem with some coatings that are too conductive is that they absorb the infrared or visible radiation that they are meant to pass. Table 6.4 gives the performance of a commercial conductive coating used for plastics that must shield against radio frequencies while passing visible light. As the surface resistance of the coating decreases from 20 to 2 Ω/❑, the visible transmission decreases from 81 to 40%. Note also that the shielding effectiveness decreases as frequency increases.

Another set of commercial coatings for ZnS or sapphire from Rafael in Israel is reported to provide >75% average transmission in the 3.5-5 μm range for a sheet resistance of 30 ± 5 Ω/❑. A coating on ZnS provides >70% average transmission in the 8-11.5 μm region for a sheet resistance of 15 ± 2 Ω/❑.

Table 6.4. Performance of an electromagnetic shielding coating for plastics*

Sheet resistance (Ω/❑)	Visible transmittance (%)	Frequency (MHz)	Attenuation (dB) for 4 Ω/❑ coating
2	40	0.01	146
3	55	0.1	128
4	62	1	108
6-8	72	10	88
10-12	78	100	65
14-20	81	1000	48

*OCC-20 coating sold by Optical and Conductive Coatings, Pacheco, California.

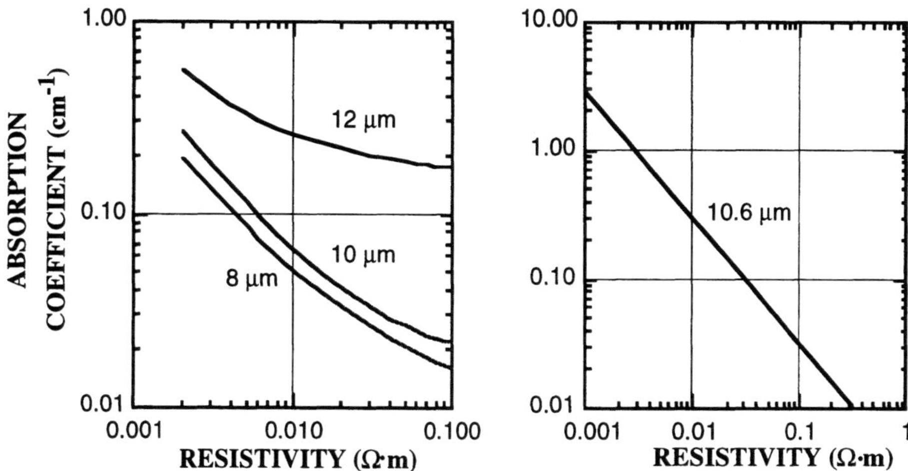

Fig. 6.15. *Left:* Effect of bulk resistivity of Ge on infrared absorption at wavelengths of 8, 10 and 12 μm. (Data from Eagle Picher Electro-Optic Materials Catalog.) *Right:* Absorption coefficient of *n*-type GaAs from Amorphous Materials, Inc. and other sources at 10.6 μm as a function of resistivity.[39]

An approach to electromagnetic shielding by infrared-transparent semiconductors is to dope the material to increase its conductivity. Germanium aircraft windows doped with antimony have low resistivity and provide good shielding. Doping increases the infrared absorption at 300 K (Fig. 6.15), but may actually decrease the infrared absorption at elevated temperature (400 K).[37,38] In Ge, *n*-type doping leads to less infrared absorption than *p*-type doping.[37] Figure 6.15 also shows the effect of doping on the absorption coefficient of *n*-type GaAs.[39] Conductive GaAs produced by Raytheon Systems Co. provides 60 dB of shielding in the 10-1000 GHz range and 30 dB in the 0.1-10 GHz range for thicknesses on the order of 1 cm. The absorption coefficient near 10 μm is 0.02 cm^{-1} and the material has low infrared absorption up to at least 200°C. Figure 6.16 shows the frequency dependence of electromagnetic attenuation by doped GaAs.

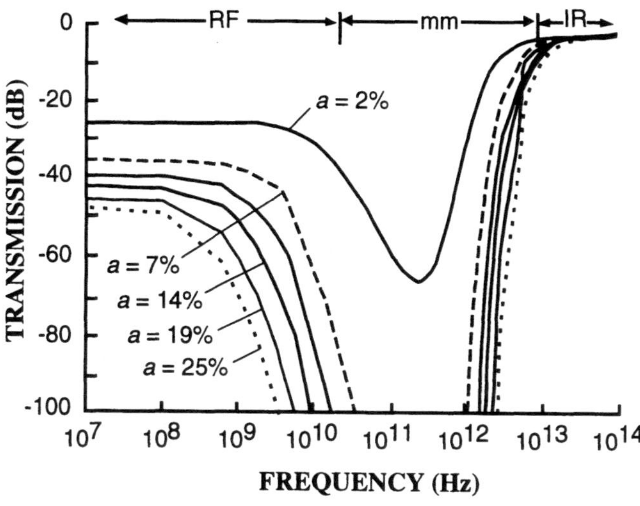

Fig. 6.16. Frequency dependence of electromagnetic shielding by doped GaAs with a thickness of 6.4 mm.[40] Absorptance labels indicate the fraction of 8-12 μm infrared radiation absorbed by the GaAs. The more conductive the material, the greater the microwave attenuation and the higher the absorption of infrared radiation.

Still another way to achieve electromagnetic shielding while retaining infrared transparency is to cover a window or dome with a conductive metal mesh created by lithographic techniques. A relatively coarse mesh blocks radio waves but allows infrared radiation to pass through the openings between the wires. The thickness and width of the wires and the size of the openings determine the electromagnetic shielding characteristics. A given design has a wavelength at which it provides optimum shielding. If the metal covers 20% of the area, the maximum usable infrared energy transmitted through a coarse mesh is 80%. In general, diffraction effects lower the fraction of usable infrared energy below the fraction of open area.

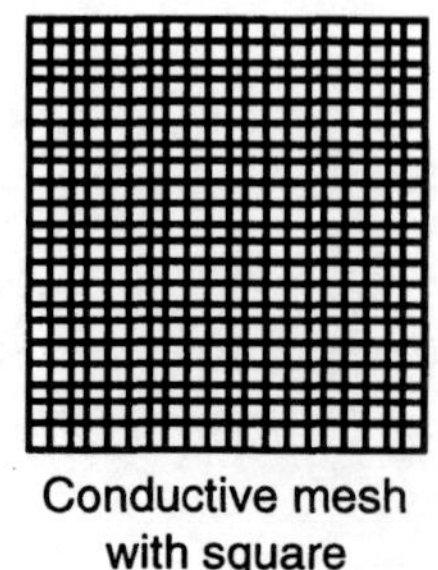
Conductive mesh with square openings

Undesirable features of meshes include blockage of the infrared signal, reflection of stray light by the mesh, diffraction, and poor rain and sand erosion resistance. The erosion problem has been addressed by burying the mesh in the window as illustrated in Fig. 6.17.[41] The Al_2O_3 overcoat is quite durable and the mesh is no longer exposed to an erosive environment.

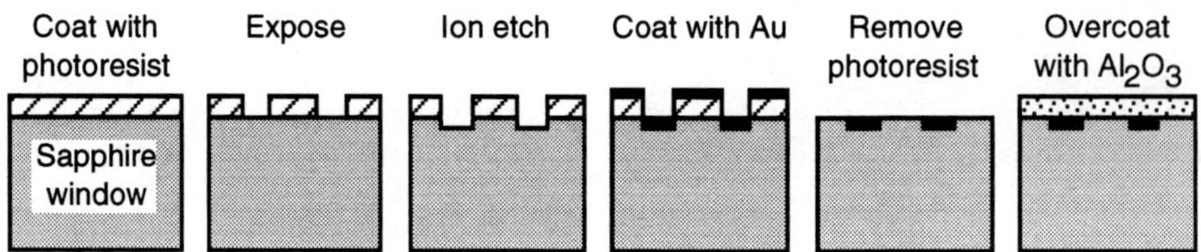

Fig. 6.17. Steps in the fabrication of a buried gold mesh.[41]

Figure 6.18 shows a *resonant mesh*,[42] in which metal covers most of the surface of a zinc sulfide window. The hexagonal array of holes whose dimensions and spacing are similar to the wavelength of infrared radiation creates interference effects in which certain infrared wavelengths are transmitted well and others are reflected. The spectrum shows a peak in transmission near 8.5 μm, with 40-50% transmission elsewhere. The shielding effectiveness is 30 dB at 10 GHz.

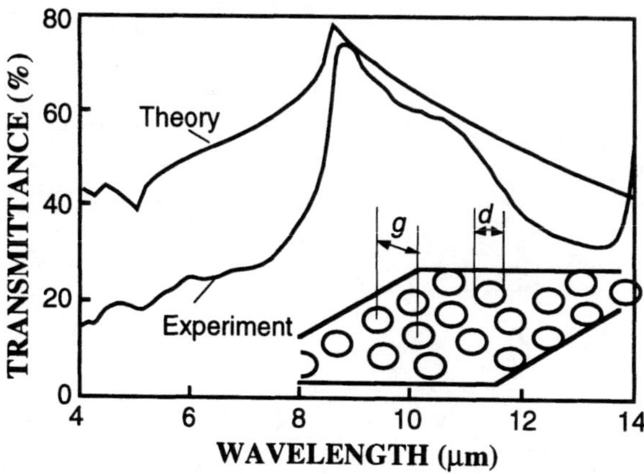

Fig. 6.18. Resonant mesh made of 0.1-μm-thick Al deposited on ZnS window.[42] The hexagonal array of circular holes (d = 4.0 μm, g = 4.5 μm) gives an infrared transmission maximum at 8.5 μm, while rejecting 99.9% of 10 GHz radiation. Transmission in this figure only refers to the mesh and does not include absorption and reflection by the ZnS.

References

1. J. D. Rancourt, *Optical Thin Films User's Handbook*, McGraw-Hill, New York (1987).
2. J. A. Dobrowolski, "Optical Properties of Films and Coatings," in *Handbook of Optics* (M. Bass, E. W. van Stryland, D. R. Williams and W. L. Wolfe, eds.), Vol I, Chap 42, McGraw-Hill, New York (1995).
3. D. E. Morton, "Optical Thin-Film Coatings," in *Handbook of Infrared Optical Materials* (P. Klocek, ed.), Marcel Dekker, New York (1991).
4. H. E. Bennett and D. K. Burge, "Simple Expressions for Predicting the Effect of Volume and Interface Absorption and of Scattering in High-Reflectance or Antireflectance Multilayer Coatings," *J. Opt. Soc. Am.*, **70**, 268-276 (1980).
5. E. Cojocaru, "Simple Relations for a Triple-Layer Antireflection Coating," *Appl. Opt.*, **31**, 2966-2968 (1992).
6. L. Li, J. A. Dobrowolski, J. D. Sankey and J. R. Wimperis, "Antireflection Coatings for Both Visible and Far-Infrared Spectral Regions," *Appl. Opt.*, **31**, 6150-6156 (1992).
7. S. F. Pellicori and E. Colton, "Fluoride Compounds for IR Coatings," *Thin Solid Films*, **209**, 109-115 (1992).
8. M. Saito, T. Gojo, Y. Kato, M. Miyagi, "Optical Constants of Polymer Coatings in the Infrared," *Infrared Phys. Technol.*, **36**, 1125-1129 (1995).
9. *Proc. SPIE*, **50**, entire volume (1974).
10. J. E. Rudisill, M. Braunstein and A. I. Braunstein, "Optical Coatings for High Energy ZnSe Laser Windows," *Appl. Opt.*, **13**, 2075-2080 (1974).
11. W. Hasan and H. T. Bui, "Broadband Durable Anti-Reflection Coating for an E-O System Window Having Multiple Wavelength Applications," *Proc. SPIE*, **1760**, 253-260 (1992).
12. A. B. Harker and J. F. DeNatale, "Diamond Gradient Index 'Moth-Eye' Antireflection Surfaces for LWIR Windows," *Proc. SPIE*, **1760**, 261-267 (1992).
13. C. G. Bernhard, "Structural and Functional Adaptation in a Visual System," *Endeavour*, **26**, 79-84 (1967).
14. S. J. Wilson and M. C. Hutley, "The Optical Properties of 'Moth Eye' Anitreflection Surfaces," *Opt. Acta*, **29**, 993-1009 (1982).
15. J. F. DeNatale, P. J. Hood, J. F. Flintoff and A. Harker, "Fabrication and Characterization of Diamond Moth Eye Antireflective Surfaces on Ge," *J. Appl. Phys.*, **71**, 1388-1393 (1992).
16. D. H. Raguin and G. M. Morris, "Antireflection Structured Surface for the Infrared Spectral Region," *Appl. Opt.*, **32**, 1154-1167 (1993).
17. P. Osbond, "Plasma Sprayed Anti-Reflection Coatings for Microwave Optical Components," *Adv. Mater.*, **4**, 807-809 (1992).
18. K. A. Epstein, D. K. Misemer and G. D. Vernstrom, "Optical Parameters of Absorbing Semiconductors from Transmission and Reflection," *Appl. Opt.*, **26**, 294-299 (1987); R. Swanepoel, "Determination of Surface Roughness and Optical Constants of Inhomogeneous Amorphous Silicon Films," *J. Phys. E: Sci. Instrum.*, **17**, 896-903 (1984); R. Swanepoel, "Determination of the Thickness and Optical Constants of Amorphous Silicon," *J. Phys. E: Sci. Instrum.*, **16**, 1214-1222 (1983); J. C. Manifacier, J. Gasiot and J. P. Fillard, "A Simple Method for the Determination of the Optical Constants n, k and the Thickness of a Weakly Absorbing Thin Film," *J. Phys. E: Sci. Instrum.*, **9**, 1002-1004 (1976).
19. S. Kaplan and L. Hanssen, "Emittance of Coated Sapphire Windows," *Proc. 7th DoD Electromagnetic Windows Symp.*, 46-52, Laurel, Maryland, May 1998.
20. C. S. Bartholomew, M. D. Morrow and N. P. Murarka, "Infrared Rugate Filters by Laser Flash Evaporation of ZnS and ZnSe," *Proc. SPIE*, **1112**, 424-432 (1989).

21. A. E. Ennos, "Stresses Developed in Optical Film Coatings," *Appl. Opt.*, **5**, 51-61 (1966).
22. W. D. Nix, "Mechanical Properties of Thin Films," *Metallurgical Trans.*, **20A**, 2217-2245 (1989).
23. E. Suhir, "An Approximate Analysis of Stresses in Multilayered Elastic Thin Films," *J. Appl. Mech.*, **55**, 143-148 (1988).
24. C. A. Klein, "Thermal Stress Modeling for Diamond-Coated Optical Windows," *Proc. SPIE*, **1441**, 488-509 (1991).
25. P. H. Kobrin and A. B. Harker, "Effect of Thin Compressive Films on Indentation Fracture Toughness Measurements," *J. Mater. Res.*, **24**, 1363-1367 (1989).
26 P. H. Kobrin and A. B. Harker, "Compressive Thin Films for Increased Fracture Toughness," *Proc. SPIE*, **683**, 139-143 (1986).
27. C. Ascheron, H. Neumann, G. Kühn and C. Haase, "Microhardness of ZnSe and its Change by Proton Implantation," *Cryst. Res. Technol.*, **24**, 1275-1279 (1989).
28. C. J. McHargue and W. B. Snyder, Jr., "Surface Modification of Sapphire for IR Window Application," *Proc. SPIE*, **2018**, 135-142 (1993).
29. C. J. McHargue, D. L. Joslin, J. M. Williams and M. E. O'Hern, "Surface Modification of Sapphire for Enhanced Infrared Window Performance," *Trans. Mater. Res. Soc. Jpn.*, **17**, 585-588 (1994).
30. C. W. White, C. J. McHargue, P. S. Sklad, L. A. Boatner and G. C. Farlow, "Ion Implantation and Annealing of Crystalline Solids," *Mater. Sci. Reports*, **4**, 41-146 (1989).
31. M. E. Ohern, C. J. McHargue, C. W. White and G. C. Farlow, "The Effect of Chromium Implantation on the Hardness, Elastic Modulus, and Residual Stress in Al_2O_3," *Nucl. Instrum. Methods Phys. Res.*, **B46**, 171-175 (1990).
32. M. Kohin, S. J. Wein, J. D. Traylor, R. C. Chase and J. E. Chapman, "Design of Transparent Conductive Coatings and Filters," *SPIE Crit. Rev.*, **CR39**, 3-34 (1992).
33. C. A. Klein, "Microwave Shielding Effectiveness of EC-Coated Dielectric Slabs," *IEEE Trans. Microwave Theory & Techniques*, **38**, 321-324 (1990).
34. C. A. Klein, "Simple Formulas for Estimating the Microwave Shielding Effectiveness of EC-Coated Optical Windows," *Proc. SPIE*, **1112**, 234-243 (1989).
35. C. I. Bright, "Electromagnetic Shielding for Electro-Optical Windows and Domes," *Proc. SPIE*, **2286**, 388-396 (1994).
36. H. Bui and W. Hasan, "Highly Durable Conductive Coating for Visible and NIR Applications," *Proc. SPIE*, **3060**, 2-10 (1997); H. T. Bui and S. Davé, "Low-Resistance Electrically Conductive Optical Coatings for Visible/Near-IR Wavelengths," *Proc. SPIE*, **2286**, 397-402 (1994).
37. K. A. Osmer, C. J. Pruszynski and J. Richter, "High Temperature IR Absorption of Low Resistivity Germanium," *Proc. SPIE*, **1112**, 83-93 (1989).
38. J. Thornton, "Absorption Characteristics of Low-Resistivity Germanium," *Proc. SPIE*, **1112**, 94-98 (1989).
39. A. R. Hilton, Sr., "Doped Gallium Arsenide External Windows," *Proc. SPIE*, **2286**, 91-98 (1994).
40. P. Klocek, T. McKenna and J. Trombetta, "Thermo-Optic, Thermo-Mechanical, and Electromagnetic Effects in IR Windows and Domes, and the Rationale for GaAs, GaP, and Diamond," *Proc. SPIE*, **2286**, 70-90 (1994).
41. J. Askinazi, "Large Aperture, Broadband Sapphire Windows for Common Aperture, Target Acquisition, Tracking and Surveillance Systems," *Proc. SPIE*, **3060**, 214-221 (1994).
42. J. P. Kurmer, J. I. Halman, K. A. Ramsey, D. L. Jones and J. McManigal, "Polarization Effects of Resonant Mesh Structures Fabricated on IR Transmitting Windows," *Proc. SPIE*, **1326**, 165-175 (1990).

Chapter 7

EROSION AND EROSION PROTECTION

One of the greatest challenges in the use of infrared windows and domes is to protect them from damage by particle impact. Collisions with raindrops are a problem at the velocities of airplanes and missiles. Figure 7.1 shows the damage to a missile dome after traversing a rainfield on the rocket sled in Fig. 7.2.[1] Bug impacts at high speed can also damage infrared windows (Fig. 7.3) Blowing sand is harmful in high or low speed collisions. This chapter describes the effects of rain and sand impact, discusses laboratory erosion tests, and reviews work on coatings used to improve window durability.

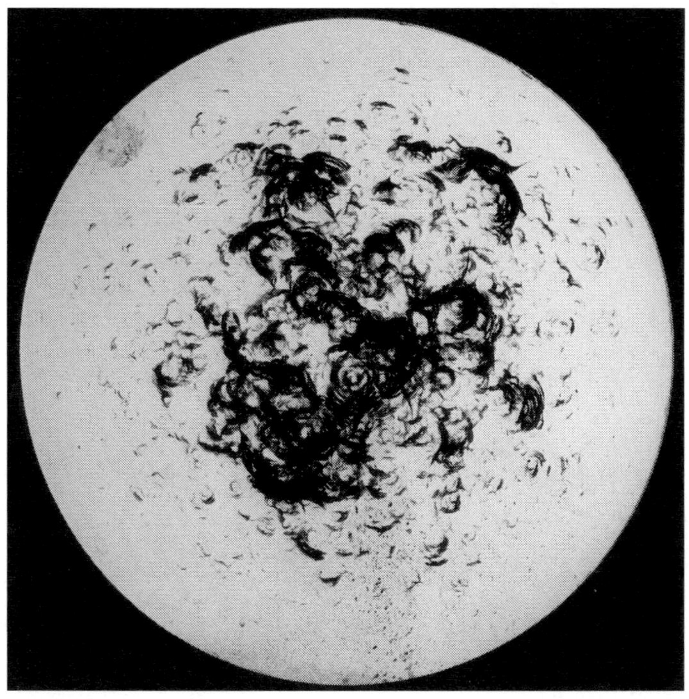

Fig. 7.1. Waterdrop impact damage on Corning 0160 glass missile dome that traversed a 760-m-long artificial rainfield (4.6 cm/h rainfall rate, average drop diameter <0.5 mm) at Mach 1.4 on the rocket sled in Fig. 7.2.[1] Notice that damage is concentrated at the center where the impact angle is near normal. [Photograph courtesy Naval Air Warfare Center, China Lake.]

Fig. 7.2. *Upper:* Dome mounted on a rocket sled at the Supersonic Naval Ordnance Research Track at the Naval Air Warfare Center at China Lake. *Lower:* Spray heads used to create artificial rainfield. [Photographs courtesy Naval Air Warfare Center, China Lake.]

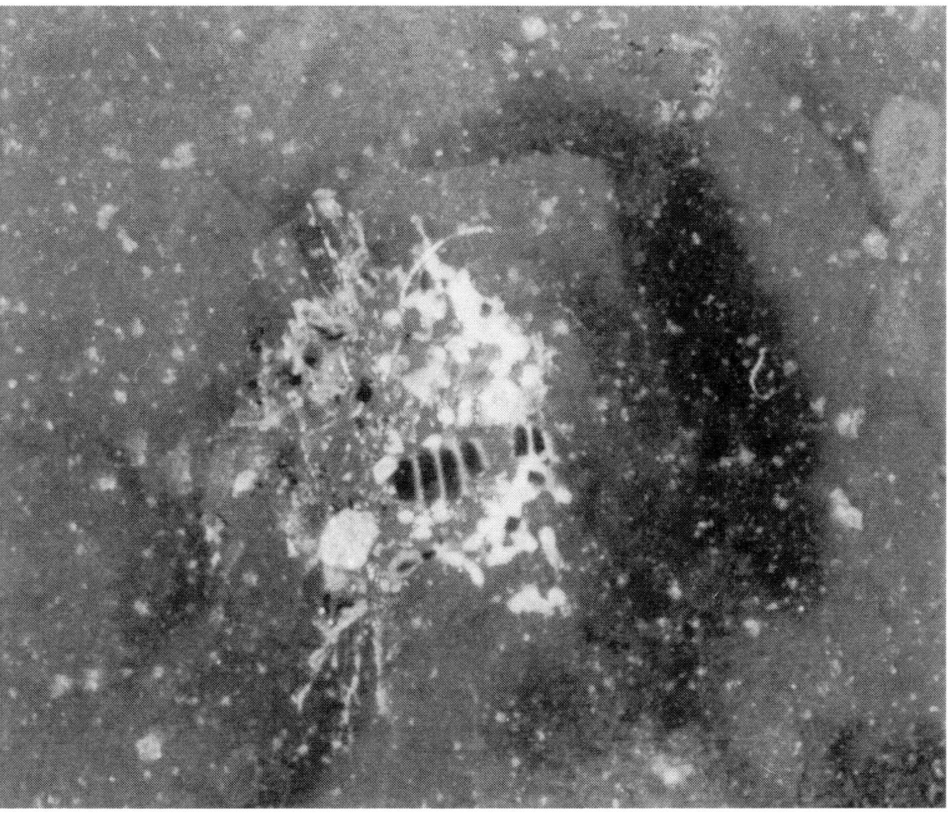

Fig. 7.3. Bug impact damage in a composite ZnS/ZnSe LANTIRN window.[2] [Courtesy Nora Osborne, University of Dayton Research Institute.]

A report on failure mechanisms in condemned LANTIRN windows is instructive.[2] LANTIRN is a *L*ow-*A*ltitude *N*avigation *T*argeting *I*nfra*R*ed *N*ight pod system carried by aircraft for navigation and targeting. The navigation pod has a trapezoidal Tuftran® window (Section 5.3.1) with approximate dimensions 119 × 150 mm consisting of 6.6 mm of ZnSe overgrown with 1.0 mm of ZnS deposited by chemical vapor deposition. There are antireflection coatings on both sides of the window. A report issued in 1992 stated that of 1100 windows delivered to the U.S. Air Force, ~200 had been condemned and removed from service. Of the condemned windows, 68 were inspected to find the cause of damage. Two thirds of the 68 windows had experienced catastrophic failure, often with multiple types of damage evident. 28% of the windows had moderate to severe sand erosion, 14% had moderate to severe rain erosion, 29% had obvious bug strikes and 13% of all failures were directly attributed to bug strikes. 42% of the windows had surface discoloration or etching attributed to attack by chemicals in the atmosphere.

The LANTIRN window has some residual stress (<10 MPa) because of the thermal expansion mismatch between ZnSe and ZnS. Clamping the window into its frame increases the stress to as much as 25 MPa. Long-term, low-level stress can cause subcritical cracks to grow (Sections 3.3.2 and 8.6) and thereby weakens the window. Rain and particle impacts can extend the subcritical cracks or can cause catastrophic failure by adding to the static stress in the window.

7.1 Rainfall characteristics

Natural rain has a distribution of drop sizes shown in Fig. 7.4.[3-5] For a rainfall rate of 2.54 cm/h, the most probable raindrop size (weighted by volume) is approximately 2.1 mm. For a rainfall rate of 2.54 mm/h the most probable drop diameter is 1.2 mm. Figure 7.4 gives us some indication why 2.0 mm is frequently chosen as a standard drop size for erosion tests.

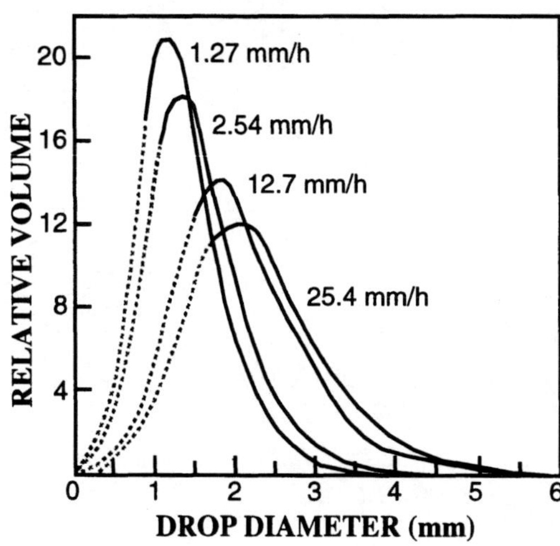

Fig. 7.4. Marshall-Palmer distribution of raindrop sizes in natural rainfall.[3-5] Dashed lines indicate range in which ideal distribution overestimates observed distribution.

An idealized distribution of raindrop sizes, taken from Marshall and Palmer,[4] follows the equation

$$N = 8000e^{-4.1d/R^{0.21}}\delta d \tag{7-1}$$

where N is the number of drops per cubic meter in the size range d to $d + \delta d$, d is the drop diameter in mm, δd is the size range in mm, and R is the rainfall rate in mm/h.

Example: Raindrop size distribution. How many drops per cubic meter are there in the size range 2.0-2.2 mm and 4.0-4.2 mm if the rainfall rate is 25.4 mm/h? Eq. (7-1) with $d = 2.0$ mm, $\delta d = 0.2$ mm and $R = 25.4$ mm/h gives

$$N = 8000e^{-4.1(2.0)/(25.4)^{0.21}}(0.2) = 25.0 \text{ drops/m}^3 .$$

For $d = 4.0$ mm, we find $N = 0.39$ drops/m^3

What is the volume of raindrops per cubic meter in the two size ranges? The volume of a sphere of diameter d is $\pi d^3/6$. Therefore the volume of drops in the 2.0-2.2 mm range is approximately (25.0 drops/m^3)[π(2.1 mm)^3/6] = 121 cubic millimeters per cubic meter, which is 0.121 cubic centimeters/m^3 or 0.121 mL/m^3. (The conversion factors are 1 mm^3 = 0.001 cm^3 and 1 cm^3 $\equiv$ 1 mL = 1 milliliter.) For this estimate, we used

2.1 mm as the mean diameter of the drop. The volume of drops in the 4.0-4.2 mm range is approximately (0.39 drops/m^3)[π(4.1 mm)^3/6] = 0.014 mL/m^3.

Now consider raindrops with diameter d mm falling at a precipitation rate of R mm/h. Each drop reaches a terminal velocity v_t m/s that is given approximately by[3] v_t (m/s) $\approx 9.65 - 10.3\ e^{-0.6\ d}$, where d is the diameter in mm. Imagine an optical window with cross sectional area A (cm^2 normal to the velocity vector) traveling through this rainfield at velocity V m/s for a period of t seconds. The number of raindrop impacts on the window is[3]

$$\text{Impacts} = \frac{R\ V\ A\ t}{6\ \pi\ d^3\ v_t} . \tag{7-2}$$

Equation (7-2) is derived by considering the number of drops in the volume swept out by the window in time t. The 6 in the denominator takes into account conversion between cm, mm, h and s in the various parameters. Table 7.1 gives representative results.

Example: Raindrop impacts in an artificial rainfield with 2 mm drop size. How many impacts per minute are expected for a 2.54-cm-diameter window traveling at 250 m/s through a 15 mm/h rainfall consisting of 2-mm-diameter raindrops? The terminal velocity for 2-mm raindrops is v_t (m/s) $\approx 9.65 - 10.3\ e^{-(0.6)(2)} = 6.55$ m/s. Equation (7-2) then gives us the number of impacts:

$$\text{Impacts} = \frac{(15)\ (250)\ (\pi\ 1.27^2)\ (60)}{6\ \pi\ 2^3\ (6.55)} = 1150 .$$

This corresponds to 3.8 impacts per second per square centimeter.

Table 7.1. Raindrop impacts[3] as a function of drop diameter for a rainfall rate of 25.4 mm/h and vehicle speed of 340 m/s

Drop diameter (mm)	Terminal velocity (m/s)	Impacts/s/cm^2*
0.25	0.945	31000
0.75	3.07	355
1.25	4.77	49.2
1.75	5.98	14.3
2.25	6.99	5.75
2.75	7.76	2.84
3.50	8.52	1.25
4.50	9.01	0.56
5.50	9.17	0.30

*Impacts per second per square centimeter of surface normal to the direction of travel. For different conditions, the number of impacts/s/cm^2 is proportional to the vehicle speed and rainfall rate [Eq. (7-2)].

Table 7.2. Raindrop impacts* at a vehicle speed of 340 m/s for natural rainfall[3]

Drop diameter (mm)	2.54 mm/h rainfall Volume %	2.54 mm/h rainfall Impacts/s/cm^2	25.4 mm/h rainfall Volume %	25.4 mm/h rainfall Impacts/s/cm^2
0-0.5	0.2	62.1	0.4	124
0.5-1	16.0	5.68	3.8	13.5
1-1.5	35.4	1.74	12.6	6.20
1.5-2	26.1	0.373	22.7	3.24
2-2.5	12.1	0.0696	23.3	1.34
2.5-3	5.2	0.0148	15.6	0.44
3-4	2.8	0.0035	16.2	0.203
4-5	0.4	0.00012	4.2	0.0234
5-6			1.1	0.0033

*Impacts per second per square centimeter of surface normal to the direction of travel.

Table 7.2 gives impact calculations that take into account the distribution of drop sizes in natural rainfall. We see that the number of collisions with relatively harmful large raindrops is much less than the number of collisions with small drops.

Example: Impacts in a natural rainfield. How many impacts per minute with 1- to 1.5-mm-diameter raindrops are expected for a 2.54-cm-diameter window traveling at 250 m/s through a 25.4 mm/h rainfall? How many 4- to 5-mm-diameter drops will strike the window in the same period? Table 7.2 tells us that the number of impacts/s/cm^2 for 1- to 1.5-mm-diameter raindrops is 6.20, if the vehicle speed is 340 m/s. For a speed of 250 m/s, the impacts will be reduced to (250/340)(6.20) = 4.56 impacts/s/cm^2. The window has an area of $\pi \times 1.27^2 = 5.07$ cm^2, which will receive $5.07 \times 4.56 = 23.1$ impacts/s or 1390 impacts/min. Table 7.2 says that the ratio of impacts by 4- to 5-mm-diameter drops compared to impacts by 1- to 1.5-mm-diameter drops is 0.0234/6.20. Therefore the number of impacts of 4- to 5-mm-diameter drops on the same window is (0.0234/6.20)(1390) = 5.2 impacts/min.

7.2 The raindrop impact event

Collisions with raindrops at sufficient velocity can damage or even break a window or dome. Long wave window materials such as ZnS and Ge are especially susceptible to damage that slowly accumulates during a lifetime of occasional encounters with clouds and rain at subsonic aircraft speeds.

Figure 7.5 shows the effect of a single 2.3-mm-diameter raindrop impacting a ZnS window at a speed of 540 m/s (Mach 1.6*). The top view shows the characteristic impact crater with rings of damage surrounding a relatively undamaged central area. The side view shows cracks radiating down into the bulk from the surface. The bottom view is a surface profile showing the crater rim and the crack system. The circular damage pattern

*Table 4.4 gives the speed of sound in the atmosphere as a function of altitude.

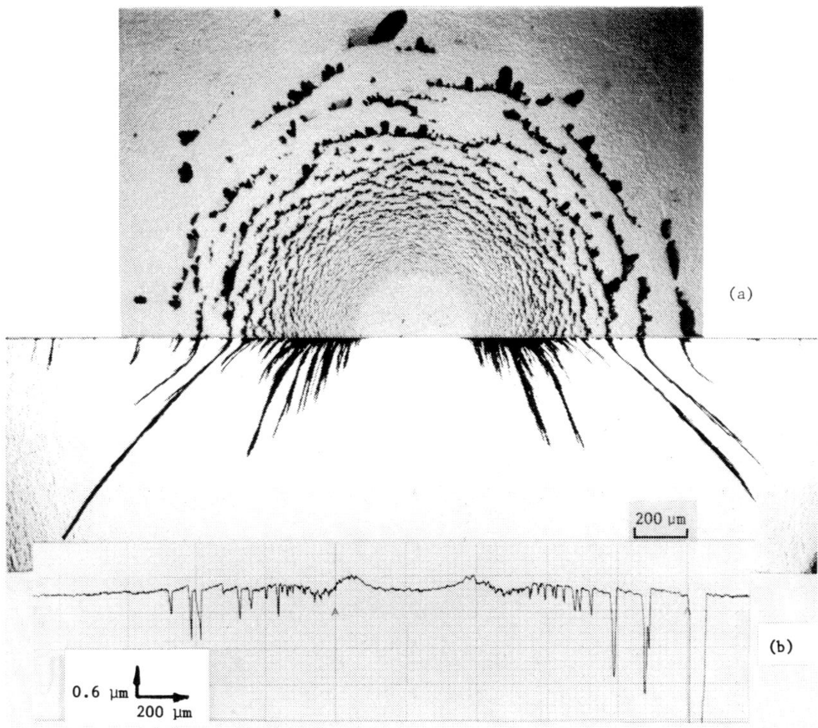

Fig. 7.5. Damage on ZnS caused by 2.3-mm-diameter raindrop with impact speed of 540 m/s.[6] Upper part of the illustration shows top view of crater seen with reflected light. Middle part is a cross sectional view and lower trace is a surface profile. (Courtesy W. F. Adler, GRC International.)

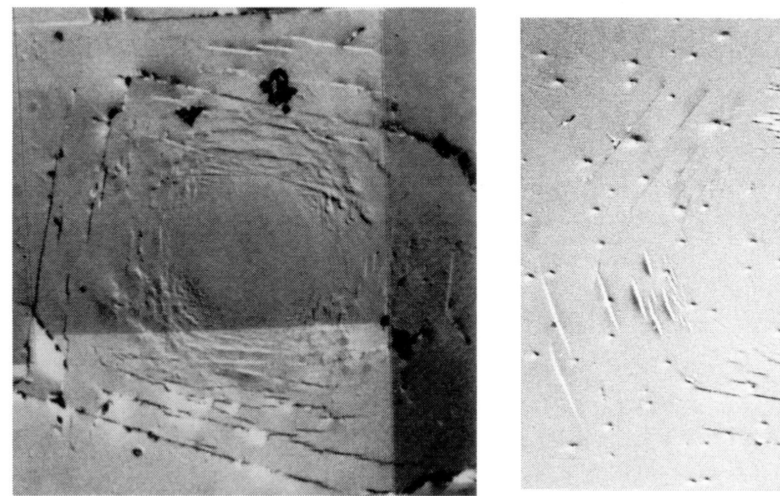

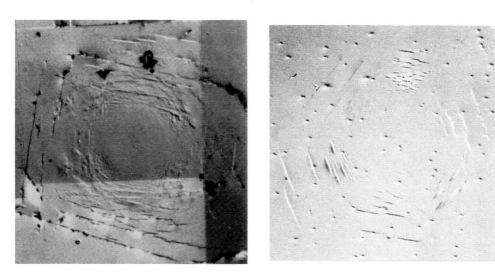

Fig. 7.6. Damage to single-crystal GaAs (*left*) and Si (*right*) caused by Mach 1 impact of 2-mm-diameter raindrops. Central crater diameter in both cases is approximately 0.2 mm. (Courtesy P. Klocek, Raytheon Systems Co.)

is typical of a polycrystalline material with small, randomly oriented grains. In Fig. 7.6 we see straight lines that follow crystal cleavage planes when single-crystal materials are damaged by raindrops.[7] By contrast, glasses may give ill-defined fracture patterns without the well-developed ring fractures of ZnS.[8]

Propagation of shock waves during the impact event is shown in Fig. 7.7.[9-11] In the initial stage of impact at the left in Fig. 7.7, compressed liquid rebounds from the solid-liquid interface and two shock waves are launched into the solid. The faster *compression wave* (also called the *longitudinal wave* or dilatational wave or dilational wave) spreads down into the solid away from the impact site. It is followed by a slower moving *shear wave*, also called a *transverse wave*. Approximately 2/3 of the collision energy goes into a surface wave, called the *Rayleigh wave*, which is not shown in Fig. 7.7. The left side of Fig. 7.7 depicts the initial ("compressible") stage in which the shock wave in the liquid trails behind the leading edge of the impact which is spreading out across the solid surface. The compressed liquid behind the leading edge creates the high "water hammer" pressure which is discussed further below. After ~0.1 μs, the shock wave in the liquid overtakes the leading edge and lateral outflow of liquid occurs at the solid-liquid interface, as shown for the "incompressible" stage at the right in Fig. 7.7. When the shock wave in the liquid reaches the leading edge, a second "release wave" propagates back into the liquid drop, reducing the pressure inside the drop.

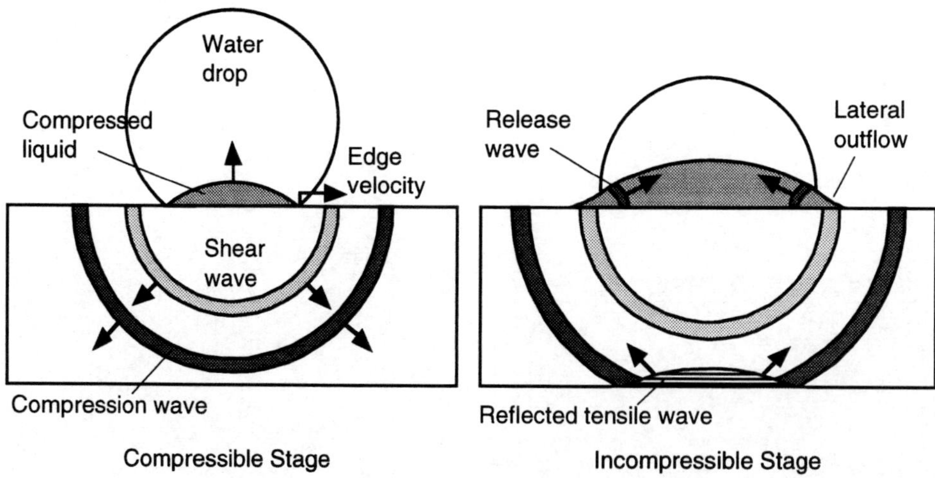

Fig. 7.7. Propagation of shock waves when a spherical water drop impacts a solid surface.[9] Equations for the wave velocities are given in the footnote beneath Eq. (7-8).

When the compression wave in the solid in Fig. 7.7 strikes the rear (lower) surface of the solid, it rebounds back into the solid and changes sign from compression to tension. If the energy in the wave is high enough, it could cause material to spall off the back (lower) surface of the solid. If the solid is sufficiently thin and the wave velocity is high enough, the reflected tensile wave can catch up with the Rayleigh wave on the front surface and intensify damage on the front surface (Fig. 7.28). This reinforcement occurs in thin ($\lesssim$1 mm) diamond windows, but not in other common materials.

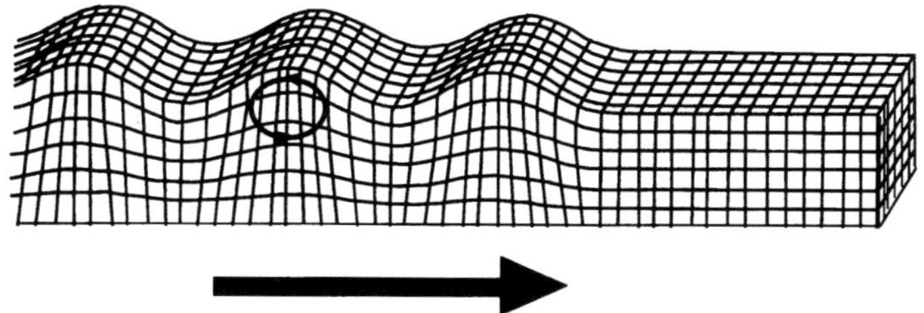

Fig. 7.8. A Rayleigh wave on the surface of a solid has both horizontal and vertical components. A point on the surface describes an ellipse as the wave passes by. (Adapted from Bolt.[12]) Equation (7-7) gives the Rayleigh wave velocity.

Rayleigh waves (Fig. 7.8) are surface waves which spread out in circles away from the impact site like ripples on a pond when you throw a rock into the water. Rayleigh waves are generated during the compressible stage of impact in Fig. 7.7. The initial wavelength of the Rayleigh wave is small because the compressible stage is very short. As the wave expands, it loses energy to microcracks in the material, so its amplitude decreases. The wave also slows down, so its wavelength increases. The increased wavelength is clearly seen in Fig. 7.5, in which the damage rings are tightly spaced near the impact and spread apart with increasing distance from the impact. The rings of damage in Fig. 7.5 correspond to tensile peaks of the Rayleigh wave in Fig. 7.8.

In the incompressible stage of impact in Fig. 7.7, lateral outflow of liquid away from the base of the water drop occurs. Figure 7.9 shows the appearance of lateral jetting as computed by a finite element model of the collision. If the surface of the solid is rough or has been roughened by damage from the Rayleigh wave or by nearby impacts, the lateral outflow can catch the raised surface and do further damage.

Fig. 7.9. Lateral outflow jetting of a spherical drop during a collision with a solid surface. (Adapted from Adler.[13])

The "water hammer" impact pressure where the raindrop strikes the surface of the window is[14]

$$\text{Pressure at center of impact} \approx \rho\, c\, \text{v} \qquad (7\text{-}3)$$

where ρ is the density of the raindrop (1000 kg/m^3), c is the shock wave speed in water ($\approx$ 1500 m/s + 2v) and v is the impact speed. The duration of this pressure is approximately $3d\text{v}/4c^2$, where d is the diameter of the raindrop. The pressure at the outer edge of the crater attains a value around three times higher than that in Eq. (7-3) for about one-tenth of the duration of the impact.[6,14] The contact diameter of the drop with the surface is approximately

Table 7.3. Impact of 2 mm diameter raindrop with rigid target

Impact speed (m/s)	Impact pressure (GPa)	Impact duration (μs)	Contact diameter (mm)
50	0.08	0.03	0.06
100	0.17	0.05	0.12
350 (Mach 1)	0.77	0.11	0.32
700 (Mach 2)	2.0	0.12	0.48
1050 (Mach 3)	3.8	0.12	0.58

$$\text{Diameter of contact } (\approx \text{crater diameter}) \approx \frac{d\,v}{c} . \tag{7-4}$$

Table 7.3 gives pressure, duration and crater diameter for a range of conditions.

7.3 Raindrop damage threshold velocity

Damage threshold velocity is the minimum impact velocity at which damage is observed. The damage threshold velocity may be defined in many ways, depending on what kind of damage is being monitored. This section discusses many of the damage thresholds that have been used. Section 7.4.3 discusses the multiple impact waterjet damage threshold velocity, which is one of the most practical thresholds to measure.

7.3.1 Threshold velocity for fracture or loss of mechanical strength

Hackworth[7] took the threshold velocity to be the lowest impact speed at which damage could be detected by examination of the surface at 165× with a Nomarski (differential interference) microscope. Using this definition, he reported single drop damage threshold velocities in Table 7.4.

In Fig. 7.5 we see two systems of cracks extending into the sample from its surface. The inner system (closer to the center of impact) makes an angle of about 65° with the surface. The outer system of longer cracks intersects the surface at 45°. Adler[6,3,15] calls the inner system Type I cracks and the outer system Type II cracks. Figure 7.10 plots the penetration depth of the cracks as a function of impact speed of the raindrop. A precarious extrapolation to zero depth provides another definition of damage impact velocity as the speed at which cracking just begins.

Table 7.4. Damage threshold velocity for single impacts of 2-mm-diameter raindrops

Material	Threshold (m/s)	Material	Threshold (m/s)
ZnS	<175	MgF_2 (hot pressed)	340-381
ZnSe	137-152	MgF_2 (single-crystal)	274-320
Si (single-crystal)	<274	Spinel (single-crystal)	<395
Sapphire	457-533	Spinel (fusion cast)	<457

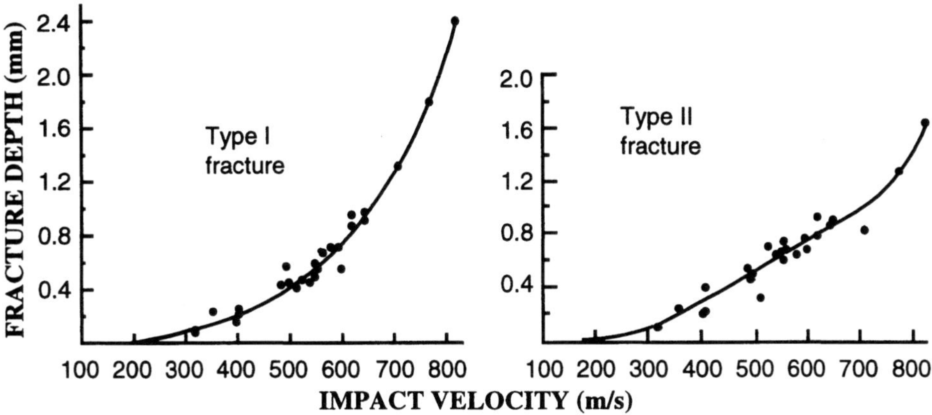

Fig. 7.10. Penetration depth for Type I and Type II fractures as a function of impact velocity for 2-mm-diameter raindrops on zinc sulfide.[3]

Another way to define a damage threshold for raindrop impact is based on loss of mechanical strength. Figure 7.11 shows a series of experiments in which the strengths of zinc sulfide disks were measured after the disks had been impacted at various speeds.[14,16] The strength of undamaged or slowly impacted materials is approximately 80 MPa. When the impact speed exceeded 125 m/s, the strength began to drop. The residual strength after impact at speeds between 150 and 600 m/s is approximately 40 MPa.

In Fig. 7.11, the strength falls abruptly when the material is impacted once at a particular speed. When zinc sulfide is subjected to random, multiple impacts in a whirling arm rainfield, the strength falls continuously for impacts in the range 120-210 m/s. The strength of zinc sulfide after 10 min of impact at 210 m/s in a 2.54 cm/h rainfield of 2-mm-diameter drops is about half of the strength of unimpacted ZnS.[17]

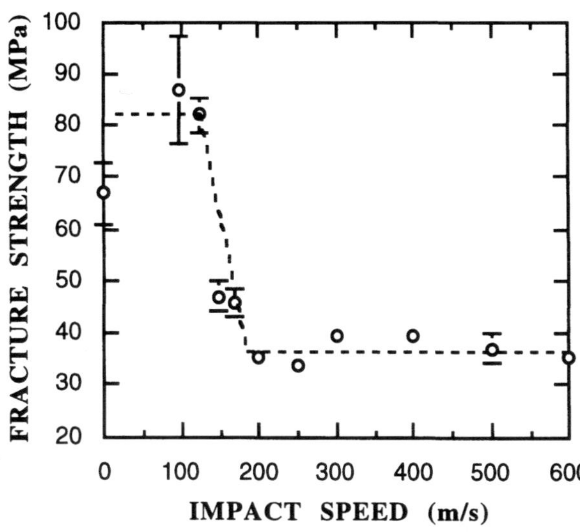

Fig. 7.11. Residual strength of zinc sulfide after impact by a waterjet with a 0.8-mm nozzle.[14,16] The low value of strength (68 MPa) of unimpacted samples in this particular experiment is not characteristic of ZnS and was attributed to poor surface finish of the unimpacted samples only.

At least up to some diameter, damage threshold velocity decreases with increasing drop size. For drops of diameter d_1 and d_2, the damage threshold velocities, v_1 and v_2, are expected to be approximately inversely proportional to the cube root of drop diameter[14]:

$$\frac{v_2}{v_1} \approx \left(\frac{d_1}{d_2}\right)^{1/3} . \tag{7-5}$$

Example: Damage threshold velocity as a function of drop size. The observed damage threshold velocity for 2.2-mm-diameter waterdrops on zinc sulfide is 160 m/s for the initial onset of fracture.[18] To predict the damage threshold velocity for a 3.2-mm-diameter drop, we use Eq. (7-5) with v_1 = 160 m/s, d_1 = 2.2 mm and d_2 = 3.2 mm:

$$\frac{v_2}{160 \text{ m/s}} \approx \left(\frac{2.2}{3.2}\right)^{1/3} \quad \Rightarrow \quad v_2 = 141 \text{ m/s} .$$

The observed threshold[3] of 140 m/s is in excellent agreement with the prediction.

At some size when the drop becomes large enough, the damage threshold velocity ceases to decrease with size, at least for zinc sulfide.[18] Figure 7.12 shows that the threshold velocity for 4.4-mm-diameter drops is essentially the same as that for 3.2-mm drops. However, above the damage threshold, the bigger drops create deeper damage.

When the waterjet experiment in Fig. 7.11 was carried out with sapphire, the damage threshold velocity was near 500 m/s for the same 0.8-mm waterjet.[14] That is, the damage threshold velocity for sapphire was approximately four times greater than that of zinc sulfide. Hot pressed silicon nitride had a threshold slightly above 600 m/s.

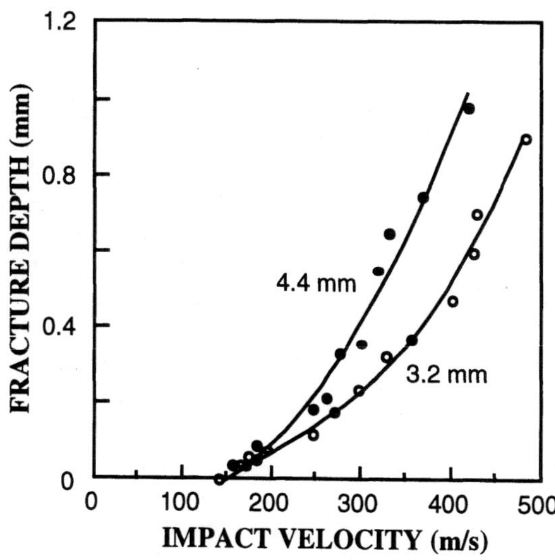

Fig. 7.12. Fracture depth as a function of impact velocity for 3.2- and 4.4-mm diameter raindrops on zinc sulfide.[18]

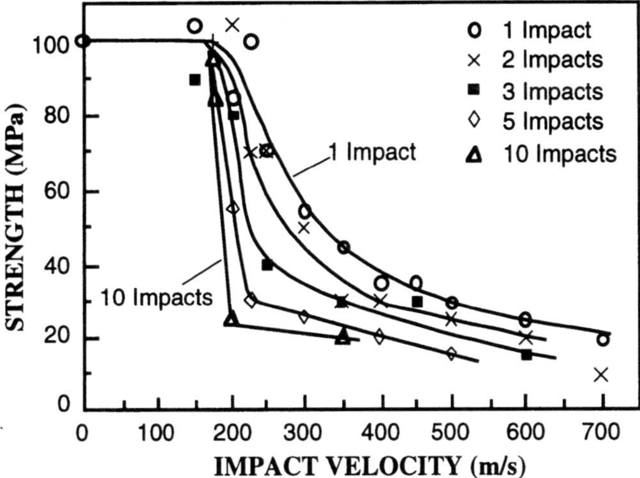

Fig. 7.13. Residual strength of soda-lime glass disks after single or multiple impacts by waterjet with 0.8-mm nozzle.[19,20]

Figure 7.13[19,20] shows the results of single and multiple impact experiments using soda-lime glass as an inexpensive, brittle substrate. The damage threshold velocity is approximately 200 m/s, but the decrease in strength is relatively gradual if only a single impact is used. The fall in strength becomes steeper as the number of impacts at the same site is increased from 1 to 10. The interpretation of this important experiment is that impacts below the threshold velocity do not damage the material. A large number of sub-threshold impacts at the same site will not weaken the material. However, every impact above the threshold velocity damages and weakens the material. The greater the number of impacts at the same site, the weaker the sample becomes, until the plateau level of approximately 10-20 MPa strength is reached.

A theoretical expression for the damage threshold velocity is[21]

$$v_{threshold} \approx 1.41 \left(\frac{K_{Ic}^2 \, c_R}{\rho_w^2 \, c_w^2 \, d_w} \right)^{1/3} \tag{7-6}$$

where K_{Ic} is the fracture toughness (Section 3.8) of the window material, c_R is the Rayleigh wave speed of the window material, ρ_w is the density of water, c_w is the compression wave speed in water (~1500 m/s) and d_w is the diameter of the raindrop. A Rayleigh wave is a surface wave on an elastic solid (Fig. 7.8). Its speed is given by[22]

$$c_R = \left(\frac{0.862 + 1.14\, \nu}{1 + \nu} \right) \sqrt{\frac{E}{2\,(1 + \nu)\,\rho}} \tag{7-7}$$

where ν is Poisson's ratio, E is Young's modulus and ρ is the density of the window material.

The agreement between measured damage threshold velocities and those calculated with Eq. (7-6) is modest, at best.[3,15] However, the numerator of Eq. (7-6) has been taken as a "damage parameter" and used to correlate observed damage threshold velocities with material properties. For example, Fig. 7.14 correlates a range of threshold data for single and multiple impacts with the damage parameter, $K_{Ic}^{2/3} c_R^{1/3}$.[23]

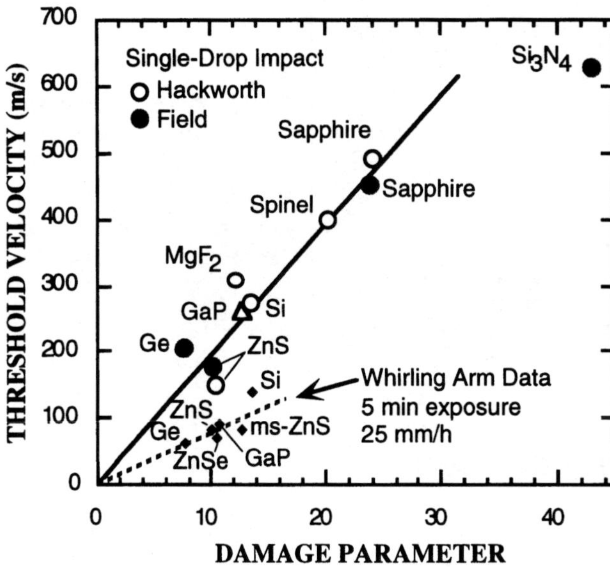

Fig. 7.14. Measured damage threshold velocity as a function of damage parameter, $K_{Ic}^{2/3}\, c_R^{1/3}$.[23] Single drop thresholds were taken from work by Hackworth or Field. GaP data point is from Ref. 104. Multiple drop thresholds are based on whirling arm experiments reported by Raytheon, involving 2000-4000 impacts on each sample.

A simpler, empirical correlation is that damage threshold velocity is proportional to the logarithm of the fracture toughness of the window material.[24] Figure 7.15 shows this correlation for waterjet impact experiments which are described in Section 7.4.3.

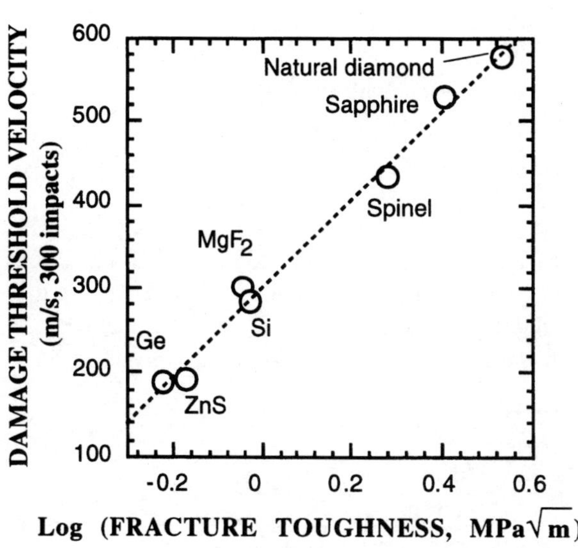

Fig. 7.15. Multiple impact waterjet damage threshold velocity (300 shots on one spot) is proportional to the logarithm of the fracture toughness of the window material.[24]

7.3.2 Threshold velocity for loss of optical transmission or contrast

In addition to crack growth and loss of strength, another manifestation of impact damage is loss of optical transmission (or increase of optical scatter). Figure 7.16 shows

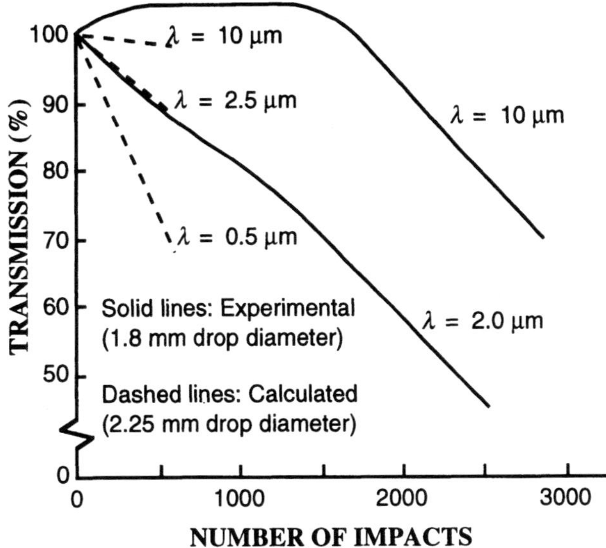

Fig. 7.16. Measured and calculated loss of transmission by ZnS in a whirling arm rainfield with impact speed of 222 m/s. The calculations are terminated when the number of overlapping impact sites becomes significant.[3,15] Transmission is normalized to its initial value, which is defined as 100%.

measured and calculated loss of transmission by ZnS after exposure to multiple raindrop impacts in a whirling arm facility.[3,15] The calculated values are based on loss of transmission as light traverses the fractured areas in Fig. 7.5. At a wavelength of 2.5 μm, there is nice agreement between the experiment and the model. At 10 μm wavelength the experimental transmission actually *increases* for a period of time before a decrease sets in. While no explanation was given, it is possible that the slightly damaged surface serves as an antireflection coating, just as the moth's eye structure in Section 6.1.1 increases transmission through the window.

Figure 7.17 show the loss of transmission in zinc sulfide when it was impacted randomly by a waterjet at 190 m/s.[24] As in Fig. 7.16, the shorter wavelength is more sensitive to transmission loss than is the longer wavelength.

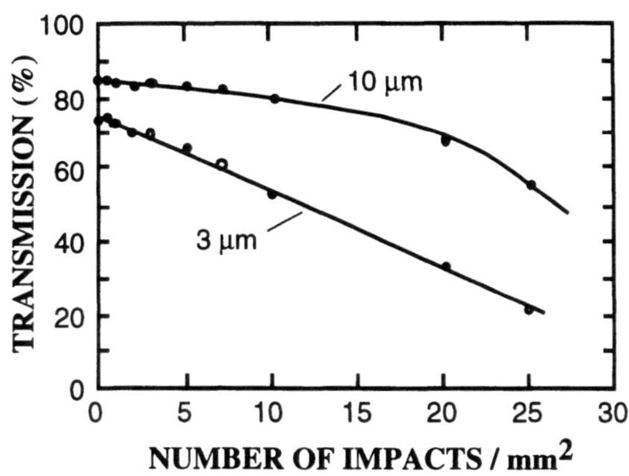

Fig. 7.17. Loss of transmission in ZnS impacted randomly by a waterjet over a 1 cm^2 area at 190 m/s from a 0.8-mm-diameter nozzle.[24]

Table 7.5. Time for 10% loss of transmission* in rainfield at 250 m/s[25]

Material	Time (s)	Material	Time (s)
Silicon	1100	Germanate glasses:	
MgF_2	500	Cortran 9754	250
ZnS (CVD)†	300	VIR6 glass	70
ZnS (PS)†	150	Germanium	40
ZnSe	9	As_2Se_5 glass	0.6

*Time needed for transmission to drop to 90% of its initial value at a selected wavelength
†CVD ZnS is 6-μm-grain-size material produced by chemical vapor deposition. PS ZnS is 1-μm-grain-size material produced by pressure sintering.

Table 7.5 compares the time required for different materials to lose 10% of their initial optical transmission in a whirling arm rainfield.[25] Notice that the order of erosion resistance is Si>MgF_2>ZnS>>ZnSe. Zinc selenide is a particularly poor material for rain erosion resistance. Notice also that silicon performs much better than MgF_2 in this experiment, whereas the damage threshold velocity of silicon and MgF_2 are essentially equal in Figs. 7.14 and 7.15.

When the percent loss in transmission of zinc sulfide was measured at 2.5 μm wavelength as a function of impact speed in a whirling arm rainfield, the rate of loss of transmission was approximately proportional to the 9*th* power of the normal impact speed.[7] In the same experiment with 1.8-mm-diameter raindrops, the erosion rate was proportional to the 14*th* power of impact speed.[7] In other experiments, the erosion rate was approximately proportional to the 11*th* power of impact speed.[25]

An empirical conclusion reached by Cassaing *et al.*[25] is that the time for 10% loss of transmission is related to the liquid water content of the rainfield (grams of water per cubic meter), the impact speed (v), the density (ρ) of the window material, and the longitudinal wave velocity‡ (c_L) in the material:

$$\text{Time for 10\% loss of transmission} \propto \frac{(c_L / \sqrt{\rho})^{4.5}}{(\text{water content}) \times v^{11}} \quad . \tag{7-8}$$

Pre-existing stress in a window has an enormous effect on rain erosion damage.[26] When a ZnSe window was slightly flexed during raindrop impact, the time needed for 10% transmission loss decreased by a factor of 100. Magnesium fluoride and Ge samples

‡For a material with modulus E, Poisson's ratio ν and density ρ, the longitudinal wave velocity is[22] $c_L = \sqrt{(\lambda+2\mu)/\rho}$, where $\lambda = E\nu/[(1+\nu)(1-2\nu)]$ and $\mu = E/(2+2\nu)$. The transverse wave velocity is $c_T = \sqrt{\mu/\rho}$.

shattered upon raindrop exposure under the mildest flexure conditions that were tested. Pre-existing stress in the LANTIRN window described at the beginning of this chapter is a significant contributor to window failure. Typically, the outer ZnS layer is in compression and the thick ZnSe window is in tension. When impact damage propagates through the ZnS and reaches the tensile ZnSe region, catastrophic failure occurs.

A more subtle manifestation of optical degradation of a window by raindrop impacts is loss of optical contrast, as measured by the modulation transfer function (MTF) defined in Section 2.3. Contrast is defined as MTF measured after rain exposure divided by MTF measured before exposure. Figure 7.18(a) shows the loss of contrast in magnesium fluoride as a function of time for different impact speeds in a whirling arm rainfield.[27] The greater the impact speed, the less time it takes to lose optical contrast.

The time required for the contrast to lose 10% of its initial value is designated $t_{0.1}$. A graph of log ($t_{0.1}$) versus log (impact speed) is a straight line with a negative slope, implying that the relationship between $t_{0.1}$ and v is

$$t_{0.1} = A\mathrm{v}^{-n} \qquad (7\text{-}9)$$

where A is a constant. For MgF_2, the exponent n is 13. For ZnSe, the exponent is 10. The dependence of optical contrast loss on drop size is not straightforward. Figure 7.18(b) shows that the time $t_{0.1}$ goes through a minimum as drop size increases.

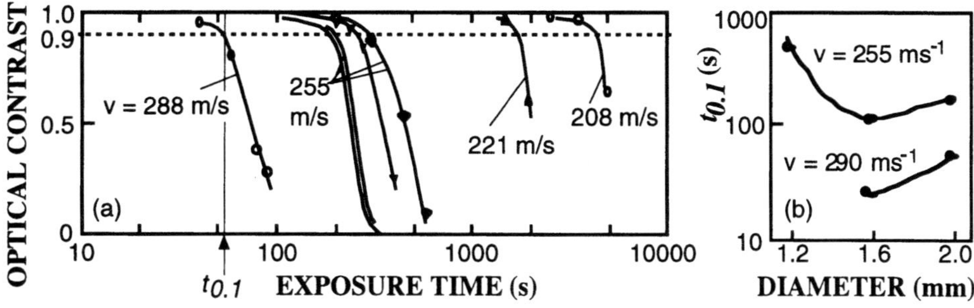

Fig. 7.18. (a) Loss of optical contrast as a function of time for MgF_2 in a whirling arm rainfield at different impact speeds.[27] MTF was measured at a spatial frequency of 66 cycles/radian in the wavelength range 2-6 μm. Optical contrast degradation is more rapid for higher spatial frequencies. (b) Dependence of $t_{0.1}$ on drop diameter.

7.3.3 Threshold velocity for loss of mass

Another definition of damage threshold is the onset of loss of mass from the material. Figure 7.19 shows mass loss by MgF_2 in a whirling arm rainfield as a function of time for different impact speeds.[27] The incubation time is defined by the extrapolation of the linear region of the curves down to zero mass loss. Incubation time follows a power law dependence on impact speed of the form time $\propto$ v^{-n}, as in Eq. (7-9) for loss of optical contrast.

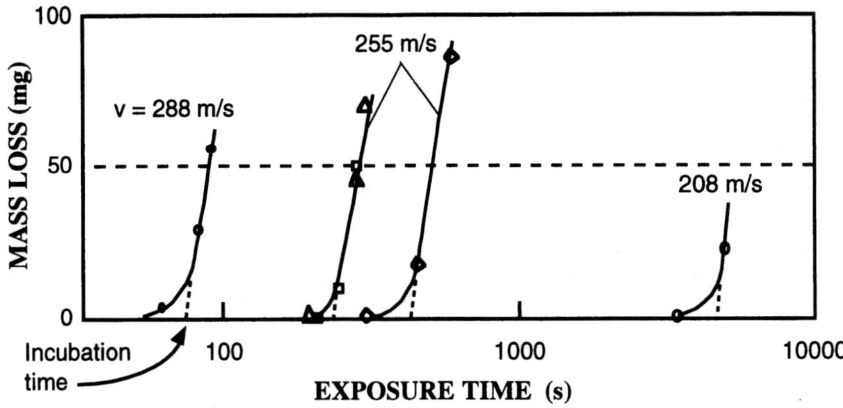

Fig. 7.19. Mass loss by MgF_2 in a whirling arm rainfield for different impact velocities.[27]

Fig. 7.20 compares the time $t_{0.1}$ for loss of 10% of optical contrast to the incubation time for mass loss.[27] In the case of MgF_2, the two times are essentially equal, which means that the effects occur concurrently. For ZnSe, loss of optical contrast comes significantly before loss of mass. It is reported[25] that for silicon, the time for 10% loss of infrared transmission (not optical contrast) is half as great as the incubation time for mass loss. For germanium, the time for 10% loss of transmission is one-tenth of the incubation time for mass loss.

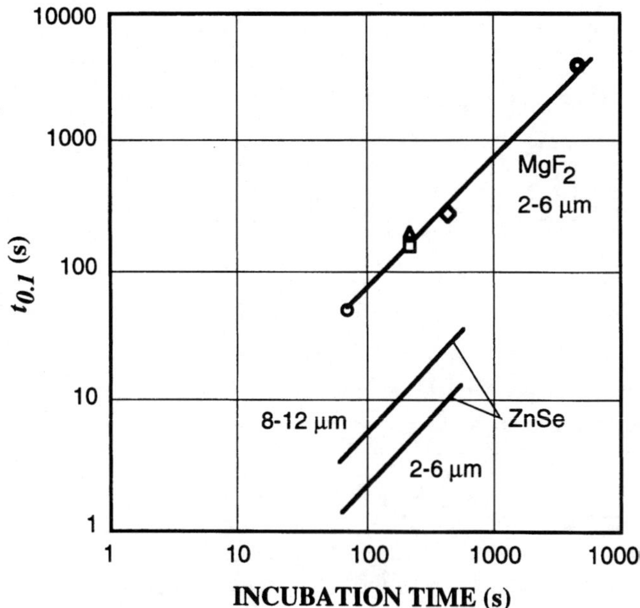

Fig. 7.20. Time for 10% optical contrast loss plotted versus incubation time for mass loss by MgF_2 and ZnSe.[27]

7.4 Rain erosion test facilities

Common rain erosion tests utilize whirling arm rain field facilities,[28,29] waterjet impact,[20,14,30] single or multiple drop impact[31,32] or a rocket sled going through a rain field (Fig. 7.2). The few existing whirling arms are national-scale facilities, whereas waterjets are more common laboratory-scale facilities. The single or multiple drop impact facilities are unique to GRC International (Santa Barbara CA).

7.4.1 Whirling arm

In the whirling arm facility shown in Fig. 7.21 a 2.5-cm-diameter flat specimen is whirled at the end of a propeller blade through an artificial rain field consisting of waterdrops falling from hypodermic needles. A common test measures the damage to a sample after whirling at 210 m/s for periods ranging from 1-20 min in a 25.4 mm/h rain field with 2-mm-diameter drops. The 2-mm size was chosen because it represents the most common drop size in Fig. 7.4 for a 25.4 mm/h rainfall rate. This choice of drop size can be criticized because the fewer, larger drops in natural rainfall are much more damaging than the 2-mm drops.[33]

Several whirling arm facilities are listed in Table 7.6. Characterization of three whirling arms using witness plates made of Plexiglas (also called poly(methyl-methacrylate), PMMA, or Lucite) to record drop imprints shows that each facility is unique.[29] Turbulence or shock waves distort the drops into different shapes at each facility. Spherical drops with an initial diameter of 2 mm may become ragged ellipsoids with equivalent diameters of 3-6 mm. This distortion makes comparison of tests between facilities very difficult (Fig. 7.22[34]). Studies at the UDRI facility showed that the actual number of waterdrop impacts on a Plexiglas substrate were less than the theoretical number for a rotor speed of 200 m/s.[33] When the rotor speed was increased, the number

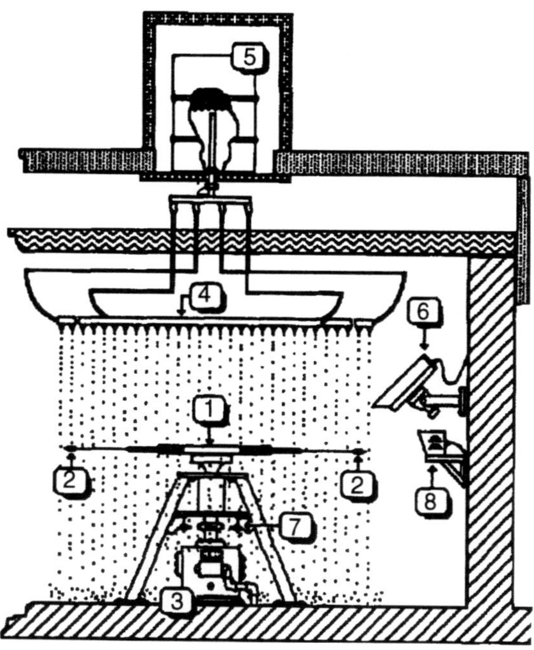

Fig. 7.21. Whirling arm rain erosion test facility run by the University of Dayton Research Institute at Wright-Patterson Air Force Base. Legend: (1) double arm blade, (2) mated test specimens, (3) vertical drive gear box and shaft, (4) curved manifold quadrant, (5) water storage tank, (6) video camera, (7) magnetic pick-ups for firing strobe light, (8) strobe light for stop-motion viewing.

Table 7.6. Whirling arm rain erosion test facilities[28]

Facility*	Maximum mach number	Arm radius (m)	Nominal drop size (mm)	Rainfall rate (mm/h)
UDRI	0.8	1.22	2	25.4
Bell Helicopter	0.75	1.22	Natural rain	76.2
DERA (formerly RAE)	0.70	1.45	2	25.4
SAAB-SCANIA	1.0	2.19	1.2-2	1.4-25
Dornier	3.0	1.2	0.5-1.7	1.8-320

*UDRI: University of Dayton Research Institute, Wright-Patterson Air Force Base, Dayton, Ohio. Bell Helicopter: Fort Worth, Texas. DERA: Defence Evaluation and Research Agency, Farnborough, United Kingdom. SAAB-SCANIA: Linköping, Sweden. Dornier: Friedrichshafen, Germany.

of impacts decreased disproportionately until, at the highest speed of 320 m/s, there were few impacts. It is therefore not possible to make an absolute comparison of erosion rate as a function of impact speed. Changing the specimen holder also changed the number of impacts when the speed was held constant. Tests run under the same nominal conditions one year apart gave significantly different results. Despite these serious limitations on quantitative experiments, side-by-side tests of different materials provide useful comparisons.

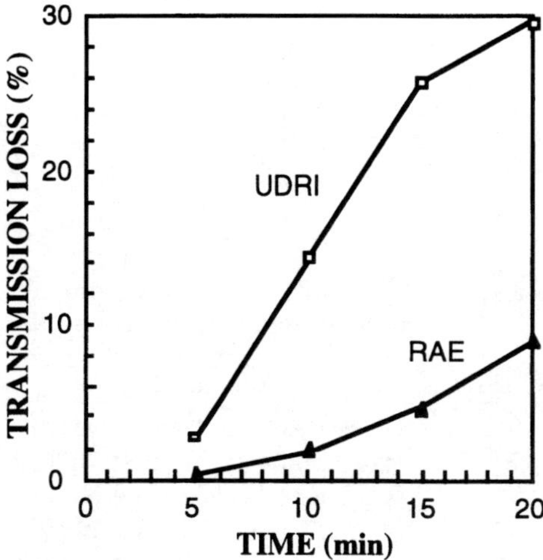

Fig. 7.22. Comparison of results obtained at whirling arm facilities at Wright-Patterson Air Force Base (UDRI) and Farnborough, Great Britain (DERA) under nominally identical conditions.[34] At DERA the distorted drops have an equivalent spherical diameter in the range 0.4 - 4.4 mm. At UDRI, most of the drops are distorted to an equivalent diameter of 4.0 - 6.6 mm, which do more damage than the smaller drops at DERA.[29]

A whirling arm is not reliable for non-normal angles of incidence.[35] As a specimen is inclined from normal incidence, the number of impacts per unit area should *decrease* if the speed and rainfall rate are constant, because the projected area of the specimen decreases. However, tilting a specimen 10 or 20° from normal at the UDRI whirling arm facility *increased* the density of impact sites compared to the number for normal incidence.

At angles of 30 and 40°, the impact density then decreased. Tilting the specimen *up* gave a different density of impact sites than tilting a specimen *down* by the same angle. The unexpected results might have been caused by turbulence or a bow shock wave from each end of the whirling arm affecting the other end.

By using a heated sample holder, it is possible to conduct rain impact experiments with windows at elevated temperature. The stress in the window is a combination of the impact stress and the thermal stress created by the contact of cold liquid with the hot surface. Figure 7.23 shows the results of whirling arm tests of heated windows.[36] The time required for ZnSe and Ge to lose 10% optical transmission decreases with increasing window temperature, as might be predicted from increased thermal stress. The time required for ZnS to lose 10% transmission *increases* with window temperature, because the mechanical strength of ZnS increases with temperature.

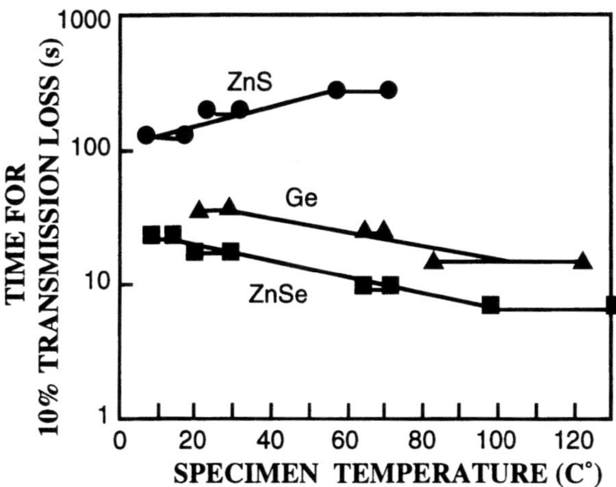

Fig. 7.23. Whirling arm erosion response of heated window materials.[36] The ordinate is the time required for the transmission in the 8-12 μm region to fall to 90% of its initial value. Horizontal bars denote the range for duplicate experiments. ZnS and Ge were tested with 2-mm-diameter raindrops at 230 m/s. ZnSe was tested with 1.2-mm drops at 210 m/s.

7.4.2 Single-impact waterjet

The single-impact waterjet apparatus in Fig. 7.24, introduced by the University of Cambridge, is relatively simple for characterizing the raindrop impact resistance of materials.[20,14] Although the equipment is inexpensive, the technique is extremely labor intensive.

The lead slug fired from an air gun in Fig. 7.24 strikes a neoprene rubber disk that extrudes a jet of water through the nozzle at 3-5 times the speed of the bullet. Only the somewhat rounded front of the jet in the photographs in Fig. 7.24 has any resemblance to a water drop. The speed is measured by the time between cutoff of two light beams perpendicular to the direction of travel. The nozzle is the critical component of the system, which must produce a waterjet with a smooth, slightly curved front that simulates the leading edge of a raindrop.

The reason why an elongated jet can crudely simulate the impact of a spherical drop is that most damage occurs during the initial stage of impact. If the front of the jet looks like the front of a sphere, the damage will be roughly similar. The waterjet can only be used for perpendicular incidence, because only the front simulates a sphere.

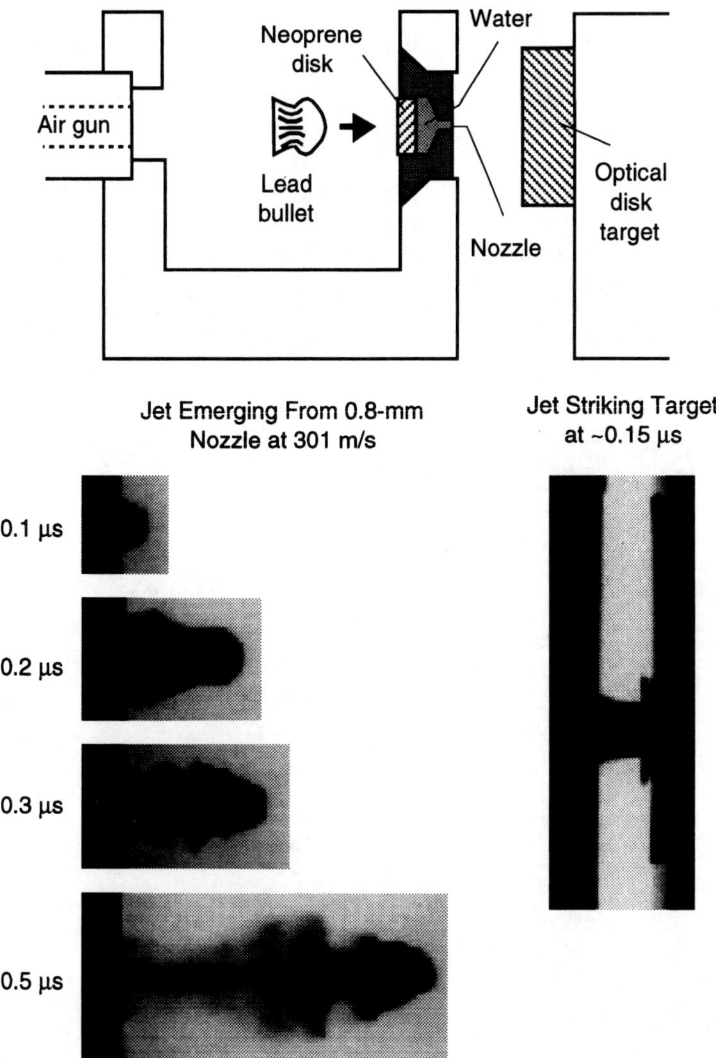

Fig. 7.24. Schematic diagram of waterjet test apparatus[14] and high speed photographs of the waterjet. The flat target of test material is placed 1 cm from the nozzle. (Photos courtesy K. Klemm, Naval Air Warfare Center.)

The damage ring produced by a waterjet striking Plexiglas can be compared to the ring produced by a spherical drop of known diameter. Such measurements produce an "equivalent drop size" for the waterjet. Figure 7.25 shows that the equivalent drop size depends on the nozzle diameter and speed. The circles in Fig. 7.25 show that direct measurements of the radius of curvature of the front of the waterjet (recorded by high speed photography) do not agree with the equivalent drop size measured by damage to witness plates.[33] In the early waterjet literature, attempts were made to correlate the actual impact condition (speed and nozzle size) to a corresponding impact speed if the jet were a 2-mm waterdrop. *The best current practice in reporting waterjet results is to list the nozzle size and actual waterjet speed and avoid any further correlation or interpretation.*

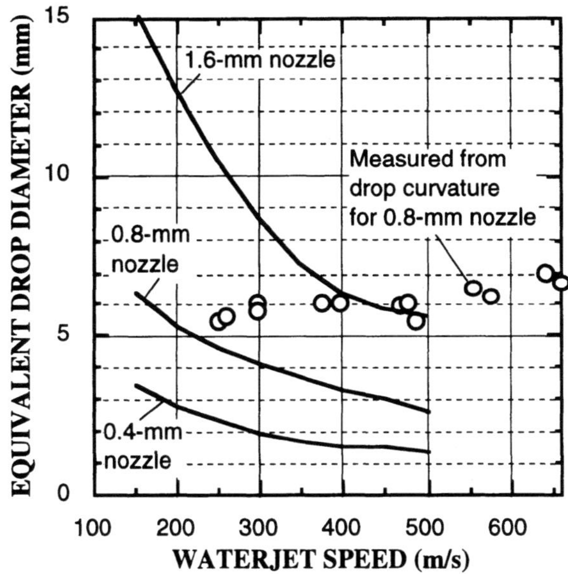

Fig. 7.25. Lines show relation between waterjet and diameter of equivalent single raindrop, based on total area of damage on Plexiglas witness plates.[14,37] Circles show measured diameter of front of waterjet from 0.8-mm nozzle.[33]

7.4.3 Multiple-impact jet apparatus (MIJA)

The multiple-impact jet apparatus, commonly abbreviated MIJA, is a modification of the single-impact jet apparatus in Fig. 7.24. Instead of firing a bullet to create the waterjet, compressed air thrusts a nylon piston against a titanium metal shaft which expels the waterjet from the nozzle. The piston is then automatically retracted and a fresh supply of water refills the nozzle. The entire sequence is automated and can fire 20 shots/min with a spread in velocity of ≤1% over the range 20-650 m/s.[9] The horizontal position of the 25- to 50-mm-diameter target specimen is governed by a computer-controlled x-y stage.

Figure 7.26 shows how the damage threshold velocity curve is measured.[24] Points on the target are impacted at speeds of 100 to 500 m/s. After each site is impacted once, the specimen is examined under a microscope with a magnification up to 50×. In the example in Fig. 7.26, a ring fracture was observed under the microscope at sites impacted at ≥440 m/s. These sites are marked by circles in the upper graph of Fig. 7.26 and are no longer impacted. Each undamaged site is impacted again, the sample is inspected, and damaged sites are marked on the graph. The middle graph in Fig. 7.26 shows the results after 7 impacts per site. The lower graph shows results after 300 impacts per site.

The "single-impact" damage threshold velocity is obtained from the first shot. In Fig. 7.26 the single-impact damage threshold is between 420 and 440 m/s and would be reported as 430 m/s. For multiple impacts, the damage threshold curve flattens out. A reasonable approximation is that 300 shots at one site is close to the limit that would be observed for an infinite number of shots at one site. The "multiple-impact" damage threshold velocity is the value observed after 300 shots, which is 310 m/s.

One caveat for Fig. 7.26 is that the damage noted on the graphs is just ring fracture around the impact site. Cracks can be induced by the waterjet near the center of the impact, but they are harder to observe and are not customarily reported. Waterjet damage threshold curves published by Cambridge University are normally for ring fracture only.

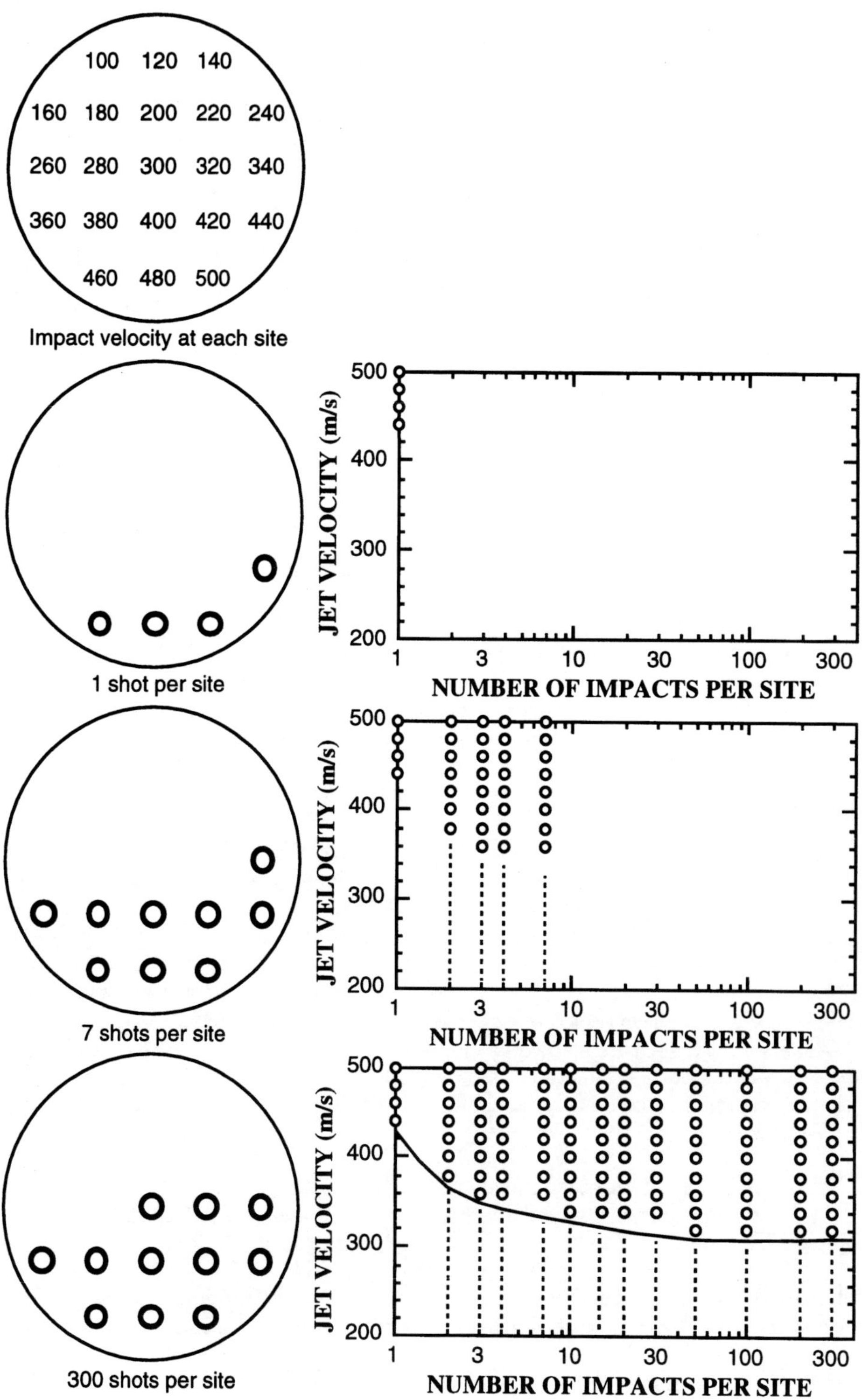

Fig. 7.26. How multiple-impact jet apparatus (MIJA) damage threshold velocity is measured.

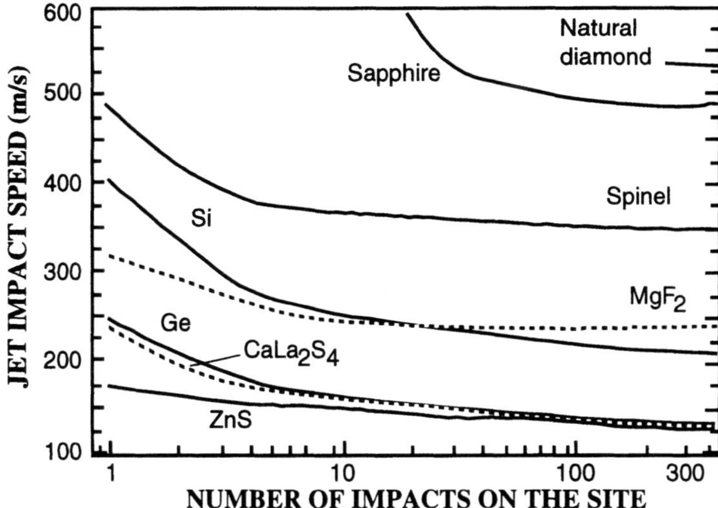

Fig. 7.27. MIJA damage threshold velocity for ring fracture of infrared materials.[24] For ALON, which is not shown on the chart, the threshold for 4 impacts is 690 m/s.[38]

For small specimens, MIJA absolute damage threshold velocity (ADTV) is measured by impacting one site 300 times at low velocity and inspecting for damage at 200× magnification. Then velocity is increased and the sample is impacted 300 more times and inspected. The process is repeated until cracking is observed.[39]

Figure 7.27 shows damage threshold velocity curves for infrared window materials.[24] Sapphire is clearly the best. Spinel is good and MgF_2 is respectable. The curve for aluminum oxynitride (ALON) is expected to lie between those of spinel and sapphire.

Specimen thickness can play a role in the outcome of a MIJA experiment. The compression wave in Fig. 7.7 moves faster than the Rayleigh wave. When the compression wave is reflected at the rear of the specimen, it becomes a tensile wave. When the tensile wave catches the Rayleigh wave moving along the front surface in Fig. 7.28, the two reinforce each other, which can produce ring fracture if the waves are still sufficiently strong.[9,40] This phenomenon is observed in thin specimens of diamond[9] and sapphire,[41] both of which have very high compression wave speeds. For sapphire, the damage threshold velocity of a 3-mm-thick disk from a 0.8-mm jet is 485 m/s for 300 impacts. The damage threshold velocity of a 1-mm-thick disk is 430 m/s.

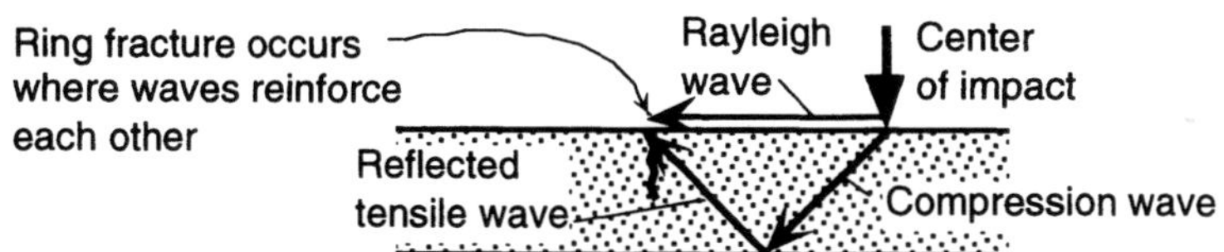

Fig. 7.28. A tensile wave reflected from the rear surface of a window reinforces the Rayleigh wave moving along the front surface. The Rayleigh wave amplitude is attenuated in proportion to $1/\sqrt{r}$, where r is the distance traveled. The compression wave (and the shear wave behind it in Fig. 7.7) are attenuated much more rapidly ($1/r^2$).[42]

7.4.4 Single-drop impact testing

Neither the whirling arm nor the waterjet provides spherical waterdrops. To accomplish this requires apparatus such as that in Fig. 7.29,[32] which is, unfortunately, expensive to run. This experiment begins when a 1.5- to 5-mm-diameter spherical waterdrop falling from a hypodermic needle near the center of the apparatus is detected by light beam 1, which triggers the gun at the right to fire. The detonation hurls an optical test specimen (up to 20 mm diameter, mounted in a holder called a sabot) toward the waterdrop at speeds up to Mach 3. When the specimen traveling from right to left toward the falling drop crosses light beam 2, a photograph of the drop is taken to document its size and shape. The time needed for the specimen to traverse the distance between light beams 2 and 3 measures the impact speed. The long arm at the left contains a graded distribution of material that slows and captures the specimen without destroying it.

Impact in Fig. 7.29 takes place in a helium atmosphere at 10 Pa (0.1 torr) to eliminate bow shock that distorts waterdrops in a whirling arm. The experiment is designed to study the response of materials to waterdrops with a simple and reproducible geometry. Impacts at angles from 90° (perpendicular) down to 20° can be produced by tilting the sample in the sabot. The sample and holder are configured to avoid damage from edge effects and from stress wave reflections at the back of the specimen.

Nylon beads produce similar damage to water drops of the same size and speed impacting ZnS.[30,43,44] (By contrast, water encapsulated in a polyamide film, fish eggs, or ice particles produced damage that is qualitatively different from waterdrops.) A multiparticle impact facility at GRC International can launch a volley of nylon beads at speeds up to at least Mach 4 at a full scale infrared dome or even a larger radome. The objective is to simulate a rainfield in which a high speed projectile encounters many raindrops in a short period. A related test method uses an exploding wire to launch single 1.5-mm-diameter nylon beads toward an optical specimen at speeds up to Mach 12.[45] Damage threshold velocities for 1.2-mm nylon beads hitting various materials at 30° from perpendicular are reported to be approximately 1600 m/s for sapphire and ALON, 1500 m/s for cubic zirconia, 1100-1200 m/s for spinel and 600 m/s for lanthana-doped yttria.[45]

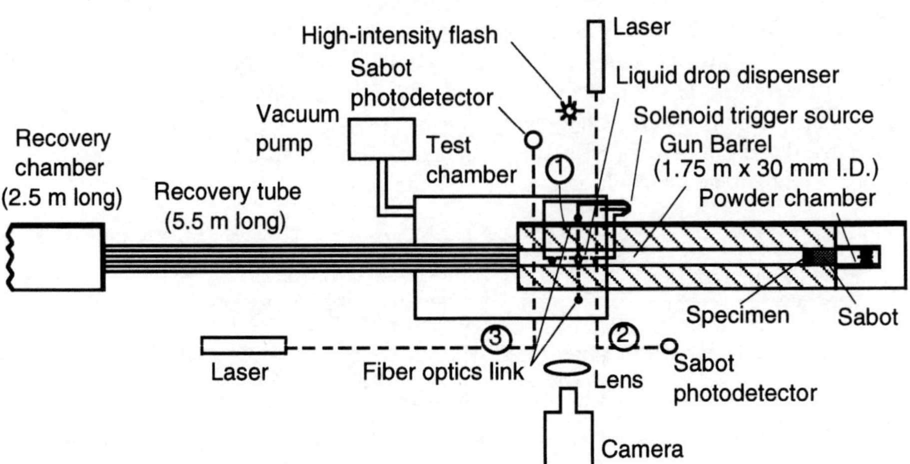

Fig. 7.29. Hydrometeor Impact Facility at GRC International hurls an optical specimen at speeds up to Mach 3 at a single, well-characterized, spherical waterdrop.[32]

7.5 Aerodynamic effects in rain erosion

A spherical raindrop is distorted as it traverses the shock wave surrounding a high speed vehicle such as a missile.[46-48] Figure 7.30 shows progressive distortion of an initially spherical waterdrop traveling at Mach 2 through an atmosphere of increasing pressure. When sufficiently distorted, the drop begins to disintegrate.

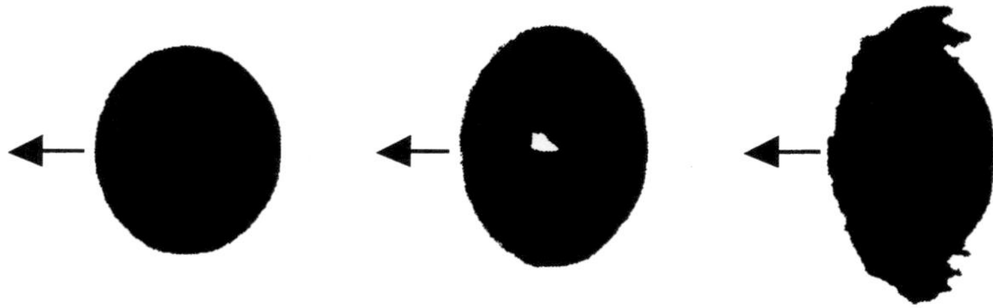

Fig. 7.30. *Left:* Slightly distorted, 5-mm-diameter waterdrop traveling at 883 m/s in the GRC Hydrometeor Impact Facility in Fig. 7.29.[48] *Center:* With some gas inside the gun barrel in Fig. 7.29, a 5-mm-diameter spherical drop traveling at 729 m/s distorts to a radius of curvature of 6.1 mm at the leading edge. *Right:* At higher gas pressure, a drop traveling at 769 m/s develops surface instability and begins to lose mass at the top and bottom of this silhouette. The radius of curvature is 6.4 mm at the leading edge.

Figure 7.31 shows the behavior of two raindrops colliding with a 15-cm-diameter hemispherical missile dome traveling at 610 m/s at an altitude of 6 km.[47] Both drops are distorted into elliptical shapes as they enter the shock wave. The larger drop deforms, but loses less than 10% of its mass prior to hitting the dome. The smaller, more distorted drop is in the process of breaking up by the moment of impact. More than 60% of its original mass has been lost.

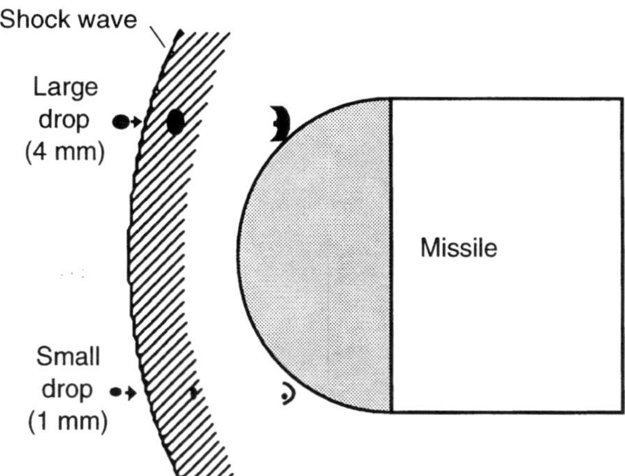

Fig. 7.31. Behavior of large and small drops as they travel approximately 8 cm through the shock wave in front of a 15-cm-diameter missile dome traveling at 610 m/s at an altitude of 6 km.[47]

Figure 7.32 shows the behavior of different drops on the same trajectory as those in Fig. 7.31, as a function of the initial drop size, D_o. The ordinate gives the fraction of initial mass remaining. An 8-mm drop loses almost no mass. A 2.5-mm drop retains 80% of its original mass at the time of impact. A 1.0-mm drop has less than 40% of its mass remaining. Drops smaller than 0.58 mm disintegrate before striking the dome.

The pictures in Fig. 7.32 show the distortion that would occur as different drops travel through the shock wave. The radius of curvature at the front of the distorted drop at the upper right is 2.5 times greater than its initial, undistorted radius. This 3.5-mm raindrop behaves as a >9-mm drop when it hits the dome. Damage to the dome is mainly a function of the effective diameter of the drop, even though the mass is unchanged.

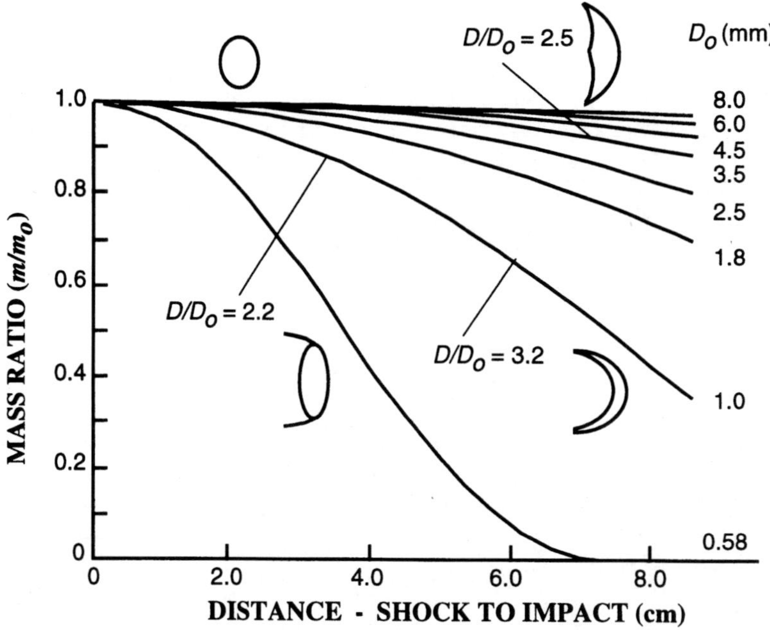

Fig. 7.32. Raindrop mass loss and distortion as it traverses the shock wave in Fig. 7.20 at a distance 6.5 cm from the centerline of the missile.[47] D_o is the initial diameter of the drop and m_o is its initial mass.

Table 7.7 confirms that impact damage increases with increasing radius of curvature of the leading edge of a waterdrop.[18] A spherical drop with a speed of 305 m/s and a diameter of 4.3 mm was distorted as in Fig. 7.30 by gas pressure in the gun barrel of the GRC Hydrometeor Impact Facility. After the drop impacted a ZnS target, the maximum fracture depth was measured. The process was repeated for six levels of distortion. Table 7.7 shows that the greater the radius of curvature, the greater the fracture depth. All drops had the same mass.

Table 7.7. Effect of waterdrop distortion on damage in zinc sulfide[18]

Radius of curvature at front of drop (mm)	Maximum fracture depth (μm)	Radius of curvature at front of drop	Maximum fracture depth
2.25	432	3.06	780
2.52	600	3.96	>1000
2.97	720	4.14	>1000

A computer code called DROPS calculates the distortion of waterdrops as they pass through the flowfield around a moving object.[48] Figure 7.33 shows two drops that pass through the shock wave of a missile traveling at Mach 2 at 3 km altitude. A slanted, flat window is located on one side of the missile. By the time they hit the window, initial 2-mm-diameter drops are distorted to a radius of curvature between 12 and 50 mm! Larger, 4-mm-diameter drops are less distorted, with radii of curvature between 22 and 33 mm when they reach the window.

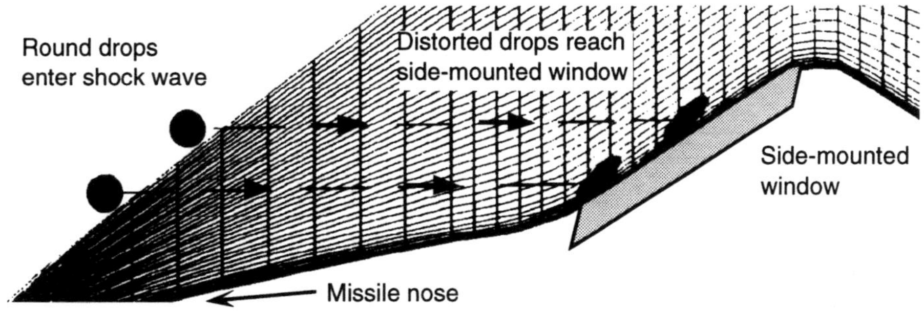

Fig. 7.33. Distortion of raindrops passing through the flowfield around a missile traveling at Mach 2 at 3 km altitude.[48]

A pod containing an infrared window below the fuselage of a supersonic aircraft provides an interesting example of design for rain erosion resistance.[47] If the pod is located sufficiently aft, the aircraft body, not the pod, creates the shock flow field through which raindrops must traverse to reach the pod. At supersonic speeds, the vast majority of raindrops disintegrate on their extended voyage through the shock wave prior to reaching the pod. By contrast, most of the raindrops that encounter the aircraft in subsonic flight survive long enough to hit the window on the pod. More damage is expected in subsonic flight than in supersonic flight.

7.6 Erosion by solid particles

At high altitudes, a vehicle is more likely to encounter ice particles than waterdrops. Ice balls do about half as much damage as waterdrops of the same size and speed in experiments with Plexiglas.[49,50] The effects of hypersonic velocity ice particle impacts on slip cast fused silica[50] (a radome material) and of wind tunnel debris on sapphire[51] have been reported. The erosion rate of fused silica[50] was approximately proportional to the *6th* power of the normal velocity of impact.

Sand particles, which are mainly SiO_2, account for a good deal of the solid particle erosion experienced by infrared windows. Figure 7.34 shows the kind of damage that sand does to zinc sulfide.[42] A plastically deformed central crater forms during the loading stage of particle impact. During the unloading stage, lateral cracks form below the crater and curve upward to intersect the front surface. The material within the circular ring of the lateral crack on the surface can easily be lost and is largely responsible for mass loss by sand erosion. Radial cracks extend deeper into the material than lateral cracks.†

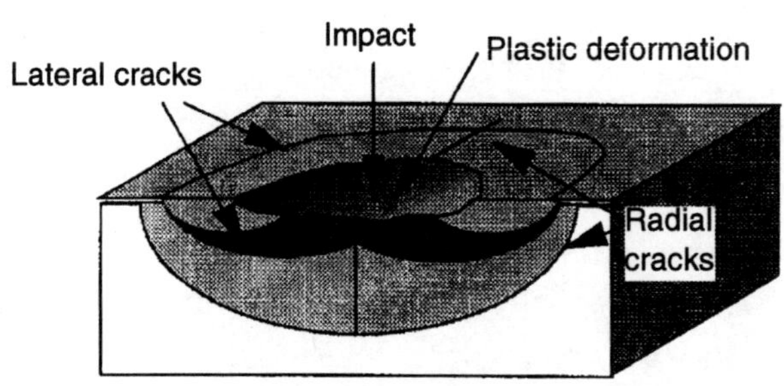

Fig. 7.34. Typical damage observed from solid particle impact on ZnS.[42]

Eroded windows lose mechanical strength.[53] When ZnS was eroded at 30 m/s by 450-μm-diameter sand particles for an unstated time, it lost 70% of its mechanical strength.[42] BK-7 glass lost 2/3 of its mechanical strength in a sand erosion test and 1/3 of its strength in a dust erosion test.[54]

An example of laboratory sand erosion apparatus is shown in Fig. 7.35. Coarse or fine sand dispensed from a conveyor belt or vibrator falls through the inlet section of a barrel through which dry air flows at high speed. After flowing through the long barrel, the particles have been accelerated to an appreciable fraction of the speed of the air flow. After leaving the barrel, the particles travel 8 cm through the air prior to striking the test specimen. Particle speed is measured prior to impact by a video camera. The Cambridge equipment provides particle velocities of 7-200 m/s and fluxes of 0.01-20 kg/m^2/s. Flux is measured by weighing the sand collected in the target chamber. A particular sand erosion rig commonly used in the United States is now located at Wright-Patterson Air Force Base. It rasters a plate holding up to sixteen 2.54-cm-diameter specimens in front of the sand spray so as to expose all specimens equally. Particle speeds can be selected in the range 30-500 m/s and angles of incidence range from 90° to 20°. Speed is measured by a laser Doppler velocimeter. Perhaps the simplest sand erosion test is based on ASTM Standard Test Method D-968, in which sand of a specified size range falls from a funnel by gravity through a distance of ~1 m before striking the test object. The impact velocity is ~4 m/s at a typical angle of incidence of 45°.

---

†Damage from a scratch scribed into a brittle solid is similar to the damage from solid particle impact. Lateral and radial cracks form when the scribe is removed.[52] Lateral cracks can be considerably wider than the plastically deformed scratch. Similar damage is expected from a particle of solid abrasive used for lapping or polishing an optical surface.

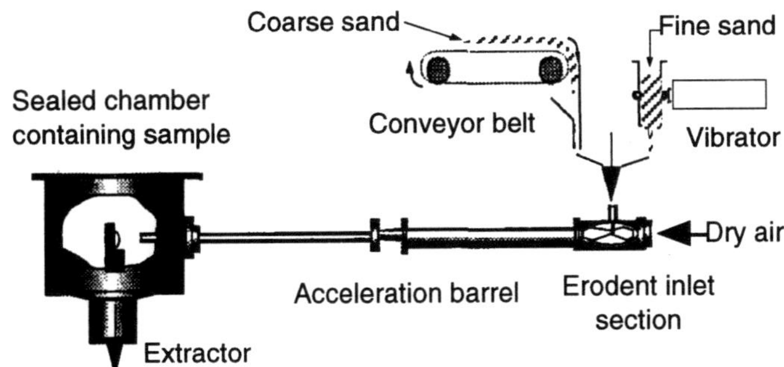

Fig. 7.35. University of Cambridge sand erosion rig.[42]

Figure 7.36 shows erosion rates of ZnS and Ge as a function of particle velocity and angle of incidence.[55] Erosion rate is measured in milligrams of target mass lost per kilogram of erodent impacting the surface. Erosion rates for ZnS and Ge increase as the angle of incidence approaches 90°. ZnS is eroded at a slower rate than Ge at low impact speeds, but the order switches at the highest speed.

Mass loss is proportional to the impact velocity, v, raised to an exponent, n:

$$\text{Erosion rate} = A\text{v}^n \tag{7-10}$$

where A is a constant. Table 7.8 gives erosion data for several infrared materials. The relatively soft materials, ZnS, Ge and calcium lanthanum sulfide, have erosion rate exponents of 2.3-3.5, arising from the elastic-plastic damage mechanism in Fig. 7.34. The low exponent of 0.6 for sapphire is thought to come about because the erodent, SiO_2, is softer than the target. Sand particles break up when they hit the sapphire.

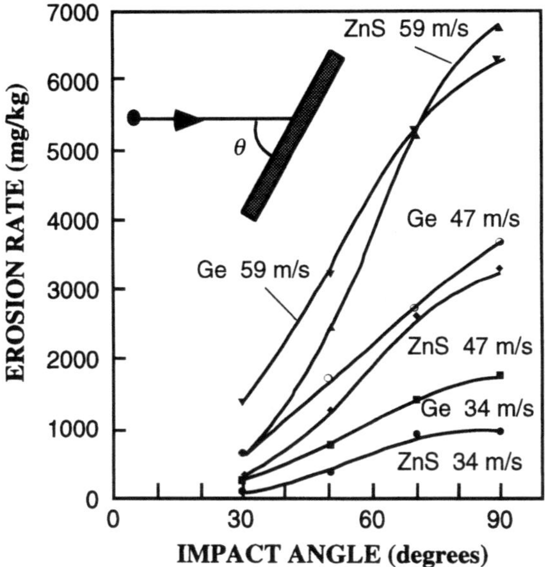

Fig. 7.36. Sand erosion rates of ZnS and Ge.[55]

Table 7.8. Sand erosion mass loss rates for infrared materials*[55,42]

Material	Erosion rate (mg/kg)	Exponent *n* in Eq. (7-10)
Zinc sulfide (34 m/s)	950	3.5
Zinc sulfide (80 m/s)	20 000	not measured
Germanium (34 m/s)	1730	2.3
Calcium lanthanum sulfide (34 m/s)	7500	2.3
Sapphire (80 m/s)	83	0.6
CVD diamond (nucleation surface) (80 m/s)	0.18	not measured
Natural diamond (100 face) (140 m/s)	0.05	not measured

*300-600 μm sand impacting at normal incidence

Sand erosion damage is commonly characterized by infrared transmission loss in the material. Figure 7.37 shows the effects of particle size and impact velocity on ZnS.[42] Transmission loss is more severe for larger particles and higher speeds.

Micrographs in Fig. 7.38 show that MgF_2 is more easily damaged than aluminum oxynitride (ALON) and spinel by sand impact. In Table 3.8, we find the Knoop hardness of the materials to be 580 kg/mm^2 for MgF_2, 1600 for spinel, and 1800 for ALON. The harder the material, the less it is damaged by the sand. In a sand erosion study of ALON, impact damage could be observed under a microscope with no measurable loss of mass and negligible change in infrared transmission.[38] Figure 7.38 also shows that two commercial antireflection coatings, which were not designed for durability, were damaged much more than the more durable substrate material. Section 7.8 discusses coatings that are designed to impart durability to windows. In general, few coatings are as erosion resistant as ALON, spinel or sapphire.

Figure 7.39 illustrates relative erosion rates of different materials measured by loss of optical transmission.[56] The order of erosion is diamond < calcium aluminate < silicon << germanium < standard ZnS < multispectral-ZnS. In a different study comparing standard ZnS to multispectral ZnS, the mass loss rate for multispectral ZnS was approximately three times greater than the rate for standard ZnS (for sand particles in the range 100-500 μm at a speed of 50 m/s and a flux of 173 g/m^2/s).[57]

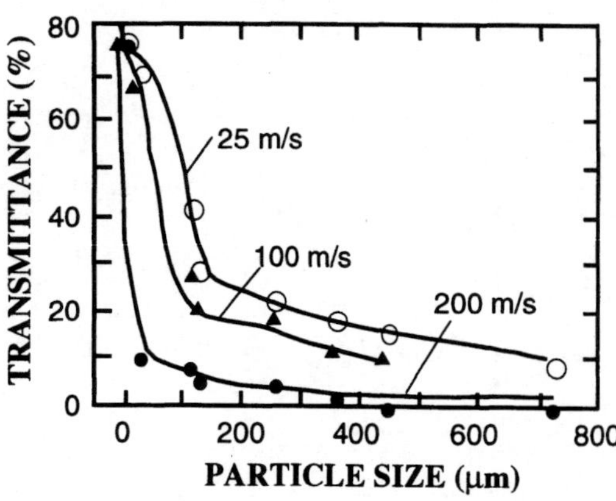

Fig. 7.37. Transmission loss in ZnS at 10 μm in a sand erosion experiment.[42]

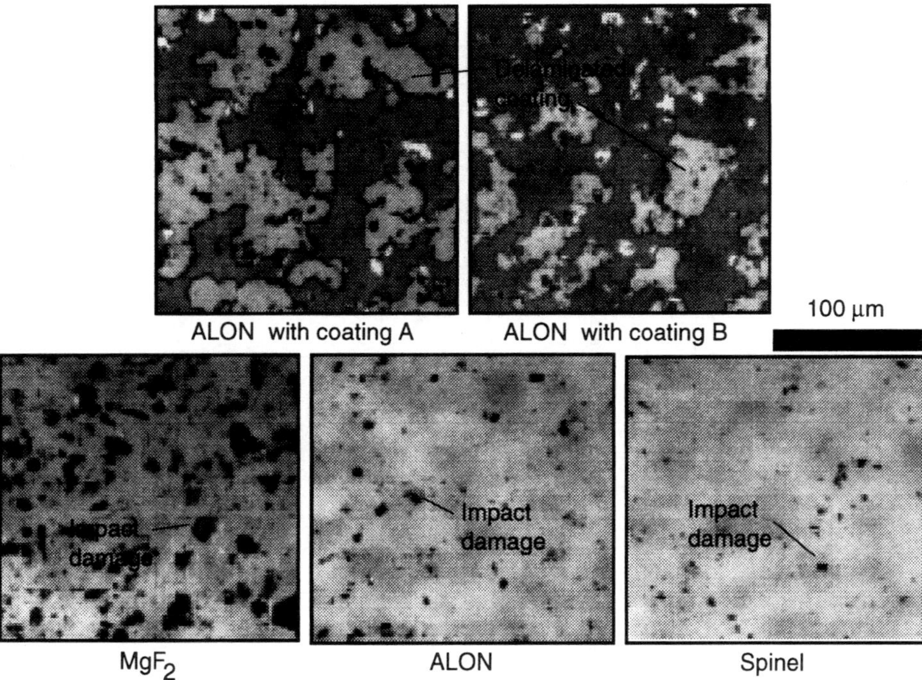

Fig. 7.38. Sand erosion damage[58] from particles of diameter ≤38 μm impacting at normal incidence at 206 m/s for a total load of 8 mg/cm^2.

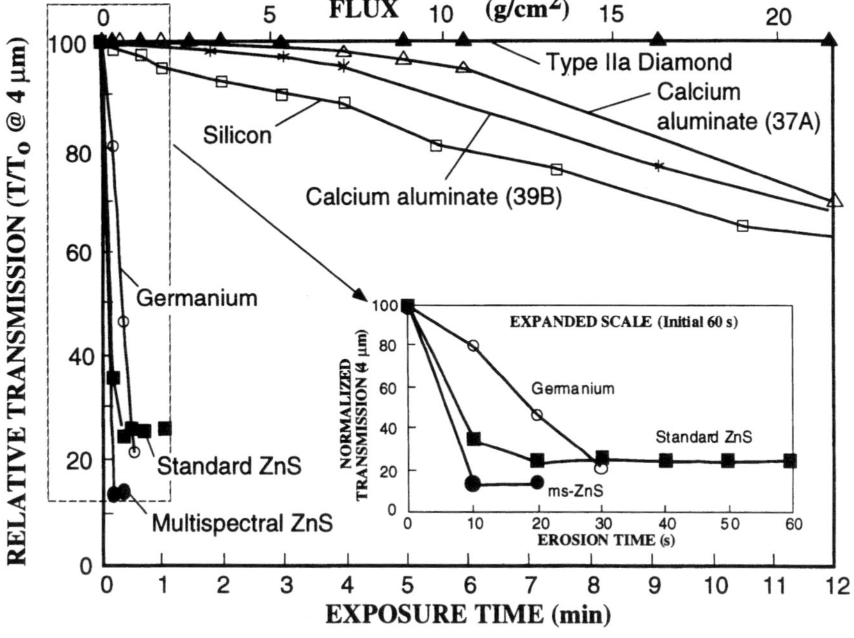

Fig. 7.39. Relative sand erosion rates of materials exposed to a commercial sand blaster at a 45° angle of incidence.[56] Diamond exhibited no loss of transmission and multispectral zinc sulfide was degraded most rapidly.

Natural diamond is the hardest known material and the most resistant to sand erosion. Some data were given in Table 7.8 and Fig. 7.39, where natural diamond is labeled "Type IIa." Although it is also extremely erosion resistant, chemical-vapor-deposited (CVD) synthetic diamond loses mass approximately 10 times faster than natural diamond in sand erosion experiments.[42] Mass loss by natural diamond by sand impact at 26 m/s was 2×10^4 times lower than the rate of mass loss by silicon nitride under impact at 47 m/s and 3×10^6 times lower than the erosion rate of alumina under impact at 34 m/s.[59] Even though diamond is extremely erosion resistant, the fact that small particles of sand at low speed (Mach 0.1) produce observable damage in diamond implies that sand erosion is a problem for all infrared window materials.

7.6.1 Combined effects of sand and rain

A sand-eroded surface is much more susceptible to rain impact damage than is a pristine surface.[60-62] When ZnS was subjected to whirling arm rain erosion (Fig. 7.40), it suffered a 2.2% loss in long wave infrared transmission.[61] Another specimen suffered a 9.7% transmission loss when exposed to sand. If ZnS was exposed to rain and then to sand, the transmission loss was just 10.3%. If ZnS was exposed to sand before rain, the transmission loss increased to 24.8%. Specimens exposed to sand before rain have thin surface cracks linking sand erosion pits. That is, pits generated by sand impact act as initiation sites for damage by raindrops. A similar effect was reported for ZKN7 glass.[60] Waterdrop impact conditions that would do no damage to a pristine surface did moderate damage to a sand-eroded surface.

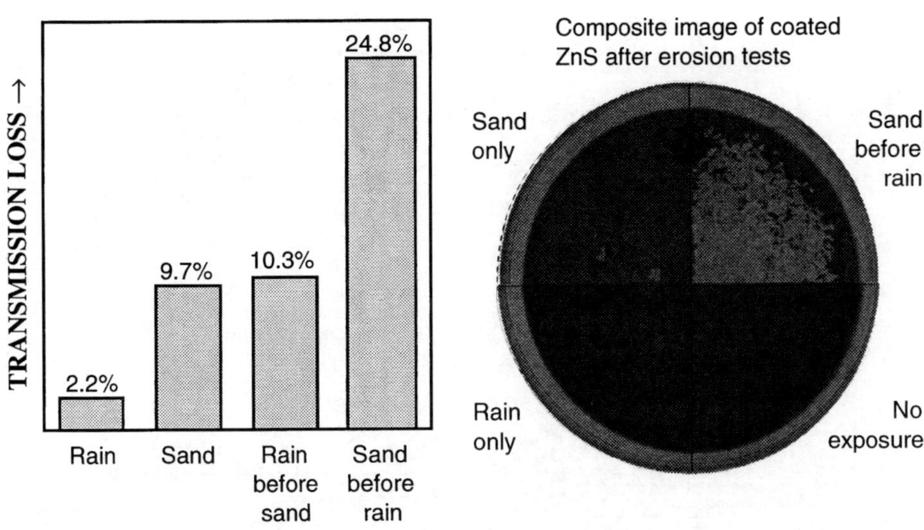

Fig. 7.40. Combined effects of sand and rain on erosion of ZnS. *Left:* Transmission loss in the 8-11.5 μm band after rain impact (5 min in whirling arm rainfield at 252 m/s with 2.54 cm/h rainfall of 2-mm-diameter drops) and sand impact (88-105 μm particles at 97 m/s for 6 min with total load of 70 mg/cm^2).[61] *Right:* Composite image of coated ZnS after erosion tests.[62] The coating is the Raytheon "DAR/REP" coating discussed in Section 7.8.4.

7.7 Effect of angle of incidence on erosion

Figure 7.1 showed a dome that had traversed a rainfield at supersonic speed. The apex was severely damaged, but the skirt was hardly affected. Impacts on the apex are near normal (90° impact angle), whereas impacts at the skirt are at a small angle of incidence. If damage were proportional to the normal component of velocity, then the damage should scale as $v \sin\theta$, where v is velocity and θ is the angle of incidence in Fig. 7.36.

7.7.1 Waterdrop impact at inclined angles

The most carefully controlled angle-of-impact experiments are single drop studies at the GRC Hydrometeor Facility in Fig. 7.29.[18] For 2.1-mm drops, the damage threshold velocities are 160 m/s for $\theta = 90°$, 262 m/s for $\theta = 60°$ and 358 m/s for $\theta = 30°$. If the damage threshold velocity is proportional to the normal component of velocity, then we could calculate the threshold velocities for 60° and 30° by using the equation

$$\text{Damage threshold velocity} = v_{\perp}/\sin\theta \qquad (7\text{-}11)$$

where $v_{\perp}$ is the damage threshold velocity for normal incidence and θ is the angle of incidence. Equation (7-11) gives the calculated values in Table 7.9. For 2.1-mm drops, calculated thresholds are lower than the observed thresholds. Therefore Eq. (7-11) is conservative: It predicts that damage should be observed at a velocity lower than the actual threshold velocity. For 3.5-mm drops, Eq. (7-11) is conservative for 60° impact and approximately correct for 30° impact. For the 4.5-mm drop, Eq. (7-11) is approximately correct for 30° impact.

What other evidence is there regarding Eq. (7-11)? Déom *et al.*[27,63,64] observed that Eq. (7-11) was obeyed by MgF_2 and Ge in the SAAB whirling arm, but not in the DERA whirling arm (Table 7.6). They were measuring the time for 10% contrast loss, as in Fig. 7.18. For the DERA whirling arm, damage at inclined angles was less severe than predicted by Eq. (7-11). In UDRI whirling arm experiments with ZnS, damage measured by transmission loss either followed Eq. (7-11) or was less severe than predicted by Eq. (7-11).[7] Whirling arm experiments with a variety of materials, not just ceramics, suggest that impact damage measured by mass loss decreases as $\sin^n\theta$, where n is in the range 3-6.[65-67] Whirling arm results must be questioned because at least some whirling arm tests are not reliable for non-normal angles of incidence (Section 7.4.1).[35] Gorham and Field reported that damage to three different materials in a waterjet test was worse for impact angles of $\theta = 10\text{-}20°$ than for $\theta = 0°$, but only at supersonic speeds.[68] They cite other studies which show that for subsonic impact the damage is worst at normal incidence ($\theta = 0°$).

Table 7.9. Single drop damage threshold velocity (m/s) for ZnS[18]

Impact angle, θ	2.1-mm-diameter drop Observed	2.1-mm-diameter drop Calculated*	3.5-mm-diameter drop Observed	3.5-mm-diameter drop Calculated*	4.5-mm-diameter drop Observed	4.5-mm-diameter drop Calculated*
90°	160	---	140	---	143	---
60°	262	185	232	162	---	---
30°	358	320	268,302	280	268	286

*Calculated from $v_{\perp}/\sin\theta$, where $v_{\perp}$ is the observed threshold velocity for normal impact.

7.7.2 Sand impact at inclined angles

In sand erosion testing, specimens can be held at their intended inclined angle in the sand spray. Fig. 7.36 showed mass loss by ZnS and Ge as a function of impact velocity and angle. The results in Fig. 7.36 can be fit with the equation

$$\text{Mass loss} = A\text{v}^n \sin^m\theta \qquad (7\text{-}12)$$

where v is the impact velocity, θ is the impact angle, A is a constant, and n and m are experimental exponents. The results for ZnS erosion are $A = 0.0126$, $n = 3.23$ and $m = 3.74$. For Ge erosion the constants are $A = 0.250$, $n = 2.48$ and $m = 2.41$. In both cases the dependence of erosion rate on angle of incidence falls off faster than $\sin\theta$.

To summarize, in both rain and sand erosion experiments, damage falls off at least as fast as the normal component of velocity, or it falls more rapidly. This observation allows us to suggest criteria for comparative erosion testing of materials.

7.7.3 Comparative erosion testing of materials

Existing standards for erosion resistance of candidate materials for a particular application are absolute standards. They specify a test condition and an allowable response. If a material is known to meet system requirements, then comparative testing could be done on new materials to see if they are "as good as" the material whose performance is known.

Hot pressed polycrystalline magnesium fluoride is a material whose performance in the field has been observed for decades on various missiles. The general experience is that MgF_2 domes do not suffer damage from raindrop impact at aircraft speeds. Experience in the Persian Gulf tells us that MgF_2 is eroded rapidly by sand impact at aircraft speeds. MgF_2 is therefore a reasonable standard for comparative rain erosion measurements of new materials, but a better material than MgF_2 needs to be qualified as a sand erosion resistance standard.

A procedure has been proposed[69] for comparative testing of materials to see if they are "as good as" a selected standard material. The angle of incidence of rain or sand on a window is a critical factor in the severity of the resulting damage. A dome material should be tested at normal incidence ($\theta = 90°$ in Figure 7.36) because the front of a dome is impacted at normal incidence. Flat windows that will be mounted at an inclined angle should be qualified for the angle at which they will be used.

The suggested protocol for comparative testing of flat samples uses the whirling arm and rejects the idea of testing at any angle other than $\theta = 0°$, because non-normal angles of incidence are not considered reliable in the whirling arm.[35] It is therefore necessary to extrapolate impact conditions for an inclined surface to impact at normal incidence. If the waterdrop has velocity v, the normal component of velocity is $\text{v}\sin\theta$. Sections 7.7.1 and 7.7.2 indicate that impact damage is proportional to $\text{v}\sin\theta$ or is less severe.

Figure 7.41 shows that for an angle of incidence θ, a window of area A moving at velocity v sweeps out a volume $A\text{v}t\sin\theta$ in time t. That is, an inclined window experiences fewer raindrop impacts than a vertical window moving at the same speed for the same time. For a vertical window to receive the same number of impacts as an inclined window, the test time should be reduced by $\sin\theta$.

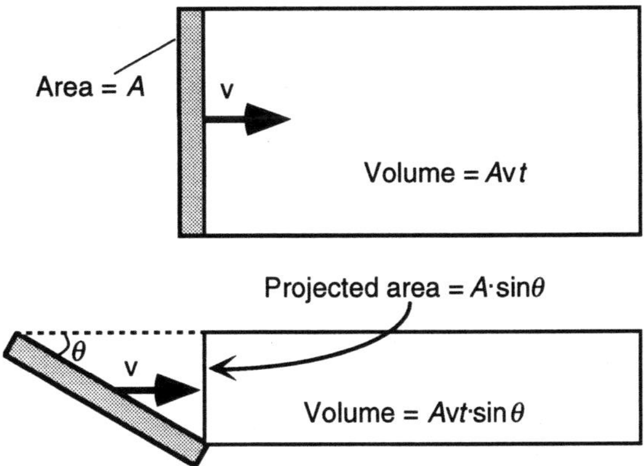

Fig. 7.41. For normal incidence, a window of area A moving at velocity v sweeps out a volume Avt in time t. If the angle of incidence is θ, the volume swept out is $Avt\sin\theta$.

To simulate the environment of an inclined window when testing a window at normal incidence, it is therefore necessary to reduce the impact velocity by $\sin\theta$ and to reduce the time of exposure by $\sin\theta$. The following scheme is proposed for comparative testing of rain erosion resistance using MgF_2 as an acceptable standard:

Suggested procedure for comparative rain erosion test with whirling arm

- Test hot pressed MgF_2 in the whirling arm at perpendicular incidence using 2-mm-diameter drops at a rainfall rate of 25.4 mm/h. The standard velocity is $v_o = 210$ m/s and the standard time is $t_o = 10$ min. Standard samples have a diameter of 25 mm and a thickness of 5 mm. The thickness should be great enough so that gross mechanical fracture does not occur.

- Expose the candidate window material (same size and thickness as MgF_2) at perpendicular incidence using 2-mm-diameter drops at a rainfall rate of 25.4 mm/h.

- Reduce the whirling arm velocity to $v = v_o\sin\theta$, where θ is the intended operational angle of incidence of the inclined window. For example, if the window will have $\theta = 30°$, then the velocity will be (210 m/s)sin30° = 105 m/s.

- Reduce the time of exposure to $t_o\sin\theta$. For $\theta = 30°$, the exposure time will be 5 min.

- Test 2-4 specimens of each type in consecutive runs in the whirling arm.

- Inspect the specimens with a microscope at a magnification of 100-200×. A material is judged to be "as good as" MgF_2 if the observed damage is "similar to or less than" the damage observed in MgF_2. The comparison is the difficult and subjective part of the procedure. At its simplest, the comparison can be based on the number of impact sites per unit area viewed under the microscope. However, different materials and coated materials can experience qualitatively different damage. The area and apparent depth of damage and extent of delamination of a coating should all be considered in making the comparison to decide if the candidate material is "as good as" MgF_2.

In sand erosion testing, specimens can be held at their intended inclined angle in the sand spray. The test below is predicated on having a "standard material" whose sand erosion resistance for normal incidence is considered acceptable for a given application.

Suggested procedure for comparative erosion test with spraying sand

- Expose the desired standard material to blowing sand at perpendicular incidence. A convenient set of conditions is 149-177 μm diameter sand particles at a velocity of 75 m/s with a total exposure of 150 mg of sand per square centimeter of sample surface. Conditions should be chosen to give a relative transmission loss ($\Delta T/T_o$) of 5-50%, where T_o is the initial transmittance and ΔT is the change in transmittance after sand exposure.

- Expose the candidate window material at its intended angle of incidence to the same blowing sand conditions used for the standard material.

- The time of exposure is the same as the time used for the standard material.

- Test 2 specimens of each type. All samples can be run simultaneously if $\theta = 90°$ or the standards and test materials can be in consecutive runs if $\theta \neq 90°$.

- Measure the relative transmittance loss in the spectral region at which the window will be used, preferably where the initial spectrum is flat.

- Inspect the specimens with a microscope at a magnification of 100-200×.

- A test material is judged to be "as good as" the standard material if the relative transmittance loss is less and the "area of the damage sites" is less than that of the standard.

- If there is not much transmission loss for either the standard or the test sample, both materials should be tested a second time to double the total exposure.

7.8 Protective coatings for erosion

Significant improvements in erosion-protection coatings for infrared windows have been made in recent years. The coating must also provide antireflection functionality, which makes its design that much more difficult. Durable coatings are now available for windows that will not exceed aircraft temperatures. Coatings for hotter systems, such as high-speed missiles, are even more difficult because they must survive higher speed impacts, have low infrared emittance, and remain bonded to the substrate during rapid heating to high temperatures.

7.8.1 Mechanisms of protection by coatings

A thin layer of rigid (high modulus) material[70,71] can protect a brittle substrate against impact damage if the coating adheres strongly to the substrate. Particle impact on the high modulus coating produces less strain in the coating than it would in the substrate. The coating transmits less stress into the substrate than direct impact on the substrate. Figure 7.42 shows the calculated reduction in stress as a function of modulus and thickness of the rigid coating. Protection increases with coating modulus and thickness, but reaches an asymptote for sufficiently thick coatings (thickness greater than 0.4 times the contact radius [Eq. (7-4)] of the impacting particle).

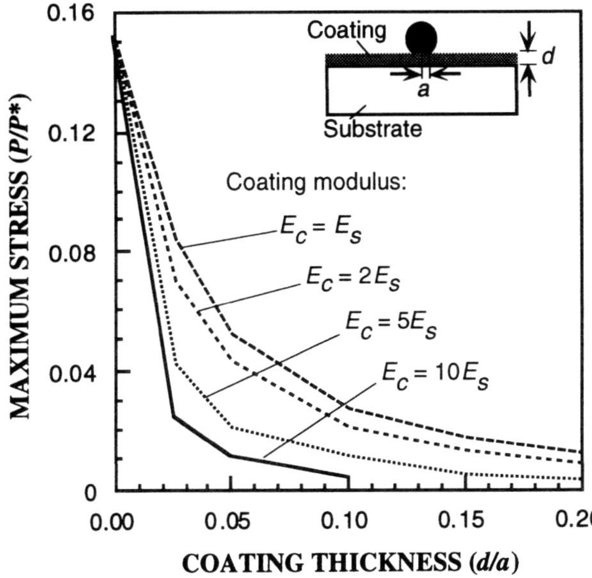

Fig. 7.42. Maximum radial tensile stress at the interface of a rigid coating (Poisson's ratio = 0.20) attached to a brittle substrate.[70] Coating modulus (E_C) ranges from one to ten times the modulus of the substrate (E_S). Stress is expressed as a multiple of the maximum stress (P^*) on the surface of the coating. Thickness (d) is expressed as a multiple of the contact radius (a) of the impacting particle.

A model for the behavior of rain erosion protective coatings rationalizes the observed performance of many coatings, as shown in Fig. 7.43.[72,73] The graph shows the relative increase in waterjet damage threshold velocity as a function of coating thickness for a variety of coatings on zinc sulfide or germanium. The damage threshold velocity increases significantly for the first 10 μm of coating thickness and then there is only slow improvement with increasing thickness.

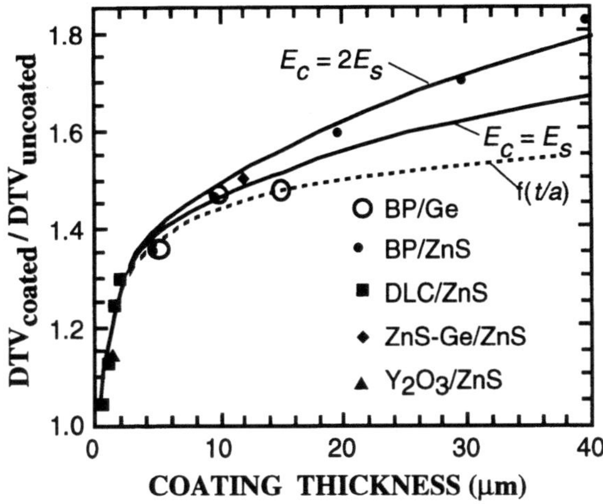

Fig. 7.43. Increase in waterjet damage threshold velocity for coatings on ZnS or Ge.[73] BP = boron phosphide, DLC = diamond-like carbon. ZnS-Ge coating contains both ZnS and Ge. Upper two curves are calculated with Eq. (7-13). Dotted line is the first term in Eq. (7-13). The flaw in Fig. 7.44 is presumed to be of length $2a$ = 50 μm for these calculations.

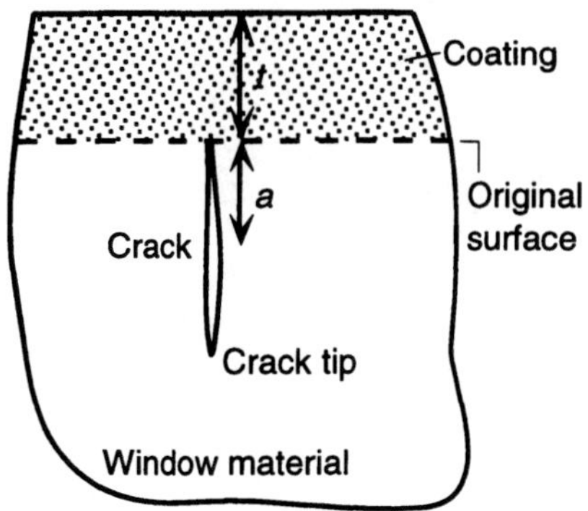

Fig. 7.44. A crack of length $2a$ initially at the surface is buried to depth t by a protective coating. The stress at the crack tip decreases as t increases because (1) the stress intensity factor decreases with increasing distance from the surface and (2) the Rayleigh wave that provides the tension to open the crack is attenuated with increasing distance below the surface.

The model on which Fig. 7.43 is based is the buried crack in Fig. 7.44. Prior to coating, a critical flaw of depth $2a$ terminates at the surface. This crack can initiate mechanical failure when the Rayleigh surface wave from a raindrop impact passes by. A coating of thickness t buries the crack below the new surface. If the coating adheres perfectly to the window, and if the coating is durable enough not to be damaged by the raindrop impact, and if the crack does not propagate into the coating (three big ifs!), then the only direction for crack propagation is downward from the crack tip in Fig. 7.44. If all these conditions are met, then the main factor governing crack propagation is the reduced stress intensity factor (Section 8.2) at the crack tip when it is buried. As the crack is buried deeper, it takes a larger and larger force at the surface to open the crack.

A second, smaller effect that makes it harder to open the buried crack is that the amplitude of the Rayleigh wave propagating along the surface is attenuated below the surface. The deeper the crack, the less it is affected by the Rayleigh wave on the surface.

Figure 7.43 showed the increase in damage threshold velocity for coated ($\mathrm{DTV_{coated}}$) versus uncoated ($\mathrm{DTV_{uncoated}}$) materials. The lines are computed from the equation

$$\frac{\mathrm{DTV_{coated}}}{\mathrm{DTV_{uncoated}}} = \mathrm{f}(t/a) + \frac{2}{3}\left(\frac{E_c}{E_s}\right)\left(\frac{c_w^2}{c_R\,\mathrm{DTV_{uncoated}}}\right)\left(\frac{t}{r}\right) . \tag{7-13}$$

The term f(t/a), which is a function of the quotient of coating thickness, t, and crack radius, a, accounts for the decreased stress intensity factor at the crack tip as the crack is buried deeper below the surface. Numerical values are given below:

t/a	f(t/a)	t/a	f(t/a)	t/a	f(t/a)
0	1	0.250	1.39	1.50	1.54
0.053	1.21	0.429	1.44	2.33	1.55
0.111	1.32	0.667	1.48	∞	1.59
0.176	1.36	1.00	1.51		

The critical flaw size for ZnS was taken as $2a$ = 50 μm to compute the curves in Fig. 7.43. The numerical values of f(t/a) say, for example, that if t/a = 0.176, which means that t = 4.4 μm, then the damage threshold velocity will be increased by 36%.

The second term in Eq. (7-13) accounts for attenuation of the Rayleigh wave with increasing depth. E_c is Young's modulus for the coating, E_s is the modulus of the substrate, c_w is the compression wave speed in water, c_R is the speed of the Rayleigh wave on the uncoated substrate, t is the coating thickness and r is the radius of the waterdrop or the effective radius of the front of a waterjet. The compression wave speed in water was given below Eq. (7-3) as $c_w \approx 1500$ m/s + 2v, where v is the impact speed (= DTV_{coated}). The Rayleigh wave speed was given in Eq. (7-7). For ZnS it is 2450 m/s.

The lowest curve in Fig. 7.43 is the term f(t/a) in Eq. (7-13). It accounts for most of the improvement in damage threshold velocity. What this means is that simply burying a crack beneath the coating increases the surface stress required to open the crack. *Much of the value of the protective coating is that it buries the crack, which decreases the stress intensity factor at the crack tip.* To perform this function, the coating must be strongly adherent and able to withstand impact stresses without failing.

By virtue of their ability to alter the stress near surface cracks, thin coatings sometimes can increase the mechanical strength of materials. For example, a 10-µm-thick layer of boron phosphide (Section 7.8.3) increased the flexure strength of germanium disks from 104 to 185 MPa.[74] A 1-µm-thick antireflection coating of Y_2O_3 on ZnS increased the flexure strength by 25%.[75] Sapphire coated at 1400°C with a 60-µm-thick glass glaze (similar to a pottery glaze) is approximately twice as strong as uncoated sapphire at 25°C and at 500°C.[76]

In contrast to the behavior of a rigid coating, a compliant coating can also protect a substrate from impact damage. The compliant coating works by spreading the stress over a greater area.[77] There is synergy between coating and substrate: The coating reduces the stress that is transmitted to the substrate, while the rigid, adherent substrate restricts the strain that can build up in the coating. For effective protection, the coating must be strongly bound to the substrate. The coating should be an elastomer with Poisson's ratio close to 0.5 so that large strains are accommodated by elastic deformation. The thickness should be approximately 20% of the contact radius of the impacting particle. Thinner coatings do not provide sufficient protection and are more likely to separate from the substrate because of high interfacial shear stresses. If the coating is too thick, reinforcement gained by attachment to the rigid substrate is lost.

Another protective structure, shown in Fig. 7.45, features a rigid *cladding* bound by a compliant, transparent, adhesive layer to the optical window. A cladding is a coating that is thick enough to be free standing. Typical coatings are <50 µm thick and claddings are typically 100 - 1000 µm thick. When a waterdrop strikes the cladding in Fig. 7.45, compression and shear waves are transmitted through the cladding (Fig. 7.7). At the interface between two materials (designated by subscripts 1 and 2 in Eq. [7-14]), waves are reflected just as light is reflected at the interface between air and glass. The fraction of energy of an incident stress wave reflected at the boundary is[78]

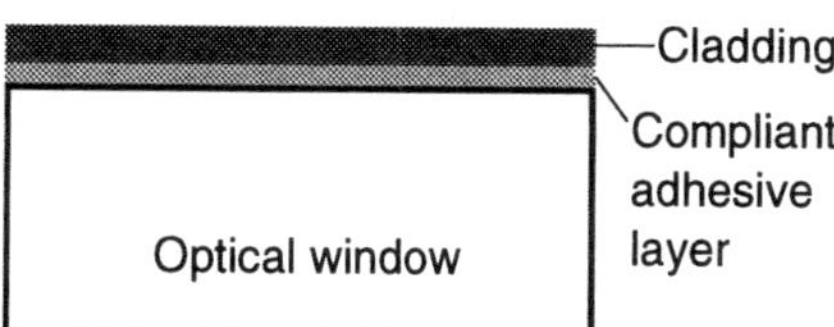

Fig. 7.45. Protective cladding attached to an optical window by a compliant adhesive layer.

$$\text{Fraction of energy reflected} = \frac{(Z_1 - Z_2)^2}{(Z_1 + Z_2)^2} \quad (7\text{-}14)$$

where Z is the *acoustic impedance* of each medium, given by

$$Z = \text{density} \times \text{speed of sound} = \sqrt{\frac{E\rho\,(1 - \nu)}{(1 - \nu)\,(1 - 2\nu)}} \quad (7\text{-}15)$$

where ρ is the density of the material, E is Young's modulus and ν is Poisson's ratio. The speed of sound is the speed of the compression wave, also called a longitudinal wave. An expression for the speed of the longitudinal wave was given in the footnote beneath Eq. (7-8). Equation (7-14) is analogous to Eq. (1-10) for reflection of light at an interface. The physics behind reflection of a shock wave and reflection of a light wave is similar.

If a cladding is directly deposited on a window, and if each has a similar acoustic impedance, then the stress from an impact passes from the cladding to the window with little reflection. A compliant interface layer has much lower acoustic impedance than a ceramic because Young's modulus for the adhesive is much lower than Young's modulus for the ceramic. The adhesive layer in Fig. 7.45 reflects a large fraction of the impact energy back into the cladding. Little of the impact energy reaches the optical window. A key requirement is that the cladding must be strong enough to absorb the impact energy without shattering and the interface must be strong enough to prohibit delamination.

Example: Reflection of stress waves in cladded structures. In the structure at the left below, a cladding of yttria is directly bonded to a ZnS window. What fraction of incident stress wave energy is reflected back into the yttria at the Y_2O_3/ZnS interface? In the structure at the right, what fraction of energy is reflected at the Y_2O_3/adhesive interface?

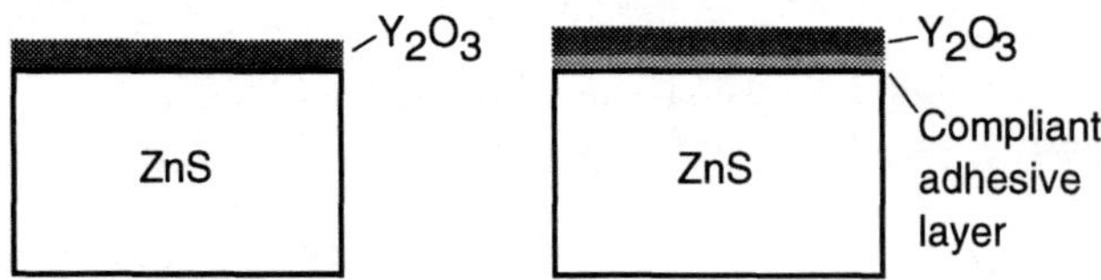

Material	E (GPa)	ρ (kg/m^3)	ν	Z (kg m^{-2} s^{-1})
Y_2O_3	173	5030	0.30	3.4×10^7
ZnS	74	4080	0.29	2.0×10^7
Adhesive	0.10	1000	0.48	9.4×10^5

E and ν are from Table 4.2. Density is from Table 3.6. Z is from Eq. (7-15). Properties of the adhesive are hypothetical.

Reflected energy is computed with Eq. (7-14). For the Y_2O_3/ZnS interface we find

$$\text{Fraction reflected} = \frac{(3.4 \times 10^7 - 2.0 \times 10^7)^2}{(3.4 \times 10^7 + 2.0 \times 10^7)^2} = 6.7\% \,.$$

For the Y_2O_3/adhesive interface, the reflected energy is

$$\text{Fraction reflected} = \frac{(3.4 \times 10^7 - 9.4 \times 10^5)^2}{(3.4 \times 10^7 + 9.4 \times 10^5)^2} = 90\% .$$

Approximately 90% of the impact energy is reflected back into the yttria cladding from the compliant layer. Only 6.7% is reflected into the coating in the 2-layer structure.

Figure 7.46 shows an impressive example of reflection of impact energy at a compliant interface.[79] The left photograph shows damage propagating from a ZnS cladding into a ZnSe window when the two layers are in direct contact in the commercial material, Tuftran.® The right photo shows that fractures do not propagate from the ZnS into the ZnSe through a compliant polymer adhesive layer. Damage in the ZnS layer, as measured by loss of transmission, is greater with the polymer adhesive, but the ZnSe is free of damage.

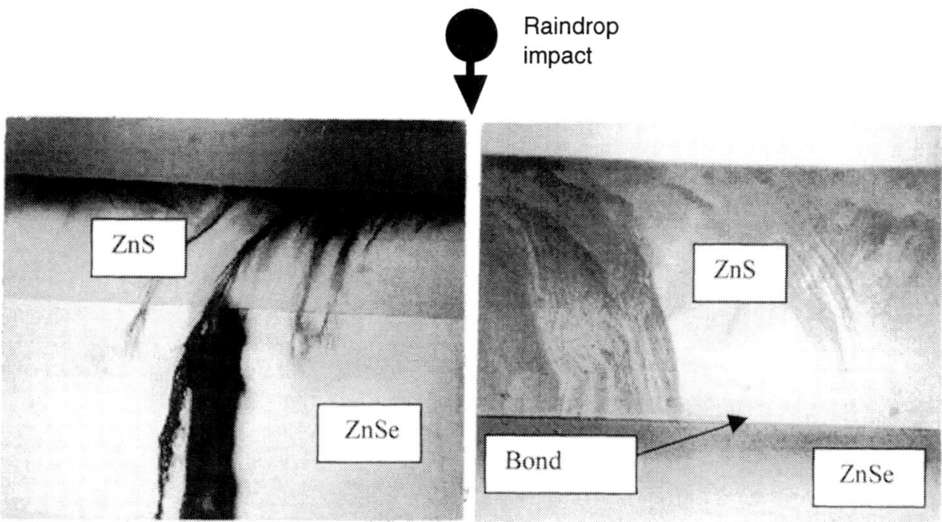

Fig. 7.46. Cross-sectional view of whirling arm waterdrop impact damage in Tuftran® ZnS/ZnSe composite (*left*) or a ZnS cladding bonded to a ZnSe substrate with a compliant polymer adhesive (*right*).[79] Damage is not transmitted through the compliant interlayer into the ZnSe substrate.

Finite element analysis has been used to model waterdrop impacts on composite windows with a compliant interlayer.[80] The maximum tensile stress in the outer cladding is likely to be from flexure of the rear surface of the cladding directly beneath the impact point.

7.8.2 Diamond-like carbon and germanium-carbon coatings

"Diamond-like carbon," often abbreviated DLC, is a class of hard, dense, amorphous hydrocarbons or pure carbon materials.[81-84] The densities are greater than graphite (2.25 g/mL), but lower than diamond (3.51 g/mL). The hydrocarbons are very rich in carbon, with compositions in the approximate range $CH_{0.3}$ to $CH_{1.5}$. The material with no hydrogen is often called "hard carbon." Diamond-like carbon can be prepared by a variety of deposition methods that have in common the presence of 10-100 electron-volt carbon ions bombarding the surface during deposition. The Knoop hardness of diamond-like carbon, 1800-2000 kg/mm^2, is similar to that of sapphire. The index of refraction of ~2.0 at 10 μm makes diamond-like carbon a useful protective/antireflection coating for germanium, which has an index of refraction of 4.0. Coatings range from pale yellow to brown and the band gap for material grown from methane in a radio frequency plasma is 1.06 eV. "Diamond-like carbon" prepared by electron cyclotron resonance chemical vapor deposition from neopentane and nitrogen is reported to have an index of refraction of 1.7 and good visible transparency when the thickness is 1.4 μm.[85]

Diamond-like carbon is almost always deposited with high compressive stress (~0.5 - 1 GPa[63,64,81]), which prevents growth of thicknesses greater than ~2 μm. Its high absorption coefficient (150-400 cm^{-1} in the 8-12 μm region) also makes it futile to grow thick optical coatings. Diamond-like carbon prepared by magnetron sputtering of graphite can be prepared with an absorption coefficient as low as 80 cm^{-1} at 10.6 μm.[86] With a thickness of ~0.1 μm, such a coating can protect KCl laser windows from moisture without being damaged by the CO_2 laser.

Diamond-like carbon is excellent for protection against abrasion (scratching) by solid particles, and is used on eyeglasses because it can be deposited on substrates close to room temperature. It adheres well to germanium, silicon and fused silica, but ZnS and ZnSe require thin intermediate layers of Ge to promote adhesion. Diamond-like carbon is not very good for rain erosion[81,83,87,88] because it cannot be grown in thick layers. With a thickness of ~1.3 μm, diamond-like carbon is reported to increase the time for 10% loss in optical contrast (Section 7.3.2) of Ge by a factor of 3 with 2-mm raindrops in a whirling arm test.[63,64] Diamond-like carbon finds its principal use as a thin abrasion-resistant outer layer in multilayer optical coatings. Another excellent abrasion-resistant coating called "amorphic diamond"[89] is made of amorphous carbon containing no hydrogen. Made by laser ablation of carbon from a graphite target, this black-colored coating is limited to thin layers because it has strong absorption across the infrared spectrum. Figure 7.47 shows the increased waterjet damage threshold velocity (Section 7.4.3) of amorphic diamond compared to diamond-like carbon.[24] Another hard carbon coating made by radio frequency plasma chemical vapor deposition is a durable material with relatively low internal stress, but high absorption coefficient.[90] The absorption problem was circumvented by making multilayer coatings containing hard carbon and silicon.

Germanium-carbon[91,92] (sometimes incorrectly called "germanium carbide"), a hard, amorphous material containing Ge, C and H, is made from GeH_4 and butane or methane by plasma-assisted chemical vapor deposition or by radio frequency sputtering of a Ge target in a hydrocarbon atmosphere. It can be grown in thick layers (>100 μm), but provides only modest rain erosion resistance. Its index of refraction can be varied between 2 and 4 by altering the Ge:C ratio. The absorption coefficient of one preparation was <10 cm^{-1} in the 3-12 μm range.[91] A more Ge-rich preparation had an absorption coefficient

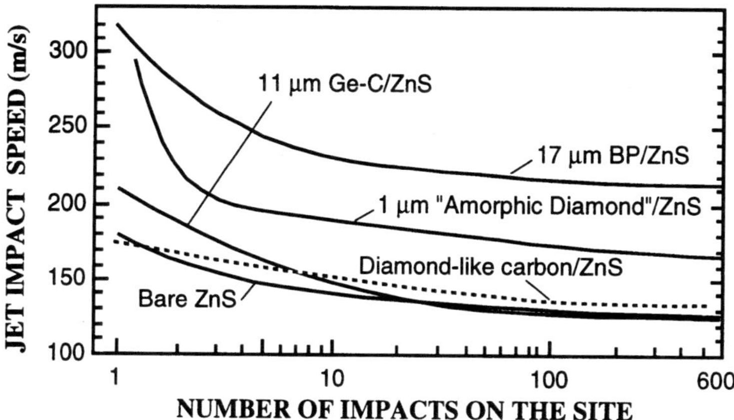

Fig. 7.47. MIJA damage threshold velocity curves for different coatings on ZnS.[24] Boron phosphide (BP) has the best erosion resistance of the coatings in this chart.

in the range 40-270 cm^{-1} at 10.6 μm.[92] Young's modulus for Ge-C is reported to be near 300 GPa[93] and the nanoindentation hardness is 14-20 GPa.[92,93] An abrasion-resistant, multilayer, dual-band (3-5 and 8-12 μm) antireflection coating called ARG6 for ZnS is based on several different layers of germanium-carbon.[91]

7.8.3 "Boron phosphide" and other phosphorus-based coatings

Boron phosphide is a coating from Pilkington Optronics (Glasgow, U.K.) that has demonstrated what is arguably the best erosion resistance of any thin coating material other than diamond.[9,24,42,69,74,94-100] Figures 7.47 - 7.49 show representative rain and sand erosion comparisons.

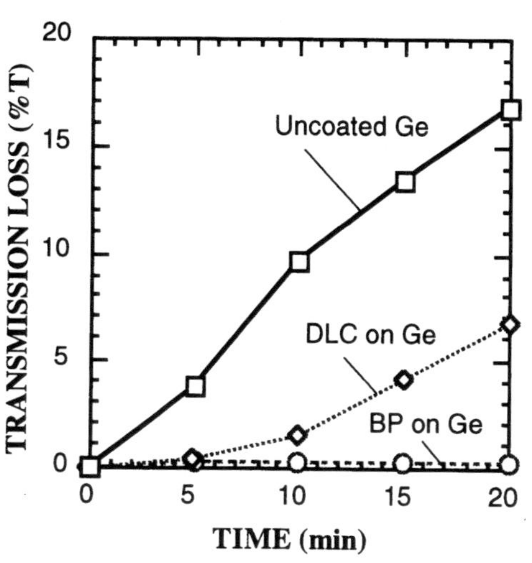

Fig. 7.48. Rain erosion resistance of Ge coated with diamond-like carbon (DLC) or "boron phosphide" (BP)[34] Whirling arm test was conducted at 210 m/s at a rainfall rate of 2.54 cm/h using 2-mm-diameter drops.

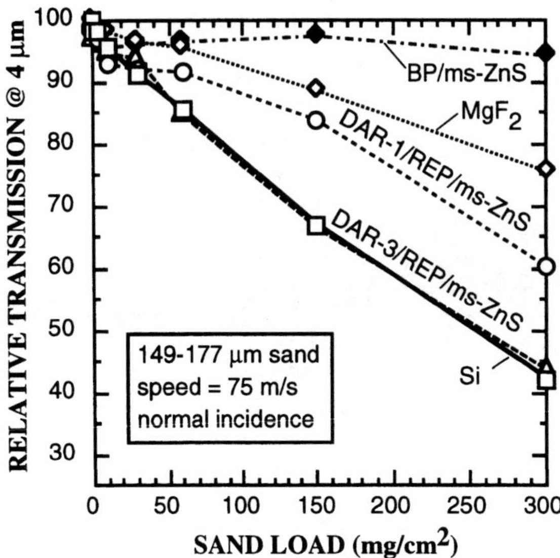

Fig. 7.49. Sand erosion resistance of multispectral ZnS with three different coatings, as well as uncoated MgF_2 and silicon.[69] DAR/REP coatings are described in Section 7.8.4. The ordinate gives the measured transmittance divided by the initial transmittance of the uneroded sample. The transmittance of the BP-coated sample was measured at 3.33 μm to avoid an absorption feature near 4 μm.

The name "boron phosphide" is a misnomer. The chemical substance, BP, is a crystalline compound containing a 1:1 atomic ratio of boron and phosphorus. The commercial coating, made by plasma-assisted chemical vapor deposition from PH_3 and B_2H_6 or BCl_3 is an amorphous phosphorus glass containing some boron, hydrogen and oxygen. A typical formula might be $BP_4O_{0.1}H_{0.4}$[100] and the P:B atomic ratio is normally in the range 2:1 to 8:1. We will use the abbreviation "BP" for the boron phosphide coating, even though it does not have the chemical composition BP. The index of refraction, ~2.9, depends on composition.[94] The fracture toughness of boron phosphide is estimated to be 1.3 $MPa\sqrt{m}$.[9] Films with a thickness of 5-17 μm on silicon are compressively stressed at 300-400 MPa.[93] The Knoop hardness of thin films, measured with a tiny load of 5 grams, is 6000 kg/mm^2.[94] This hardness compares to GaP films at 4000 and germanium-carbon films at 2000 kg/mm^2 with the same indentor load.

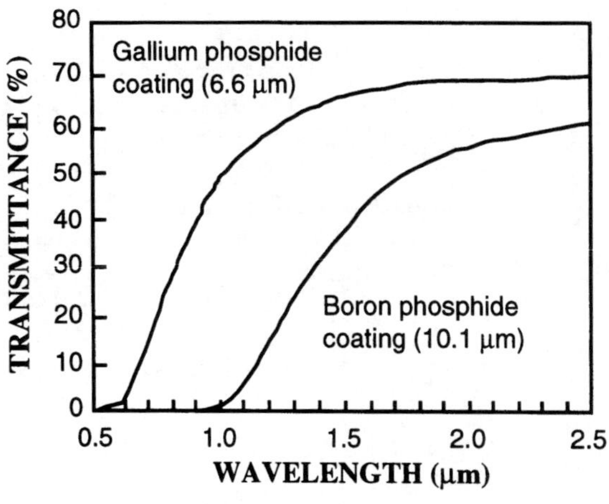

Fig. 7.50. Near-infrared and visible transmission curves of boron phosphide and gallium phosphide coatings.[94]

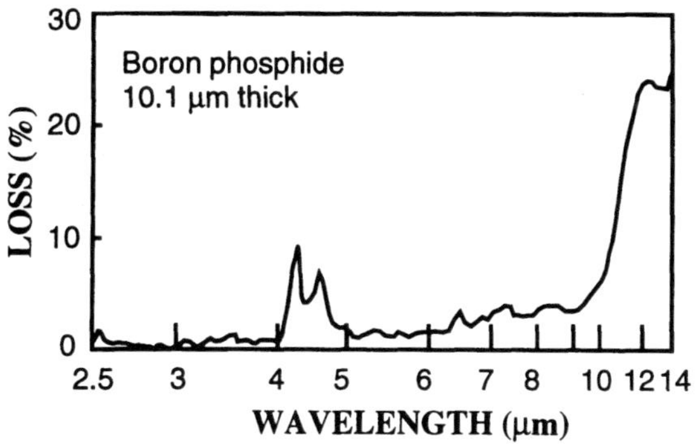

Optical transmission loss (%/μm)

λ (μm)	BP	GaP
9.24	0.48 ± 0.17	0.05 ± 0.05
10.21	0.54 ± 0.16	0.01 ± 0.05
10.59	0.80 ± 0.20	0.18 ± 0.07

Fig. 7.51. Infrared loss in boron phosphide film.[94] The table gives losses expressed as percent transmission per micrometer of film thickness.

Boron phosphide coatings have a black, lustrous appearance. Figure 7.50 shows the short wavelength cutoff of boron phosphide near 1 μm, with negligible visible transmittance.[94] Properties of a gallium phospide (GaP) coating are also shown in Fig. 7.50. BP films have infrared absorptions at 4-5 μm and beyond 10 μm (Fig. 7.51).

Up to a point, thick boron phosphide coatings offer more rain erosion protection than thinner coatings (Fig. 7.52).[34] The optimum thickness is thought to be in the 10-15 μm range. However, absorption at wavelengths longer than 10 μm limits the thickness for practical coatings. In sand erosion experiments with 220-500 μm sand particles at a speed of 30 m/s and a flux of 500 g/m^2/s, the optimum thickness of boron phosphide to protect ZnS or Ge was 15 μm. The optimum is thought to arise because thicker coatings reduce the impact stress at the substrate, but the intrinsic stress within the coating also increases with thickness. There is an intermediate thickness where these two opposing tendencies have the optimum balance.[57,101] The optimum required coating thickness increases in proportion to the impact speed and increases in a quadratic manner with the diameter of the sand particles.[101]

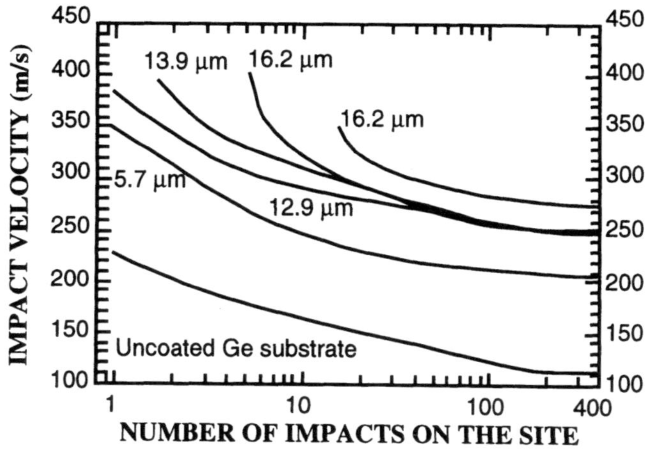

Fig. 7.52. Multiple impact jet apparatus damage threshold velocity curves for different thicknesses of boron phosphide coating on germanium.[24]

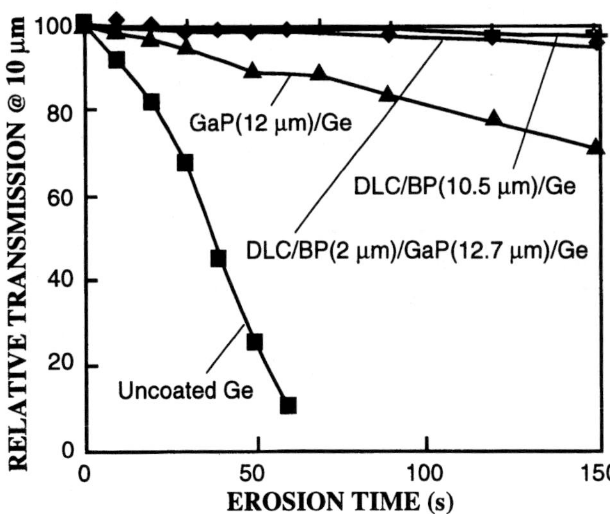

Fig. 7.53. Sand erosion resistance of Ge coated with boron phosphide, gallium phosphide, or a combination of the two.[95] Transmission loss was measured at 10 µm. Sand in the 45-80 µm size range was traveling at 100 m/s with a flux of 110 $mg/cm^2/s$ at 45° incidence.

GaP coatings have much lower absorption than boron phosphide (see table in Fig. 7.51), so composite structures with a thin layer of BP on top of a thicker layer of GaP have been developed. The BP is overcoated with a thin layer of diamond-like carbon as an abrasion-resistant antireflection coating. Figure 7.53 shows that the composite structure has almost as much sand erosion resistance as a thicker BP coating. Figure 7.54 compares the optical performance of a thick BP coating to that of the BP/GaP composite. The two lower curves at the right side of the spectrum show that the composite structure has significantly higher transmittance. Figure 7.55 compares the emittance of several coatings at 500°C. Both BP and the BP/GaP composite have too much emittance to be used at elevated temperature.

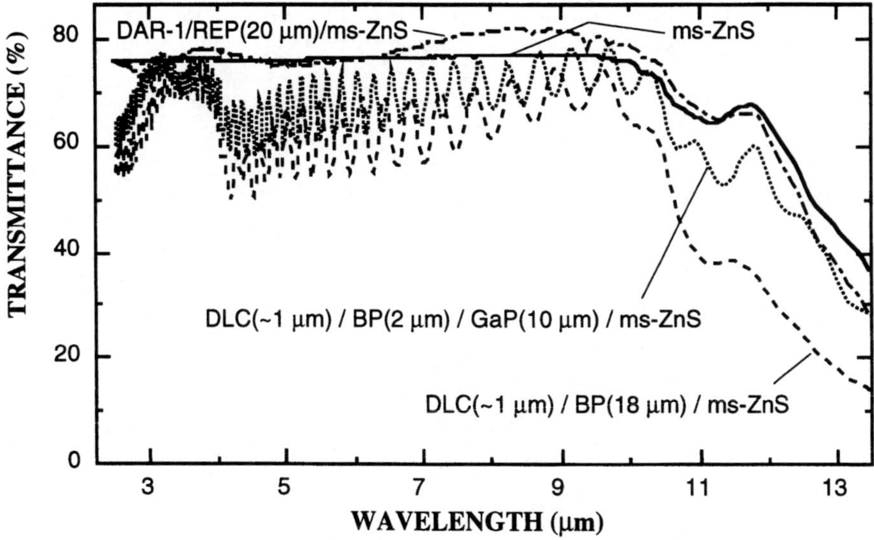

Fig 7.54. Infrared transmission of multispectral zinc sulfide with different coatings.[69] Transmission would be increased if the back side had an antireflection coating.

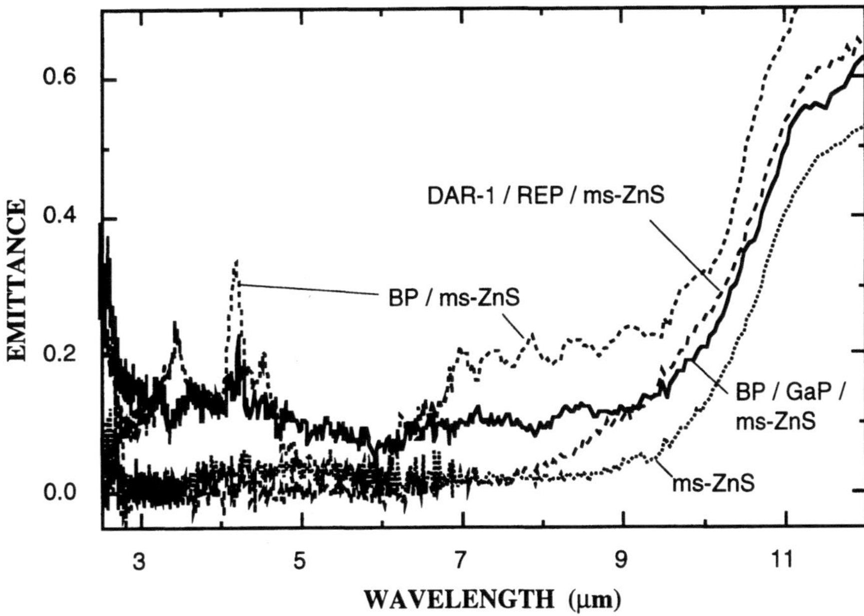

Fig. 7.55. Emittance of infrared window materials at 500°C measured at Rockwell Science Center. The diamond-like carbon outer antireflection coating burned off the BP/GaP/ms-ZnS sample at 500°C prior to the measurement. The diamond-like carbon coating probably also burned off the BP/ms-ZnS sample at 500°C.

Several studies found that waterdrop impact damage occurs in a ZnS substrate prior to damage being observed in a boron phosphide coating.[24,69] Seward *et al.*[24] observed damage to the ZnS substrate before there was any damage in a 17-µm-thick BP coating. For an 11-µm-thick coating, damage to the coating and substrate were concurrent. The damage threshold velocity curve for BP on standard ZnS in Fig. 7.56 shows a dual threshold behavior.[9] The "coating threshold curve" involves damage to both the substrate and the coating. The "substrate threshold curve" is observed after the coating has been stripped away for the substrate. To try to determine the "intrinsic" impact resistance of BP, it was coated on sapphire, which is extremely durable. A 10-µm-thick BP coating on sapphire began to fail after ~100 impacts on one site at a velocity of 490 m/s.[24]

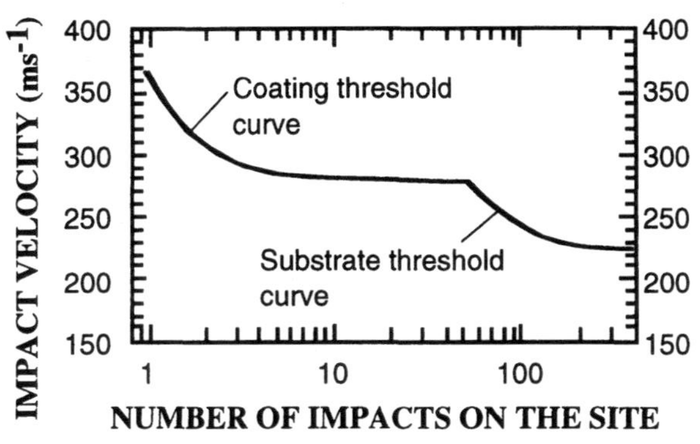

Fig. 7.56. Dual threshold behavior of 14.9-µm-thick BP coating on standard ZnS in multiple jet impact apparatus with 0.8-mm nozzle.[9]

Chemical-vapor-deposited (CVD) gallium phosphide with a thickness of ~20 μm provides moderate rain erosion resistance on GaAs and Ge.[102,103] A diamond-like carbon outer layer serves as an antireflection coating and provides abrasion resistance. An advantage of GaP is its low absorption throughout the midwave and long wave infrared regions. In a study of coated, polycrystalline germanium, it was observed that grains in which the (111) crystal plane was impacted had 6 times less damage from waterdrops than grains in which the (100) crystal plane was impacted.[103] The single-impact damage threshold velocity for bulk, polycrystalline, CVD GaP for a 0.8-mm waterjet is ~200 m/s.[104] This is similar to the performance of germanium-carbon on ZnS in Fig. 7.47.

A promising erosion-resistant coating with broadband capability is aluminum gallium phosphide (abbreviated AlGaP).[105] Transmission spectra of 18-μm-thick coatings on multispectral ZnS are shown in Fig. 7.57. The Knoop hardness of the coating, measured with a 25 g load, is 1300 kg/mm^2. The hardness of bare multispectral ZnS is 180 kg/mm^2. When multispectral ZnS was coated with AlGaP and overcoated with a multilayer, hard carbon antireflection coating, it survived whirling arm rain erosion tests (210 m/s, 20 min, 2-mm drops, 25.4 mm/h rainfall) and separate sand erosion tests (54-75 μm sand, 77 m/s, normal incidence, 20 mg/cm^2) with <2% transmission loss.

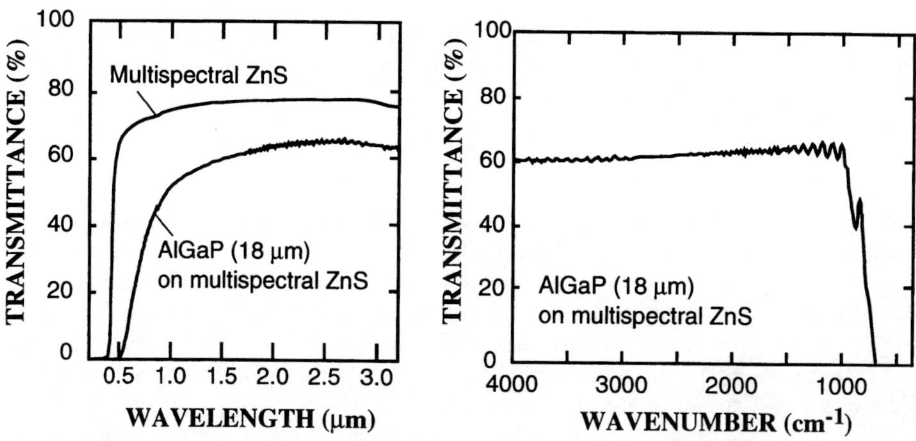

Fig. 7.57. Transmission spectra of 18-μm-thick aluminum gallium phosphide on multispectral ZnS.[105] Transmission would be increased to ~85% in selected bands by 2-sided antireflection coatings.

7.8.4 "REP" coating

Raytheon demonstrated that a sputtered layer of ZnS on a bulk ZnS window provides rain impact protection.[73] The protection might arise from a combination of burying flaws on the original ZnS (Fig. 7.44) and, perhaps, from intrinsic compressive stress in the coating. The sputtered ZnS coating was given the imaginative name "REP" coating, which stands for "rain erosion protection." Because ZnS has virtually no abrasion resistance, a durable oxide-based antireflection coating was applied over the REP coating. The oxide coating is called "DAR" for "durable antireflection coating." Several different oxide coatings are designated DAR-1, DAR-2, *etc.*

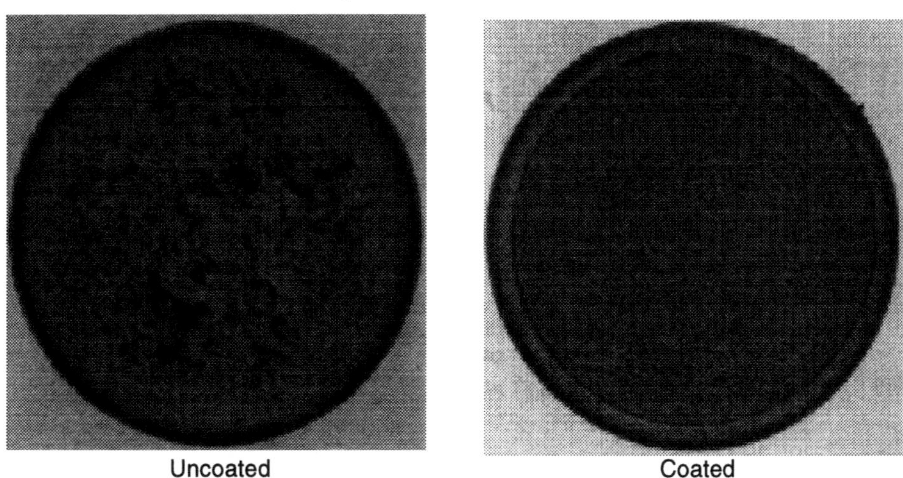

Fig. 7.58. Appearance of uncoated and REP-coated standard ZnS after 10 min in a 25.4 mm/h whirling arm rainfield at 210 m/s.[73,106]

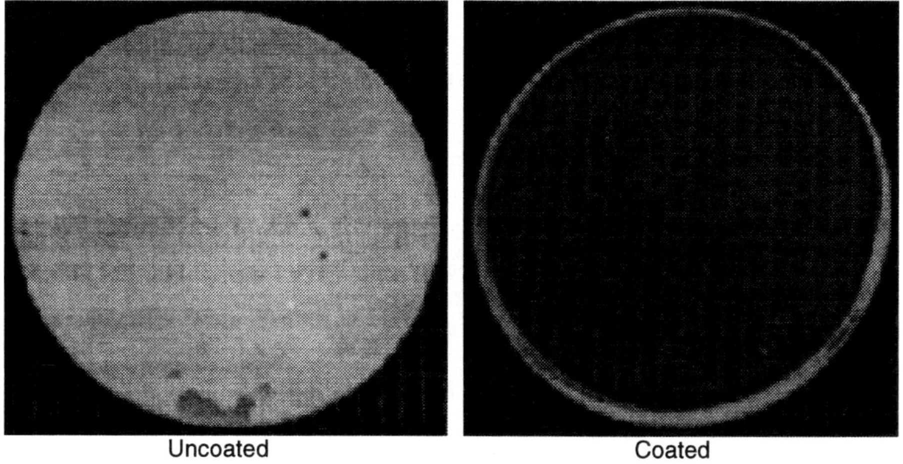

Fig. 7.59. Appearance of uncoated and DAR/REP-coated standard ZnS after 1 min of heating in air at 1000°C.[73] The DAR coating protects the ZnS from reaction with air at elevated temperature.

A REP coating thickness of 25 μm increases the waterjet damage threshold velocity of ZnS by 60%, which is consistent with the solid curve in Fig. 7.43. Figure 7.58 shows the surfaces of coated and uncoated ZnS following a whirling arm rain exposure. The protection offered by the coating is obvious. An added bonus of the DAR antireflection coating is that it protects ZnS from oxidation. Figure. 7.59 shows the appearance of uncoated and coated specimens after exposure to 1000°C in air for 1 min. The opaque appearance of the uncoated specimen is likely the result of surface oxidation.

The DAR/REP coating provides better sand erosion resistance than many standard antireflection coatings, as seen in the second trace from the top in Fig. 7.63. However, Fig. 7.49 showed that the sand erosion resistance of DAR/REP-coated multispectral ZnS is poorer than MgF_2 and much poorer than boron phosphide-coated multispectral ZnS. In side-by-side rain erosion tests, the DAR/REP coating did not protect multispectral ZnS as well as a boron phosphide coating.[69]

Figure 7.54 shows that the long wave infrared transmission of DAR/REP-coated ZnS is similar to that of uncoated material. However, the DAR/REP coating has more emittance than uncoated material at wavelengths beyond 9 μm in Fig. 7.55. This is most likely a result of the DAR oxide coating.

7.8.5 Claddings

A *cladding* is a relatively thick layer of material used to protect a window. A cladding is thick enough to be free standing if it were not attached to the window. Although there is no recognized definition, a cladding might be considered to be a layer that is ≥100 μm in thickness, whereas a *coating* is thinner. In Section 5.3.1 and Fig. 7.46 we encountered the commercial material, Tuftran®, which consists of a ZnSe window with a 1-mm-thick ZnS cladding. The cladding provides some erosion resistance for the very soft ZnSe. In Tuftran, the ZnS is deposited directly on top of the ZnSe. Now we describe protective claddings made of silicon or polymer bonded to a window.

Silicon has moderate rain erosion resistance (Fig. 7.15), but poor sand erosion resistance (Fig. 7.49). A 1-mm-thick layer of polycrystalline silicon was demonstrated to provide rain erosion protection for ZnSe in 1980.[107] The Si cladding was bonded to the ZnSe with a rigid, infrared-transparent glass braze.†[108] One disadvantage of this design is that silicon has weak absorptions in the long wave infrared region where the window could be used (Fig. 1.30). (However, the Si-O absorption band near 9 μm in current polycrystalline silicon from Raytheon is only about one fourth as intense as the corresponding band in most single-crystal silicon.) A second disadvantage is that stress is created by differential thermal expansion of Si and ZnSe. As the composite cools down ~250°C from the brazing temperature, the cladding contracts less than the underlying window. The cladding ends up in compression and the ZnSe in tension.

The concept of protecting a gallium arsenide window with a layer of silicon was granted a U.S. patent in 1993.[109] When a 75-μm coating of Si was deposited on GaAs, the damage threshold velocity for 4-mm waterdrops increased from 130 m/s to 210 m/s . It was proposed that doping the Si to make it electrically conductive could be used to provide protection from electromagnetic interference or to allow de-icing capability.

The particularly promising structure in Fig. 7.60 employs a polymer adhesive layer to bind 2-mm-thick ZnS or 1-mm-thick polycrystalline silicon claddings to ZnSe windows.[110] The window and cladding are separately polished and antireflection coated prior to assembly. Figure 7.46 showed that a compliant layer is extremely effective at reflecting impact shock waves back into the cladding. When the cladding is damaged, it can be removed and replaced without having to replace the more expensive ZnSe window.

---

†A braze made of arsenic, sulfur and selenium softens below 250°C and is useful for this purpose. The braze is called a chalcogenide (pronounced "kal **ko** jen ide") braze because the elements S and Se belong to the family of elements called chalcogenides.

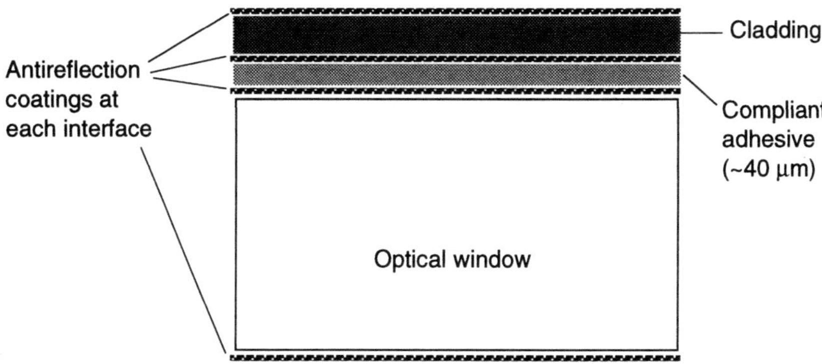

Fig. 7.60. Structure of optical window with protective cladding.[110]

Figure 7.61 shows the optical performance of two composite windows. The fine-grain polycrystalline silicon cladding, which is somewhat stronger than single-crystal silicon,† is made by chemical vapor deposition. It cannot readily fracture on crystal cleavage planes the way single-crystal silicon does when impacted by waterdrops (Fig. 7.62).

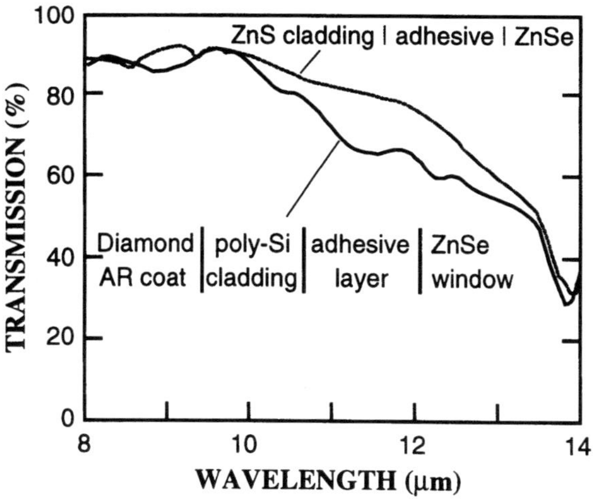

Fig. 7.61. Measured transmission of ZnSe windows with protective claddings.[110]

†Raytheon flexure strength data: chemical-vapor-deposited polycrystalline Si: 169 MPa; single-crystal (100) Si: 138 MPa; single-crystal (111) Si: 119 MPa. Coarse-grain polycrystalline silicon grown from melts is no stronger than single-crystal silicon.

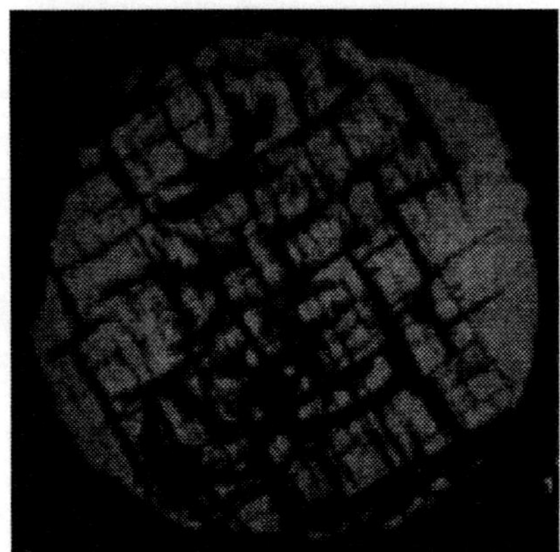

Single-crystal silicon (100)
290 m/s for 2 min

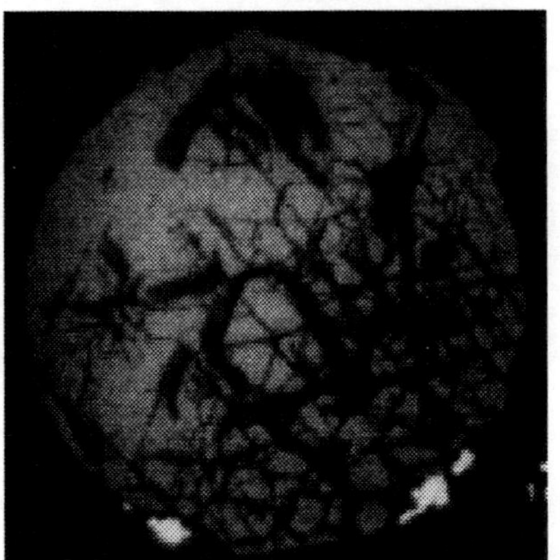

Fig. 7.62. Undamaged polycrystalline silicon and heavily damaged single-crystal silicon after whirling arm rain impact tests. Single-crystal materials fracture on crystal cleavage planes. [Courtesy R. Korenstein, Raytheon.]

Figure 7.63 illustrates the improvement in sand erosion resistance attained by coatings.[110] A conventional antireflection coating on ZnS loses transmission rapidly. We saw in Fig. 7.49 that uncoated single-crystal silicon has poor abrasion resistance. A diamond-like carbon coating on polycrystalline silicon still provides only modest sand erosion resistance in Fig. 7.63. A zirconia coating on ZnS performs better. By far, the best results were observed with a 1-μm-thick antireflection coating made of chemical-vapor-deposited diamond on polycrystalline silicon. A cladded structure like the one in Fig. 7.60 with a compliant adhesive, a polycrystalline silicon cladding, and an outermost layer of diamond is extremely durable.

Another cladding worthy of mention is the polymer described in Section 5.7.[111] This material has a high modulus in the plane of the cladding and a low modulus perpendicular to the plane. It is easily deformed by waterdrop impact (20 min at 210 m/s in 2.54 cm/h whirling arm rainfield), but does not readily transmit strain in a lateral direction. Its resistance to sand erosion is excellent. The polymer has a transmission window from 8 to 13 μm. Figure 7.64 shows a composite window structure and Fig. 7.65 shows a deep dimple left by the impact of a 2.2-mm waterdrop at 337 m/s (Mach 1.0). The cladding is damaged, but there is no damage in the underlying window. The theory is that the polymer cladding could be removed and replaced when it is worn out.

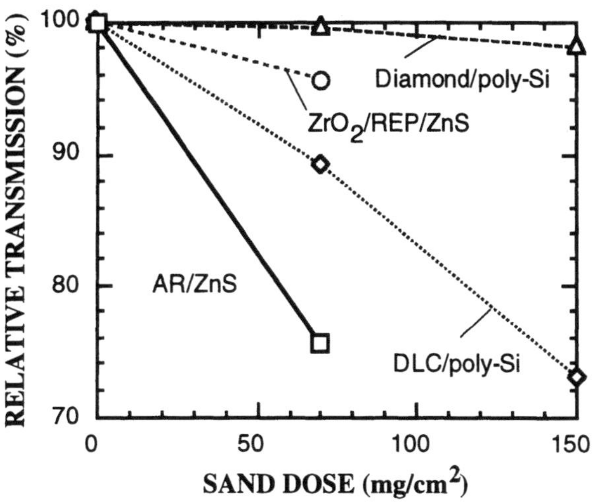

Fig. 7.63. Sand erosion experiments showing relative transmission (T/T_o) for surfaces subjected to 88-105 μm sand particles at 97 m/s.[110]

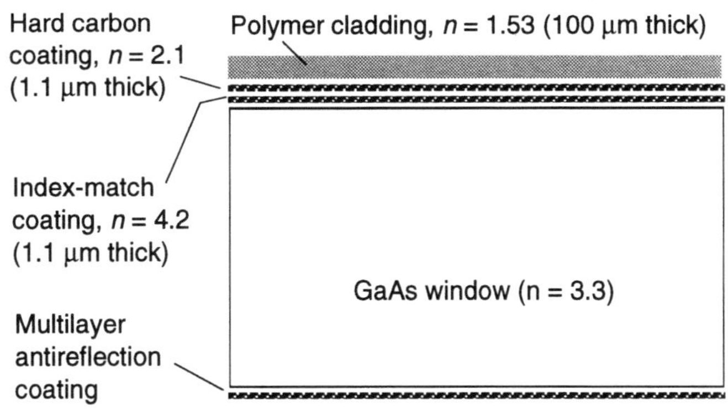

Fig. 7.64. Composite window with polymer cladding.[111]

Fig. 7.65. Round dimple just below center is impact damage from 2.2-mm waterdrop at 337 m/s on polymer cladding on 19-mm-diameter GaAs.[111] Damage is confined to the cladding.

7.8.6 Diamond coatings

Diamond coatings up to several microns thick can be grown directly on silicon by chemical vapor deposition. Thicker layers do not survive cooldown after growth because diamond has significantly lower thermal expansion than silicon. Stresses developed during cooldown of thick layers result in delamination and/or fracture. Diamond cannot be grown directly on ZnS because the hydrogen plasma used to grow diamond attacks the ZnS. Instead, infrared-transparent interlayers are used to protect the substrate from the plasma, to provide an intermediate thermal expansion, and to promote adhesion.[112,113] By this means, continuous diamond films greater than 1 μm in thickness have been deposited on ZnS. Adhesion is enhanced by lithographically patterning the ZnS with a grid structure prior to interlayer deposition (Fig. 7.70).[112,113]

To deposit diamond on germanium, the Ge surface is first patterned by lithography and ion milling to create a series of "hilltops" that promote adhesion of the diamond.[114] An expansion-matching interlayer is then applied prior to diamond deposition. Conditions that maintain a grain size ≤1 μm are required to minimize optical scatter from the diamond layer.[115] Based on published transmission curves, the optical scatter of the small-grain diamond is negligible in the long wave infrared region and ~10% in the midwave region. A thin, transparent diamond layer has also been applied to sapphire with the use of an interlayer.[115] Table 7.10 shows the improvement in waterjet damage threshold velocity for various materials coated with diamond.

Table 7.10. Waterjet (MIJA) damage threshold velocity* for diamond coatings[9,116]

Substrate	Diamond thickness (μm)	Uncoated (m/s)	Coated (m/s)
ZnS	5	125	213
Ge	variable	130	313
Si	unstated	210	325
Sapphire	7	425	565

*Threshold for 300 shots from 0.8-mm nozzle.

Instead of growing a thin coating directly on a substrate, diamond can be attached with an optical braze (Fig. 7.66).[117,108] First, a thin diamond layer is grown on an optically smooth silicon substrate. The diamond facing the substrate is optically smooth, but the growth surface of the diamond is relatively rough and therefore scatters light. Ordinarily, this rough surface would require polishing before it could be used in an optical system. Instead, a sandwich is made in which a glass bonding layer is placed between the rough surface of the diamond and the smooth surface of a zinc sulfide or zinc selenide window. The bonding layer is an arsenic-selenium-sulfur glass whose refractive index is close to that of diamond (which, coincidentally, is nearly equal to those of ZnS or ZnSe).

The sandwich is then hot pressed (Fig. 7.66, *center*) above the softening temperature of the glass, so that the glass flows into the crevices of the rough diamond surface and glues the diamond to the zinc selenide. In the final step (Fig. 7.66, *right*), the silicon is dissolved in hydrofluoric acid, exposing the smooth diamond surface. A finished window is shown in Fig. 9.1.

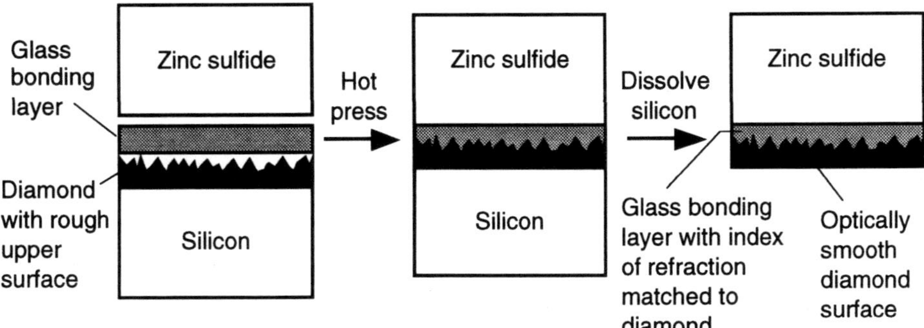

Fig. 7.66. Westinghouse optical brazing process[117,108] for attachment of diamond to ZnSe or ZnS using an infrared-transparent glass bonding layer whose index of refraction is matched to that of diamond.

Figure 7.67 shows the effect of brazing on infrared optical scatter. The lower curve gives the transmission of an 8-μm-thick diamond layer, while the upper curve shows the transmission of the same diamond brazed to zinc selenide. The upper curve is level near the expected transmittance of a nonabsorbing, nonscattering layer of diamond on zinc selenide. The difference between the two curves represents scatter from the rough diamond surface. By embedding the diamond surface in a medium of the same refractive index, the roughness is effectively eliminated and scatter is reduced to a negligible level.

The damage threshold velocity for a 2-mm waterdrop on ZnS is 175 m/s. An optically brazed 20-μm diamond layer increased the threshold velocity to 290-335 m/s.[118] For ZnSe, the threshold velocity increased from 140 to 350 m/s. Doubling the thickness of the diamond did not increase the damage threshold velocity, nor did increasing the thickness of the braze layer to as much as 35 μm.

The dramatic micrographs in Fig. 7.68 show the damage in ZnS impacted above the damage threshold velocity. The photo at the left, taken in reflected light, shows two ring

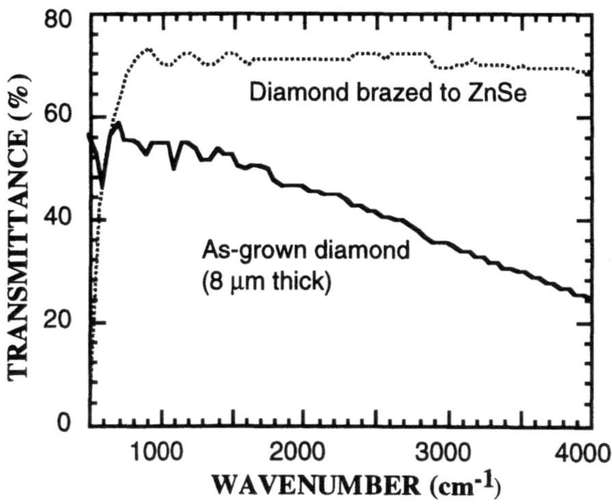

Fig. 7.67. Effect of optical brazing on infrared transmission.

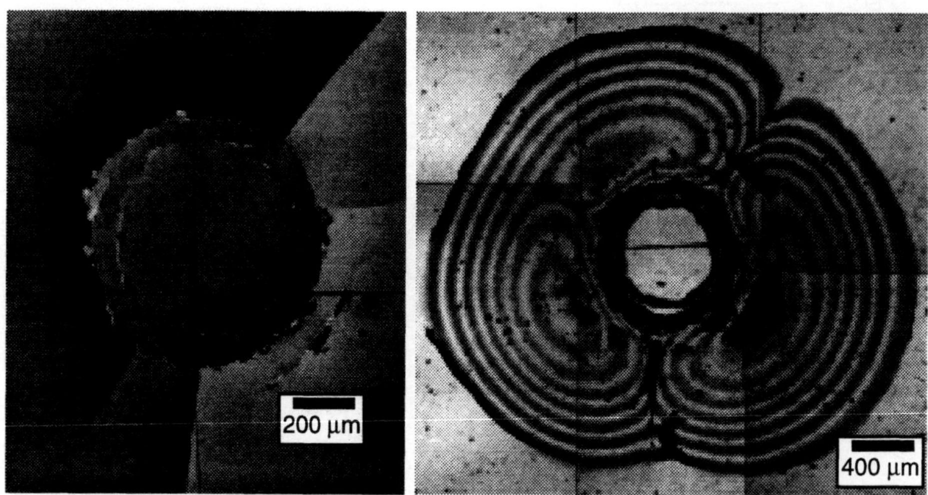

Fig. 7.68. Single 2-mm waterdrop impact at 390 m/s on diamond-coated ZnS (20-μm-thick diamond bonded with chalcogenide glass braze, Fig. 7.66).[118] *Left:* Nomarski micrograph showing only surface features in reflected light. *Right:* Internal damage seen with transmitted light.

fractures with several radial fractures propagating away from the center of the impact. The photo at the right (in which the scale is increased by a factor of 2) shows the same impact site in transmitted light, which highlights internal damage in the ZnS. You can see that the damage inside the ZnS covers a much wider area than the ring fracture on the surface.

Rain erosion resistance of diamond has not been as great as originally expected, because the chemical-vapor-deposited material is significantly weaker than natural diamond. However, chemical-vapor-deposited diamond is as hard as natural diamond, so its abrasion resistance is excellent (although not as good as natural diamond[55]).

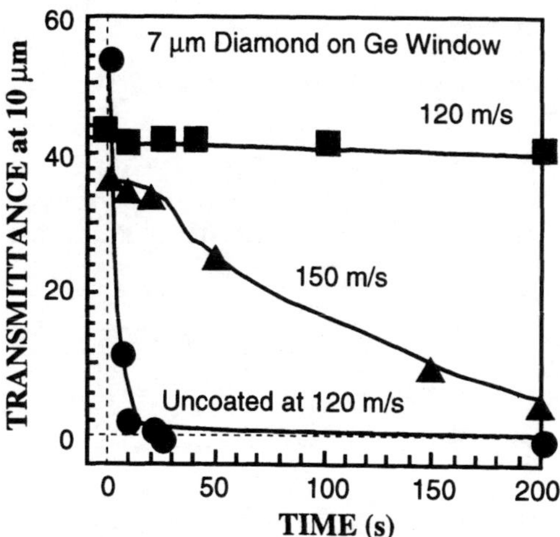

Fig. 7.69. Sand erosion response of Ge with 7-μm-thick CVD diamond coating from GEC Marconi.[42,114] Sand particles were 20-40 μm in size at a flux of 50 mg/cm²/s.

Figure 7.69 shows that a sand erosion condition which obliterates the surface of germanium in 10 s does essentially no damage in 200 s if the Ge is protected by 7 μm of diamond.[42,114] When the sand velocity is increased from 120 to 150 m/s, the diamond is damaged. The sand load after 200 s in Fig. 7.69 is equivalent to flying through light haze for more than 2000 h or flying through a dust storm for 37 h.[114] Detailed studies of sand abrasion of diamond coatings show that the first step is formation of ring cracks, followed by debonding under more severe conditions.[55,119]

The grid structure patterned into ZnS in Fig. 7.70 promotes diamond adhesion and enhances sand erosion resistance.[112] The gently rounded bumps provided better erosion resistance than did a flat surface or a grid with sharp peaks. A diamond thickness of 1.0 μm on ZnS was more effective than thicker diamond.

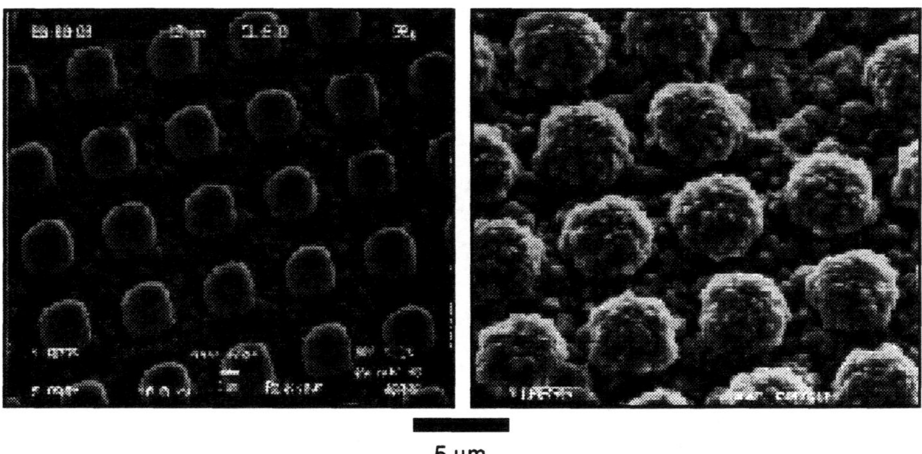

Fig. 7.70. *Left:* Lithographically patterned ZnS surface designed to promote adhesion of diamond layer.[112] The ZnS surface has been coated with a proprietary interlayer. *Right:* Appearance of surface after depositing 1 μm of diamond in a microwave plasma. [Photos courtesy R. Korenstein, Raytheon.]

References

1. R. E. Fowler and J. W. Joyner, *Rain Erosion Tests of AIM-9B Dome Materials and Coatings at Supersonic Speeds*, Naval Ordnance Test Station Report TP 4336, China Lake, California, April 1967.
2. N. Osborne, G. Graves, K. Alexander, D. Simerlink and R. Haren, "LANTIRN Infrared Window Failure Analysis," *Proc. SPIE*, **2286**, 444-455 (1992).
3. W. F. Adler, "Development of Design Data for Rain Impact Damage in Infrared-Transmitting Windows and Radomes," *Opt. Eng.*, **26**, 143-151 (1987).
4. J. S. Marshall and W. McK. Palmer, "The Distribution of Raindrops with Size," *J. Meteorology*, **5**, 165-166 (1948).

5. J. O. Laws and D. A. Parsons, "The Relation of Raindrop Size to Intensity," *Trans. Am. Geophys. Union*, **24**, 452-460 (1943).
6. W. F. Adler and T. W. James, "Analysis of Water Impacts on Zinc Sulfide," in *Fracture Mechanics of Ceramics* (R. C. Bradt, A. G. Evans, D. P. H. Hasselman and F. F. Lange, eds.), Plenum Press, New York (1983).
7. J. V. Hackworth, "Damage of Infrared-Transparent Materials Exposed to Rain Environments at High Velocities," *Proc. SPIE*, **362**, 123-136 (1982).
8. W. F. Adler, "Liquid Drop Collisions on Deformable Media," *J. Mater. Sci.*, **12**, 1253-1271 (1977).
9. E. J. Coad and J. E. Field, "The Liquid Impact Resistance of CVD Diamond and Other Infrared Materials," *Proc. SPIE*, **3060**, 169-180 (1997).
10. W. F. Adler, "The Mechanics of Liquid Impact," in *Treatise on Materials Science and Technology* (C. M. Preece, ed.), Academic Press, New York (1979).
11. N. K. Bourne, T. Obara and J. E. Field, "High-Speed Photography and Stress Gauge Studies of Jet Impact Upon Surfaces," *Phil. Trans. Royal Soc. Lond. A*, **355**, 607-623 (1997).
12. B. A. Bolt, *Earthquakes*, 4th ed., W. H. Freeman, New York (1999).
13. W. F. Adler, "Waterdrop Impact Modeling," *Wear*, **186-187**, 341-351 (1995).
14. J. E. Field, R. J. Hand and C. J. Pickles, "Strength and Rain Erosion Studies of I.R. Materials," *Proc. SPIE*, **1112**, 306-315 (1989). The origin of Eq. (7-5) is Eq. (12) in R. J. Hand, J. E. Field and D. Townsend, "The Use of Liquid Jets to Simulate Angled Drop Impact," *J. Appl. Phys.*, **70**, 7111-7118 (1991).
15. W. F. Adler, "Development of Design Data for Rain Impact Damage in Infrared-Transmitting Materials," *Proc. SPIE*, **297**, 143-154 (1981).
16. S. van der Zwaag and J. E. Field, "Liquid Jet Impact Damage on Zinc Sulfide," *J. Mater. Sci.*, **17**, 2625-2636 (1982).
17. J. A. Detrio, "Rain Impact Testing of Durable Coatings on Water Clear ZnS," *Proc. SPIE*, **1760**, 243-252 (1992).
18. W. F. Adler, "Recent Observations of Waterdrop Impacts on Infrared Windows," *Proc. 6th DoD Electromagnetic Windows Symposium*, Huntsville, Alabama, October 1995, pp. 204-213.
19. S. van der Zwaag and J. E. Field, "Rain Erosion Damage in Brittle Materials," *Eng. Frac. Mech.*, **17**, 367-379 (1983).
20. C. R. Seward, C. S. J. Pickles and J. E. Field, "Single and Multiple Impact Jet Apparatus and Results," *Proc. SPIE*, **1326**, 280-290 (1990).
21. A. G. Evans, Y. M. Ito and M. Rosenblatt, "Impact Damage Thresholds in Brittle Materials Impacted by Water Drops," *J. Appl. Phys.*, **51**, 2473-2482 (1980).
22. J. D. Achenbach, *Wave Propagation in Elastic Solids*, North-Holland, Amsterdam (1973).
23. R. W. Tustison and R. L. Gentilman, "Current and Emerging Materials for LWIR External Windows," *Proc. SPIE*, **968**, 25-34 (1988).
24. C. R. Seward, E. J. Coad, C. S. J. Pickles and J. E. Field, "The Rain Erosion Resistance of Diamond and Other Window Materials," *Proc. SPIE*, **2286**, 285-300 (1994).
25. J. J. Cassaing, A. A. Déom, A. M. Bouveret and D. L. Balageas, "IR Materials Rain Damage Prediction and Test Results," *Proc. SPIE*, **1112**, 295-305 (1989).
26. A. A. Déom, D. L. Balageas, F. G. Laturelle, G. D. Gardette and G. J. Freydefont "Sensitivity of Rain Erosion Resistance of Infrared Materials to Environmental Conditions Such as Temperature and Stress," *Proc. SPIE*, **1326**, 301-309 (1990).
27. A. A. Déom, D. L. Balageas, G. Gardette and G. Gauffre, "Optical Damage Characterization of Rain-Eroded IR Materials," Office National d'Etudes et de Recherches Aérospatiales (ONERA) TP 1985-161, Châtillon, France; *Proc. SPIE*, **590**, 157-162 (1985).

28. W. F. Adler, "Supersonic Waterdrop Impacts on Materials," *Proc. 6th European Electromagnetic Windows Conf.*, Friedrichshafen, FRG (1991), pp. 237-245.
29. W. F. Adler, "Rain Erosion Testing," *Proc. SPIE*, **1112**, 275-294 (1989).
30. P. N. H. Davies and J. E. Field, "Multiple Impact Jet Apparatus (MIJA): Application to Rain Erosion Studies," *Proc. SPIE*, **1112**, 316-327 (1989).
31. W. F. Adler and P. L. Boland, "Multiparticle Supersonic Impact Test Program," *Proc. SPIE*, **1326**, 268-279 (1990).
32. "Projectile Impact Test Capabilities," General Research Corporation, Santa Barbara, California (1989).
33. W. F. Adler, "Rain Impact Retrospective and Vision for the Future," *Wear*, **233-235**, 25-38 (1999).
34. B. C. Monachan, D. Morrison, E. M. Waddell, D. R. Gibson, A. D. Wilson and K. Lewis, "Boron Phosphide I.R. Coatings," in *Infrared Thin Films, SPIE Critical Reviews*, **CR39**, 91-123 (1992).
35. J. A. Detrio, "Non-Normal Angle of Incidence Rain Impact Testing with Whirling Arm Apparatus, " *Proc. SPIE*, **2286**, 561-569 (1994).
36. A. A. Déom, A. Luc, C. Flamand, R. Gouyon and D. L. Balageas, "Are Classical Rain Erosion Experiments of Infrared Materials Used in High Velocity Seekers Representative?" *Proc. SPIE*, **3060**, 130-141 (1997).
37. C. R. Seward, C. S. J. Pickles, R. Marrah and J. E. Field, "Rain Erosion Data on Window and Dome Materials," *Proc. SPIE*, **1760**, 280-290 (1992).
38. E. A. Maguire, J. K. Rawson and R. W. Tustison, "Aluminum Oxynitride's Resistance to Impact and Erosion," *Proc. SPIE*, **2286**, 26-32 (1994).
39. C. F. Kennedy, R. H. Telling and J. E. Field, "Liquid Impact and Fracture of Free-Standing CVD Diamond," *Proc. SPIE*, **3705** (1999).
40. F. P. Bowden and J. E. Field, "The Brittle Fracture of Solids by Liquid Impact, by Solid Impact, and by Shock," *Proc. Royal Soc. London*, **A282**, 331-352 (1964).
41. Data from C. R. Seward, University of Cambridge (1994).
42. R. H. Telling, G. H. Jilbert and J. E. Field, "The Erosion of Aerospace Materials by Solid Particle Impact," *Proc. SPIE*, **3060**, 56-67 (1997).
43. W. F. Adler and P. L. Boland, "Multiparticle Supersonic Impact Test Program," *Proc. SPIE*, **1326**, 268-279 (1990).
44. W. F. Adler, "Laboratory Rain Impact Evaluation of a Full-Scale Radome," *Proc. 7th DoD Electromagnetic Windows Symp.*, 223-232, Laurel, Maryland, May 1998.
45. W. F. Adler and J. A. Cox, "High Velocity Rain Erosion Effects on Optical Sensor Windows," IRIS Materials Specialty Group Meeting, Washington, DC. (June 1987).
46. A. Wierzba and K. Takayama, "Experimental Investigation of the Aerodynamic Breakup of Liquid Drops," *AIAA J.*, **26**, 1329-1335 (1988).
47. W. F. Adler and D. J. Mihora, "Aerodynamic Effects on Raindrop Impact Parameters," in *Proc. 5th European Electromagnetic Windows Conf.*, Antibes-Juan-Les-Pins, France (1989), pp. 157-164.
48. W. F. Adler and D. J. Mihora, "Infrared-Transmitting Window Survivabilty in Hydrometeor Environments," *Proc. SPIE*, **1760**, 291-302 (1992).
49. W. F. Adler and J. W. Flavin, "Comparison of Water Drop and Ice Ball Impacts on Materials," General Research Corp. Report CR-88-1069, Santa Barbara, California (1988).
50. F. B. Weiskopf, J. S. Lin, R. A. Drobnick and B. K. Feather, "Erosion Modelling and Test of Slip Cast Fused Silica," *Proc. SPIE*, **1326**, 310-320 (1990).
51. T. S. Blackwell and D. A. Kalin, "High Velocity, Small Particle Impact Erosion of Sapphire Windows," *Proc. SPIE*, **1326**, 291-300 (1990).
52. V. H. Bulsara and S. Chandrasekar, "Direct Observation of Contact Damage Around Scratches in Brittle Solids," *Proc. SPIE*, **3060**, 76-88 (1997).

53. A. G. Evans, "Strength Degradation by Projectile Impacts," *J. Am. Ceram. Soc.*, **56**, 405-409 (1973).
54. E. R. Fuller, Jr., S. W. Freiman, J. B. Quinn, G. D. Quinn and W. C. Carter, "Fracture Mechanics Approach to the Design of Glass Aircraft Windows: A Case Study," *Proc. SPIE*, **2286**, 419-430 (1994).
55. J. E. Field, Q. Sun and H. Gao, "Solid Particle Erosion of Infrared Transmitting Materials," *Proc. SPIE*, **2286**, 301-306 (1994).
56. S. P. McGeoch, D. R. Gibson and J. A. Savage, "Assessment of Type IIa Diamond as an Optical Material for Use in Severe Environments," *Proc. SPIE*, **1760**, 122-142 (1992).
57. E. J. Coad, C. S. J. Pickles, C. R. Seward, G. H. Jilbert and J. E. Field, "The Erosion Resistance of Infrared Transparent Materials," *Proc. Roy. Soc. Lond. A*, **454**, 213-238 (1998).
58. D. C. Harris, *Comparative Sand and Rain Erosion Studies of Spinel, Aluminum Oxynitride (ALON), Magnesium Fluoride, and Germanate Glass*, Naval Air Warfare Center Report TP 8147, China Lake, California, August 1993.
59. I. P. Hayward and J. E. Field, "The Solid Particle Erosion of Diamond," *J. Hard Mater.*, **1**, 53-64 (1990).
60. W. F. Adler, "Particulate Impact Damage Predictions," *Wear*, **186-187**, 35-44 (1995).
61. E. S. Kelly, R. J. Ondercin, J. A. Detrio and P. R. Greason, "Environmental Testing of Long Wave Infrared (LWIR) Windows," *Proc. SPIE*, **3060**, 68-75 (1997).
62. J. A. Detrio, P. Greason, R. Bertke, E. S. Kelly and R. J. Ondercin, "Durability Testing and Characterization of LWIR Window Materials," *Proc. 7th DoD Electromagnetic Windows Symp.*, 206-214, Laurel, Maryland, May 1998.
63. A. A. Déom, D. L. Balageas, T. M. Mackowski and P. Robert, "Rain Damage of Diamond-Like Coated Germanium," Office National d'Etudes et de Recherches Aérospatiales (ONERA) TP 1985-160, Châtillon, France.
64. A. A. Déom, D. L. Balageas, T. M. Mackowski and P. Robert, "Rain Damage of Diamond-Like Coated Germanium," *Proc. 2nd Int. Symp. Opt. & Electro-Opt. Appl. Sci. Eng.*, Cannes, November 1985.
65. R. J. Hand, J. E. Field and D. Townsend, "The Use of Liquid Jets to Simulate Angled Drop Impact," *J. Appl. Phys.* **70**, 7111-7118 (1991).
66. H. Busch, G. Hoff and G. Langbein, *Philos. Trans. Roy. Soc. London*, **A260**, pt. 1110, 168 (1966).
67. D. W. C. Baker, J. K. H. Jolliffe and D. Pearson, *Philos. Trans. Roy. Soc. London*, **A260**, pt. 1110, 193 (1966).
68. D. A. Gorham and J. E. Field, "Anomalous Behaviour of High Velocity Oblique Liquid Impact," *Wear*, **41**, 213-222 (1977).
69. D. C. Harris, "Side-by-Side Comparison of Erosion-Resistant Coatings," *Proc. SPIE*, **3060**, 17-29 (1997).
70. S. van der Zwaag and J. E. Field, "The Effect of Thin Hard Coatings on the Hertzian Stress Field," *Phil. Mag. A*, **46**, 133-150 (1982).
71. R. J. Hand, J. E. Field and S. van der Zwaag, "High Modulus Layers as Protective Coatings for 'Window' Materials," *Proc. SPIE*, **1112**, 120-128 (1989).
72. L. M. Goldman, J. K. Rawson and R. W. Tustison, "A Phenomenological Model of Rain Protection," *Proc. 5th DoD Electromagnetic Windows Symposium*, Boulder, Colorado, 219-224, October 1993.
73. L. M. Goldman and R. W. Tustison, "High Durability Infrared Transparent Coatings," *Proc. SPIE*, **2286**, 316-324 (1994).
74. E. M. Waddell, D. R. Gibson and M. Wilson, "Broadband IR Transparent Rain and Sand Erosion Protective Coating for the F14 Aircraft Infra-Red Search and Track Germanium Dome," *Proc. SPIE*, **2286**, 376-385 (1994).

75. J. M. Wahl and R. W. Tustison, "Mechanical Enhancement of LWIR Materials via Coatings," *Proc. SPIE,* **1326**, 128-136 (1990).
76. R. L. Gentilman, E. A. Maguire, H. S. Starrett, T. M. Hartnett and H. P. Kirchner, "Strength and Transmittance of Sapphire and Strengthened Sapphire," *J. Am. Ceram. Soc.*, **64**, C116-C117 (1981).
77. M. J. Matthewson, "The Effect of a Thin Compliant Protective Coating on Hertzian Contact Stress," *J. Phys. D*, **15**, 237-249 (1982).
78. R. Gentilman, "Preventing Rain Erosion Damage to ZnS and ZnSe Window with a Compliant Interlayer," *Proc. 5th DoD Electromagnetic Windows Symposium*, Boulder, Colorado, 203-210, October 1993.
79. R. J. Ondercin, A. B. Harker and L. M. Goldman, "Development of a Repairable Composite IR Window," *Proc. SPIE,* **3705** (1999).
80. W. F. Adler and D. J. Mihora, "Analysis of Waterdrop Impacts on Layered Window Constructions," *Proc. SPIE,* **2286**, 264 (1994).
81. D. M. Swec and M. J. Mirtich, "Diamondlike Carbon Protective Coatings for Optical Windows," *Proc. SPIE,* **1112**, 162-173 (1989).
82. J. A. Savage, *Infrared Optical Materials and Their Antireflection Coatings*, Adam Hilger, Bristol (1985), pp. 218-220.
83. R. W. Tustison, "Protective, Infrared Transparent Coatings," in *Infrared Thin Films, SPIE Critical Reviews*, **CR39** (1992).
84. J. Robertson, "Diamond-Like Carbon," *Pure Appl. Chem.*, **66**, 1789-1796 (1994).
85. M. B. Moran and L. F. Johnson, "Diamond-Like Carbon Films Synthesized by Electron Cyclotron Resonance Chemical Vapor Deposition, " *Proc. SPIE,* **3060**, 42-54 (1997).
86. F. X. Lu, B. X. Yang, D. G. Cheng, R. Z. Ye, W. X. Yu and J. B. Sun, "Low Hydrogen Content Diamond-Like Carbon Coatings of KCl Optics for High Power Industrial CO_2 Lasers," *Thin Solid Films*, **212**, 220-225 (1992).
87. S. van der Zwaag and J. E. Field, "Indentation and Liquid Impact Studies of Coated Germanium," *Phil. Mag. A*, **48**, 767-777 (1983).
88. R. D. Harris and A. W. Towch, "Window Evaluation Programme for an Airborne FLIR System: Environmental and Optical Aspects," *Proc. SPIE,* **1112**, 244-257 (1989).
89. C. B. Collins, F. Davanloo, E. M. Juengerman, D. R. Jander and T. J. Lee, "Amorphic Diamond Films Grown With a Laser-Ion Source," *Proc. SPIE,* **1112**, 192-198 (1989).
90. W. Hasan and S. H. Propst, "Durability Testing of Hard Carbon Coatings for Be and ZnS Substrates," *Proc. SPIE,* **2286**, 354-363 (1994).
91. A. H. Lettington, C. J. H. Wort and B. C. Monachan, "Development and IR Applications of GeC Thin Films," *Proc. SPIE,* **1112**, 156-161 (1989).
92. J. M. Mackowski, B. Cimma, R. Pignard, P. Colardelle and P. Laprat, "Rain Erosion Behavior of Germanium Carbide (GeC) Films Grown on ZnS Substrates," *Proc. SPIE,* **1760**, 201-209 (1992).
93. E. D. Nicholson, C. S. J. Pickles and J. E. Field, "The Mechanical Properties of Thin Films for Aerospace Applications," *Proc. SPIE,* **2286**, 275-284 (1994).
94. D. R. Gibson, W. M. Waddell, S. A. D. Wilson and K. Lewis, "Ultradurable Phosphide-Based Antireflection Coatings for Sand and Rain Erosion Protection," *Opt. Eng.*, **33**, 957-966 (1994).
95. D. R. Gibson, E. M. Waddell and K. Lewis, "Advances in Ultradurable Phosphide-Based Broadband Anti-Reflection Coatings for Sand and Rain Erosion Protection of Infrared Windows and Domes," *Proc. SPIE,* **2286**, 335-346 (1994).
96. E. M. Waddell, D. R. Gibson and J. Meredith, "Sand Impact Testing of Durable Coatings on FLIR ZnS Relevant to the LANTIRN E-O System Window," *Proc. SPIE,* **2286**, 364-375 (1994).

97. D. R. Gibson, E. M. Waddell, J. W. Kerr, A. D. Wilson and K. Lewis, "Ultradurable Phosphide-Based Anti-Reflection Coatings for Sand and Rain Erosion Protection," *Proc. SPIE*, **1760**, 178-200 (1992).
98. E. M. Waddell and B. C. Monachan, "Rain Erosion Protection of IR Materials Using Boron Phosphide Coatings," *Proc. SPIE*, **1326**, 144-156 (1990).
99. B. C. Monachan, C. J. Kelly and E. M. Waddell "Ultra-Hard Coatings for IR Materials," *Proc. SPIE*, **1112**, 129-143 (1989).
100. K. L. Lewis, C. J. Kelly and B. C. Monachan, "Recent Progress in the Development of Boron Phosphide as a Robust Coating Material for Infra-Red Transparencies," *Proc. SPIE*, **1112**, 407-4163 (1989).
101. G. H. Jilbert and J. E. Field, "Optimum Coating Thickness for the Protection of Zinc Sulfide and Germanium Substrates from Solid Particle Erosion," *Wear*, **217**, 15-23 (1998).
102. P. Klocek, J. T. Hoggins and M. Wilson, "Broadband IR Transparent Rain Erosion Protection Coating for IR Windows," *Proc. SPIE*, **1760**, 210-223 (1992).
103. M. Wilson, M. Thomas, I. Perez and D. Price, "Impact Damage as a Function of Crystal Orientation in Ge IR Windows Employing Durable Phosphide Coatings," *Proc. SPIE*, **2286**, 108-119 (1994).
104. J. M. Wahl and R. W. Tustison, "Optical, Mechanical and Water Drop Impact Characteristics of Polycrystalline GaP," *J. Mater. Sci.*, **29**, 5765-5772 (1994).
105. J. Askinazi and A. Narayanan, "Protective Broadband Window Coatings," *Proc. SPIE*, **3060**, 356-365 (1997).
106. L. M. Goldman, P. E. Cremin, T. E. Varitimos and R. W. Tustison, "Damage Resulting from Single and Multiple Waterdrop Impacts on Coated and Uncoated LWIR Substrates," *Proc. SPIE*, **1760**, 224-242 (1992).
107. P. A. Miles and R. W. Tustison, "Erosion Resistant Infrared Windows: Thermal, Mechanical and Optical Aspects of Composite Designs," *Proc. SPIE*, **204**, 108-110 (1980).
108. R. H. Hopkins, W. E. Kramer, G. B. Brandt, J. S. Schruben, R. A. Hoffman, K. B. Steinbruegge and T. L. Peterson "Fabrication and Evaluation of Erosion-Resistant Multispectral Optical Windows," *J. Appl. Phys.*, **49**, 3133-3139 (1978).
109. A. R. Hilton, Sr., "Doped Gallium Arsenide External Windows," *Proc. SPIE*, **2286**, 91-98 (1994).
110. R. J. Ondercin, L. M. Goldman and A. B. Harker, "Development of a Repairable IR Composite Window," *Proc. SPIE*, **3705** (1999).
111. N. Brette and P. Klocek, "Engineered Polymeric IR-Transparent Protective Coatings," *Proc. SPIE*, **2286**, 325-334 (1994).
112. R. Korenstein, L. Goldman, R. Hallock, R. Ondercin and E. Kelly, " Diamond Coated ZnS for Improved Erosion Resistance," *Proc. SPIE*, **3060**, 181-195 (1997).
113. D. A. Tossell, M. C. Costello and C. J. Brierley, "Diamond Layers for the Protection of Infrared Windows," *Proc. SPIE*, **1760**, 268-278 (1992).
114. M. D. Hudson, C. J. Brierley, A. J. Miller and A. E. J. Wilson, "Fabrication and Testing of Diamond Coatings on Infrared Windows for the Harrier GR7 and AV8-B Systems," *Proc. SPIE*, **3060**, 196-202 (1997).
115. C. J. Brierley, M. C. Costello, M. D. Hudson and T. J. Bettles, "Diamond Coatings for Large Area IR Windows," *Proc. SPIE*, **2286**, 307-315 (1994).
116. E. J. Coad, C. S. J. Pickles, G. H. Jilbert and J. E. Field, "Aerospace Erosion of Diamond and Diamond Coatings," *Diamond and Related Materials*, **5**, 640-643 (1996).

117. W. D. Partlow, R. E. Witkowski and J. P. McHugh, "CVD Diamond Coatings for the Infrared by Optical Brazing," in *Applications of Diamond Films and Related Materials* (Y. Tzeng, M. Yoshikawa, M. Murakawa and A. Feldman, eds.), Elsevier, Amsterdam (1991), pp 163-168.
118. W. F. Adler, *Initial Assessment of the Waterdrop Impact Resistance of Diamond Coated Zinc Sulfide and Zinc Selenide*, General Research Corp. Report CR-91-1263, Santa Barbara, California, November 1991; W. F. Adler, *Waterdrop Impact Resistance of Diamond Coated Zinc Sulfide and Zinc Selenide*, General Research Corp. Report CR-92-1313, Santa Barbara, California, September 1992.
119. Z. Feng, Y. Tzeng and J. E. Field, "Solid Particle Impact of CVD Diamond Films," *Thin Solid Films*, **212**, 35-42 (1992).

Chapter 8

PROOF TESTING

The range of strengths of a set of identically manufactured ceramic windows is usually quite large. For example, in Table 3.1 the weakest member of a set of 13 zinc sulfide disks had about half the strength of the strongest member. The median strength of ZnS windows with the Weibull distribution in Fig. 3.18 is 94 MPa. If you were asked to design a system with 99.9% reliability, you would have to limit the stress on the window to ≤28 MPa. Such a limitation severely inhibits the use of ceramic components. A *proof test* removes the weakest members of a population of components so that the remaining population will function reliably at a higher level of stress.

Suppose that you must provide ZnS windows that will withstand a service stress of 50 MPa with 99.9% reliability. One way that you might accomplish this is to subject each window to a proof stress of, say 60 MPa. Figure 8.1 shows that 5.8% of the

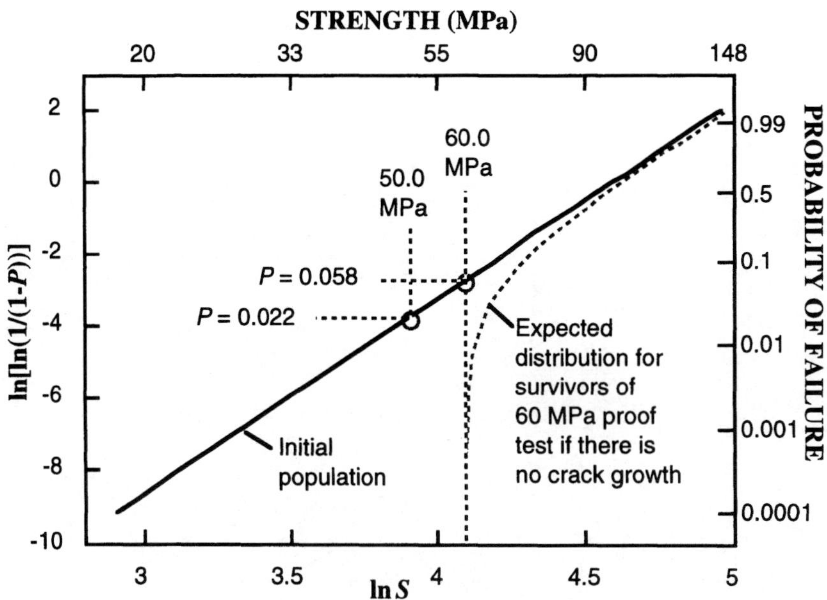

Fig. 8.1. Strengths of ZnS windows based on Weibull parameters from Fig. 3.18 (solid line). P is the probability of failure and S is strength. The dashed line is the expected distribution of strengths of windows that survive a proof test at 60 MPa if there is no slow crack growth during the proof test to weaken the windows. Note that the Weibull distribution is a straight line when plotting $\ln(\ln[(1-P)^{-1}]$ vs. $\ln S$.

windows are expected to break during the proof test. After sweeping the broken pieces from the factory floor, you would like to be able to assure your customer with some confidence that the remaining windows will survive when they are reloaded to 50 MPa.

Unfortunately, it is possible for pre-existing microcracks to grow larger during a proof test. If this happens, then the surviving proof-tested windows would not be as strong as we expect in Fig. 8.1. In the worst case, it is possible for a proof-tested set of windows to be weaker than the original untested distribution. Let's examine a case study before considering the analytical details of proof tests.

8.1 Case study: proof testing of zinc selenide

When chemical vapor deposited zinc selenide was first developed, its properties were studied for application as a laser window.[1] In one experiment, rectangular bars were cut from several plates of ZnSe that were grown in an identical manner. The bars were polished and their strengths were measured in a 4-point flexure test. The median failure stress was 47 MPa: Half of the bars failed below 47 MPa and half failed above 47 MPa.

A proof test was conducted in the following manner to select bars whose strength was ≥47 MPa:[1]

1. A fresh set of 41 polished bars was selected and each was loaded at a constant rate (crosshead speed = 0.508 mm/min) until it broke or until the stress was 47 MPa. Results are shown by (•) in Fig. 8.2: 20 of the 41 bars broke with an average strength of 42 ± 6 MPa.

2. The load on each surviving bar was relaxed to 1 MPa and then the bar was reloaded until it broke. Results shown by (x) in Fig. 8.2 indicate that the 21 survivors from step 1 had an average strength of 49 ± 3 MPa. Four of the 21 broke slightly below the proof stress of 47 MPa, with the weakest breaking at 45.5 MPa.

The curves in Fig. 8.2 are the expected Weibull distribution based on earlier experiments. Bars that failed the proof test agree with the expected distribution. The strengths of the survivors are slightly below the predicted values. We conclude that the proof test weakened the ZnSe slightly. However, if we want each bar to survive reloading once to a stress of, say, 40 MPa, then a proof test at 47 MPa is reasonable.

A second proof test was carried out by loading each of 21 ZnSe bars 20 consecutive times to a stress of 28 MPa.[1] One bar broke during the proof test. The survivors were loaded until they failed, with results shown in Fig. 8.3. The average strength of the survivors was 41 ± 3 MPa, which is 87% of the average strength of the virgin population of ZnSe prior to proof testing. However, the repetitive loading proof test eliminated the part of the population with a strength below 28 MPa.

Figs. 8.2 and 8.3 show that proof testing does weaken ZnSe somewhat. Would ZnSe eventually break after many consecutive cycles of loading and unloading at a fairly low stress? Table 8.1 shows the results of one such *dynamic fatigue test* in which 9 specimens of ZnSe from the population with a median strength of 47 MPa were each loaded 10 million times. Specimens loaded up to 38.5 MPa survived 10^7 loading cycles and then had a mean strength of 42.9 ± 5.2 MPa. This residual strength is not significantly different from the residual strength after 20 loadings to 28 MPa in Fig. 8.3.

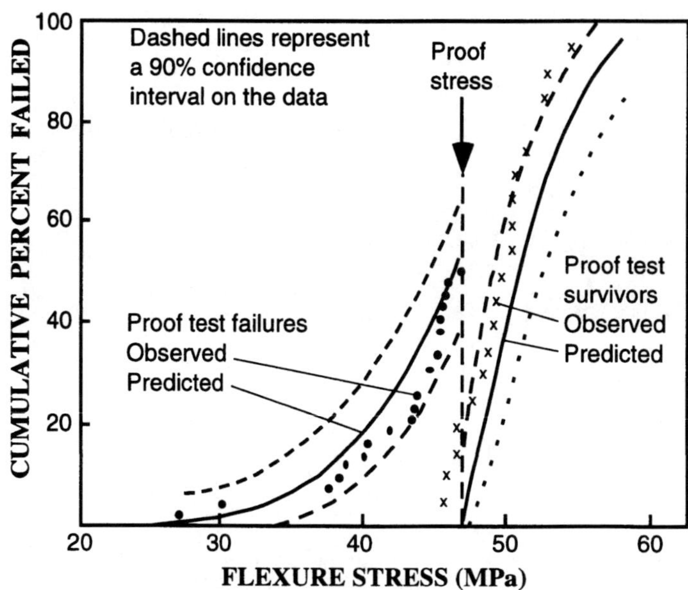

Fig. 8.2. Cumulative failure distribution of ZnSe proof tested at 47 MPa.[1]

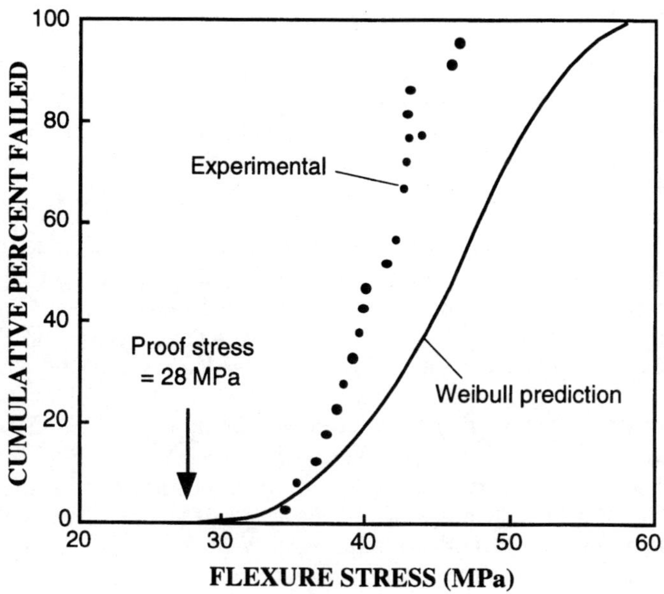

Fig. 8.3. Failure distribution of ZnSe repetitively loaded at a proof stress of 28 MPa.[1]

Of the two specimens loaded at 39.9 MPa, one survived 10^7 cycles and one broke after 174 000 cycles. Two bars loaded above 40 MPa broke during the first cycle. In this experiment it appears that 10^7 loading cycles are not more harmful than 20 cycles.

Table 8.1. Dynamic fatigue test of ZnSe[1]

Maximum cyclic stress (MPa)	Cycles to failure	Post-test flexure strength (MPa)
14.1	survived 10^7	39.6
21.2	survived 10^7	37.4
28.3	survived 10^7	40.6
33.6	survived 10^7	47.5
38.5	survived 10^7	49.4
39.9	survived 10^7	44.9
39.9	174 000	---
40.7	1	---
42.1	1	---

One more set of tests was carried out with ZnSe from the same population with a median strength of 47 MPa to see if slow crack growth occurs under *static fatigue.* Flexure bars were loaded to the stresses in Table 8.2 and left until they broke (or until the experimenter got tired). Three bars loaded at 26, 38 and 39 MPa did not fail after many days. The remaining samples loaded at 35-39 MPa broke in times ranging from 0 to 600 min. The spread of behavior in Table 8.2 is daunting if you are concerned about ceramic reliability. Subsequent tests indicated that slow crack growth occurs in ZnSe during static loading under the influence of atmospheric moisture.[2] The growth of cracks under the combined influence of stress and a chemical agent, called *stress corrosion*, is common to many ceramics. Water is the most common corrosive agent for many ceramics.

Table 8.2. Static fatigue test of ZnSe[1]

Static stress (MPa)	Failure time (min)	Comments
26	>15 900	no failure
35	4	---
37	4	---
38	0	failed on loading
38	0	failed on loading
38	0	failed on loading
38	4	---
38	600	---
38	>6600	no failure
39	240	---
39	>7200	no failure

8.1.1 An example of an unsuccessful proof test

If a proof test is not designed properly, the strength of the surviving specimens can be the same[3] or even lower[4,5] than the strength of the original specimens. Fig. 8.4 shows an example in which the survivors of a mechanical test have essentially the same strength distribution as the initial population.

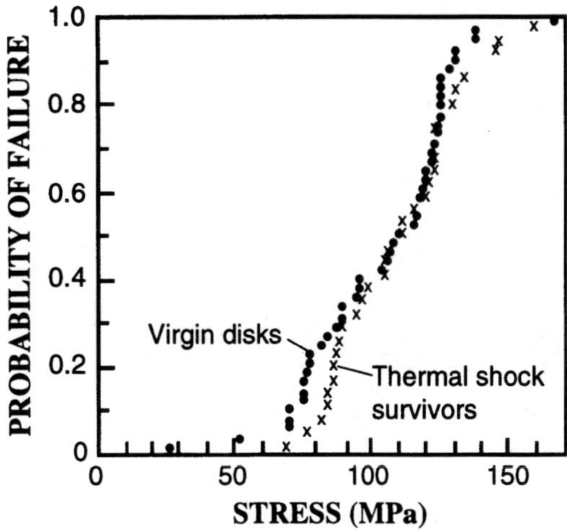

Fig. 8.4. Strength distribution of virgin glass disks and a set of similar disks that had survived a thermal shock test.[3] The population of thermal shock survivors is no stronger than the initial population, even though the thermal shock test destroyed 25-89% of the disks that were tested.

In this experiment, glass disks were subjected to a thermal shock test by contact with a hot brass rod using oil on the contact surface to facilitate heat transfer. In different sets of tests 25 to 89% of the disks broke when rapidly heated by the brass rod. Ring-on-ring flexure testing of 47 virgin disks that had not been subjected to thermal shock gave a mean strength of 104 MPa with a standard deviation of 26 MPa and a Weibull modulus of 4. Flexure testing of 33 disks that survived the thermal shock test gave a mean strength of 110 ± 23 MPa with a Weibull modulus of 5. The similarity of the two strength distributions in Fig. 8.4 suggests that the thermal shock test caused subcritical cracks to grow during the test, thus weakening the survivors.

One way to verify that a mechanical proof test is not weakening the survivors is to show that specimens that pass the test once can pass repeatedly. Without such assurance, the proof test is not trustworthy.

8.2 What is the stress intensity factor?

In order to discuss slow crack growth during a proof test, we need to describe the stress distribution near a crack tip. Figure 8.5 shows a thin crack of length $2a$ in an infinite plane. You can think of this as a small, half-penny shaped crack of radius a intersecting the surface of a ceramic window. Relative to the size of the crack (tens of micrometers), the window is an infinite plane. If there were no crack in the plane, then when the stress $\sigma_{applied}$ is imposed across the plane, all points would feel the uniform stress $\sigma_{applied}$ in the y direction. If the plane has a crack in it, the cracked portion is not bearing a load, so the uncracked portion must bear some extra load. The extra stress is concentrated near the tips of the crack. Even though force is only applied in the y direction in Fig. 8.5, the presence of the crack generates stress in the x direction and slight shear stress in the xy plane.

The stresses in the element of material whose coordinates are (r, θ) in Fig. 8.5 are σ_y in the y direction, σ_x in the x direction, and the shear τ_{yx} $(= \tau_{xy})$ in the xy plane. All three types of stress depend on the coordinates r and θ in the following general manner:[6]

$$\text{Stress} = \frac{K_I}{\sqrt{2\pi r}} f(\theta) \tag{8-1}$$

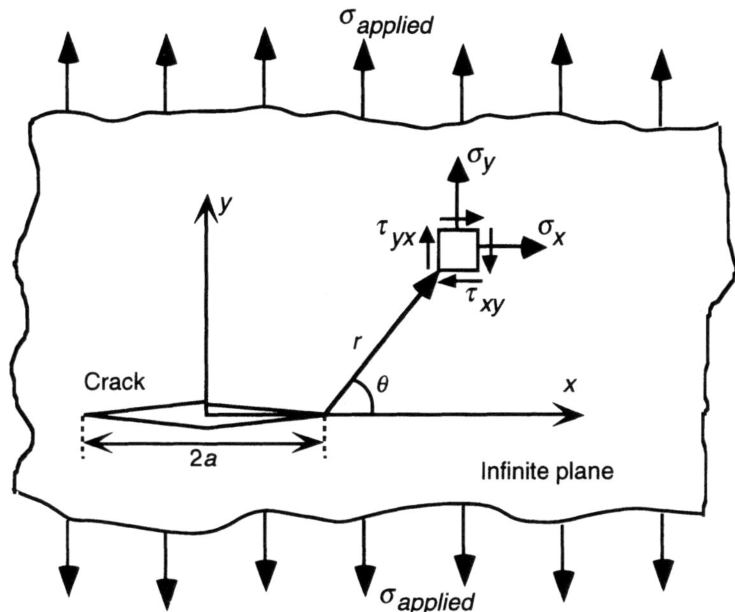

Fig. 8.5. A crack of length $2a$ in an infinite plane subjected to a stress $\sigma_{applied}$. At any point in the plane with coordinates (r, θ), there is a stress in the y direction (σ_y), a stress in the x direction (σ_x), and a shear stress ($\tau_{yx} = \tau_{xy}$).

where the constant K_I, called the *stress intensity factor* is determined by the geometry of the problem. Common units for K_I are $\text{MPa}\sqrt{\text{m}}$. The angular function, $f(\theta)$, is different for σ_x, σ_y, and τ_{xy}, but does not depend on the geometry of the problem. The entire stress field in Fig. 8.5 is specified if we know the constant K_I.

For a crack in an infinite plane, it turns out that the stress intensity factor is $K_I = \sigma_{applied}\sqrt{\pi a}$. Figure 8.6 shows stress contours for σ_x and σ_y near the crack tip of Fig. 8.5. The shapes of the contours are described by $f(\theta)$ in Eq. (8-1). The magnitudes of the stresses are governed by K_I. Stress is proportional to $1\sqrt{r}$ as the distance from the crack tip increases. The shear stresses are much smaller than σ_x and σ_y and are confined to a region too close to the crack tip to be shown in this diagram.

According to Eq. (8-1), stress increases to infinity at the crack tip. Obviously, this cannot happen. When stress becomes great enough immediately adjacent to the crack tip, material flows in a plastic manner instead of the elastic manner described by Eq. (8-1).

8.3 Slow crack growth

The classic method for measuring slow crack growth in a ceramic is with the double cantilever beam specimen in Fig. 8.7.[7,8,9] A small crack is intentionally introduced and the constant force P is then applied to pull the beam apart and cause the crack to grow along the direction of the guiding notch. The position of the crack tip is monitored optically as the crack grows to lengths of 2 to 4 cm at typical rates of 10^{-10} to 10^{-4} m/s.

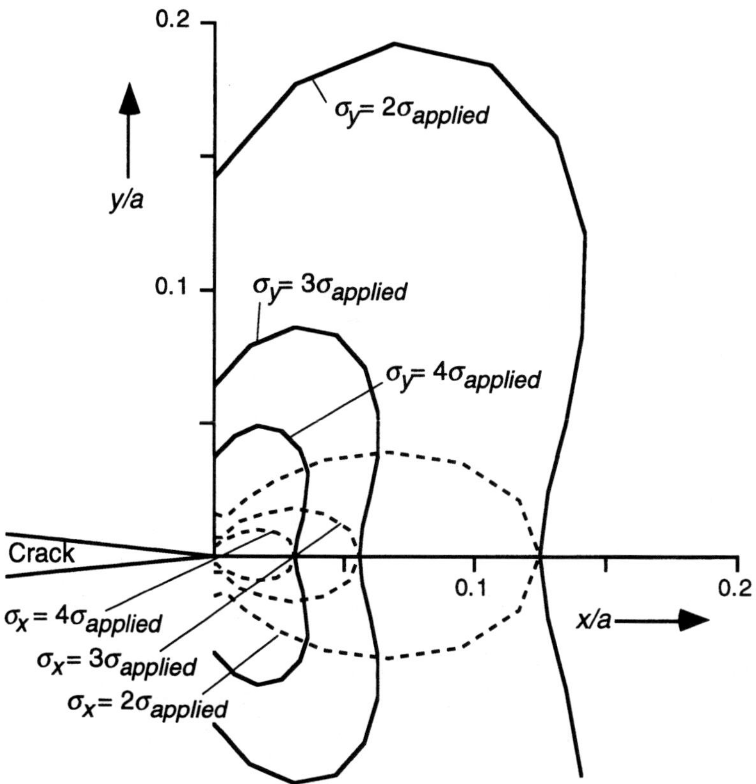

Fig. 8.6. Enlarged view of the region near the crack tip in Fig. 8.5 showing σ_y (solid curves) and σ_x (broken curves). Stresses are expressed as multiples of the external stress, $\sigma_{applied}$. Distances along the Cartesian axes are expressed as multiples of the crack radius, a. That is, the length of the x axis is just two tenths of the crack radius.

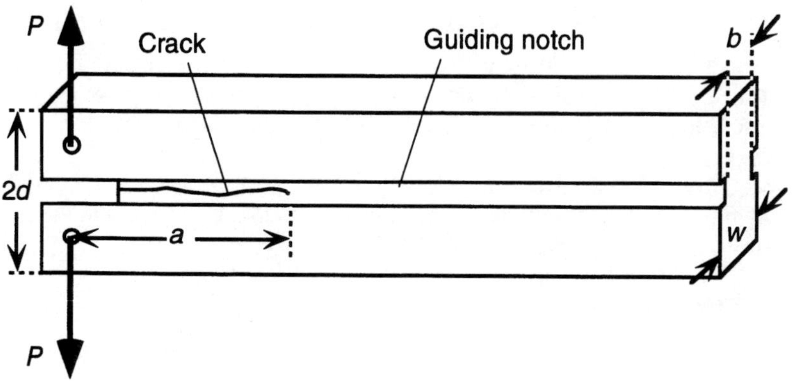

Fig. 8.7. Double cantilever beam specimen for measuring slow crack growth. An improved experimental design applies a constant bending moment to the specimen, rather than the constant load shown in this figure.[10]

Crack velocity is the rate of change of crack length (a) with time (t):

$$\text{Crack velocity} = v = da/dt\,. \tag{8-2}$$

The stress intensity factor in Eq. (8-1) governing the stress field at the tip of the crack for the double cantilever specimen is[11,12]

$$K_I = \frac{Pa}{d^{3/2}\sqrt{wb}}\left(3.467 + 2.315\,\frac{d}{a}\right) \tag{8-3}$$

where the dimensions d, w and b are defined in Fig. 8.7. The stress intensity factor increases as the applied force (P) increases and as the length of the crack increases.

An extensive study of crack propagation in glass is summarized in Fig. 8.8.[8,13] In this illustration, the observed crack velocity is plotted as a function of the stress intensity factor calculated with Eq. (8-3). The greater the stress intensity factor, the faster the rate of crack growth. In much of the graph, crack velocity increases by one or more

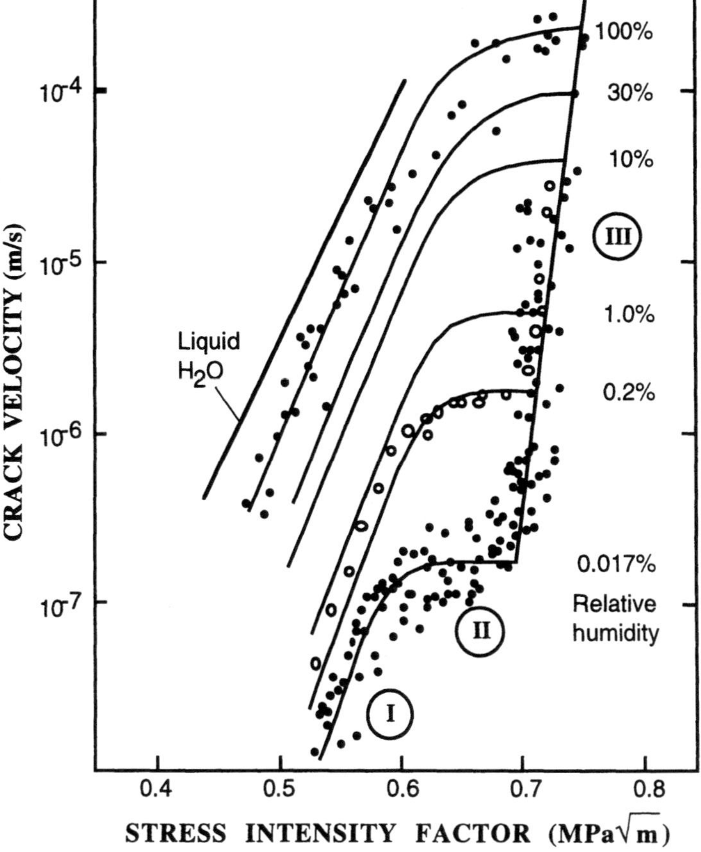

Fig. 8.8. Crack velocity in soda lime silicate glass in humid nitrogen atmospheres and in liquid water. Relative humidities labeled on the graph range from 0.017% to 100%.[8,13] Points are shown for three of the conditions to indicate the spread of the data.

orders of magnitude when the stress intensity factor increases by just 10%. The crack velocity curves can be divided into three general regions:

1. In region I of Fig. 8.8 (the lower left portion of each curve), crack velocity increases as the humidity in the test atmosphere increases. Crack velocity is greatest when the specimen is immersed in liquid water (upper left curve). The rate of crack growth in region I is governed by the rate of reaction of water with the glass.

2. In region II the crack velocity is independent of the stress intensity factor, but is still strongly dependent on humidity. Crack velocity in region II is governed by the rate at which water diffuses to the crack tip to react with the glass.

3. In region III, crack velocity again increases very rapidly with increasing stress intensity factor, but is independent of the concentration of water in the environment. The rate of crack growth is controlled by the chemical composition of the glass. Crack velocity may be so rapid in region III that fracture is essentially instantaneous.

The ability of water to enhance crack propagation under an applied stress is called *stress corrosion.* A possible mechanism by which water reacts with the silicate lattice of glass at the tip of the growing crack is shown in Fig. 8.9.[14] In region I of Fig. 8.8, the crack velocity increases with increasing temperature and with increasing pH (increasing OH^- concentration) if the glass is immersed in aqueous solution.

In regions I and III of Fig. 8.8 the dependence of crack velocity on stress intensity is:

$$\text{Crack velocity} = \text{v} = AK_I^n \tag{8-4}$$

where A and n are constants. If v is given in m/s and K_I is expressed as MPa$\sqrt{\text{m}}$, the

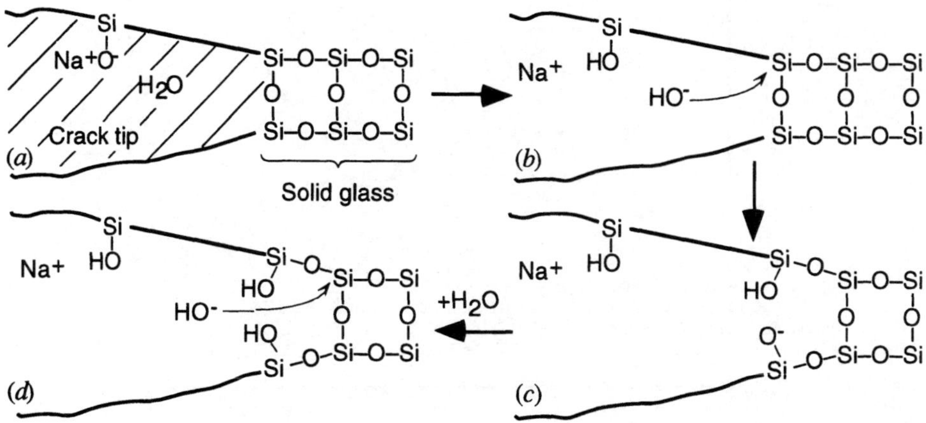

Fig. 8.9. Proposed mechanism for stress corrosion in glass.[14] In going from step (*a*) to step (*b*), water reacts with a $-SiO^-Na^+$ group on the surface of the glass to produce –SiOH and Na^+OH^-. In step (*c*) the OH^- breaks a –Si–O–Si– bond at the crack tip. In step (*d*) another molecule of H_2O reacts with the $-SiO^-$ group generated in step (*c*) to create a new molecule of OH^- that can attack another –Si–O–Si– bond at the crack tip.

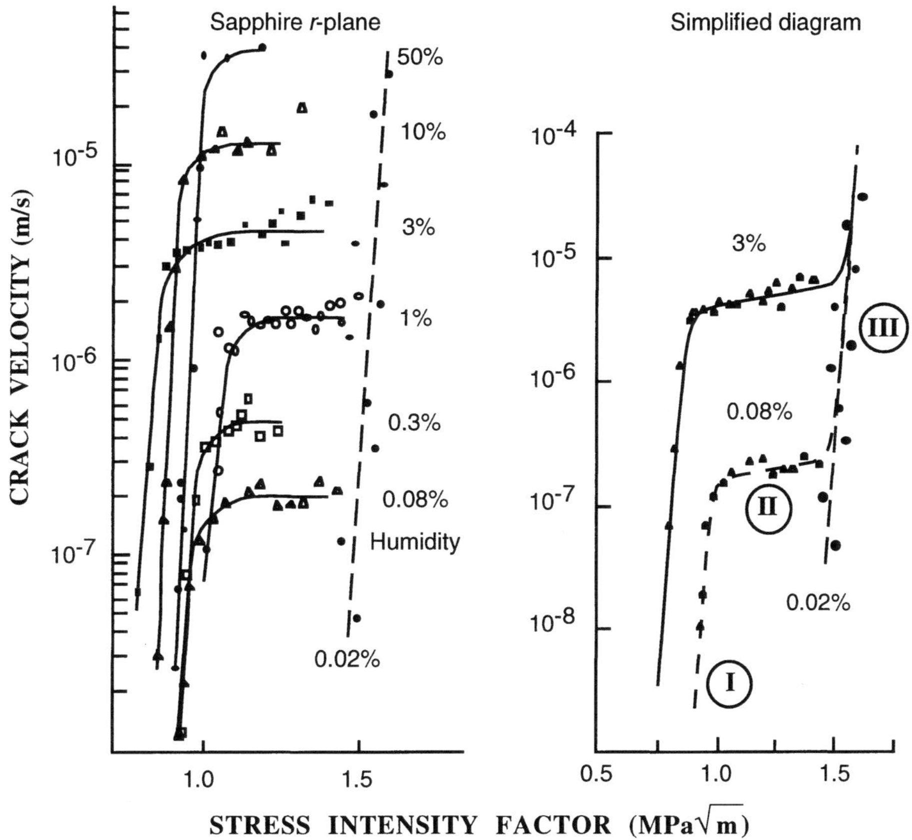

Fig. 8.10. Crack velocity on the r crystal plane in sapphire at relative humidities from 0.02% to 50% demonstrates that stress corrosion occurs.[9,13,15]

exponent n is in the range 15-50 for glasses,[8,12] near 50 for sapphire (Fig. 8.10),[13,15] 40 for ZnSe[2] and 170 for ZnS in water.[16] Equation (8-4) simply restates that crack velocity increases extremely rapidly with increasing stress intensity.

Example: Effect of humidity on crack growth rate in sapphire. Region I crack growth for the r-plane of sapphire in Fig. 8.10 obeys Eq. (8-4) with the following numerical constants:

0.08% humidity: $\quad v\ (\text{m/s}) = (1.5 \times 10^{-6})\ K_I^{63.2}$

3.0% humidity: $\quad v\ (\text{m/s}) = (3.5 \times 10^{-3})\ K_I^{49.9}$

when K_I is expressed in the units MPa$\sqrt{\text{m}}$. How much faster does a crack grow in 3.0% humidity compared to 0.08% humidity if the stress intensity factor is 0.85 MPa$\sqrt{\text{m}}$?

Plugging in for K_I in the rate equations gives

0.08% humidity: $v = (1.5 \times 10^{-6})(0.85^{63.2}) = 5.2 \times 10^{-11}$ m/s

3.0% humidity: $v = (3.5 \times 10^{-3})(0.85^{49.9}) = 1.7 \times 10^{-6}$ m/s .

Increasing the humidity from 0.08% to 3.0% increases the rate of crack growth by a factor of $(1.7 \times 10^{-6}$ m/s$)/(5.2 \times 10^{-11}$ m/s$) = 3 \times 10^4$ in region I in Fig. 8.10. In region II (from $K_I = 1.0$ to 1.4 MPa$\sqrt{\text{m}}$), the crack growth rate in 3.0% humidity is only 24 times greater than the growth rate at 0.08% humidity. In region III the crack growth rate is independent of humidity.

Equation (8-4) provides a simple fit for crack velocity in regions I and III of Figs. 8.8 and 8.10. Another equation that has been used in regions I and III is $v = v_o e^{\beta K_I}$, where v_o and β are empirical constants.[17] A more complex equation that fits crack velocity in all three regions has also been given.[18]

Temperature can have a very strong influence on the rate of crack propagation. A study of crack growth on the *r*-plane of sapphire in vacuum at temperatures from 200° to 600°C produced the following approximate relation between crack velocity, v, and temperature:[19]

$$v\ (\text{m/s}) = v_o e^{(-E^* + 0.423K_I)/(RT)} \qquad (8\text{-}5)$$

where K_I is expressed in Pa$\sqrt{\text{m}}$, E^* is an activation energy equal to 8.9×10^5 J/mol, R is the gas constant (8.3145 J/[mol·K]) and T is in kelvins. The value of v_o is $e^{19.3}$ m/s. We see in Fig. 8.11 that increasing the temperature from 473 K to 873 K increases the rate of crack propagation by a factor of 10^{10}.

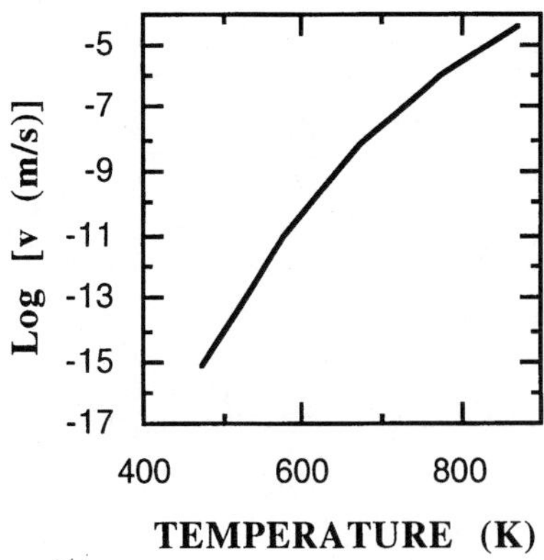

Fig. 8.11. Crack propagation rate versus temperature for the *r*-plane of sapphire in vacuum based on Eq. (8-5) for a stress intensity of 1.6×10^5 Pa$\sqrt{\text{m}}$.

8.4 The theory of proof testing

Our goal now is to derive an equation that describes how the strength of a ceramic changes during a proof test. Our approach follows that of Fuller *et al.*[20] which, in turn, builds on work of Evans and Fuller.[21] Three simple equations relating flaw size, stress intensity factor, crack velocity and strength serve as the starting point.

First, we note that the stress intensity factor for a crack of radius a (diameter = $2a$) in a solid of finite dimensions under the externally applied stress σ has the general form

$$\text{Stress intensity factor} = K_I = \sigma Y \sqrt{a} \tag{8-6}$$

where Y is a geometric factor that depends on the geometry of the loading and the crack configuration. The second key equation was (8-4) which related crack velocity to the stress intensity factor:

$$\text{Crack velocity} = \text{v} = \frac{da}{dt} = AK_I^n \tag{8-4}$$

where t is time and A and n are parameters measured in an experiment such as the one in Figs. 8.7 and 8.8. In many cases, slow crack growth is only significant in region I of Fig. 8.8. In regions II and III, crack growth is so fast that failure is very rapid. Although we will only analyze crack growth in region I, Fuller *et al.*[20] give results for the general case in which regions I, II and III are significant.

The third key equation is the relation between the strength (S) of a ceramic part and the critical stress intensity factor, K_{Ic}:

$$\text{Strength} = S = \frac{K_{Ic}}{Y\sqrt{a}}. \tag{8-7}$$

Equation (8-7) is equivalent to Eq. (3-35). It says that when the stress intensity factor at the crack tip reaches the critical value K_{Ic}, the crack grows so fast that the part fails essentially instantaneously. The strength of the part is the applied stress at which failure occurs. Equation (8-7) is just a rearranged form of Eq. (8-6) at the instant of failure.

Now consider the proof test in Fig. 8.12. Stress is applied to a ceramic component at the constant rate $\dot{\sigma}_\ell$ until the proof stress σ_p is reached. Then stress is then relieved at the constant rate $\dot{\sigma}_u$. The symbol $\dot{\sigma}_\ell$ stands for the time rate of change of the *l*oading

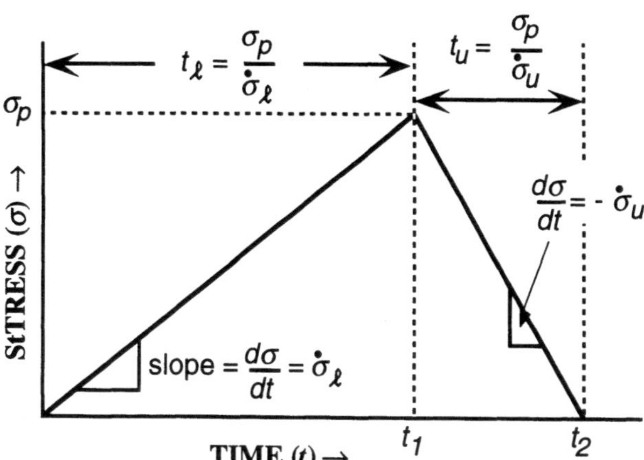

Fig. 8.12. Stress profile for a proof test. In general, there might be a period added to the middle of the test during which the stress is held constant at σ_p. This more general case is treated by Fuller *et al.*[20]

stress (MPa/s). Similarly, $\dot{\sigma}_u$ is the magnitude of the rate (MPa/s) at which the component is *u*nloaded. ($\dot{\sigma}_u$ is a positive number.) Other stress profiles can certainly be used in proof tests, but we will only analyze the case in Fig. 8.12.

8.4.1 How strength changes during a proof test

If we want to know how strength changes with time during a proof test, an obvious start is to differentiate the equation for strength (8-7) with respect to time:

$$\frac{dS}{dt} = -\frac{K_{Ic}}{2Y}a^{-3/2}\frac{da}{dt} . \tag{8-8}$$

But $da/dt = AK_I^n$ in Eq. (8-4) and we can find an expression for K_I by combining Eqns. (8-6) and (8-7):

$$\frac{S}{\sigma} = \frac{K_{Ic}/Y\sqrt{a}}{K_I/Y\sqrt{a}} = \frac{K_{Ic}}{K_I} \Rightarrow K_I = K_{Ic}\left(\frac{\sigma}{S}\right) . \tag{8-9}$$

Therefore $da/dt = AK_I^n = AK_{Ic}^n(\sigma/S)^n$. Also, we know from Eq. (8-7) that $a^{-3/2} = Y^3S^3/K_{Ic}^3$. Substituting for da/dt and $a^{-3/2}$ in Eq. (8-8) produces the result

$$\frac{dS}{dt} = (-\tfrac{1}{2}AY^2K_{Ic}^{n-2})S^{3-n}\sigma^n . \tag{8-10}$$

Next, we rearrange and integrate Eq. (8-10) to find strength as a function of time:

$$\int_{S_i}^{S_f} S^{n-3}\,dS = (-\tfrac{1}{2}AY^2K_{Ic}^{n-2})\int_0^t \sigma^n\,dt$$

$$\frac{1}{n-2}(S_f^{n-2} - S_i^{n-2}) = (-\tfrac{1}{2}AY^2K_{Ic}^{n-2})\int_0^t \sigma^n\,dt \tag{8-11}$$

where S_i is the initial strength at time $t = 0$ and S_f is the final strength at time t. For the loading and unloading cycle in Fig. 8.12, the integral on the right side of Eq. (8-11) is broken into two portions, one from time 0 to t_1 and the second from t_1 to t_2. In the first time interval, there is a constant loading rate $d\sigma/dt = \dot{\sigma}_\ell$, for which we can write $dt = d\sigma/\dot{\sigma}_\ell$. In the second time interval, we can write $dt = -d\sigma/\dot{\sigma}_u$. Making these substitutions for dt on the right side of Eq. (8-11) gives

$$\frac{1}{n-2}(S_f^{n-2} - S_i^{n-2}) = (-\tfrac{1}{2}AY^2K_{Ic}^{n-2})\left[\int_0^{\sigma_p}\sigma^n\frac{d\sigma}{\dot{\sigma}_\ell} - \int_{\sigma_p}^{0}\sigma^n\frac{d\sigma}{\dot{\sigma}_u}\right] . \tag{8-12}$$

Integrating Eq. (8-12) gives an expression for the strength at the end of the proof test:

Strength at end of proof test: $$S_f^{n-2} = S_i^{n-2} - \left(\frac{n-2}{2n+2}\right) A Y^2 K_{Ic}^{n-2} \left(\frac{1}{\dot{\sigma}_\ell} + \frac{1}{\dot{\sigma}_u}\right) \sigma_p^{n+1} . \quad (8\text{-}13)$$

If the proof test involved a stress profile different from that in Fig. 8.12, we would use the appropriate function, $\sigma(t)$, in the integral at the right side of Eq. (8-11) to find how strength changes during the proof test.

8.4.2 A theoretical example: proof testing of sapphire

Figure 8.13 shows the behavior of sapphire predicted with Eq. (8-11) in a proof test in which stress is applied at a constant rate until reaching a peak of 200 MPa, and then the stress is removed at the same constant rate, as shown by the dotted lines. In parts (a) and (c), the loading and unloading rates are 1 MPa/s. In parts (b) and (d) the rates are 1000 MPa/s. We suppose that stress is applied in such a manner that crack growth occurs only on the *m*-plane of the crystal, for which we know crack growth parameters from Fig. 8.10. For parts (a) and (b), the humidity is 3% and for parts (c) and (d) the humidity is 0.08%.

Let's examine individual curves in Fig. 8.13(a), for which the humidity is 3% and the loading rate is slow (1 MPa/s). In curve 1, the initial strength of the test coupon is 600 MPa and application of a 200 MPa proof stress does not degrade the coupon: The final strength after testing is still 600 MPa. If the coupon has an initial strength of 550 MPa (curve 2), the final strength is 550 MPa. However, if the initial strength is 525 MPa (curve 3), slow crack growth occurs enough during the proof test to degrade the strength slightly to 523 MPa. When the initial strength is around 507 MPa, crack growth enters an unstable regime in which imperceptible changes in initial strength lead to huge changes in strength degradation. For example, curves 4-7 behave as follows:

	Initial strength (MPa)	Final strength (MPa)
Curve 4	506.680 000 000	429
Curve 5	506.676 385 732	310
Curve 6	506.676 385 731	broke at 126
Curve 7	506.676 000 000	broke at 166

The remaining cases in Fig. 8.13(a), for which the initial strength is below 500 MPa, all result in mechanical failure during the proof test. Note that *a peak proof stress of 200 MPa is breaking test coupons with initial strengths up to 506 MPa* because of slow crack growth during the proof test.

For the calculations in Fig. 8.13(b), the only change is that the rates of loading and unloading have been increased from 1 MPa/s to 1000 MPa/s. This gives less time for slow crack growth to occur during the proof test, so significant strength degradation occurs only for samples whose initial strengths are somewhat lower than those that are degraded in Fig. 8.13(a). For example, curve 9 shows that a sample with an initial strength of 500 MPa is not degraded in strength, even though a sample with the same initial strength failed during the slower proof test in Fig. 8.13(a). *The quicker the test, the less strength degradation occurs.* The divergent behavior exhibited in curves 10 and 11 of Fig. 8.13(b) occurs when the initial strength is 438.63 MPa.

In Figs. 8.13(c) and (d), the humidity is reduced from 3% to 0.08%, thereby decreasing stress corrosion and decreasing the rate of crack growth during proof testing. Mechanical failure (curve 13) does not occur above initial strengths of 424.11 MPa at a

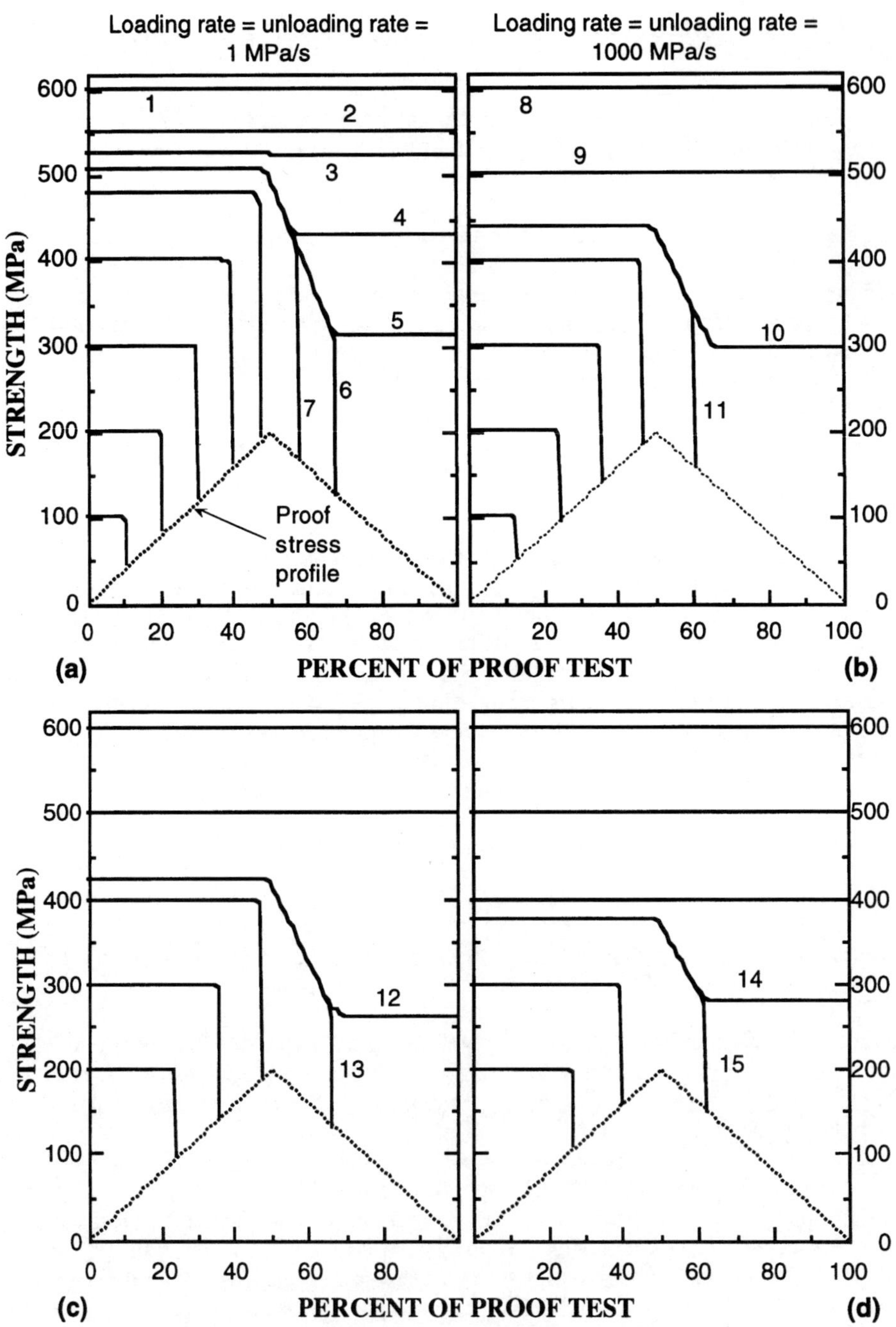

Fig. 8.13. Predicted variation in strength of sapphire during a proof test. Calculations are based on Eq. (8-11) for crack growth on the *m*-plane of a crystal in 3% humidity [(a) and (b)] and 0.08% humidity [(c) and (d)]. Crack growth parameters are $A = 0.0035$ and $n = 49.9$ for 3% humidity and $A = 1.5 \times 10^{-6}$ and $n = 63.2$ for 0.08% humidity. K_{Ic} is assumed to be 2 MPa$\sqrt{\text{m}}$. The peak applied proof stress is 200 MPa. For (a) and (c), $\dot{\sigma}_\ell = \dot{\sigma}_u = 1$ MPa/s; for (b) and (d), $\dot{\sigma}_\ell = \dot{\sigma}_u = 1000$ MPa/s.

Table 8.3. Minimum initial strength to survive 200 MPa proof test in Fig. 8.13

Humidity	Loading and unloading rate 1 MPa/s	1000 MPa/s
3%	507	439
0.08%	424	379

loading rate of 1 MPa/s and above an initial strength of 378.84 MPa (curve 15) at a loading rate of 1000 MPa/s. *The lower the humidity in the test atmosphere, the less strength degradation occurs.* (If we were testing material that is not subject to stress corrosion, the humidity would not be important.) Table 8.3 summarizes the effects of humidity and load rate for the sapphire proof test in Fig. 8.13.

The best proof tests will use the most rapid feasible unloading rate (to minimize strength degradation after reaching the peak proof stress) and will control the environment to minimize strength degradation.[20,21] Immersing the material in a dry liquid such as toluene or heptane[22] or in a dry atmosphere reduces slow crack growth and strength degradation in materials that exhibit moisture-assisted stress corrosion.

8.5 Designing a proof test for the space shuttle window[17]

When the space shuttle was being designed in the 1970's, an ultra-low expansion glass was chosen for the windows to limit thermal stress during the hot re-entry into the atmosphere from space. A proof test was sought to guarantee a desired minimum lifetime (such as a month or a year) under the anticipated stresses that the window would see in service. The proof test would subject the window to a stress σ_p that is higher than the anticipated applied service stress, σ_a. We begin by deriving an expression for the minimum time to failure of a part that survives the proof test.

8.5.1 Minimum time to failure after a proof test[23]

If a is the length of a crack, the crack growth velocity, v, is $\mathrm{v} = da/dt$, where t is time. The relation between stress intensity, K_I, the applied stress, σ_a, and crack length is

$$K_I = \sigma_a Y \sqrt{a}, \quad \Rightarrow \quad K_I^2 = \sigma_a^2 Y^2 a \tag{8-14}$$

where Y is a factor ($\approx \sqrt{\pi}$) that depends on the crack geometry. Let's consider how the crack length changes under the application of a constant applied stress. Taking the derivative of both sides of Eq. (8-14) with respect to time, we find

$$2K_I \frac{dK_I}{dt} = \sigma_a^2 Y^2 \frac{da}{dt} = \sigma_a^2 Y^2 \mathrm{v} \quad \Rightarrow \quad \frac{2}{\sigma_a^2 Y^2}\left(\frac{K_I}{\mathrm{v}}\right) dK_I = dt . \tag{8-15}$$

The time, t_f, required for a crack to grow from its initial size to the critical size for failure is obtained by integrating Eq. (8-15) from $t = 0$ to $t = t_f$ and from $K_I = K_{Ii}$ to $K_I = K_{If}$:

$$t_f = \frac{2}{\sigma_a^2 Y^2} \int_{K_{Ii}}^{K_{If}} \left(\frac{K_I}{\mathrm{v}}\right) dK_I . \tag{8-16}$$

In the crack growth diagram in Fig. 8.8, region II comes at sufficiently high crack velocity for most ceramics that crack propagation time is controlled mainly by the slower crack growth in region I. Equation (8-4), v = AK_I^n, related crack growth velocity to stress intensity in region I. The constants A and n depend on the material and the environment. Substituting AK_I^n for v in Eq. (8-16) allows us to evaluate the integral to obtain

$$t_f = \frac{2(K_{Ii}^{2-n} - K_{If}^{2-n})}{(n - 2)A\sigma_a^2 Y^2} \quad . \tag{8-17}$$

Now we make some simplifications. First, we equate K_{If} with K_{Ic}, since failure occurs when the stress intensity reaches K_{Ic}. Since the exponent, n, is typically in the range 9 to 50 and the initial stress intensity K_{Ii} is substantially less than K_{Ic}, we can ignore the term K_{Ic}^{2-n} in comparison to K_{Ii}^{2-n}. Therefore Eq. (8-17) reduces to

$$t_f \approx \frac{2K_{Ii}^{2-n}}{(n - 2)A\sigma_a^2 Y^2} \quad . \tag{8-18}$$

Eq. (8-18) tells us the time to failure if we know the initial stress intensity, K_{Ii}, at the largest flaw and if we have measured A and n in a separate crack growth experiment.

When a proof-tested component is subjected in service to an initial stress intensity K_{Ii} at an applied stress, σ_a, the value of K_{Ii}/σ_a is equal to the maximum value of K_{Iproof}/σ_p achieved during the proof test, which cannot have exceeded the value K_{Ic}/σ_p or the part would have broken. Substituting the inequality $K_{Ii}/\sigma_a < K_{Ic}/\sigma_p$ into Eq. (8-18) gives

$$t_f \lesssim \frac{2(\sigma_p/\sigma_a)^{n-2} K_{Ic}^{2-n}}{(n - 2)A\sigma_a^2 Y^2} \quad \Rightarrow \quad t_{min} = \frac{2(\sigma_p/\sigma_a)^{n-2} K_{Ic}^{2-n}}{(n - 2)A\sigma_a^2 Y^2} \tag{8-19}$$

where t_{min} is the minimum time to failure when the proof-tested part is put into service at a constant stress σ_a.

8.5.2 Crack growth parameters for space shuttle window material

To use Eq. (8-19) to predict the minimum time to failure for the space shuttle window, we need to know the crack growth parameters A and n in Eq. (8-4). This was done in two ways.[17] In one set of experiments, crack velocity was measured as a function of the stress intensity factor using the double cantilever beam specimen in Fig. 8.7. To assess worst-case crack growth, the specimen was either immersed in water or exposed to 100% humidity. A graph of crack velocity versus stress intensity factor was linear in the range $10^{-11} < v < 10^{-4}$ m/s and $0.32 < K_I < 0.50$ MPa$\sqrt{\text{m}}$ and gave the best fit values $A = 6 \times 10^6$ and $n = 35.8$ at 25°C if K_{Ic} is expressed in units of MPa$\sqrt{\text{m}}$ in Eq. (8-4). These values describe slow crack growth in region I of Fig. 8.8 for the space shuttle window material. When crack growth was measured at 60°C, the parameters were $A = 1 \times 10^5$ and $n = 28.7$.

An independent check on the validity of the crack growth parameters was obtained by measuring fracture strength in ring-on-ring flexure tests as a function of the loading rate from 0.0074 to 74 MPa/s. The tensile surface of the disk was immersed in water during the test to attain worst case subcritical crack growth during the test. The procedure described in Ref. 17 based on equations in Ref. 24 was then used to estimate the crack growth parameters $A = 3 \times 10^3$ and $n = 27$ at 25°C, which gave crack growth velocities in adequate agreement with those measured with the double cantilever beam specimen.

Subcritical crack growth did not occur when the glass was maintained in a dry nitrogen atmosphere or in vacuum. The fracture toughness, $K_{Ic} = 0.70$ MPa$\sqrt{m}$ measured by indentation in vacuum, was independent of temperature from 25 to 600°C.

8.5.3 Proof test design

With values for A, n, K_{Ic} and Y ($\approx \sqrt{\pi}$), we can use Eq. (8-19) to design a proof test. For a required service applied stress, σ_a, and desired minimum time to failure, t_{min}, Eq. (8-19) specifies what proof stress, σ_p, is needed. Fig. 8.14 shows t_{min} as a function of service stress, σ_a, for different ratios of σ_p/σ_a. For a given proof stress level, the predicted lifetime is lower at 60° than at 25°C because cracks grow faster at the higher temperature.

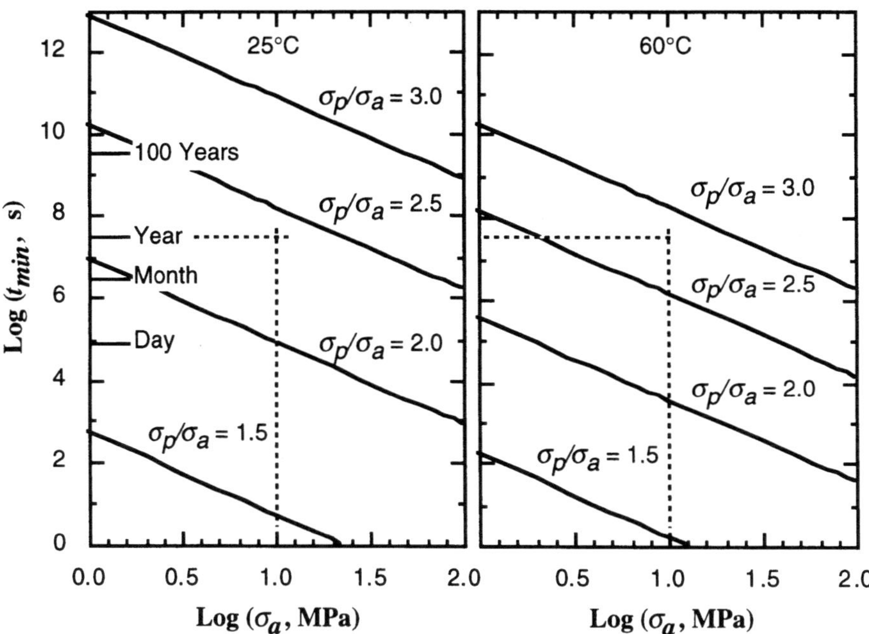

Fig. 8.14. Proof test design chart for space shuttle window showing minimum time to failure as a function of service stress. Predictions are based on Eq. (8-19) using the crack growth rates v (m/s) = $(6 \times 10^6)K_I^{35.8}$ at 25°C and $(1 \times 10^5)K_I^{28.7}$ at 60°C. In these equations, the units of K_I are MPa$\sqrt{m}$.

Example: Selecting the space shuttle window proof stress. Suppose that the minimum time to failure is required to be 1 year at a service stress of 10 MPa at 25°C. What proof stress is required to guarantee 1 year of service life? In the graph at the left side of Fig. 8.14, draw a horizontal line at 1 year and a vertical line at log (10 MPa) = 1. The intersection is between the curves for $\sigma_p/\sigma_a = 2.0$ and $\sigma_p/\sigma_a = 2.5$, from which we estimate $\sigma_p/\sigma_a = 2.4$. That is, the proof stress should be 2.4 times the service stress: $\sigma_p = 2.4\sigma_a = (2.4)(10\text{ MPa}) = 24$ MPa. A proof stress of 24 MPa should guarantee a minimum time to failure of 1 year at a service stress of 10 MPa at 25°C.

If the operating temperature were 60°C instead of 25°C, the graph at the right in Fig. 8.14 predicts that $\sigma_p/\sigma_a = 2.8$, so a proof stress of 28 MPa would be required.

The period when subcritical cracks can grow in the space shuttle window is during launch and initial orbit, when moisture from the atmosphere is still present at crack tips. After sufficient exposure to the vacuum of space, the moisture evaporates and slow crack growth ceases. Upon re-entry into the atmosphere, the windows get hot, but crack growth should be negligible because there is no moisture at the crack tips.

Extensions of the ideas presented in this section have been used to design reliable double-pane glass aircraft windows.[25,26] The goal is to ensure that one pane of the window can survive for a required time under a certain stress if the other pane breaks. Required crack growth parameters were obtained by creating a known surface flaw with a Vickers indentor and measuring the strength at different stress rates (from 0.01 to 100 MPa/s). Glass that had been subjected to dust and sand erosion was characterized as part of the study to see how erosion would reduce the lifetime of the window. The resulting lifetime predictions for the full scale window are shown in Fig. 8.15.

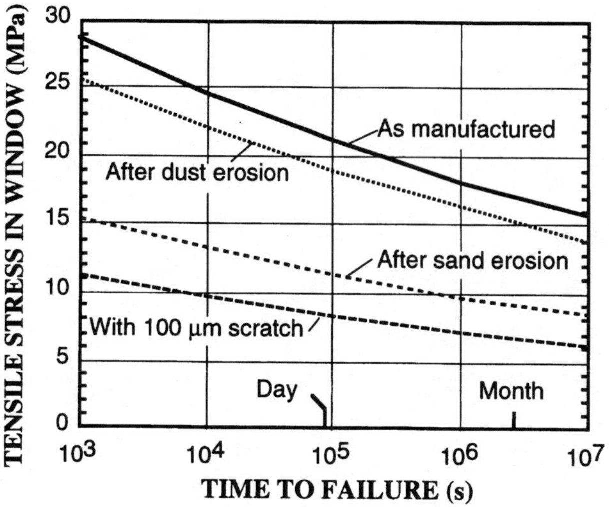

Fig. 8.15. Predicted time to failure for BK-7 glass aircraft window.[26]

8.6. Fatigue

When a ceramic is subjected to stress over a period of time under conditions in which stress corrosion can occur, subcritical cracks grow and the material becomes weaker. Fig. 8.16 shows an example of 3-point flexure specimens made of 99% dense alumina left under a static stress until they failed.[27] When tested rapidly in an inert atmosphere, the flexure strength of the material was 505 ± 40 MPa. The graph shows that when the same bars were left in the air (at unstated humidity and temperature) at a constant stress of 250 MPa, they broke in ~10^4 s. The greater the static stress, the shorter the time to failure.

A theoretical equation for the time to failure under the stress, σ, is[28]

$$\text{Time to failure} = t_f \quad \frac{2}{\sigma^n (n-2) A Y^2} \left(\frac{S_i}{K_{Ic}} \right)^{n-2} \tag{8-20}$$

where n and A are the crack growth parameters in Eq. (8-4), Y is a geometric factor ($\approx \sqrt{\pi}$), S_i, is the initial strength of the material measured by rapid application of stress in an inert atmosphere, and K_{Ic} is the fracture toughness of the material. The exponent, n, is typically in the range 9 to 50. A graph of log t_f versus log σ should have a slope of $-n$. If you measure the mean failure time, t_{f1}, of a series of identical samples under the stress σ_1, then the mean failure time under the stress σ_2 is expected to be

$$\frac{t_{f2}}{t_{f1}} = \left(\frac{\sigma_1}{\sigma_2} \right)^n . \tag{8-21}$$

Example: Dependence of fatigue life on stress. If the exponent in Eq. (8-21) is n = 30, by what factor does the time to failure decrease when the stress is increased by 10%?

$$\frac{t_{f2}}{t_{f1}} = \left(\frac{\sigma_1}{\sigma_2} \right)^n = \left(\frac{1}{1.1} \right)^{30} = 0.057$$

The lifetime decreases to 5.7% of its initial value when the stress is increased by 10%.

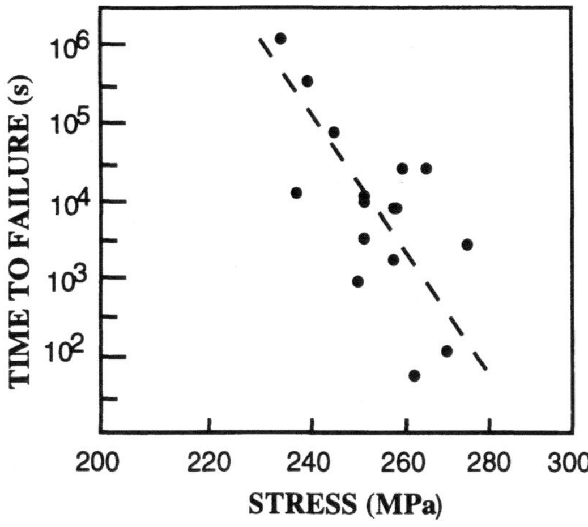

Fig. 8.16. Fatigue in alumina flexure bars left under static stress.[27] The dashed line is a guide for the eye, not a fit to the data points.

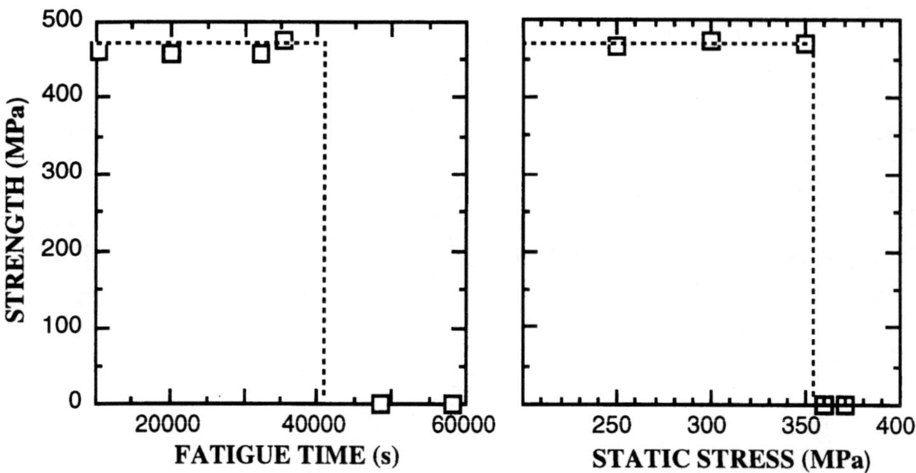

Fig. 8.17. Residual strength of alumina 3-point flexure specimens after being left under a static stress of 350 MPa for various times (*left*) or under different static stresses for a constant time of 35 000 s (*right*).[29] The 3-point flexure strength was measured in a rapid test under an inert atmosphere at the end of the fatigue period.

Fig. 8.17 shows how the residual strength of the alumina 3-point flexure bars from the previous figure changes in two different static fatigue tests.[29] In one case, the bars were left under a static flexure stress of 350 MPa for different periods of time and then their residual strength was measured by rapidly breaking each bar. The graph at the left shows that the residual strength is essentially constant before the bar breaks. In the second case, the bars were left for 35 000 s under a fixed flexure stress. As the fixed stress was varied from 250 to 350 MPa, the residual strength measured at the end of the experiment was essentially constant. When the bar was left at 360 MPa, it broke. The theoretical expression for the residual strength, S_t, after time, t, at the fatigue stress, σ, is

$$S_t^{n-2} = S_i^{n-2} - \frac{1}{2}\sigma^n t\,(n-2) A Y^2 K_{Ic}^{\;n-2} \tag{8-22}$$

where the symbols have the same meaning as in Eq. (8-20). For a large value of n, this expression changes very little until the failure time is closely approached. Because of the exponential rate of subcritical crack growth, materials subject to static fatigue retain their strength for a long time prior to a rapid demise.

References

1. J. C. Wurst and T. P. Graham, *Thermal, Electrical, and Physical Measurements of Laser Window Materials*, Air Force Materials Laboratory Report AFML-TR-75-28, April 1975.
2. S. W. Freiman, J. J. Mecholsky, Jr., R. W. Rice and J. C. Wurst, "Influence of Microstructure on Crack Propagation in ZnSe," *J. Am. Ceram. Soc.*, **58**, 406-409 (1975).

3. W. P. Rogers and A. F. Emery, "Contact Thermal Shock Test of Ceramics," *J. Mater. Sci.*, **27**, 146-152 (1992).
4. J. E. Ritter, Jr., P. B. Oates, E. R. Fuller, Jr., and S. M. Wiederhorn, "Proof Testing of Ceramics. Part 2: Experiment," *J. Mater. Sci.*, **15**, 2275-2281 (1980).
5. J. E. Ritter, Jr., "Assessment of Reliability of Ceramic Materials," in *Fracture Mechanics of Ceramics*, (R. C. Bradt, A. G. Evans, D. P. H. Hasselman and F. F. Lange, eds.), Vol. 5, pp. 227-251, Plenum Press, New York (1983).
6. D. Broek, *Elementary Engineering Fracture Mechanics*, 4th ed., Chap. 1, Martinus Nijhoff Publishers, Dordrecht, The Netherlands (1986).
7. J. J. Gilman, "Direct Measurements of the Surface Energies of Crystals," *J. Appl. Phys.*, **31**, 2208-2218 (1960).
8. S. M. Wiederhorn, "Influence of Water Vapor on Crack Propagation in Soda-Lime Glass," *J. Am. Ceram. Soc.*, **50**, 407-414 (1967).
9. S. M. Wiederhorn, "Subcritical Crack Growth in Ceramics," in *Fracture Mechanics of Ceramics*, (R. C. Bradt, D. P. H. Hasselman and F. F. Lange, eds.), Vol. 2, pp. 613-645, Plenum Press, New York (1974) .
10. S. W. Freiman, D. R. Mulville and P. W. Mast, "Crack Propagation Studies in Brittle Materials," *J. Mater. Sci.*, **8**, 1527-1533 (1973).
11. S. M. Wiederhorn, A. M. Shorb and R. L. Moses, "Critical Analysis of the Theory of the Double Cantilever Method of Measuring Fracture-Surface Energies," *J. Appl. Phys.*, **39**, 1569-1572 (1968).
12. S. M. Wiederhorn and L. H. Bolz, "Stress Corrosion and Static Fatigue in Glass," *J. Am. Ceram. Soc.*, **53**, 543-548 (1970).
13. S. M. Wiederhorn, "Moisture Assisted Crack Growth in Ceramics," *Int. J. Fracture Mech.*, **4**, 171-177 (1968).
14. R. J. Charles, "Static Fatigue of Glass. I," *J. Appl. Phys.*, **29**, 1549-1553 (1958); Static Fatigue of Glass. II," *Ibid.*, **29**, 1554-1560 (1958).
15. S. D. Brown, "A Multibarrier Rate Process to Subcritical Crack Growth" in *Fracture Mechanics of Ceramics*, (R. C. Bradt, D. P. H. Hasselman and F. F. Lange, eds.), p. 606, Vol. 4, Plenum Press, New York (1978).
16. C. S. J. Pickles and J. E. Field, "The Dependence of the Strength of Zinc Sulfide on Temperature and Environment," *J. Mater. Sci.*, **29**, 1115-1120 (1994).
17. S. M. Wiederhorn, A. G. Evans, E. R. Fuller and H. Johnson, "Application of Fracture Mechanics to Space-Shuttle Windows," *J. Am. Ceram. Soc.*, **57**, 319-323 (1974).
18. S. D. Brown, "Multibarrier Kinetics of Brittle Fracture: I, Stress Dependence of the Subcritical Crack Velocity," *J. Am. Ceram. Soc.*, **62**, 515-524 (1979); *Ibid.*, p. 635.
19. S. M. Wiederhorn, B. J. Hockey and D. E. Roberts, "Effect of Temperature on the Fracture of Sapphire," *Philos. Mag.*, **28**, 783-796 (1973).
20. E. R. Fuller, Jr., S. M. Wiederhorn, J. E. Ritter, Jr. and P. B. Oates, "Proof Testing of Ceramics. Part 2: Theory," *J. Mater. Sci.*, **15**, 2282-2295 (1980).
21. A. G. Evans and E. R. Fuller, "Proof Testing — The Effects of Slow Crack Growth," *Mater. Sci. Eng.*, **19**, 69-77 (1975).
22. J. E. Ritter, Jr., K. Jakus, G. M. Young and T. H. Service, "Effect of Proof-Testing Soda-Lime Glass in Heptane," *J. Am. Ceram. Soc.*, **65**, C134-C135 (1982).
23. A. G. Evans and S. M. Wiederhorn, "Proof Testing of Ceramic Materials – An Analytical Basis for Failure Prediction," *Int. J. Fracture Mech.*, **10**, 379-392 (1974).
24. A. G. Evans, "Slow Crack Growth in Brittle Materials Under Dynamic Loading Conditions," *Int. J. Fracture Mech.*, **10**, 251-259 (1974).
25. E. R. Fuller, Jr., S. W. Freiman, J. B. Quinn, G. D. Quinn and W. C. Carter, "Fracture Mechanics Approach to the Design of Glass Aircraft Windows: A Case Study," *Proc. SPIE*, **2286**, 419-430 (1994).

26. J. W. Pepi, "Failsafe Design of an All BK-7 Glass Aircraft Window," *Proc. SPIE*, **2286**, 431-443 (1994).
27. S. E. Park, B. S. Hahn and H. L. Lee, "Static and Cyclic Fatigue Behaviour in Alumina Ceramics," *J. Mater. Sci. Lett.*, **14**, 1688-1690 (1995).
28. J. B. Wachtman, *Mechanical Properties of Ceramics*, Chap. 8, Wiley, New York (1996).
29. S. E. Park and H. L. Lee, "Prediction of Static Fatigue Life in Ceramics," *J. Mater. Sci. Lett.*, **16**, 1352-1353 (1997).

Chapter 9

OPTICAL-QUALITY CVD DIAMOND

A combination of extraordinary thermal, mechanical and optical properties places natural diamond in a class by itself among materials.[1-3] Diamond is the only material that offers long wave infrared (8-14 μm) transmission along with great strength and resistance to thermal shock and erosion. Diamond is also an excellent window for microwave, visible and ultraviolet radiation. Diamond absorbs midwave infrared radiation, so its utility for midwave windows is limited to a role as a thin, protective film. The extraordinary properties of diamond have been known for decades, but only the smallest of optical windows could be made because large crystals of natural diamond are extremely rare. The key technological advance that now makes diamond windows possible is the growth of chemical-vapor-deposited (CVD) diamond over large areas.[4-7] CVD diamond has most of the properties of natural diamond, but its mechanical strength remains modest.

Figure 9.1 shows two ways in which diamond will find use as an optical material.[8] As we discussed in Section 7.8.6, diamond can serve as a thin, durable, transparent coating for more fragile infrared window materials. Alternatively, thick pieces may serve as free-standing windows or domes. This chapter describes the properties of diamond and some of the technology that allows us to capitalize on diamond's extraordinary capabilities.

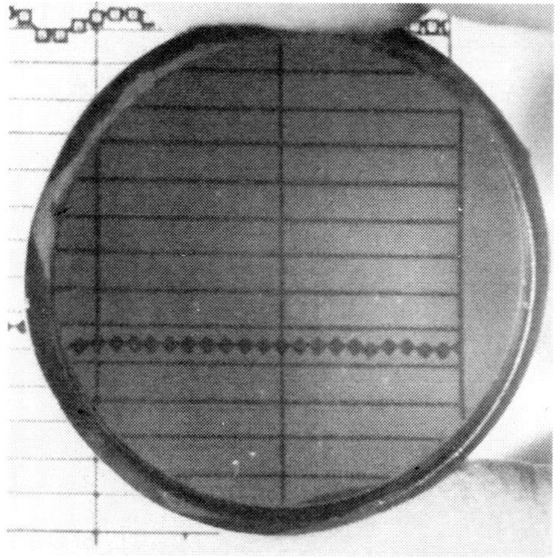

Fig. 9.1. *Left*: Diamond coating (7 μm thick) attached to 38-mm-diameter zinc selenide window by optical brazing (Fig. 7.66) at Westinghouse. *Right*: Transparent, polished diamond window from Raytheon is 0.75 mm thick.

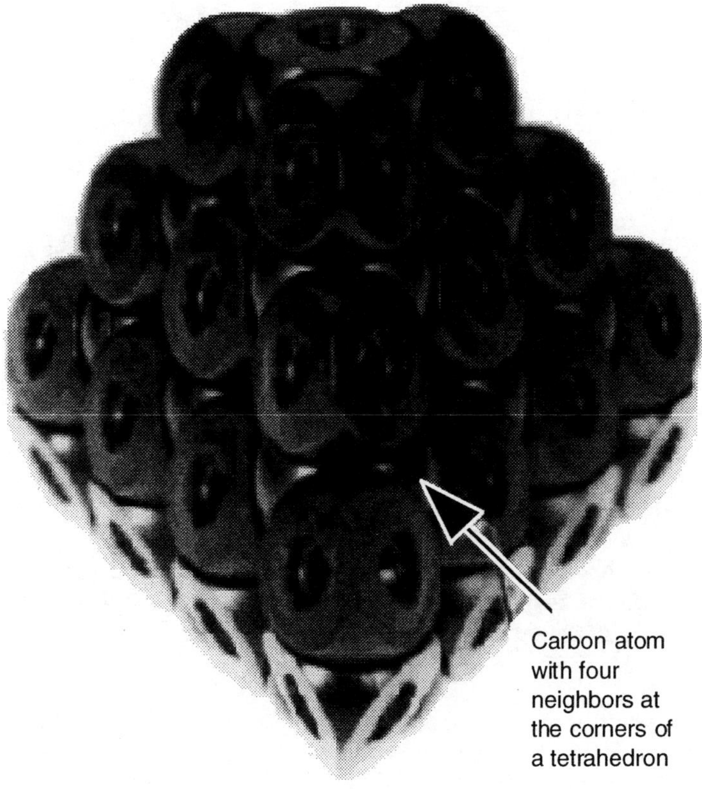

Fig. 9.2. Model of diamond crystal lattice. Every atom inside the crystal has the same tetrahedral arrangement of four nearest neighbors as the one shown. Atoms on the surface of the crystal, which lack one or two bonds to other carbon atoms, may be capped by hydrogen atoms.

9.1 What is diamond and how is it made?

Diamond and graphite are the two most common crystalline forms of carbon. Diamond owes its extraordinary properties to a tightly packed, three-dimensional array of atoms in the crystal (Fig. 9.2).* Each atom is strongly bonded to its four nearest neighbors at the corners of a regular tetrahedron. Diamond contains the greatest number density of atoms (1.76×10^{23} atoms/cm^3) of any material. The tight packing of strongly bonded atoms makes diamond very strong and very stiff. Single-crystal natural diamond is approximately 4 times stronger than silicon carbide and sapphire (Table 3.3), and 2-3 times as stiff (2-3 times higher Young's modulus).

Natural diamond is classified into four types. *Type Ia*, which accounts for 98% of natural diamond, contains up to 0.1% nitrogen in small aggregates or platelets that strongly absorb ultraviolet light below 320 nm. The thermal conductivity of this diamond is less than 900 W/m·K at 25°C and its electrical resistivity is $>10^{14}$ Ω·m. Almost all industrial synthetic diamond is *Type Ib*, which contains up to 0.2% nitrogen in the crystal lattice. Low concentrations of nitrogen give diamond a yellow color, while higher concentrations lead to shades of green. Colorless *Type IIa* diamond, which is almost free of nitrogen, has the best optical and thermal properties. It transmits light

*In contrast, graphite consists of layers of carbon atoms that are tightly bonded within each layer, but loosely bonded between layers. As a result, the layers can easily slide past each other, making graphite a good lubricant.

above 225 nm and has a thermal conductivity near 2000 W/(m·K) at 25°C. *Type IIb* diamond is also free of nitrogen, but contains small quantities of boron (typically < 1 ppm) that impart a blue color and *p*-type conductivity (resistivity = 0.1-10 Ω·m) to the crystal. Only Type IIa is suitable as an infrared window. Nitrogen gives rise to infrared absorption bands in the 7.5-25 μm region, while boron impurities yield sharp absorptions at 3.58 and 4.07 μm, and weak absorption in the long wave window.[1,9]

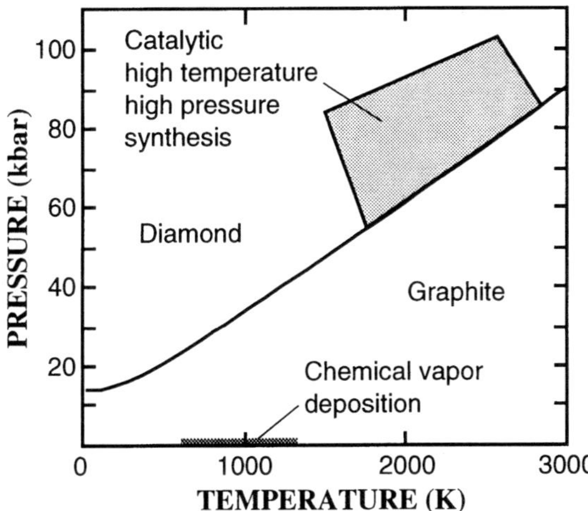

Fig. 9.3. Phase diagram of carbon, showing regions of stability of diamond and graphite. Pressure is in kbar (1 bar = 10^5 Pa = 0.9869 atm). Temperature and pressure regimes in which synthetic diamond is made are shaded.

The phase diagram of carbon in Fig. 9.3 shows that graphite is stable at relatively low pressure, while diamond is stable at high pressure. Although graphite is the thermodynamically stable form of carbon at atmospheric pressure, a temperature above 1500°C is required before the rate of conversion of diamond to graphite is significant.[10,11] At 2100°C, the rate of this *graphitization* process is such that a 0.1 carat (1 carat = 0.2 g) octahedral crystal of diamond is converted to graphite in less than 3 min.

High-temperature, high-pressure processes are employed for industrial synthesis of diamond.[1,2] Figure 9.3 shows a region of temperature and pressure in which a mixture of graphite with molten metals such as nickel, cobalt or iron produces diamond. Diamond is the thermodynamically stable form of carbon under these conditions. Carbon atoms dissolved in the metal under these conditions have sufficient mobility to recrystallize into the diamond structure. Industrial diamond synthesis generally produces micrometer-size powders, but carefully controlled conditions can yield single-crystal, gem-quality diamond up to 5 carats in mass.

Another form of synthetic diamond, called *polycrystalline diamond*, is widely used in cutting tools.[1] Polycrystalline diamond is made by sintering diamond powder in the presence of cobalt at pressures of 50-100 kbar and temperatures of 1200-1600 K. The diamond crystallites grow together to form a compact mass with a grain size of 2-25 μm in which the interstices are filled with residual cobalt. Cutting tools with a thick (0.7-mm) layer of polycrystalline diamond on a base of tungsten carbide are sold under such trade names as Compax® (General Electric) and Syndite® (De Beers).

9.1.1 Chemical vapor deposition of diamond

A revolution in diamond technology occurred in the 1980s, when continuous polycrystalline diamond films were grown over several square centimeters by low pressure chemical vapor deposition.[4-7] This process operates at lower temperature than high-temperature, high-pressure diamond synthesis (Fig. 9.3) and, more importantly, at pressures below 1 atm. Typical conditions might be 950°C and 0.2 bar — easily attained with conventional equipment. Note in Fig. 9.3 that chemical vapor deposition of diamond takes place in the region of the phase diagram in which graphite, not diamond, is the stable form of carbon. Hydrogen atoms in the reactor are essential for the growth of diamond and also stabilize diamond relative to graphite.[12-14]

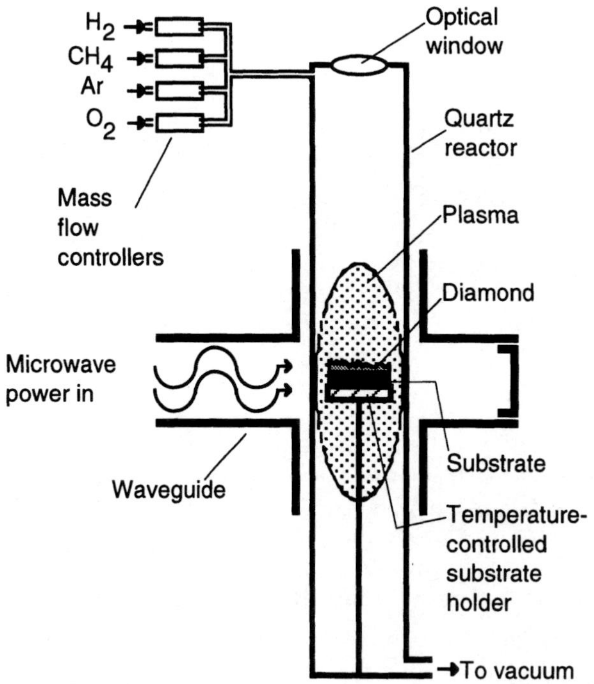

Fig. 9.4. Schematic diagram of microwave plasma chemical vapor deposition of diamond.

Figure 9.4 shows a *microwave plasma reactor* in which diamond is grown on a silicon substrate. A mixture that is typically 99 volume percent hydrogen and 1 percent methane, possibly with additives such as argon, oxygen[15] or parts per million of nitrogen,[16] is passed through a quartz tube inside a waveguide. Microwave radiation partially dissociates the gas into a plasma containing hydrogen atoms, methyl radicals (CH_3), high energy electrons[17] and other reactive species such as hydroxyl radicals (OH). Diamond crystallites nucleate on the substrate and grow into a continuous polycrystalline film. The diamond growth surface in Fig. 9.5 is rough, but the underside facing the substrate is as smooth as the original silicon surface. Diamond with a thickness greater than 1 mm can be grown over areas greater than 100 cm^2 with growth rates of 0.1-10 μm/h by this technique. The maximum growth rate reported for good optical-quality diamond is ~3 μm/h.

Fig. 9.5. Microstructure of growth surface of diamond from a microwave plasma.

The principal requirements for diamond deposition are dissociation of the mixture of hydrogen and methane into atoms and radicals at elevated temperature, and crystal growth at lower temperature. A variety of techniques can be used to activate the gas phase. Instead of microwaves, the apparatus in Fig. 9.6 uses an arc discharge (a spark) between two electrodes to partially dissociate H_2 gas.[18] The *dc torch* produces diamond at rates of several hundred μm/h over small areas, or at lower rates over larger areas.[19,20] Diamond may be grown on such substrates as silicon, molybdenum, copper, tungsten, nickel and stainless steel. In the dc torch, the substrate must be cooled to approximately 900°C so that the temperature is not too high for diamond growth. Figure 9.7 shows an industrial scale dc torch diamond reactor.

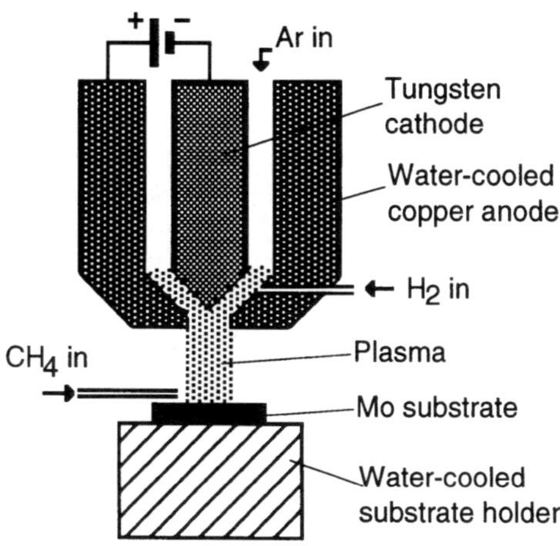

Fig. 9.6. Schematic diagram of direct current torch (plasma jet) used for chemical vapor deposition of diamond.

Two other common means of activating the gas phase for diamond deposition are a hot filament and an acetylene torch. In a *hot filament reactor*, the gas is heated to approximately 2200°C by a tungsten (light bulb) filament held 5-10 mm from the substrate, whose temperature may be 700-1000°C. At a pressure of 25 mbar, approximately half of the hydrogen molecules in the 1% methane/99% hydrogen mixture

Fig. 9.7. Industrial scale dc arc jet diamond deposition reactor that uses a magnetic field to help shape the plasma. Jet is seen impinging on the substrate at the center of the photo. (Courtesy Norton Co.)

are dissociated into hydrogen atoms.[21] An advantage of this type of reactor is that it is readily scaled up to large sizes and irregular shapes. A disadvantage is the relatively slow growth rate of 0.1-1 µm/h. Also, the hot filament method does not grow the highest optical-quality material. An *acetylene welder's torch* with a 1:1 mole ratio of acetylene and oxygen has also been used to grow diamond. The use of acetylene instead of methane illustrates that almost any organic compound[22,23] may be a source of carbon for diamond growth. Methane is not a necessary ingredient, although methyl radicals formed in the gas phase are thought to be the immediate precursor to diamond growth.[24-27]

Molybdenum and silicon are the most common substrates for diamond growth. Conditions are also known for deposition on transition metals such as Cu, Cr, Mn, Fe, Co and Ni.[28] Thin diamond coatings have been grown directly on fused silica and sapphire by metal-induced nucleation of diamond.[29] This process uses a source of Cr, Ti or Ni near the growth surface to promote nucleation and adherence of diamond and gives a growth rate of 0.4 µm/h at 650°C. Thick layers cannot be produced because thermal expansion mismatch causes delamination upon cooldown. Diamond films up to 1.5 µm in thickness can be grown on glass.[30] If a sputtered interlayer such as silica, hafnia, silicon carbide or aluminum nitride is deposited on a ceramic, diamond can be grown directly on the interlayer.[29,31] The interlayer improves adhesion by binding well to both the diamond and the substrate. The interlayer also encapsulates the substrate to protect it from attack by the plasma and creates a diffusion barrier between diamond and the substrate. Figure 7.70 showed a patterned surface etched into zinc sulfide to promote adhesion of a thin diamond layer to an interlayer deposited on the ZnS.

Fig. 9.8. Preferential appearance of (100) square diamond crystal faces in the absence of oxygen (*left*) and (111) triangular faces (*right*) in the presence of oxygen in a microwave reactor. [Photos courtesy C. E. Johnson, Naval Air Warfare Center.]

Preferential orientation of CVD diamond crystallites can be controlled by deposition conditions. For example, Fig. 9.8 shows that square crystal faces are selectively deposited in a microwave plasma under particular conditions in the absence of oxygen. When oxygen is added to the feed gas, triangular faces are preferentially deposited.

Because of the much lower thermal expansion coefficient of diamond compared to most substrates, catastrophes are known to occur during cooldown after a CVD growth run. If the diamond adheres tightly to the substrate, the diamond and/or the substrate are likely to fracture. If there is no fracture, serious bowing can occur. Proprietary release layers are applied to the substrate prior to growth of diamond to aid the separation of the diamond from the substrate. One patented release layer is a coating of niobium metal on a molybdenum substrate.[32] As the molybdenum-niobium-diamond sandwich cools down in a hydrogen atmosphere, niobium absorbs hydrogen (from the side of the sandwich) to form a weak, embrittled niobium hydride which releases the diamond without fracture. On the subject of patents, it is of interest to note that General Electric Co. has two patents which claim the rights to all transparent CVD diamond more than 50 μm thick.[33]

9.1.2 The two surfaces of CVD diamond

The first diamond grains that nucleate on a substrate are submicron in size. The grains tend to grow in conical shapes (Fig. 9.9) as the thickness of the deposit increases.[34] Therefore, exposed grains on the growth surface tend to become larger and larger and the deposit becomes rougher as the diamond becomes thicker (Fig. 9.10).

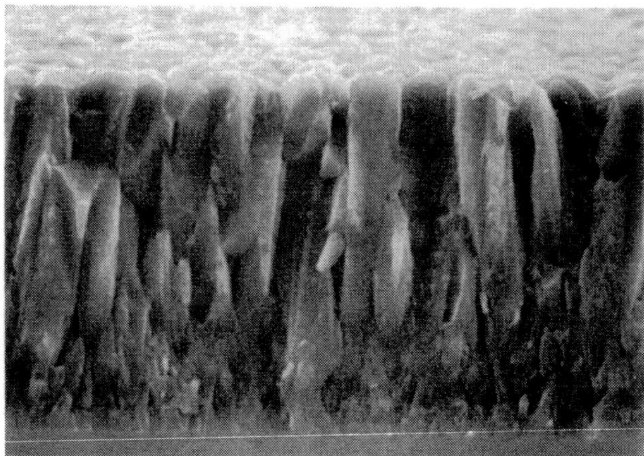

Fig. 9.9. Cross section of diamond film showing columnar growth.[34]

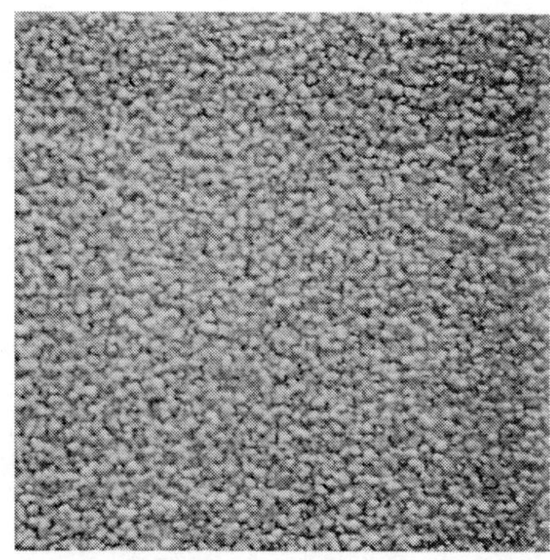

3.5 μm thick

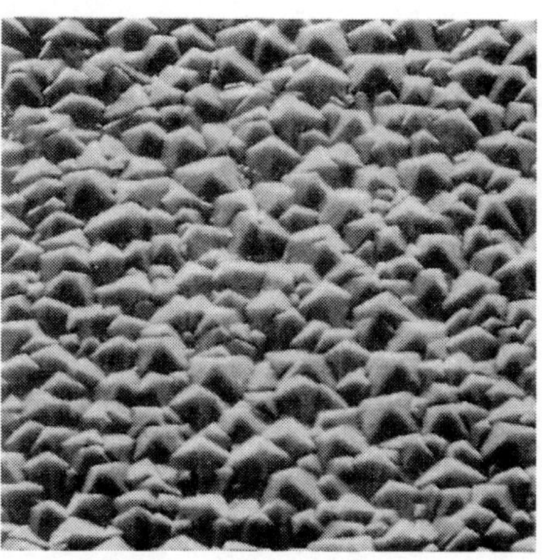
105 μm thick

160 μm thick

Fig. 9.10. Scanning electron micrographs showing grains on the growth surface of hot-filament CVD diamond films of increasing thickness.[34] Both grain size and surface roughness increase with thickness.

Fig. 9.11. Nomarski micrographs showing substrate and growth surfaces of 1-mm-thick polished, optical-quality diamond.

Polished, optical diamond windows normally have two easily distinguished surfaces. The substrate surface (also called the nucleation surface) has tiny grains and the growth surface has large grains, as seen in Fig. 9.11. In Section 3.8.1 we saw that mechanical strength tends to scale as $1/\sqrt{\text{grain size}}$. The large-grain growth surface has much larger flaws than the small-grain substrate surface. The growth surface typically has half the tensile strength of the substrate surface.

9.2 Mechanical and thermal properties of diamond

The comparison of properties of diamond with those of zinc sulfide and sapphire in Table 9.1 indicates why diamond is so promising as an infrared window or dome material.

Table 9.1. Comparison of properties of diamond, zinc sulfide and sapphire near 300 K

Property	CVD diamond	Zinc sulfide	Sapphire
Long wave infrared cutoff (μm)	none	11	5
Hardness (kg/mm^2)	9000	230	1600
Strength (MPa)	200	100	300-1000
Young's modulus (GPa)	1143	74	344
Poisson's ratio	0.07	0.29	0.27
Expansion coefficient (ppm/K)	0.8	7.0	5.3
Thermal conductivity [W/(m·K)]	2000	19	36
Thermal shock figure of merit, R' [kW/m, Eq. (4-7)]	410	2.6	4.3
Microwave dielectric constant	5.7	8.35	9.39 ($E \perp c$) 11.58 ($E \parallel c$)
Microwave loss tangent	<0.0004	0.0024	0.00005 ($E \perp c$) 0.00006 ($E \parallel c$)

Zinc sulfide is the most durable of the currently used long wave (8-14 μm) infrared materials and sapphire is the most durable midwave (3-5 μm) material. Diamond is much harder, much stiffer, has lower thermal expansion, has extraordinarily higher thermal conductivity and thermal shock resistance, and has better microwave dielectric properties (for radar transmission) than sapphire or zinc sulfide.

9.2.1 Hardness, toughness and elastic properties

As the hardest known material (Table 3.7), diamond is especially resistant to abrasion by solid particles. In a study of sand erosion at an impact speed of 26 m/s, the rate of loss of mass by natural, single-crystal diamond was 2×10^4 times lower than the mass loss rate of silicon nitride (at an impact speed of 47 m/s) and 3×10^6 times lower than the mass loss rate of alumina (at an impact speed of 34 m/s).[35] The hardness of chemical-vapor-deposited diamond is equal to that of natural diamond.[36] Table 9.2 gives the hardness of natural diamond measured on different crystal planes[37] and Fig. 9.12 shows the temperature dependence of the hardness of CVD diamond.[38] There is a significant drop in hardness around 700°C. The great hardness of diamond makes it especially valuable for coating ceramic cutting tools.[39]

Table 9.2. Vickers hardness of natural diamond on different crystal planes[37]

Crystal plane	(001)	(001)	(110)	(110)	(111)	(111)
Direction	<100>	<110>	<001>	<110>	<110>	<112>
Type Ia:						
kg/mm^2	10000	8500	11000	9000	5700	6400
GPa	98	83	108	88	56	63
Type IIa						
kg/mm^2	10500	9300	11700	9600	7800	11200
GPa	103	91	115	94	76	110

1 kg load = 9.8 N at 20°C

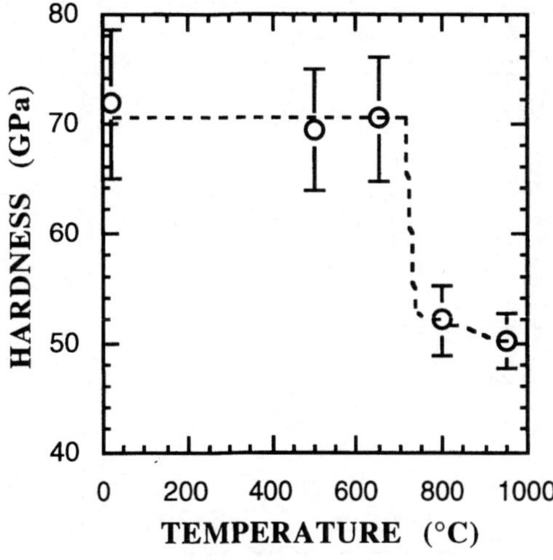

Fig. 9.12. Temperature dependence of Vickers hardness of 400-μm-thick CVD diamond with grain sizes in the range 2-20 μm.[38] Crystallites have a preferential {100} form with (111) perpendicular to the plane of the film. Measurements were made under Ar to avoid oxidation. In another study of CVD diamond, the hardness at 20°C was 81 ± 18 GPa.[40]

Surveying various measurements of the fracture toughness (K_{Ic}) of small crystals of natural diamond, Field estimated that K_{Ic} for natural diamond is ~5 MPa$\sqrt{m}$.[41] Finding the toughness of CVD diamond by indenting it with diamond is problematic, because the test material is as hard as the indentor. A more meaningful method for diamond is the double-torsion method, in which a starter flaw is machined into a rectangular specimen for 4-point bending.[42] The fracture toughness is derived from the applied load at the onset of fracture. Four specimens with thicknesses from 0.43 to 0.88 mm had fracture toughness values of 5.7, 6.0, 6.9 and 8.7 MPa$\sqrt{m}$.[43] In another study, Vickers indentation of CVD diamond gave an average fracture toughness of 5.3 ± 1.3 MPa$\sqrt{m}$.[40] Indentation of low-optical-quality (nearly black) CVD diamond gave K_{Ic} = 8 ± 2 MPa$\sqrt{m}$.[44] Identification of the critical flaw size in flexure disks of the same diamond gave K_{Ic} = 8 ± 1 MPa$\sqrt{m}$.

Elastic constants of single-crystal diamond are listed in Table E.1 in Appendix E. From these constants, Klein and Cardinale calculated that Young's modulus for CVD diamond with randomly oriented crystals is E = 1143 MPa and Poisson's ratio is ν = 0.069.[45] Young's modulus and Poisson's ratio are anisotropic, so CVD diamond with preferential crystallite orientations will have values different from that of randomly oriented material. Bulge tests of CVD diamond are consistent with the values of E and ν derived from single-crystal diamond.[45] The bulk modulus calculated for randomly oriented CVD diamond is 442 MPa and the shear modulus is 553 MPa.[46] The pressure and temperature dependence of the elastic constants of diamond have been measured.[47]

9.2.2 Mechanical strength

Field[48] states that the mechanical strength of natural diamond is highly variable, but an approximate value is 3 GPa. Single crystals normally cleave on (111) planes whose

Table 9.3. Flexure strength of CVD diamond

Specimen description	Strength (MPa)
Disk burst test:	
6.4 mm diameter × 0.25 mm thick[50]	230-410 (4 disks)
10-20 mm diameter × 0.18-0.30 mm thick[51]	740-1140 (64 disks)
20-25 mm diameter × 0.5 mm thick[52]	280 ± 50 (14 disks) (coarse-grain side)
25 mm diameter × 1.07 mm thick[52]	220, 240 (coarse-grain side)
Ring-on-ring disk flexure[53,54]	
20 mm diam. × 0.6-1.2 mm thick, disks from 3 lots	180-270 (25 disks)
Ring-on-ring disk flexure at 1000°C[53,54]	
20 mm diameter × 0.5-0.9 mm thick	410 ± 150 (3 disks)
4-point flexure bars (25 mm × 5 mm × 0.13 mm thick)[55]	570 ± 120 (15 bars)*
3-point flexure bars (28 mm × 2 mm × 1 mm thick)[56]	
Thermal conductivity grade (1320 W/[m·K]))	280-370 (coarse-grain side) 650-780 (fine-grain side)
Thermal conductivity grade (1000-1200 W/[m·K])	440, 510 (coarse-grain side) 950, 980 (fine-grain side)

*Weibull modulus = 5.1 according to Harris analysis of data in the paper.

cleavage energy is 5.5 ± 0.15 J/m^2 (Section 3.3.1). The compressive strength of diamond grit is ~10 GPa. The 3-point flexure strength of 5 specimens of Type IIa natural diamond (2 × 20 × 0.2 mm) was in the range 2.0 - 3.0 GPa.[49] For comparison, the 3-point flexure strength of 15 specimens of synthetic single-crystal diamond with a thickness of 0.35 mm was in the range 0.9 - 2.4 GPa with a mean value near 1.5 GPa.

CVD diamond in Table 9.3 is weaker than single-crystal diamond because the CVD material is laced with internal flaws. Material from De Beers[51] might be stronger than other CVD diamond, perhaps because De Beers has found a way to minimize the conical spreading of grains as they grow. Table 9.3 shows that the fine-grain side of CVD diamond is twice as strong as the coarse-grain side and the strength of diamond at 1000°C in an inert atmosphere is at least as great as the strength at 20°C.

Figure 9.13 shows the strength of optical-quality diamond from De Beers measured in 3-point bending tests. The substrate (nucleation) side of the diamond is about twice as

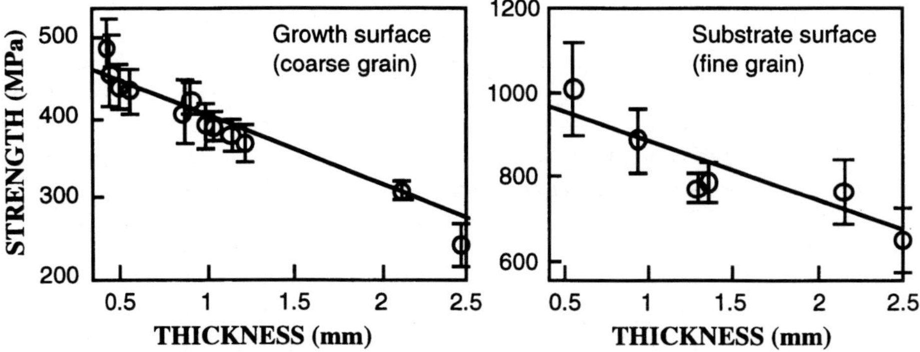

Fig. 9.13. Strength of as-grown (unpolished) De Beers optical-quality diamond measured in 3-point bending test with 18 × 2 mm bars.[57] Each point is an average of 5-10 measurements and error bars represent the standard deviation.

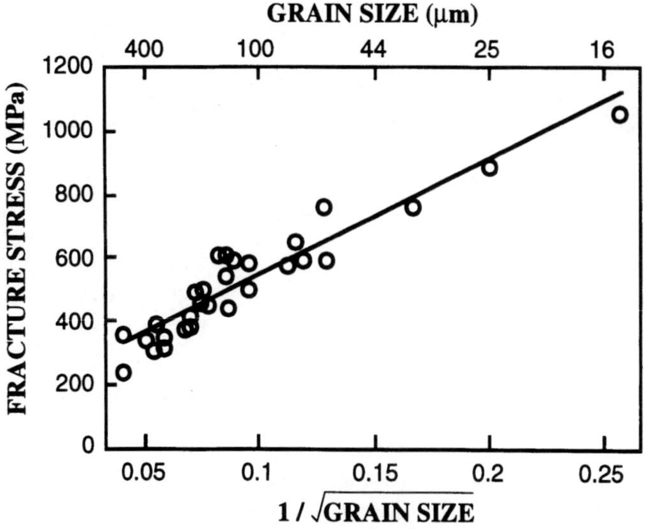

Fig. 9.14. Strength of De Beer's diamond (18 × 2 mm 3-point bend bars) as a function of grain size.[57] Data come from both the growth and substrate surfaces. The substrate surface has been polished.

strong as the growth side. For both surfaces, strength decreases with increasing thickness of the diamond. The explanation is that the flaw size tends to increase with the thickness of the diamond. Figure 9.14 shows the direct correlation of strength with grain size. The data follow the Petch equation (3-36) with the intercept $\sigma_o \approx 200$ MPa. There is a relatively small spread of strengths in DeBeers diamond in the 3-point bending test. The Weibull modulus for the data in Fig. 9.13 is 23 on the growth surface and 11 on the substrate surface.

Figure 9.15 shows ring-on-ring disk flexure data (Section 3.2.2) from Raytheon that are in qualitative disagreement with the De Beers data in Fig. 9.13. In agreement with De Beers, the strength of the substrate surface decreases with increasing thickness of the diamond. However, in disagreement with De Beers, the strength of the growth surface appears to be independent of thickness. A pair of data points designated MP3-58 show the effect of polishing the growth surface. The strength prior to polishing was 132 ± 20 MPa and it increased to 213 ± 10 MPa after polishing. It is hard to rationalize the independence of strength from thickness on the growth surface.

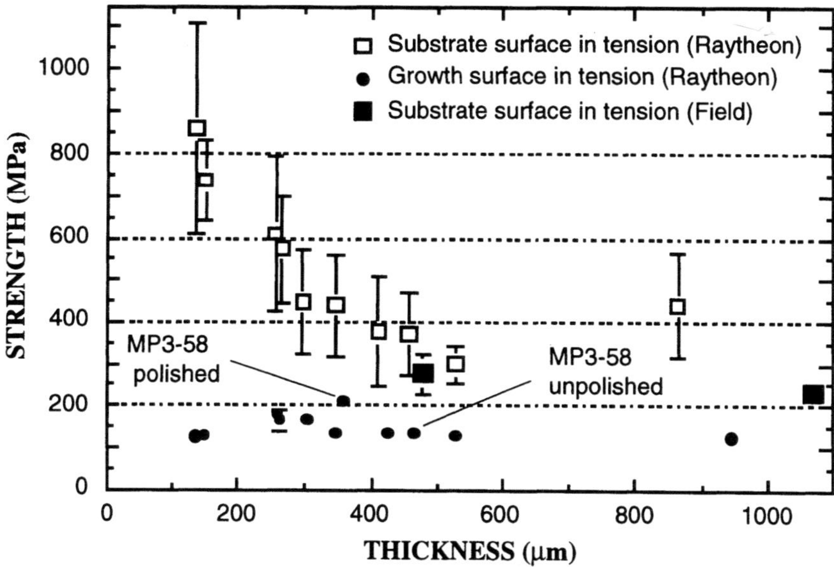

Fig. 9.15. Ring-on-ring disk flexure strength of as-grown (unpolished) optical-quality CVD diamond from seven diamond depositions at Raytheon.[58] Two additional data points are from Field.[52] Disks with a diameter of 17 mm were tested with a load diameter of 7 mm and support diameter of 14 mm. Each point is an average of 3-9 specimens, with error bars showing the standard deviation. For most of the growth surface specimens, the standard deviation lies within the diameter of the black circles.

Raytheon diamond in Fig. 9.15 appears to be less than half as strong as De Beers diamond in Fig. 9.13. This might be largely a result of area scaling: The area under stress in the 3-point bend test of De Beers is much smaller than the area under stress in the ring-on-ring test at Raytheon. Figure 3.20 showed that the apparent strength of alumina doubled in going from a ring-on-ring test to a ball-on-ring test.

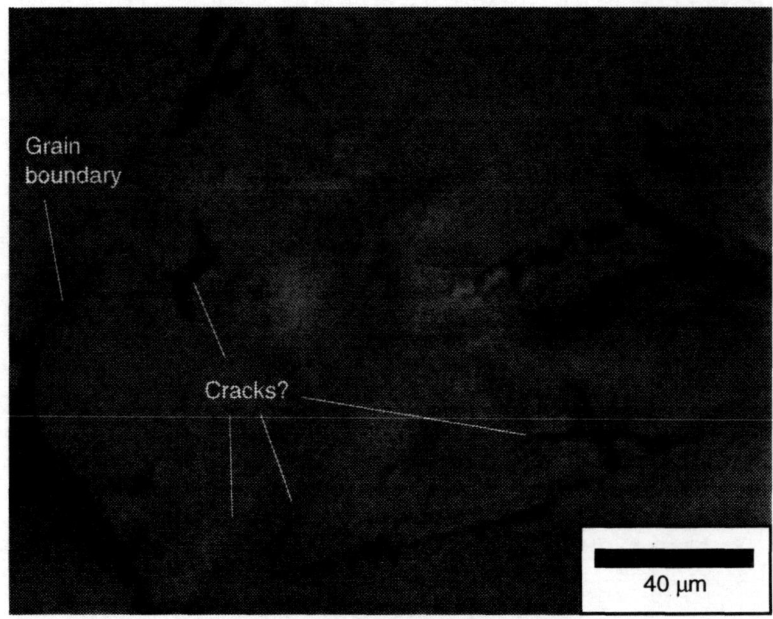

Fig. 9.16. Cracks extending into a single grain on the polished growth surface of CVD diamond.

Whereas small crystals of natural diamond can be nearly perfect, CVD diamond is laced with obvious defects when viewed under a microscope. Figures 9.16 to 9.18 show some of these defects, any of which might limit the strength of CVD diamond.

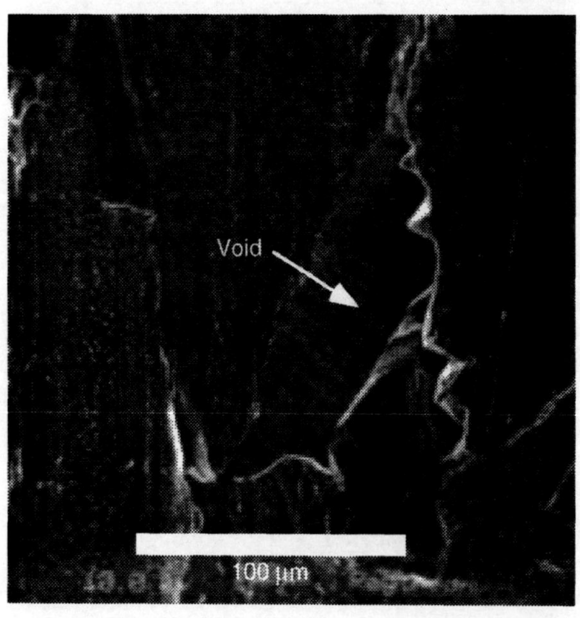

Fig. 9.17. Large void observed in laser-cut CVD diamond.[58] [Raytheon photo courtesy C. B. Willingham.]

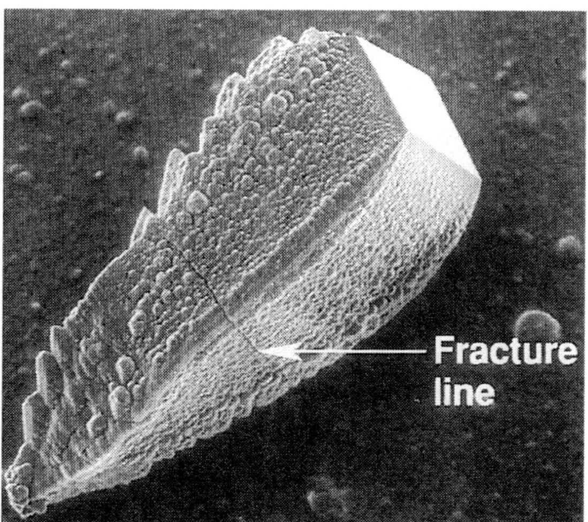

Fig. 9.18. Spontaneous fracture of a single grain of diamond. Tiny spherical nodules of carbon were first grown by a microwave plasma torch on a molybdenum surface with a $CH_4/H_2/Ar$ mixture. Then O_2 was added to the gas to initiate growth of single (100) diamond grains from the carbon nodules at 1000°C. When the diamond cooled down, isolated, fractured crystals were observed. [Photo courtesy A. Harker, Rockwell Science Center.]

100 μm

The spontaneous fracture of a single grain in Fig. 9.18 most likely arises from intrinsic stress in the CVD diamond. Numerous studies found large, intrinsic stresses in diamond that vary systematically with growth conditions such as methane concentration and substrate temperature.[59] In addition to significant average stresses in CVD diamond, local stresses within grains can vary between compressive and tensile with magnitudes on the order of 1 GPa.[60-62]

Example: Critical flaw size in diamond. The fracture toughness of diamond is approximately $K_{Ic} = 6.0 \text{ MPa}\sqrt{\text{m}}$ for both single-crystal and CVD diamond. What is the size of the critical flaw in each material if the strength of single-crystal diamond is 2.5 GPa and the strength of CVD diamond is 300 GPa? To estimate the critical flaw size, we use Eq. (3-35):

$$\text{Strength} = S = \frac{K_{Ic}}{1.24\sqrt{r}} \quad \Rightarrow \quad r = \left(\frac{K_{Ic}}{1.24\,S}\right)^2$$

where r is the radius of the critical flaw. Inserting $K_{Ic} = 6.0 \text{ MPa}\sqrt{\text{m}}$ and $S = 2500$ MPa for single-crystal diamond gives $r = 3.7$ μm or a flaw diameter of $2r \approx 7$ μm. For CVD diamond with a strength of 300 GPa, the flaw diameter is $2r \approx 520$ μm = 0.52 mm.

9.2.3 Thermal expansion

The low thermal expansion of diamond in Table 9.4[63] makes it difficult to retain a diamond coating on most substrates, because the substrate material expands at a greater rate than diamond, leading to delamination or fracture from significant temperature excursions. The data in Table 9.4 are for natural diamond. However, the thermal expansion of Raytheon[54] and Diamonex[64] CVD diamond are similar to that of natural diamond in the range 123-523 K and 50-1100 K, respectively. A polynomial fit to the data for natural diamond in Table 9.4 over the range 100-1600 K is:

Table 9.4. Comparison of thermal expansion coefficients (10^{-6} K^{-1})[63]

Temp. (K)	Diamond	ZnS	ZnSe	GaAs	GaP	Ge	Si	SiC
100	0.0					2.2	-0.4	0.3
200	0.4	5.1	6.0			5.0	1.4	1.2
300	1.0	6.8	7.3	5.7	4.6	5.9	2.6	2.8
400	1.8	7.4	7.9		5.0	6.4	3.2	3.9
500	2.7				5.3	6.8	3.5	4.3
600	3.2				5.5	7.2	3.8	4.6
800	3.8				6.0	7.9	4.0	5.0
1000	4.4			7.3		8.5	4.2	5.3
1200	4.9						4.3	5.5
1400	5.4						4.4	5.6
1600	5.9							5.65

$$\text{Expansion coefficient} = \alpha\,(\text{ppm/K}) = \frac{1}{L}\frac{dL}{dT}$$
$$= -1.144 + 9.535 \times 10^{-3}\,T - 5.149 \times 10^{-6}\,T^2 + 1.204 \times 10^{-9}\,T^3 \qquad (9\text{-}1)$$

where L is the length of the specimen and T is temperature (K).

9.2.4 Thermal conductivity and heat capacity

Diamond has the greatest thermal conductivity of any material near 300 K — 5 times higher than copper and 2-3 *orders of magnitude* greater than that of other infrared window materials in Table 4.1. Figure 9.19 compares the thermal conductivity of Type IIa natural diamond[65-68] to that of copper. Two data points in Fig. 9.19 for high quality CVD diamond[69] lie on the curve for Type IIa diamond. Table 9.5 lists some numerical values of thermal conductivity. For the temperature range 500-1200 K, the thermal conductivity of Type IIa diamond[65] can be fit by the equation

$$k\left(\frac{\text{W}}{\text{m}\cdot\text{K}}\right) = \frac{2.833 \times 10^6}{T^{\,1.245}} \qquad (9\text{-}2)$$

in which there is some uncertainty in the second digit of k.

Table 9.5. Thermal conductivity of diamond (W/[m·K])

Temperature (K)	Best Type IIa[70]	Type IIa[71]	Type Ia[70]	Excellent CVD[69]
100	12500	---	1200-5200	---
200	5500	---	1000-3000	---
300	2500	2000-2200	500-1800	2000
450	1300	1250	400-1200	---
480	---	---	---	1200
700	900			
900	600			
1200	450			

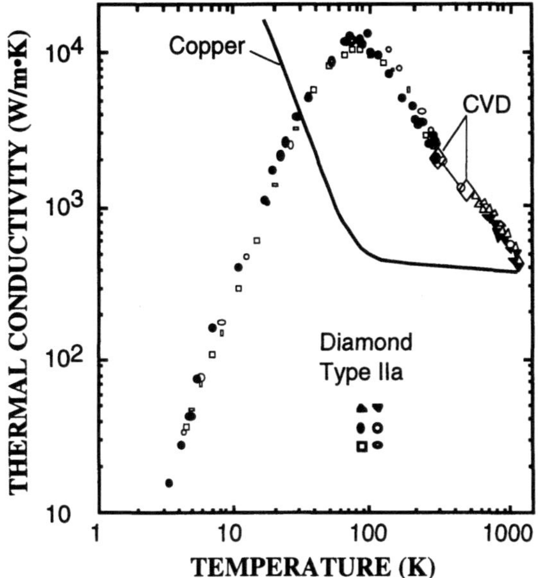

Fig. 9.19. Comparison of thermal conductivity of Type IIa single-crystal diamond[65-68] and copper. Two diamond symbols show the conductivity of high quality CVD diamond.

We saw in Fig. 9.9 that CVD diamond has a columnar microstructure. Furthermore, the grains are very fine near the substrate surface and coarse at the growth surface (Fig. 9.11). This anisotropic microstructure leads to significant anisotropy in thermal conductivity.[72-75] Conduction within single grains of CVD diamond is similar to conduction in single-crystal natural diamond. Therefore the thermal conductivity of CVD diamond normal to the surface (parallel to the columnar grains) is very high. By contrast, there is significant thermal resistance at the grain boundaries, so conductivity parallel to the surface is relatively low. Furthermore, conductivity on the coarse-grain growth surface is higher than conductivity on the fine-grain substrate surface because there are fewer grain boundaries per unit length on the growth surface. The net result is shown in Fig. 9.20, in which the conductivity increases from a low value on the substrate surface nearly to the single-crystal value on the growth surface.

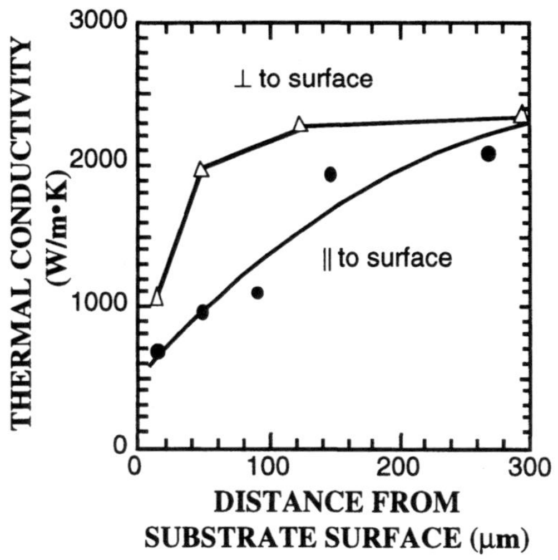

Fig. 9.20. Anisotropic thermal conductivity of high quality CVD diamond near 300 K.[72]

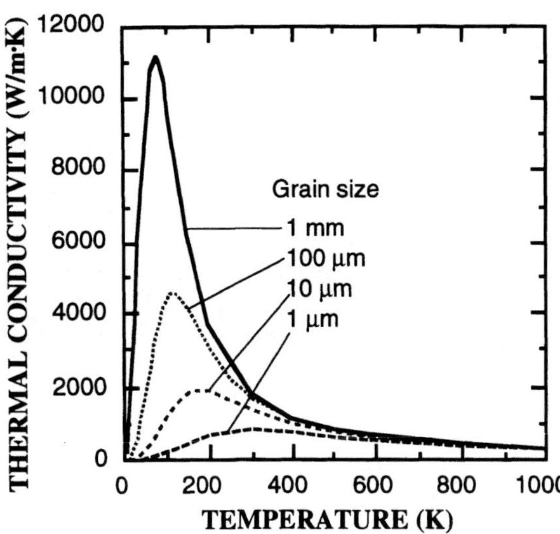

Fig. 9.21. Dependence of thermal conductivity of diamond on grain size, as calculated by the method of Morelli *et al.*[76] (Calculations courtesy C. Robinson, Raytheon Co.) Curves show qualitative trends, but are not necessarily quantitatively accurate.

Figure 9.21 shows the predicted[76] effect of grain size on thermal conductivity of diamond. Since heat is carried through the electrically insulating solid by atomic vibrations (phonons), scattering of the phonons at the boundaries of small grains decreases the thermal conductivity.[76-78] While grain size is a significant factor at low temperatures (100 K), there is little dependence on grain size at elevated temperature (500 K), where thermal conductivity is limited by phonon-phonon interactions (called *Umklapp* processes[79]) instead of grain boundary scattering.

Foreign atoms in a crystal lattice, such as ^{13}C in a ^{12}C lattice also give rise to phonon scattering. When the ^{13}C in single-crystal diamond was reduced from its natural abundance of 1.1% down to 0.07%, the thermal conductivity increased from 22.3 W/(m·K) to 33.2 W/(m·K) near room temperature and this relative difference persists from 200 to 500 K.[80-82] If the ^{13}C content is increased from its natural abundance, the conductivity goes down.[83,84] The thermal conductivity of diamond is sensitive to the quality of the crystal growth, the microstructure, and the presence of defects such as nitrogen atoms in the lattice.[85-88]

Compared to other materials near 300 K, diamond has an unusually low heat capacity (Table 9.6). The heat capacity of CVD diamond with a range of quality from transparent to dark is equal to that of natural diamond.[91] The density of CVD diamond is within 1% of the density (3.5146-3.5155 g/mL) measured on a range of natural diamonds.[91]

Table 9.6. Heat capacity of natural diamond (C_p, J/[mol·K])[89,90]

Temperature (K)	Heat capacity	Temperature	Heat capacity
200	2.37	700	18.04
300	6.19	800	19.50
400	10.23	900	20.64
500	13.56	1000	21.60
600	16.12	1100	22.51

9.2.5 Commercial grades of CVD diamond

De Beers manufactures *optical, thermal,* and *mechanical* grades of diamond.[92] Most of the properties described so far apply to optical grade material, which grows slowest and has the best optical transmission and highest thermal conductivity. The grain size on the growth surface of optical quality material with thicknesses of 1-2 mm is 100-250 μm. Thermal grade material is a compromise between high thermal conductivity and faster growth rate. Its conductivity is in the range ~1200-1800 W/(m·K) at 300 K, compared to a conductivity of ≥2000 W/(m·K) for optical grade material. At elevated temperature, the conductivity of optical and thermal grades are nearly equal. At cryogenic temperatures, the optical quality material has much greater conductivity than the thermal grade.

Mechanical grade material, which is nearly opaque, has a finer grain size, higher strength, and lower thermal conductivity than other grades. The grain size on the growth surface of 1-mm-thick material is ~50 μm. The strength of the growth surface is 50% greater than that of optical material. The strengths of the fine grain, substrate surface of all three materials are equal, because the grain size is very fine in all cases.

9.3 Optical properties of diamond

The highest quality CVD diamond has optical properties similar to those of Type IIa single-crystal diamond. Figure 9.22 shows that both natural diamond and CVD diamond transmit from the ultraviolet through the visible and infrared regions, with the exception of weak absorptions in the midwave infrared. Fig. 1.43 showed an expansion of the absorption bands with 1-phonon, 2-phonon and 3-phonon regions labeled. The perfect tetrahedral symmetry of the carbon atoms in diamond (Fig. 9.2) forbids the absorption of infrared radiation by the 1-phonon vibrations centered around 1000 cm^{-1}. If it were not

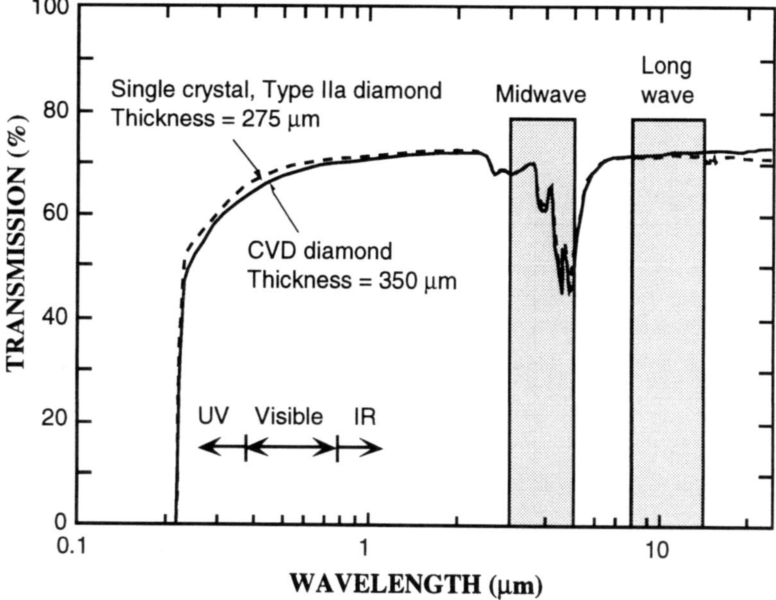

Fig. 9.22. Comparison of transmission of Type IIa natural diamond and Raytheon optical-quality CVD diamond.[58]

for this symmetry, diamond would absorb strongly in the long wave infrared window. The 2- and 3-phonon absorptions in the midwave region are weakly allowed. The flat transmission level of ~70% arises from reflection, not absorption.

9.3.1 Absorption and scatter

Figure 9.23 shows the long wave infrared absorption of Type IIa diamond at 295 and 771 K.[93,94] The extinction coefficient in Fig. 9.23 includes contributions from both absorption and scatter, but most is from absorption. Table 9.7 gives direct measurements of the absorption coefficient near 10 μm from laser calorimetry. De Beers CVD diamond made in 1999 is reported to have only about 20% greater absorption than Type IIa diamond near 10 μm. All of the CVD diamond in Table 9.7 would be suitable for infrared window applications at elevated temperature. Figure 9.24 shows a direct measure of emittance from a De Beers CVD diamond window made around 1993. Figure 9.25 compares the long wave extinction of two specimens of CVD diamond to that of Type IIa diamond. The weak peaks between 1100 and 600 cm^{-1} appear to be real.[98] Assignments of the diamond phonon frequencies in this and other regions have been reported.[99-101]

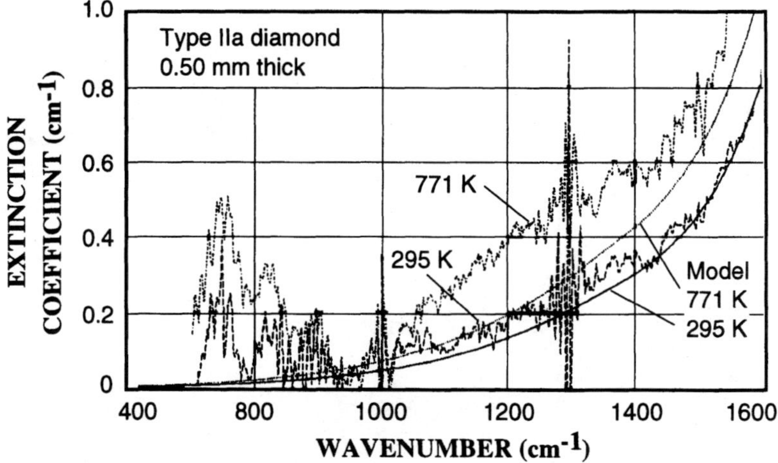

Fig. 9.23. Long wave infrared extinction of Type IIa diamond.[93,94] Extinction includes both absorption and scatter. Smooth curves are from a multiphonon model.

Table 9.7. Laser calorimetry of diamond at ~293 K

Specimen	Wavelength (μm)	Absorption coefficient (cm^{-1})
Type IIa specimen 1	10.59	0.033[95]
Type IIa specimen 2	10.59	0.042[95]
Type IIa specimen 3	10.59 and 9.27	0.047 and 0.062[a]
Type IIa (multiple specimens)	10.59	0.029 ± 0.010[96]
CVD specimen 4[53,54]	10.59	0.07[a], 0.11[b], 0.23[c]
CVD specimen 5[53,54]	10.59 and 9.27	0.13 and 0.19[a]
CVD specimen 6[53,54]	10.59 and 9.27	0.23 and 0.28[a]
De Beers optical grade CVD diamond	~10.59	≤0.07,[92] 0.05[97]

[a]Raytheon data [b]Texas Instruments data [c]China Lake data

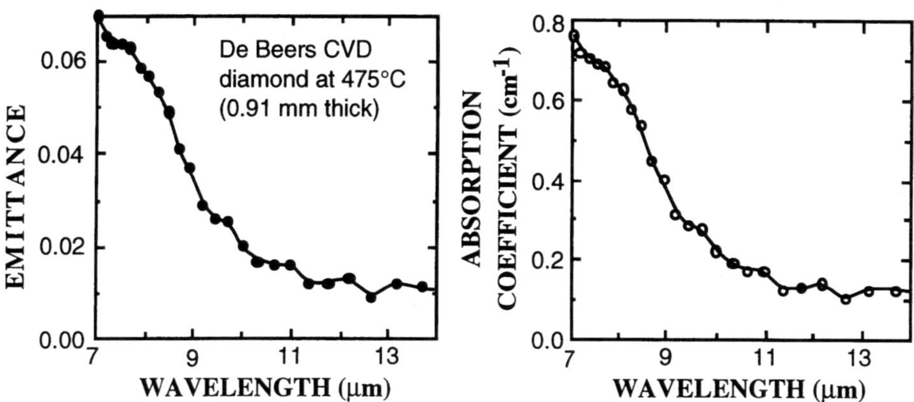

Fig. 9.24. Emittance of De Beers CVD diamond at 475°C and absorption coefficient calculated from the emittance. [Data from A. B. Harker, Rockwell Science Center.]

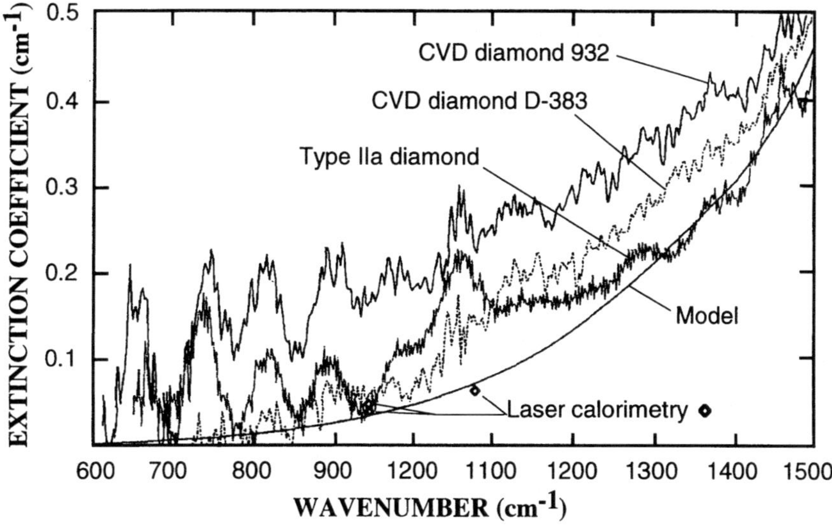

Fig. 9.25. Comparison of long wave infrared extinction coefficient of Type IIa diamond and two specimens of CVD diamond.[93,94] Specimen 932 is 0.64 mm thick and D-383 is 0.74 mm thick. The Type IIa diamond is 0.50 mm thick.

In a study of Type IIa diamond, laser calorimetry was performed on several 1-mm- and 0.5-mm-thick samples (Table 9.8).[96] The difference between the two thicknesses allowed an assignment of contributions from bulk and surface absorption. At 10.59 μm in Table 9.8, the bulk absorption coefficient of $\alpha = 0.029\ cm^{-1}$ in a specimen of thickness $b = 0.10$ cm corresponds to an internal absorptance of $1 - e^{-\alpha b} = 1 - e^{-(0.029\ cm^{-1})(0.10\ cm)} = 0.29\%$. The surface absorption contribution was 0.63% per surface, or 1.26% total. The surfaces accounted for 81% of the total absorptance. The

Table 9.8. Bulk and surface contributions to Type IIa diamond absorption*[96]

	As-received samples		After vacuum heating at 700°C	
Wavelength (μm)	Bulk absorption coefficient (cm^{-1})	Absorptance at each surface (%)	Bulk absorption coefficient (cm^{-1})	Absorptance at each surface (%)
9.24	-0.002 ± 0.014	1.50 ± 0.22	0.004 ± 0.020	0.71 ± 0.14
10.21	0.001 ± 0.011	0.70 ± 0.13	not measured	not measured
10.59	0.029 ± 0.010	0.63 ± 0.18	0.027 ± 0.011	0.11 ± 0.10
10.67	0.033 ± 0.014	0.68 ± 0.15	not measured	not measured

*Based on measuring three 1-mm-thick specimens and two 0.5-mm-thick specimens

specimens were then heated to 700°C in vacuum and calorimetry was repeated. The surface absorption was reduced by the heat treatment and the bulk absorption remained essentially unchanged. At 10.59 μm in Table 9.8, the surface absorptance decreased to 0.11% per surface. The total surface contribution to absorptance fell from 81% prior to vacuum heating to 45% after heating. It was surmised that surface species derived from optical finishing or cleaning were removed by heating.

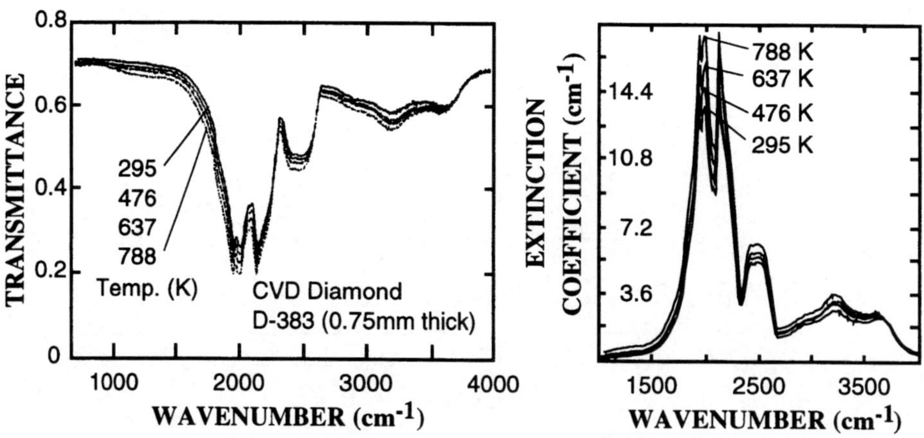

Fig. 9.26. Temperature dependence of transmittance and absorption coefficient of CVD diamond D-383.[93,102]

Figure 9.26 shows the temperature dependent transmittance and absorption coefficient of diamond in the 2- and 3-phonon regions, where high quality CVD diamond and Type IIa diamond are essentially identical. Band-integrated absorption coefficients for diamond in the multiphonon region are listed in Table 9.9. These data are useful for estimating the emittance of diamond at elevated temperature. In the 1-phonon region where the transitions are forbidden, the magnitude of the absorption depends on defects in the diamond.[103] Therefore the best CVD diamond just approaches the behavior of Type IIa diamond.

Table 9.9. Effect of temperature on band-averaged absorption of diamond*[53]

Absorption type	Spectral range (cm^{-1})	Average absorption coefficient (cm^{-1}) 298 K	522 K	656 K	753 K	771 K	973 K
2-phonon	1660-2660	6.0 [5.6]	6.7	7.4	[7.2]	8.5	[8.5]
3-phonon	2660-3890	0.8 [1.4]	0.8	1.1	[1.7]	1.4	[2.4]
4-phonon	3890-5028	0.2	0.3	0.6		0.7	

*Data not in brackets were measured on 0.50-mm-thick Type IIa diamond by M. E. Thomas, Johns Hopkins University Applied Physics Laboratory.[98,102] Data in brackets were obtained with 0.51-mm-thick CVD diamond by J. Trombetta, Texas Instruments.

Example: Emittance and reflectance of diamond. Many people would like to use diamond as a midwave window in the 3-phonon region (2660-3890 cm^{-1}, 2.57-3.76 μm) where the transmittance looks pretty good in Fig. 9.26. What would be the emittance of a 0.15-cm-thick window in this band at 771 K? Table 9.9 tells us that the band-averaged absorption coefficient (α) is 1.4 cm^{-1} in the 3-phonon region. We can use Eq. (1-27) to find the emittance. The refractive index of diamond in the infrared region (Table 9.11) is 2.38, so the single-surface reflectance in Eq. (1-11) is $R = (1-2.38)^2/(1+2.38)^2 = 0.167$. Using $\alpha = 1.4$ cm^{-1} and b = thickness = 0.15 cm in Eq. (1-27) gives the emittance:

$$\varepsilon = \frac{(1 - R)(1 - e^{-\alpha b})}{1 - Re^{-\alpha b}} = \frac{(1 - 0.167)(1 - e^{-(1.4)(0.15)})}{1 - 0.167e^{-(1.4)(0.15)}} = 0.18 .$$

The emittance of 18% is close to what we would have found from the approximation $\varepsilon \approx \alpha b = (1.4)(0.15) = 0.21$. Inspection of Fig. 2.14 shows that 18% emittance is grossly unacceptable for a hot window in the midwave region. If the diamond had been just 10 μm thick, the emittance would have been 0.14%. *A thin film of diamond (~10 μm thick) has acceptable emittance in the midwave region, but a thick window does not.*

To continue this example, let's compute the total infrared reflectance of diamond for normal incidence with Eq. (1-12):

$$\text{Total reflectance} = r = \frac{2R}{1 + R} = \frac{2 \times 0.167}{1 + 0.167} = 0.286.$$

The transmittance in the absence of absorption or scatter is therefore 1 - 0.286 = 71.4%. This is the baseline transmittance level in Fig. 9.22 in the long wave infrared region.

High quality CVD diamond in Table 9.10 has ~100 times as much optical scatter as Type IIa diamond at 10.6 μm and ~20 times as much scatter at 0.63 μm. Even so, the infrared scatter of high quality CVD diamond is below 1%, which is less than that of polycrystalline magnesium fluoride (Fig. 2.5). There is no change in the scatter of Type IIa or CVD diamond up to 500°C.[102] Infrared microscopy at Texas Instruments showed that visible and midwave infrared scatter originates predominantly at grain boundaries, not at grain centers.[54]

Table 9.10. Total integrated scatter of high quality diamond[53,54]

Material	Thickness (mm)	Scatter (% forward / % backward) @ 0.63 μm	@ 3.39 μm	@ 10.6 μm
Type IIa[102]	0.50	0.2 / --		0.004 / --
CVD diamond	0.75	4.2 / 1.9	1.0 / --	0.4 / 0.8
CVD diamond	0.35	4.5 / 2.2	0.8 / 0.4	0.1 / 0.4
CVD diamond	0.75	4.2 / --		0.1 / --
CVD diamond	0.35	2.3 / 1.3	0.6 / 0.2	0.9 / 0.2
CVD diamond	0.51	24 / 7.8	4.5 / 1.3	0.6 / 1.0
CVD diamond	0.91	9.2 / --		0.6 / --

Three specimens of De Beers CVD diamond had optical scatter at 0.5 μm of 3-16%.[104]

The ultraviolet absorption of diamond increases exponentially near the electronic band edge near 0.22 μm in Fig. 9.22. In the region 0.225 to 0.35 μm, the absorption coefficient (α, cm^{-1}) of Type IIa diamond is fit by the equation $\alpha \approx 0.28\, e^{0.45\, E}$, where E is the photon energy in electron volts.[93] Electron volts are equal to the energy in joules divided by the elementary charge (e). In terms of wavelength, the relationship is $eV = hc/(e\lambda)$, where h is Planck's constant, c is the speed of light and λ is wavelength. Impurities in diamond can give rise to ultraviolet absorption not seen in Fig. 9.22. For example, the nitrogen in Type Ia diamond gives a strong absorption centered near 0.39 μm.[9]

Early in the development of CVD diamond, it was common to observe strong infrared absorption bands in the C-H stretching region at ~2800-3000 cm^{-1} (~3.4 μm) superimposed on the diamond 3-phonon absorption in the same region. Figure 9.27 shows two very weak C-H absorptions arising from approximately 8 CH_2 groups per

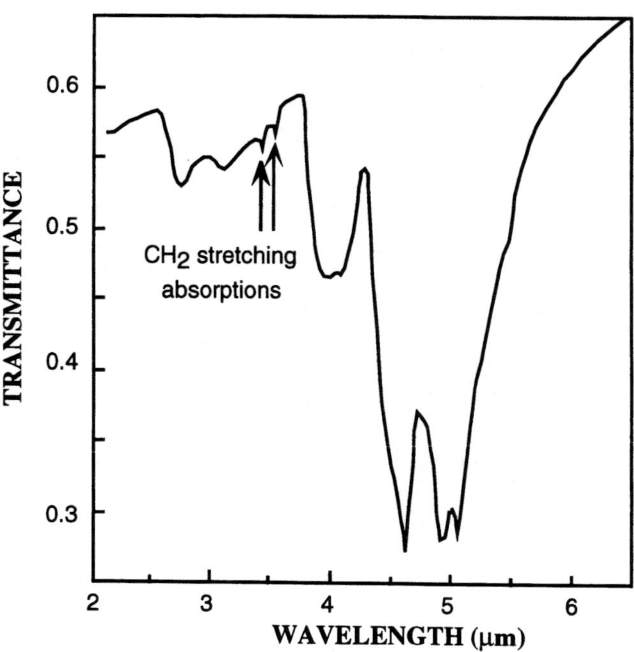

Fig. 9.27. Weak CH_2 stretching absorptions superimposed on the 3-phonon absorption in CVD diamond.[54] The concentration of hydrogen in this sample is 16 H per 10^6 C atoms.

million carbon atoms in CVD diamond.[54] Other bands at somewhat shifted frequencies in other specimens of diamond arise from CH_3 groups and CH groups in various environments. A given set of conditions for each different CVD diamond growth method has its own "signature" in the CH stretching region. The strength of the CH absorption near 3.4 μm is strongly correlated with absorption in the 1-phonon region of CVD diamond near 8 μm: The more CH impurity in the diamond, the more long wave infrared absorption in the diamond.[105-109] Hydrogen in CVD diamond is thought to reside mainly at grain boundaries. Grain boundaries can also be rich in graphite-like, amorphous carbon.[110]

Loss of thermal conductivity is also correlated with increasing hydrogen content in diamond.[108,110] However, many defects in diamond are correlated with each other and it has been argued that hydrogen itself is not responsible for lowering the thermal conductivity.[72] Figure 9.28 shows that increasing visible transmission in diamond is correlated with increasing thermal conductivity.[111]

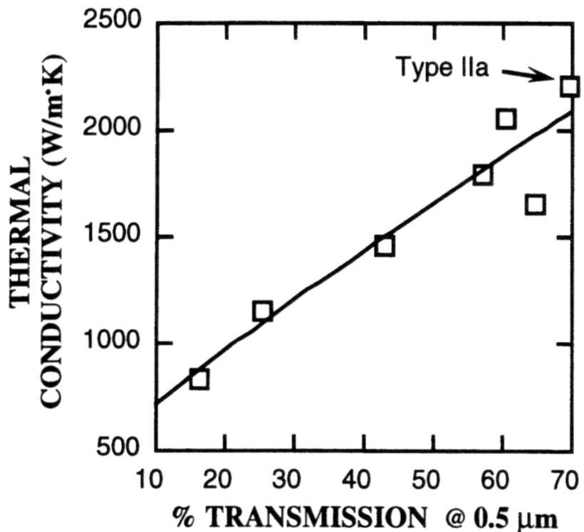

Fig. 9.28. Correlation of increasing thermal conductivity with increasing visible (and ultraviolet) transmission in diamond.[111]

9.3.2 Refractive index

Values of the refractive index (n) of diamond are listed in Table 9.11.[112] Further values for a wide range of wavelengths can be found in the *Handbook of Optical Constants of Solids*.[113] A dispersion equation providing n as a function of wavelength is given in Appendix C.[114] The optical constants of diamond have been modeled over the energy range 0.06 to 30 eV.[115]

Based on measurements at audio frequencies of 10^2-10^4 Hz, the dependence of refractive index on temperature is $dn/dT = 9.6 \times 10^{-6}$ K^{-1} and the dependence on pressure is $dn/dP = -0.86 \times 10^{-12}$ Pa^{-1} near 300 K.[117] A similar value of $dn/dT = 1.0 \times 10^{-5}$ K^{-1} is cited at a wavelength of 0.587 μm.[118] The temperature dependence of dn/dT at 0.633 μm over the range 25-1200°C for Type IIa diamond is[119]

Table 9.11. Refractive index of Type IIa diamond[112]

Wavelength (μm)	Refractive Index	Wavelength (μm)	Refractive Index	Wavelength (μm)	Refractive Index
0.2265	2.7151[116]	0.600	2.4159	5.00	2.3767
0.250	2.6333	0.700	2.4062	10.00	2.3756
0.300	2.5407	2.50	2.3786	15.00	2.3752
0.400	2.4641	3.00	2.3782	20.00	2.3750
0.500	2.4324	4.00	2.3773	25.00	2.3749

$$\frac{1}{n}\frac{dn}{dT} = 3.2 \times 10^{-6} + 3.76 \times 10^{-8}T + 3.78 \times 10^{-11}T^2 + 1.50 \times 10^{-14}T^3 \qquad (9\text{-}3)$$

where T is °C. At 25°C Eq. (9-3) gives $dn/dT = 1.0 \times 10^{-5}$ K^{-1}. An etalon of CVD diamond (Section 1.3.2) gave a mean value of $dn/dT = 1.56 \times 10^{-5}$ K^{-1} in the wavelength region 6.7-12.5 μm over the temperature range 295-784 K.[93,94]

A 300-μm-thick sample of high quality CVD diamond had the same refractive index at 10 μm as Type IIa diamond.[111] For thin films of diamond of modest quality, the situation is different. Whereas the refractive index of Type IIa diamond is 2.4-2.5 at visible wavelengths, the refractive index of ~3-μm-thick CVD diamond is in the range ~2.0-2.3.[119] The difference might arise from non-diamond carbon or voids in the films.

9.3.3 Microwave properties of diamond

Table 2.1 showed that diamond is one of the best microwave window materials among infrared materials, because of its relatively low dielectric constant (ε). For materials such as diamond with low absorption, the dielectric constant is the square of the refractive index ($\varepsilon = n^2$). A low dielectric constant means there is low reflection loss. One of the important potential uses of CVD diamond windows is for high power Gyrotron tubes that generate microwave energy with typical frequencies in the range 110-170 GHz at powers of the order of 1 MW.[57] Low absorption of the electromagnetic energy and high thermal conductivity to avoid large temperature gradients are other critical attributes of diamond that make it an excellent high-power microwave window.

Diamond has a nearly constant refractive index and dielectric constant from long wave infrared wavelengths all the way through microwave and radio wavelengths. For Type IIa diamond, the dependence of microwave dielectric constant on temperature in the range 220-340 K is[117]

$$\varepsilon = 5.70111 - 5.35167 \times 10^{-5}T + 1.6603 \times 10^{-7}T^2 \qquad (9\text{-}4)$$

where T is in kelvins. The temperature and pressure derivatives at 300 K are

$$\frac{1}{\varepsilon_{300}}\frac{d\varepsilon}{dT} = 8.09 \times 10^{-6}\ \text{K}^{-1} \qquad \frac{1}{\varepsilon_{300}}\frac{d\varepsilon}{dP} = -0.72 \times 10^{-12}\ \text{Pa}^{-1} \qquad (9\text{-}5)$$

where ε_{300} is the dielectric constant at 300 K. Figure 2.19 showed the change in

Table 9.12. Microwave loss tangent and absorption coefficient of CVD diamond[57]

Frequency (GHz)	Loss tangent	Absorption coefficient
36	$(80 - 150) \times 10^{-6}$	0.0027 - 0.014
72	73×10^{-6}	0.0026
144	$(100 - 265) \times 10^{-6}$	0.0072 - 0.0190
145	$(100 - 200) \times 10^{-6}$	0.0072 - 0.0145
145	$(15 - 50) \times 10^{-6}$	0.0011 - 0.0036

Data for other specimens are given in Ref. 104.

dielectric constant of Type IIa diamond at 35 GHz as a function of temperature from 18 to 525°C. The equation of the curve in Figure 2.19 is

$$\frac{\Delta\varepsilon}{\varepsilon_{18}} = -2.6445 + 0.68861T + 3.8313 \times 10^{-5}T^2 + 1.1853 \times 10^{-6}T^3 \qquad (9\text{-}6)$$

where ε_{18} is the dielectric constant at 18°C and T is in °C.

For high quality CVD diamond, the microwave dielectric constant is somewhat variable, and has been quoted to be in the range 5.68 ± 0.15.[57] Table 9.12 cites loss tangents (Eq. [2-8]) and absorption coefficients for high quality CVD diamond.[57]

9.4 Diamond windows and domes

Large, optical-quality diamond windows and domes have been fabricated. Figure 9.1 showed a clear, polished window and Fig. 9.29 illustrates the excellent long wave infrared characteristics of a larger, thicker window. The 70-mm-diameter optical-quality dome in Fig. 9.30 has been polished on both surfaces.

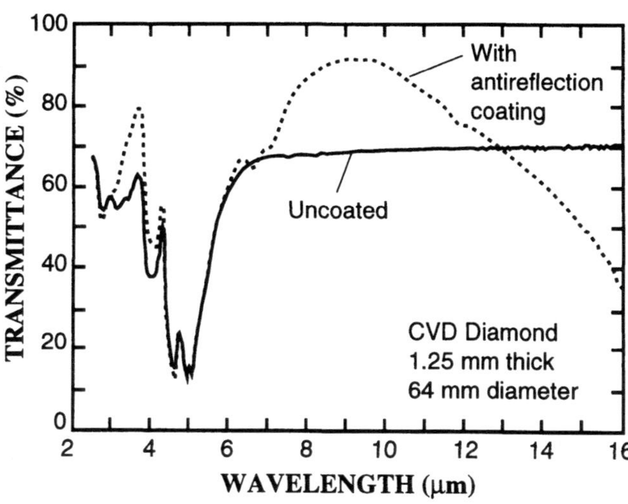

Fig. 9.29. Infrared transmission of a large disk of optical-quality diamond grown by microwave plasma chemical vapor deposition at Raytheon.

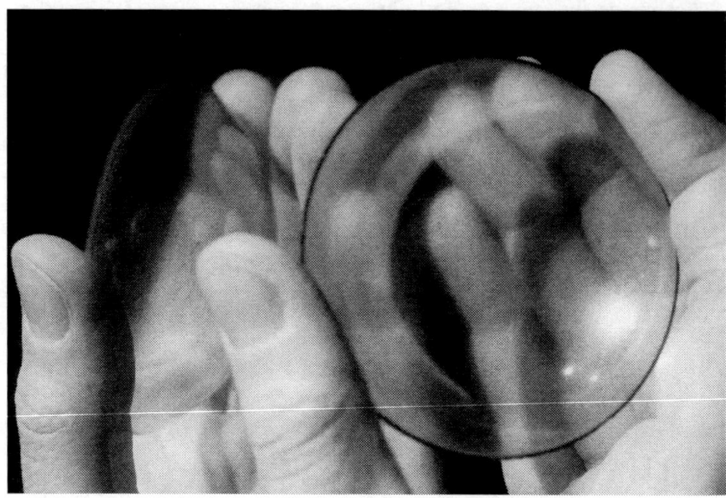

Fig. 9.30. Polished CVD diamond domes (1 mm thick) from De Beers.[97] [Photo courtesy C. J. H. Wort, De Beers Industrial Diamond Division.]

As mentioned in Section 9.3.3, diamond has outstanding properties for high power microwave windows. Diamond is also comparable to the best available materials for high power laser windows.[120] Its dielectric breakdown damage threshold electric field strength is comparable to that of other good laser window materials. Other useful attributes of diamond for laser windows are its high thermal conductivity, reasonable mechanical strength, low *dn/dT*, and weak nonlinear optical properties. The breakdown field strength of excellent CVD diamond is close to that of Type IIa diamond at 0.532 and 1.064 μm wavelength.[121] In comparison to zinc selenide for CO_2 laser windows, diamond has 100 times more absorbance at the 10.6-μm laser wavelength. However, the antireflection coatings required for either window account for much of the total absorbance.[92] With its high thermal conductivity, low *dn/dT*, and ability to be fabricated in thinner windows, thermal lensing in diamond is 100 times less than that in ZnSe.[97]

The high thermal conductivity, low *dn/dT*, and low thermal expansion of diamond windows and domes make them virtually immune to thermo-optical image distortions in rapid-heating environments. In one set of simulations, heating of domes was computed for a severe missile trajectory.[122] The relative root-mean-square wavefront errors of several dome materials after 3 seconds of flight were: zinc sulfide - 3.1, magnesium oxide - 1.0, yttria - 0.8, sapphire - 0.6, and diamond - 0.02.

9.4.1 Polishing diamond

Two challenges for making large optical components are the low growth rate of optical-quality CVD diamond (not exceeding 3 μm/h) and great difficulty of polishing the hardest known material. Conventional abrasive polishing with diamond grit requires months of effort to produce an optical finish on a large surface. De Beers developed proprietary technology to reduce the time to fabricate the dome in Fig. 9.30 from a rough blank to a finished article in ~3 days. The quality of the optical surface produced at De Beers is indicated by the modulation transfer function of a large, flat window in Fig. 9.31, which is within 3% of the diffraction limited value.

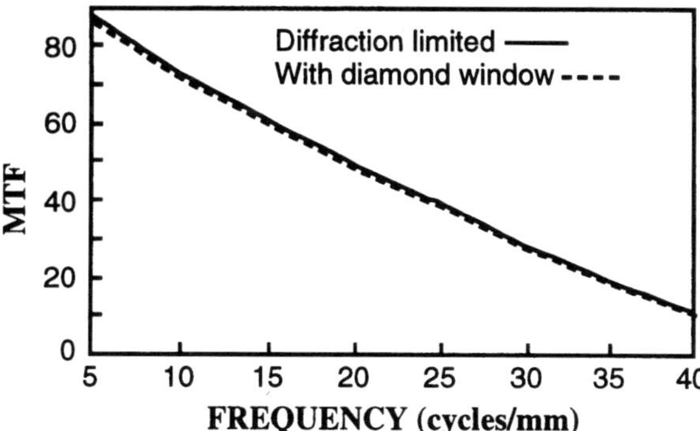

Fig. 9.31. Long wave infrared modulation transfer function (MTF) over a 35-mm aperture in a 106-mm-diameter CVD diamond window from De Beers.[97]

Raytheon uses a computer-controlled 1.06-μm Nd:YAG laser at grazing incidence to machine the outer surface of a 2-mm-thick dome blank from its initial roughness of ~100 μm down to a roughness of ~1 μm (Fig. 9.32).[58] However, final polishing to an optical finish with the required optical figure (geometric shape) requires significant effort after laser machining.

Many methods have been reported for material removal and shaping of CVD diamond. When diamond is brushed against hot iron or nickel surfaces at ~650-950°C, carbon dissolves in the metal and the diamond surface becomes smoother.[123-128] The rate mainly depends on the rate at which carbon diffuses from diamond into the metal. Reactive atmospheres containing H_2[127] or hydrogen atoms[125] are effective in removing

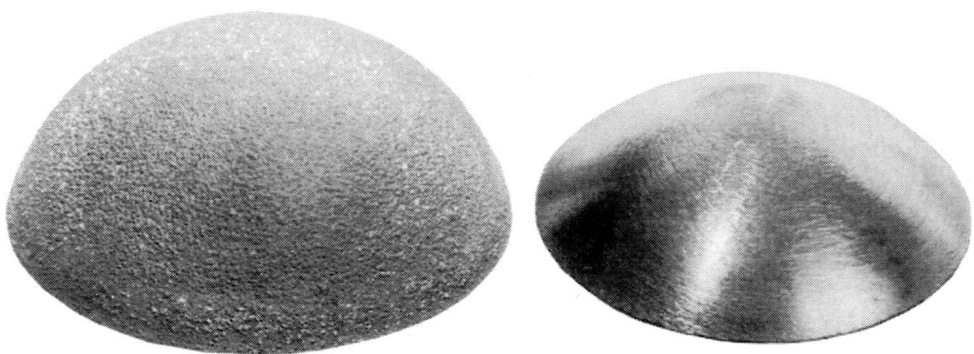

Fig. 9.32. Rough as-grown outer surface of CVD diamond dome and appearance after machining to a roughness of 1 μm with a Nd:YAG laser. The laser-machined dome appears black because the surface is converted to a thin layer of graphite which can be easily removed by subsequent polishing. [Raytheon photos courtesy C. B. Willingham.]

carbon in the form of CH_4 from the metal surface. Molten lanthanum or cerium metal dissolves carbon quickly from the diamond surface at 920°C under Ar[129] and the molten lanthanum-nickel eutectic dissolved diamond at rates of 36-215 μm/h over the temperature range 600-900°C.[130] Precise shaping of diamond with molten metals would be rather difficult. Oxygen[124,131,132] or argon[132] ion beams have also been used to smooth diamond surfaces. The oxide ion conductor, yttria-stabilized zirconia, acting as a cathode in a solid electrochemical cell with diamond and a silicon anode, smoothes diamond by oxidizing it to CO and/or CO_2 in air at 350°C.[133] Attempts have also been made to polish diamond by an abrasive liquid jet.[134]

9.4.2 Mechanical and erosion performance

The mechanical strength of CVD diamond deposited as a hemispheric dome is equivalent to the strength of flat deposits. De Beers used a laser to cut flexure bars out of dome-shaped deposits. The strength of the bars in 3-point flexure tests was equivalent to the strengths of flat pieces shown in Fig. 9.13.[57] The error in measuring the strength is negligible provided the radius of curvature of the bar is more than 8 times the thickness of the bar.

Type IIa diamond has the highest waterjet damage threshold velocity of any infrared window material. Figure 7.27 showed that diamond has a higher multiple impact damage threshold velocity than sapphire. In an experiment with 1-mm-thick specimens, ring fracture was observed on the front surface of sapphire after 190-230 impacts at 420-440 m/s from a 0.8 mm jet.[96,135] Type IIa diamond of the same thickness did not exhibit ring fracture, but finally shattered after 170 impacts at 530 m/s. A second 1-mm-thick specimen of Type IIa diamond withstood 300 impacts at 500 m/s and fractured after 9 impacts at 525 m/s.[136]

Unlike thicker specimens of other materials, thin free-standing diamond windows (~1 mm thick) generally fail at the rear surface when impacted by a sufficiently fast waterjet (Fig. 9.33).[136] Thicker materials almost always exhibit ring fracture on the front surface, rather than tensile failure on the rear surface. Figure 7.28 explained why the front surface ring fracture in diamond, when it finally does occur, is greatly enlarged in comparison to ring fractures on other materials such as zinc sulfide.

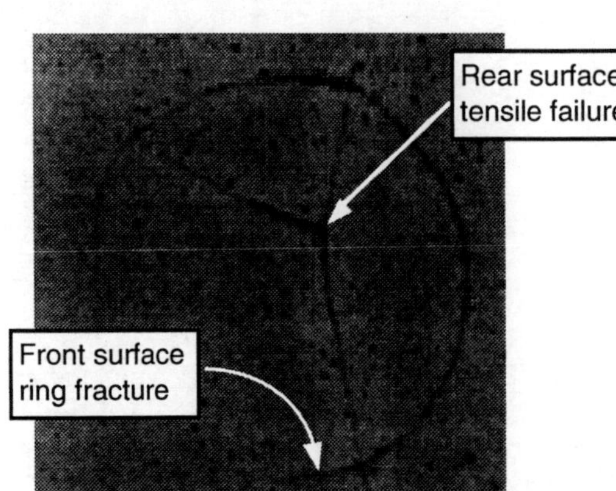

Fig. 9.33. Front and rear surface damage to CVD diamond from waterjet impact. The diameter of the ring fracture is 4.1 mm.[136]

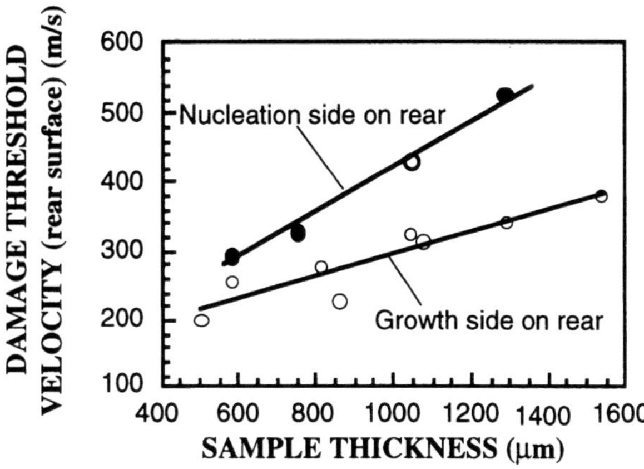

Fig. 9.34. Waterjet damage threshold velocity (0.8-mm nozzle) for rear surface tensile failure in CVD diamond.[136]

Figure 9.34 shows that the threshold velocity for rear surface tensile failure in CVD diamond is higher for damage on the fine-grain nucleation surface than for the coarse-grain growth surface.[136] This is not surprising, since the nucleation surface has a greater tensile strength than the growth surface (Fig. 9.13). Figure 9.35 shows that damage in thin diamond occurs first on the rear surface.[136] If the diamond is sufficiently thick, front surface ring fracture occurs at a lower threshold velocity than rear surface tensile failure. The reason is that the tensile stress on the rear surface decreases with the square of the thickness of the window.

A possible antireflection surface on diamond is the moth eye structure described in Section 6.1.1. Erosion of a diamond moth eye has not been reported, but there has been a study of a silicon moth eye.[137,138] The multiple impact jet apparatus (MIJA) damage threshold velocity for 300 shots from a 0.8 mm jet for flat silicon was 205 m/s. With a silicon moth eye structure, the damage threshold velocity was reduced to 70 m/s and most damage was attributed to lateral jetting of the waterdrop after impact (Fig. 7.9). It remains to be seen whether a diamond moth eye will perform better.

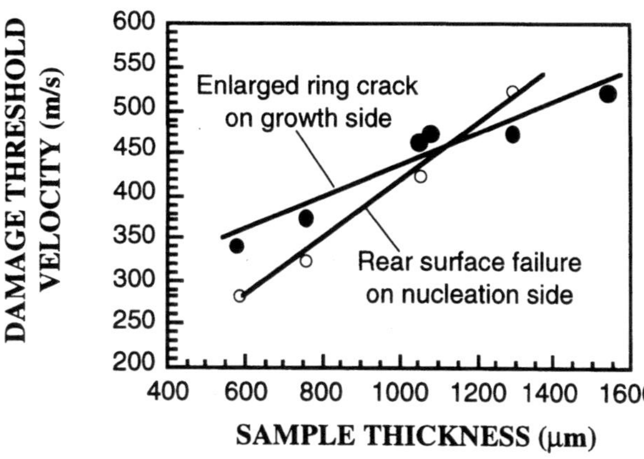

Fig. 9.35. Transition from initial rear surface tensile failure to front surface ring fracture as the thickness of CVD diamond increases.[136]

Table 9.13. Erosion of diamond by 300-600 sand particles[42]

Material	Grain size (μm)*	Sand impact velocity (m/s)	Erosion rate (mg/kg)
Optical-quality CVD diamond:			
<311>/<111> preferential orientation	38 ± 4 n	100	2.63 ± 0.05
<311>/<111> preferential orientation	280 ± 80 g	100	1.47 ± 0.05
<111> preferential orientation	30 ± 8 n	100	4.0 ± 0.5
<111> preferential orientation	190 ± 50 g	100	1.89 ± 0.2
"Mechanical-grade" CVD diamond:			
<110> preferential orientation	32 ± 4 g	100	0.29 ± 0.05
Natural diamond Type Ia <100>	single-crystal	140	0.05
Sapphire (*c*-plane)	single-crystal	100	92 ± 2
Zinc sulfide	60 ± 10	100	30 000 ± 1 000

*n = nucleation surface; g = growth surface

Type IIa diamond has extreme sand erosion resistance. Mass loss is only observed after cracks on crystal planes intersect and chipping begins.[35] Table 9.13 compares erosion rates of CVD diamond with Type IIa diamond, sapphire, and zinc sulfide.[42] In CVD diamond, there was no evidence of failure along grain boundaries. Figure 9.36 shows that the coarse-grain growth surface of 1-mm-thick CVD diamond is eroded more rapidly than the fine-grain nucleation surface.[139] (In 0.5-mm-thick diamond, the erosion rates of both surfaces were similar to each other.)

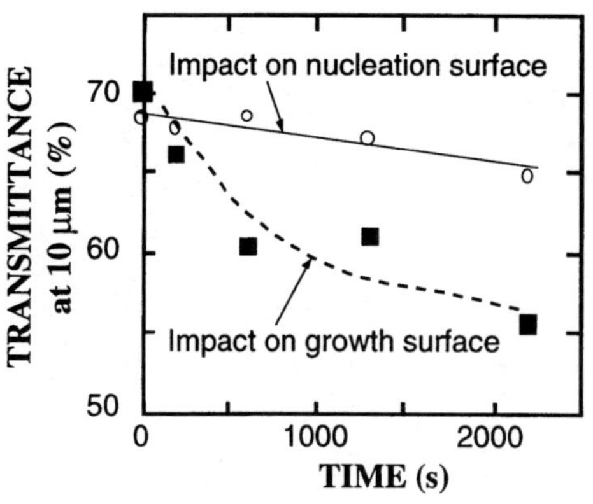

Fig. 9.36. Sand erosion of the two surfaces of 1-mm-thick CVD diamond by 300-600 μm particles at 60 m/s at 0.7 $kg/m^2/s$.[139] Under the same conditions, sapphire lost 60% transmittance in about 550 s.

9.4.3 Oxidation of diamond

Diamond is made of carbon, which burns in the air when the temperature is high enough. Figure 9.37 shows that when diamond is heated in air, the mass remains constant up to 700°C and then oxidation begins.[140,141] The sample is gone by 800°C.

Oxidation: $C\ (diamond) + O_2(g) \rightarrow CO_2(g)$

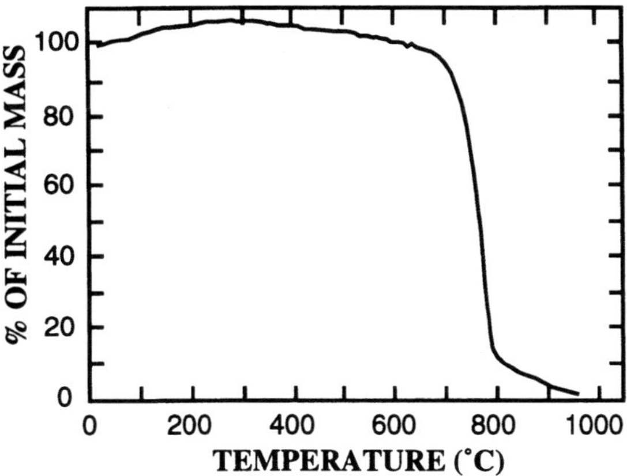

Fig. 9.37. Thermogravimetric experiment in which the mass of a chemical-vapor-deposited diamond film is measured while the diamond is heated in air at a rate of 20°C/min. (Courtesy C. E. Johnson, Naval Air Warfare Center.)

Figure 9.38 shows an experiment designed to assess the endurance of a polished diamond window in the air at high temperature. The upper trace shows the initial infrared transmittance of the disk. The disk was then placed in a preheated furnace open to the air at 700°C for 75 s. The temperature of the diamond was not measured, but it was presumed that it came close to 700°C for a large fraction of a minute. The transmittance was almost imperceptibly degraded. When the same procedure was repeated at 800°C, significant degradation occurred and electron microscopy showed etching (oxidation) at the grain boundaries and, perhaps, at polishing scratches on the surface of the diamond. The conclusion is that an uncoated CVD diamond window or dome could survive ~1 min in air at 700°C, but not at 800°C. Type IIa diamond showed virtually no loss in infrared transmission when heated for 75 or 255 s at 800°C in air. A loss of 7% in transmission at 10.56 μm was noted after a total heating time of 540 s at 800°C.

Oxidation-resistant, antireflection coatings should allow diamond to be used at higher temperatures than 700°C. Two different coatings protected diamond from oxidation at temperatures of 1000-1300°C for ~6 s.[143] However, it should be kept in mind that any

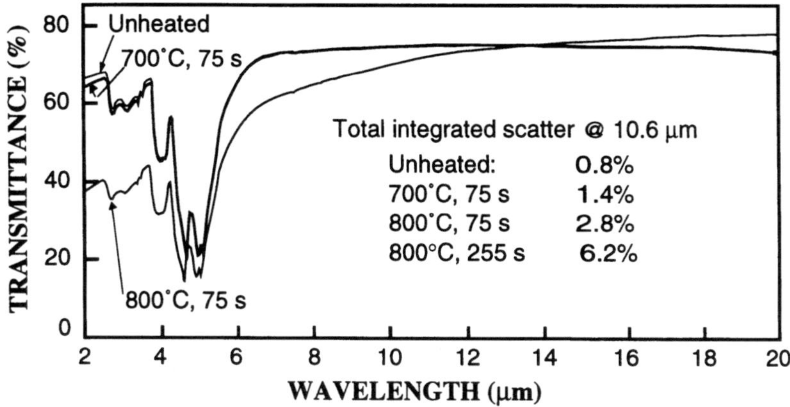

Fig. 9.38. Transmittance of polished CVD diamond disk after 75 s exposure to 700° or 800°C in air.[142]

coating on diamond is less erosion resistant than diamond. Attempts to reduce the susceptibility to oxidation by ion implantation[143] or incorporating fluorine into the diamond were unsuccessful.[144] Boron doping at a level of 10^{19} B/cm^3 decreased the rate of oxidation of CVD diamond by 50% at 700°C,[145] but boron may not be compatible with long wave infrared optical performance requirements.

9.4.4 Prospects

Diamond is likely to remain the most expensive infrared window material. However, it will perform jobs that no other material can do. It will withstand very rapid heating when all other materials would fail. It induces the least thermo-optic distortion. Unlike other thermal-shock-resistant materials, such as sapphire and gallium phosphide, diamond provides a long wave window, not a midwave window. For process control applications, diamond is the most chemically inert window. It might be the ultimate microwave or laser window for very-high-power devices. Optical applications of diamond are likely to be small compared to use for cutting tools and heat spreading in electronic circuits. Each application will help reduce the price for new applications. As with other new materials, several decades are likely to pass before a range of applications reaches maturity.

References

1. J. Wilks and E. Wilks, *Properties and Applications of Diamond*, Butterworth Heineman, Oxford (1991).
2. J. E. Field, ed., *The Properties of Natural and Synthetic Diamond*, Academic Press, London (1992).
3. G. Davies, ed., *Properties and Growth of Diamond*, INSPEC (Institute of Electrical Engineers), London (1994).
4. R. C. DeVries, A. Badzian and R. Roy, "Diamond Synthesis: The Russian Connection," *Mater. Res. Soc. Bull.*, 65-75, February 1996.
5. K. E. Spear, "Diamond -- Ceramic Coating of the Future," *J. Am. Ceram. Soc.*, **72**, 171-191 (1989).
6. J. C. Angus, "History and Current Status of Diamond Growth at Metastable Conditions," *Electrochem. Soc. Proc.*, **89-12**, 1-13 (1989).
7. M. N. R. Ashfold, P. W. May, C. A. Rego and N. M. Everitt, "Thin Film Diamond by Chemical Vapor Deposition Methods," *Chem. Soc. Rev.*, 21-30 (1994).
8. A. Feldman and S. Holly, *Diamond Optics, Proc. SPIE*, **969**, **1146**, **1325**, **1534**, **1759** (1988, 1989, 1990, 1991, 1992).
9. M. Seal and W. J. P. van Enckevort, "Applications of Diamond in Optics," *Proc. SPIE*, **969**, 144-152 (1988).
10. T. Evans and P. F. James, "A Study of the Transformation of Diamond to Graphite," *Proc. Roy. Soc. London*, **A277**, 260-269 (1964).
11. G. Davies and T. Evans, "Graphitization of Diamond at Zero Pressure and at a High Pressure," *Proc. Roy. Soc. London*, **A328**, 413-428 (1972).
12. K. E. Spear and M. Frenklach, "High Temperature Chemistry of CVD (Chemical Vapor Deposition) Diamond Growth," *Pure Appl. Chem.*, **66**, 1773-1782 (1994).
13. J. E. Butler and R. L. Woodin, " Thin Film Diamond Growth Mechanisms," *Phil. Trans. Roy. Soc. Lond.* **A342**, 209-224 (1993).
14. W. A. Yarbrough, "Vapor-Phase-Deposited Diamond — Problems and Potential," *J. Am. Ceram. Soc.*, **75**, 3179-3199 (1992).
15. P. K. Bachmann, D. Leers and H. Lydtin, "Towards a General Concept of Diamond Chemical Vapor Deposition," *Diamond and Related Mater.*, **1**, 1-12 (1991).

16. W. Müller-Sebert, E. Wörner, F. Fuchs, C. Wild and P. Koidl, "Nitrogen Induced Increase of Growth Rate in Chemical Vapor Deposition of Diamond," *Appl. Phys. Lett.*, **68**, 759-760 (1996).
17. F. M. Cerio and W. A. Weimer, "Electrostatic Probe Measurements for Microwave Plasma-Assisted Chemical Vapor Deposition of Diamond," *Appl. Phys. Lett.*, **59**, 3387-3389 (1991).
18. F. M. Cerio and W. A. Weimer, "Construction of an Inexpensive dc Plasma Jet for Diamond Deposition Using Commercially Available Components," *Rev. Sci. Instrum.*, **63**, 2065-2068 (1992).
19. K. Kurihara, K. Sasaki and M. Kawarada, "Diamond-Film Synthesis Using DC Plasma Jet CVD," *Fujitsu Sci. Tech. J.*, **25**, 44-51 (1989).
20. N. Ohtake and M. Yoshikawa, "Diamond Film Preparation by Arc Discharge Plasma Jet Chemical Vapor Deposition in the Methane Atmosphere," *J. Electrochem. Soc.*, **137**, 717-722 (1990).
21. K.-H. Chen, M.C.-Chuang, C. M. Penney and W. F. Banholzer, "Temperature and Concentration Distribution of H_2 and H Atoms in Hot-Filament Chemical-Vapor Deposition of Diamond," *J. Appl. Phys.*, **71**, 1485-1493 (1992).
22. Y. Hirose and Y. Terasawa, "Synthesis of Diamond Thin Films by Thermal CVD Using Organic Compounds," *Jap. J. Appl. Phys.*, **25**, L519-L521 (1986).
23. W. A. Weimer, F. M. Cerio and C. E. Johnson, "Examination of the Chemistry Involved in Microwave Plasma Assisted Chemical Vapor Deposition of Diamond," *J. Mater. Res.*, **6**, 2134-2144 (1991).
24. D. G. Goodwin and G. G. Gavillet, "Numerical Modeling of the Filament-Assisted Diamond Growth Environment," *J. Appl. Phys.*, **68**, 6393-6400 (1990).
25. F. G. Celii and J. E. Butler, "Direct Monitoring of CH_3 in a Filament-Assisted Diamond Chemical Vapor Deposition Reactor," *J. Appl. Phys.*, **71**, 2877-2883 (1992).
26. M. P. D'Evelyn, C. J. Chu, R. H. Hange and J. L. Margrave, "Mechanism of Diamond Growth by Chemical Vapor Deposition: Carbon-13 Studies," *J. Appl. Phys.*, **71**, 1528-1530 (1992).
27. C. E. Johnson, W. A. Weimer and F. M. Cerio, "Efficiency of Methane and Acetylene in Forming Diamond by Microwave Plasma Assisted Chemical Vapor Deposition," *J. Mater. Res.*, **7**, 1427-1431 (1992).
28. W. Zhu, P. C. Yang, J. T. Glass and F. Arezzo, "Diamond Nucleation and Growth on Reactive Transition-Metal Substrates," *J. Mater. Res.*, **10**, 1455-1460 (1995).
29. M. B. Moran, L. F. Johnson and K. A. Klemm, "Method for Growth of CVD Diamond on Thin Film Refractory Coatings and Glass Ceramic Materials,"*Proc. SPIE*, **2286**, 205-216 (1994).
30. K. E. Nariman, D. B. Chase and H. C. Foley, "Growth of Free-Standing Diamond Films on Glass," *Chem. Mater.*, **3**, 391-394 (1991).
31. C. J. Brierley, M. C. Costello, M. D. Hudson and T. J. Bettles, "Diamond Coatings for Large Area IR Windows," *Proc. SPIE*, **2286**, 307-315 (1994).
32. P. G. Kosky and T. R. Anthony, "Fracture-Free Release of CVD Diamond," *Diamond and Related Mater.*, **5**, 1313-1317 (1996); P. G. Kosky and T. R. Anthony, U.S. Patent 5,474,306, 23 June 1995.
33. T. R. Anthony and J. F. Fleischer, "Substantially Transparent Free Standing Diamond Films," U.S. Patent 5,273,731, 28 Dec 1993; Transparent Diamond Films and Method for Making," U.S. Patent 5,110,579, 5 May 1992.
34. G. F. Cardinale and C. J. Robinson, "Fracture Strength Measurements of Filament Assisted CVD Polycrystalline Diamond Films," *J. Mater. Res.*, **7**, 1432-1437 (1992).
35. I. P. Hayward and J. E. Field, "The Solid Particle Erosion of Diamond," *J. Hard Mater.*, **1**, 53-64 (1990).

36. C. J. McHargue, "Mechanical Properties of Diamond and Diamond-Like Films," in *Applications of Diamond Films and Related Materials* (Y. Tzeng, M. Yoshikawa, M. Murakawa and A. Feldman, eds.), Elsevier, Amsterdam (1991).
37. C. A. Brookes and E. J. Brookes, "Diamond in Perspective: A Review of Mechanical Properties of Natural Diamond," *Diamond and Related Materials*, **1**, 13-17 (1991).
38. D. R. Mumm, K. T. Faber, M. D. Drory and C. F. Gardinier, "High-Temperature Hardness of Chemically Vapor-Deposited Diamond," *J. Am. Ceram. Soc.*, **76**, 238-240 (1993).
39. H. Itoh, S. Shimura, K. Sugiyama, H. Iwahara and H. Sakamoto, "Improvement of Cutting Performance of Silicon Nitride Tool by Adherent Coating of Thick Diamond Film," *J. Am. Ceram. Soc.*, **80**, 189-196 (1997); P. M. Stephan, "Diamond Films Enhance Machining with Ceramics," *Am. Ceram. Soc. Bull.*, **71**, 1623-1629 (1992).
40. M. D. Drory, C. F. Gardiner and J. S. Speck, "Fracture Toughness of Chemically Vapor-Deposited Diamond," *J. Am. Ceram. Soc.*, **74**, 3148-3150 (1991).
41. Ref. 2, p. 486.
42. R. H. Telling and J. E. Field, "Fracture in CVD Diamond," *Int. J. Refractory Metals and Hard Materials*, **16**, 269-276 (1998).
43. C. F. Kennedy, R. H. Telling and J. E. Field, "Liquid Impact and Fracture of Free-Standing CVD Diamond," *Proc. SPIE*, **3705** (1999).
44. L. P. Hehn, Z. Chen, J. J. Mecholsky, Jr., P. Klocek, J. T. Hoggins and J. M. Trombetta, "Fracture Surface Analysis of Free-Standing Diamond Films," *J. Mater. Res.*, **9**, 1540-1545 (1994).
45. C. A. Klein and G. F. Cardinale, "Young's Modulus and Poisson's Ratio of CVD Diamond, " *Diamond and Related Materials*, **2**, 918-923 (1993).
46. Ref. 2, p. 677.
47. H. J. McSkimin and P. Andreatch, Jr., "Elastic Moduli of Diamond as a Function of Pressure and Temperature," *J. Appl. Phys.*, **43**, 2944-2948 (1972).
48. Ref. 2, pp. 678-679.
49. De Beers data reported at 3rd International Applied Diamond Conference, Gaithersburg, Maryland (1995).
50. D. S. Olson, G. J. Reynolds, G. F. Virshup, F. I. Friedlander, B. G. James and L. D. Partain, "Tensile Strength of Synthetic Chemical-Vapor-Deposited Diamond," *J. Appl. Phys.*, **78**, 5177-5179 (1995).
51. T. J. Valentine, A. J. Whitehead, R. S. Sussmann, C. J. H. Wort and G. A. Scarsbrook, "Mechanical Property Measurements of Bulk Polycrystalline CVD Diamond," *Diamond and Related Materials*, **3**, 1168-1172 (1994).
52. J. E. Field and C. S. J. Pickles, "Strength, Fracture and Friction Properties of Diamond," *Diamond and Related Materials*, **5**, 625-634 (1996).
53. D. C. Harris, "Properties of Diamond for Window and Dome Applications," *Proc. SPIE*, **2286**, 218-228 (1994).
54. D. C. Harris, *Development of Chemical-Vapor-Deposited Diamond for Infrared Optical Applications. Status Report and Summary of Properties*, Naval Air Warfare Center Weapons Division Report TP 8210, China Lake, California, July 1994.
55. T. E. Steyer, K. T. Faber and M. D. Drory, "Fracture Strength of Free-Standing Chemically Vapor-Deposited Diamond Films," *Appl. Phys. Lett.*, **66**, 3105-3107 (1995).
56. K. L. Jackson, D. L. Thurston, P. J. Boudreaux, R. W. Armstrong and C. CM. Wu, "Fracturing of Industrial Diamond Plates," *J. Mater. Sci.* **32**, 5035-5045 (1997).
57. J. A. Savage, C. J. H. Wort, C. S. J. Pickles, R. S. Sussmann, C. G. Sweeney, M. R. McClymont, J. R. Brandon, C. N. Dodge and A. C. Beale, " Properties of Free-Standing CVD Diamond Optical Components," *Proc. SPIE*, **3060**, 144-159 (1997).

58. C. B. Willingham, T. M. Hartnett, R. P. Miller and R. B. Hallock, "Bulk Diamond for IR/RF Windows and Domes," *Proc. SPIE*, **3060**, 160-168 (1997).
59. H. Windischmann, G. F. Epps, Y. Cong and R. W. Collins "Intrinsic Stress in Diamond Films Prepared by Microwave Plasma CVD," *J. Appl. Phys.*, **69**, 2231-2237 (1991).
60. A. B. Harker, D. G. Howitt, S. J. Chen, J. F. Flintoff and M. R. James, "Residual Stress Measurements on Polycrystalline Diamond," *Proc. SPIE*, **2286**, 254-261 (1994).
61. M. S. Haque, H. A. Naseem, A. P. Malshe and W. D. Brown, "A Study of Stress in Microwave Plasma Chemical Vapor Deposited Diamond Films Using X-Ray Diffraction," *Chemical Vapor Deposition*, **3**, 129-135 (1997).
62. I. I. Vlasov, V. G. Ralchenko, E. D. Obraztsova, A. A. Smolin and V. I. Konov, "Stress Mapping of Chemical-Vapor-Deposited Diamond Film Surface by Micro-Raman Spectroscopy," *Appl. Phys. Lett.*, **71**, 1789-1791 (1997).
63. G. A. Slack and S. F. Bartram, "Thermal Expansion of Some Diamondlike Crystals," *J. Appl. Phys.*, **46**, 189-198 (1975).
64. D. J. Pickrell, K. A. Kline and R. E. Taylor, "Thermal Expansion of Polycrystalline Diamond Produced by Chemical Vapor Deposition," *Appl. Phys. Lett.*, **64**, 2353-2355 (1994).
65. J. W. Vandersande, C. B. Vining and A. Zoltan, "Thermal Conductivity of Natural Type IIa Diamond," *NASA Technical Brief*, Vol. 16, No. 12, Item #5, December 1992.
66. R. Berman, F. E. Simon and J. S. Ziman, "The Thermal Conductivity of Diamond at Low Temperatures," *Proc. Roy. Soc. A*, **220**, 171-183 (1953).
67. R. Berman, P. R. W. Hudson, M. Martinez, "Nitrogen in Diamond: Evidence from Thermal Conductivity," *J. Phys. C, Solid State Phys.*, **8**, L430-L434 (1975).
68. E. A. Burgemeister, "Thermal Conductivity of Natural Diamond Between 320 and 450 K," *Physica B&C*, **93**, 165-179 (1978).
69. C. J. H. Wort, C. G. Sweeney, M. A. Cooper, G. A. Scarsbrook and R. S. Sussman "Thermal Properties of Bulk Polycrystalline CVD Diamond," *Diamond '93 Conference*, Portugal, September 1993.
70. J. W. Vandersande, "Thermal Conductivity of Natural-Isotope Diamond," in Ref. 3, pp. 33-35.
71. Ref. 2, p. 681.
72. J. E. Graebner, J. A. Mucha and F. A. Baiocchi, "Sources of Thermal Resistance in Chemically Vapor Deposited Diamond," *Diamond and Related Materials*, **5**, 682-687 (1996).
73. J. E. Graebner, S. Jin, G. W. Kammlott, B. Bacon, L. Seibles and W. Banholzer, "Anisotropic Thermal Conductivity in Chemical Vapor Deposition Diamond," *J. Appl. Phys.*, **71**, 5353-5356 (1992).
74. J. E. Graebner, S. Jin, G. W. Kammlott, J. A. Herb and C. F. Gardinier, "Unusually High Thermal Conductivity in Diamond Films," *Appl. Phys. Lett.*, **60**, 1576-1578 (1992).
75. O. W. Käding, M. Rösler, R. Zachai, H.-J. Fußer and E. Matthias, "Lateral Thermal Diffusivity of Epitaxial Diamond Films," *Diamond and Related Materials*, **3**, 1178-1182 (1994).
76. D. T. Morelli, C. P. Beetz and T. A. Perry, "Thermal Conductivity of Synthetic Diamond Films," *J. Appl. Phys.*, **64**, 3063-3066 (1988).
77. D. T. Morelli, T. M. Hartnett and C. J. Robinson, "Phonon-Defect Scattering in High Thermal Conductivity Diamond Films," *Appl. Phys. Lett.*, **59**, 2112-2114 (1991).
78. D. T. Morelli, C. Uher and C. J. Robinson, "Transmission of Phonons Through Grain Boundaries in Diamond Films," *Appl. Phys. Lett.*, **62**, 1085-1087 (1993).

79. R. Berman, *Thermal Conduction in Solids*, Clarendon Press, Oxford (1976).
80. T. R. Anthony, W. F. Banholzer, J. F. Fleischer, L. Wei, P. K. Kuo, R. L. Thomas and R. W. Pryor, "Thermal Diffusivity of Isotopically Enriched ^{12}C Diamond," *Phys. Rev. B*, **42**, 1104-1111 (1990).
81. J. R. Olson, R. O. Pohl, J. W. Vandersande, A. Zoltan, T. R. Anthony and W. F. Banholzer, "Thermal Conductivity of Diamond Between 170 and 1200 K and the Isotope Effect," *Phys. Rev. B*, **47**, 14850-14856 (1993).
82. D. G. Onn, A. Witek, Y. Z. Qui, T. R. Anthony and W. F. Banholzer, "Some Aspects of the Thermal Conductivity of Isotopically Enriched Diamond Single Crystals," *Phys. Rev. Lett.*, **68**, 2806-2809 (1992).
83. K. Belay, Z. Etzel, D. G. Onn and T. R. Anthony, "The Thermal Conductivity of Polycrystalline Diamond Films: Effects of Isotope Content," *J. Appl. Phys.*, **79**, 8336-8340 (1996).
84. W. F. Banholzer and T. R. Anthony, "Diamond Properties as a Function of Isotopic Composition," *Thin Solid Films*, **212**, 1-10 (1992).
85. E. Wörner, C. Wild, W. Müller-Sebert, R. Locher and P. Koidl, "Thermal Conductivity of CVD Diamond Films: High-Precision, Temperature-Resolved Measurements," *Diamond and Related Materials*, **5**, 688-692 (1996).
86. E. Wörner, J. Wagner, W. Müller-Sebert, C. Wild and P. Koidl, "Infrared Raman Scattering as a Sensitive Probe for the Thermal Conductivity of Chemical Vapor Deposited Diamond Films," *Appl. Phys. Lett.*, **68**, 1482-1484 (1996).
87. H. Verhoeven, A. Flöter, H. Reiß, R. Zachai, D. Wittorf and W. Jäger, "Influence of the Microstructure on the Thermal Properties of Thin Polycrystalline Diamond Films," *Appl. Phys. Lett.*, **71**, 1329-1331 (1997).
88. M. Reichling, T. Klotzbücher and J. Hartmann, "Local Variation of Room-Temperature Thermal Conductivity in High-Quality Polycrystalline Diamond," *Appl. Phys. Lett.*, **73**, 765-758 (1998).
89. A. C. Victor, "Heat Capacity of Diamond at High Temperature," *J. Chem. Phys.*, **36**, 1903-1911 (1962).
90. D. L. Burk and S. A. Frieberg, "Atomic Heat of Diamond from 11° to 200°K," *Phys. Rev.*, **111**, 1275-1282 (1958).
91. J. E. Graebner, "Measurements of Specific Heat and Mass Density in CVD Diamond," *Diamond and Related Mater.*, **5**, 1366-1370 (1996).
92. R. S. Sussann, J. R. Brandon, S. E. Coe, C. S. J. Pickles, C. G. Sweeney, A. Wasenczuk, C. J. H. Wort and C. N. Dodge, "CVD Diamond: A New Engineering Material for Thermal, Dielectric and Optical Applications," *Industrial Diamond Review*, March 1998, pp. 69-77.
93. M. E. Thomas, W. J. Tropf and A. Szpak, "Optical Properties of Diamond," *Diamond Films & Technol.*, **5**, 159-180 (1995).
94. M. E. Thomas and W. J. Tropf, "Optical Properties of Diamond," *Proc SPIE*, **2286**, 144-151 (1994).
95. K. Harris, G. L. Herrit, C. J. Johnson, S. P. Rummel and D. J. Scatena, "Infrared Optical Characteristics of Type 2A Diamonds," *Appl. Opt.*, **30**, 5015-5017 (1991). See correction in *Appl. Opt.*, **31**, 4342 (1992).
96. S. P. McGeoch, D. R. Gibson and J. A. Savage, "Assessment of Type IIa Diamond as an Optical Material for Use in Severe Environments," *Proc. SPIE*, **1760**, 122-142 (1992).
97. C. J. H. Wort, C. S. J. Pickles, A. C. Beale, C. G. Sweeney, M. R. McClymont, R. J. Saunders, R. S. Sussmann and K. L. Lewis, "Recent Advances in the Quality of CVD Diamond Optical Components," *Proc. SPIE*, **3705** (1999).
98. M. E. Thomas and W. J. Tropf, "Optical Properties of Diamond," *Johns Hopkins APL Technical Digest*, **14** [1], 16-23 (1993).

99. C. A. Klein, T. M. Hartnett and C. J. Robinson, "Critical-Point Phonon Frequencies of Diamond," *Phys. Rev. B*, **45**, 12854-12863 (1992).
100. J. M. Trombetta, J. T. Hoggins, W. F. Banholzer and T. R. Anthony, "^{13}C Isotope Identification of Defect-Induced One-Phonon Absorption in CVD Diamond," *Diamond Films Technol.*, **5**, 129-139 (1995).
101. J. R. Hardy and S. D. Smith, "Two-Phonon Infra-Red Lattice Absorption in Diamond," *Phil. Mag.*, **6**, 1163-1172 (1961).
102. R. E. Clausing, J. R. McNeely, M. B. McIntosh, W. B. Snyder, Jr., M. E. Thomas, M. J. Linevsky and W. J. Tropf, *High Temperature Optical Characterization of CVD Diamond and Natural Type IIa Diamond*, Oak Ridge National Laboratory/U. S. ArmySpace and Strategic Defense Command Report MMES/USASDC/SDIO/TP-92-12, 31 October 1992.
103. I. Kiflawi, C. M. Welbourn and G. S. Woods, "One-Phonon Absorption by Type IIa Diamonds," *Solid State Commun.*, **85**, 551-552 (1993).
104. R. S. Sussmann, C. J. H. Wort, C. G. Sweeney, J. L. Collins, C. N. Dodge, and J. Savage, "Optical and Dielectric Properties of CVD Polycrystalline Diamond Plates," *Proc. SPIE*, **2286**, 229-238 (1997).
105. J. M. Trombetta, J. T. Hoggins, P. Klocek and T. M. McKenna, "Optical Properties of DC Arc-Discharge Plasma CVD Diamond," *Proc. SPIE*, **1534**, 77-88 (1991).
106. K. M. McNamara, D. H. Levy, K. K. Gleason and C. J. Robinson, "Nuclear Magnetic Resonance and Infrared Absorption Studies of Hydrogen Incorporation in Polycrystalline Diamond," *Appl. Phys. Lett.*, **60**, 580-582 (1992).
107. K. M. McNamara, B. E. Scruggs and K. K. Gleason, "The Effect of Impurities on the IR Absorption of Chemically Vapor Deposited Diamond," *Thin Solid Films*, **253**, 157-161 (1994).
108. K. M. McNamara Rutledge, B. E. Skruggs and K. K. Gleason, "Influence of Hydrogenated Defects and Voids on Thermal Conductivity of Polycrystalline Diamond," *J. Appl. Phys.*, **77**, 1459-1462 (1995).
109. K. McNamara Rutledge and K. K. Gleason, "Hydrogen in CVD Diamond Films," *Chemical Vapor Deposition*, **2**, 37-43 (1996).
110. L. M. Brown and P. Fallon, "Comment on Thin Film Diamond Growth Mechanisms," *Phil. Trans. Roy. Soc. Lond.* **A342**, 224 (1993).
111. C. J. H. Wort, C. G. Sweeney, M. A Cooper, G. A. Scarsbrook and R. S. Sussman, "Thermal Properties of Bulk Polycrystalline CVD Diamond," *Diamond and Related Mater.*, **3**, 1158-1167 (1994).
112. A. Feldman, E. N. Farabaugh and L. H. Robins, *Chemical Vapor Deposited Diamond*, National Institute of Standards and Technology Report No. 11 (1991).
113. D. F. Edwards and H. R. Philipp, "Cubic Carbon (Diamond)," in *Handbook of Optical Constants of Solids* (E. D. Palik, ed.), p. 673, Academic Press, Orlando, Florida (1985).
114. D. F. Edwards and E. Ochoa, "Infrared Refractive Index of Diamond," *J. Opt. Soc. Am.*, **71**, 607-608 (1981).
115. A. B. Djurišić and E. H. Li, "Modeling the Optical Constants of Diamonds from 0.06 to 30 eV," *Appl. Opt.*, **37**, 7273-7275 (1998).
116. F. Peter, "Uber Brechungsindizes und Absorptionskonstanten des Diamanten zwischen 644 und 226 mμ," *Z. Phys.*, **15**, 358-368 (1923).
117. J. Fontanella, R. L. Johnston, J. H. Colwell and C. Andeen, "Temperature and Pressure Variation of the Refractive Index of Diamond," *Appl. Opt.*, **16**, 2949-2951 (1977).
118. M. J. Dodge, "Refractive Index," in *CRC Handbook of Laser Science and Technology* (M. J. Weber, ed.), Vol. IV, CRC Press, Boca Raton, Florida (1986).

119. Z. Yin, H. S. Tan and F. W. Smith, "Determination of the Optical Constants of Diamond Films with a Rough Growth Surface," *Diamond and Related Mater.*, **5**, 1490-1496 (1996).
120. C. A. Klein, "Pulsed Laser-Induced Damage to Diamond," *Diamond Films & Technol.*, **5**, 141-158 (1995).
121. C. A. Klein and R. DeSalvo, "Thresholds for Dielectric Breakdown in Laser-Irradiated Diamond," *Appl. Phys. Lett.*, **63**, 1895-1898 (1993).
122. T. M. Hartnett, R. B. Hallock, R. P. Miller and C. B. Willingham, "CVD Diamond for IR and RF Dome and Window Applications: Status Update," *Proc. 7th DoD Electromagnetic Windows Symp.*, pp. 89-94, Laurel, MD (1998).
123. M. Yoshikawa, "Development and Performance of a Diamond Film Polishing Apparatus with Hot Metals." *Proc. SPIE*, **1325**, 210-221 (1990).
124. A. B. Harker, J. Flinthoff and J. F. DeNatale, "The Polishing of Polycrystalline Diamond Films," *Proc. SPIE*, **1325**, 222-229 (1990).
125. T. P. Thorpe, A. A. Morrish, L. M. Hanssen, J. E. Butler and K. A. Snail, "Growth, Polishing, and Optical Scatter of Diamond Thin Films," *Proc. SPIE*, **1325**, 230-237 (1990).
126. X. H. Wang, L. Pilione, W. Zhu, W. Yarbrough, W. Drawl and R. Messier, "Infrared Measurements of CVD Diamond Films," *Proc. SPIE*, **1325**, 160-163 (1990).
127. H. Tokura, C.-F. Yang and M. Yoshikawa, "Study on the Polishing of Chemically Vapor Deposited Diamond Film," *Thin Solid Films*, **212**, 49-55 (1992).
128. S. Jin, J. E. Graebner, G. W. Kammlott, T. H. Tiefel, S. G. Kosinski, L. H. Chen and R. A. Fastnacht, "Massive Thinning of Diamond Films by a Diffusion Process," *Appl. Phys. Lett.*, **60**, 1948-1950 (1992).
129. S. Jin, J. E. Graebner, M. McCormack, T. H. Tiefel, A. Katz and W. C. Dautremont-Smith, "Shaping of Diamond Films by Etching with Molten Rare-Earth Metals," *Nature*, **362**, 822-824 (1993).
130. C. E. Johnson, "Chemical Polishing of Diamond," *Surface and Coatings Technol.*, **68/69**, 374-377 (1994).
131. B. G. Bovard, T. Zhao and H. A. Macleod, "Oxygen-Ion Beam Polishing of a 5-cm-Diameter Diamond Film, " *Appl. Opt.*, **31**, 2366-2369 (1992).
132. A. Hirata, H. Tokura and M. Yoshikawa, "Smoothing of Chemically Vapor Deposited Diamond Film by Ion Beam Irradiation," *Thin Solid Films*, **212**, 43-48 (1992).
133. J. E. Yehoda and J. J. Cuomo, "Solid State Ionic Polishing of Diamond," *Appl. Phys. Lett.*, **66**, 1750-1752 (1995).
134. M. Hashish and D. H. Bothell, "Polishing of CVD Diamond Films with Abrasive-Liquidjets — An Exploratory Investigation," *Proc. SPIE*, **1759**, 97-105 (1992).
135. C. R. Seward, C. S. J. Pickles and J. E. Field, "Liquid and Solid Erosion Properties of Diamond," *Diamond and Related Mater.*, **2**, 606-611 (1993).
136. E. J. Coad and J. E. Field, "The Liquid Impact Resistance of CVD Diamond and Other Infrared Materials," *Proc. SPIE*, **3060**, 169-180 (1997).
137. C. R. Seward, E. J. Coad, C. S. J. Pickles and J. E. Field, "The Rain Erosion Resistance of Diamond and Other Window Materials," *Proc. SPIE*, **2286**, 285-300 (1994).
138. E. J. Coad, C. S. J. Pickles, G. H. Jilbert and J. E. Field, "Aerospace Erosion of Diamond and Diamond Coatings," *Diamond and Related Mater.*, **5**, 640-643 (1996).
139. R. H. Telling, G. H. Jilbert and J. E. Field, "The Erosion of Aerospace Materials by Solid Particle Impact," *Proc. SPIE*, **3060**, 56-67 (1997).

140. C. E. Johnson, W. A. Weimer and D. C. Harris, "Characterization of Diamond Films by Thermogravimetric Analysis and Infrared Spectroscopy," *Mater. Res. Bull.*, **24**, 1127-1134 (1989).
141. C. E. Johnson, M. A. S. Hasting and W. A. Weimer, "Thermogravimetric Analysis of the Oxidation of CVD Diamond Films," *J. Mater. Res.*, **5**, 2320-2325 (1990).
142. C. E. Johnson, J. M. Bennett and M. P. Nadler, "Oxidative Degradation of CVD Diamond," *Proc. SPIE*, **2286**, 247-253 (1994).
143. K. A. Klemm, H. S. Patterson, L. F. Johnson and M. B. Moran, "Protective Optical Coatings for Diamond Infrared Windows," *Proc. SPIE*, **2286**, 347-353 (1994).
144. K. J. Grannen, D. V. Tsu, R. J. Meilunas and R. P. H. Chang, "Oxidation Studies of Fluorine Containing Diamond Films," *Appl. Phys. Lett.*, **59**, 745-747 (1991).
145. K. Miyata and K. Kobashi, "Air Oxidation of Undoped and B-Doped Polycrystalline Diamond Films at High Temperature," *J. Mater. Res.* **11**, 296-304 (1996).

Appendix A

PHYSICAL CONSTANTS AND CONVERSION FACTORS

Table A.1. Physical constants*

Quantity	Symbol	Value
Elementary charge	e	$1.602\ 177\ 33\ (49) \times 10^{-19}$ C
Speed of light in vacuum	c	$2.997\ 924\ 58 \times 10^{8}$ m/s
Planck's constant	h	$6.626\ 075\ 5\ (40) \times 10^{-34}$ J·s
Boltzmann's constant (= R/N)	k	$1.380\ 658\ (12) \times 10^{-23}$ J/K
Electron rest mass	m_e	$9.109\ 389\ 7\ (54) \times 10^{-31}$ kg
Proton rest mass	m_p	$1.672\ 623\ 1\ (10) \times 10^{-27}$ kg
Permittivity of free space	ε_o	$8.854\ 187\ 817 \times 10^{-12}$ $C^2/(N{\cdot}m^2)$
Permeability of free space	μ_o	$4\pi \times 10^{-7}$ H/m
Avogadro's number	N	$6.022\ 136\ 7\ (36) \times 10^{23}$ mol^{-1}
Gas constant	R	8.314 510 (70) J/(mol·K)
Faraday constant (= Ne)	F	$9.648\ 530\ 9\ (29) \times 10^{4}$ C/mol
Gravitational constant	G	$6.672\ 59\ (85) \times 10^{-11}$ $m^3/(s^2{\cdot}kg)$
Gravitational acceleration (standard)	g	9.806 65 m/s^2
Bohr magneton (= $eh/2m_e$)	μ_B	$9.274\ 015\ 4\ (31) \times 10^{-24}$ J/T
Electron magnetic moment	μ_e	$9.284\ 770\ 1\ (31) \times 10^{-24}$ J/T
Pi	π	3.141 592 653 589 793...
Basis of natural logarithm	e	2.718 281 828 459 045...

*Numbers in parentheses are one-standard-deviation uncertainties in last digits. From E. R. Cohen and B. N. Taylor, *J. Res. Nat. Bur. Stand.*, **92**, 85 (1987).

Table A.2. Conversion factors*

To convert from	Into	Multiply by
Pounds (mass)	Kilograms	0.453 592 37
Pounds (force)	Newtons	4.448 223
Inches	Meters	2.54×10^{-2}
Ångströms	Meters	10^{-10}
Pounds per square inch (psi)	Pascals (N/m^2)	$6.894\ 76 \times 10^3$
Ksi (1000 psi)	MPa (10^6 Pa)	6.894 76
Foot-pounds	Joules	1.355 82
Atmospheres	Pascals	$1.013\ 25 \times 10^5$
Torr (mm Hg)	Pascals	$1.333\ 22 \times 10^2$
Bars	Pascals	10^5
Dynes	Newtons	10^{-5}
Liters	Cubic meters	10^{-3}
Ergs	Joules	10^{-7}
Electron volts	Joules	$1.602\ 177\ 33 \times 10^{-19}$
Calories (thermochemical)	Joules	4.184
British thermal units (BTU)	Joules	$1.055\ 06 \times 10^3$
BTU / ($ft^2 \cdot s$)	W/cm^2	1.135 66
Horsepower	Watts	$7.457\ 00 \times 10^2$

*Example: To convert 3.00 pounds into kilograms, multiply pounds by 0.453 592 37: 3.00 lb × 0.453 592 37 kg/lb = 1.36 kg.

Appendix B

SUPPLIERS OF INFRARED MATERIALS AND SOURCES OF INFORMATION

This list is by no means exhaustive. We apologize in advance to suppliers who are not represented and welcome submissions for inclusion on future lists. More listings of manufacturers and suppliers and optical coating laboratories can be found at the home page of *Laser Focus World* at http://www.lfw.com/ on the World Wide Web and in *The Photonics Buyers' Guide* published by Laurin Publishing Co., P.O. Box 4949, Pittsfield, MA 01202-4949, Phone: 413-499-0514. See also the *SPIE Buyers Guide* at http://optics.org/search/advsearch.html. Another source is the *American Precision Optics Manufacturers Association Buyers Guide* available from the Center for Optics Manufacturing, 240 East River Road, Rochester NY 14623-1212 (Phone: 716-275-1093; http://www.opticam.rochester.edu)

Amorphous Materials
3130 Benton Street
Garland TX 75042-7410
972-494-5624
GaAs, AMTIR-1, AMTIR-3, As_2S_3

Battelle Pacific Northwest Laboratory
Richland WA 99352
509-375-2076
coatings, electromagnetic shielding

Bicron (formerly Harshaw)
6801 Cochran Road
Solon OH 44139
440-248-7400
BaF_2, CaF_2, CsBr, CsI, PbF_2, LiF, MgF_2, KBr, KCl, KF, KI, AgBr, AgI, NaCl, NaI, SrF_2, TlBr, KRS-5, TlCl, KRS-6

Bicron (formerly Union Carbide)
750 S. 32nd Street
Washougal WA 98671
206-835-8566
Sapphire

Ceramiques Techniques Desmarquest
48 rue des Vignerons
94685 Vincennes Cedex France
FAX: 33 1 48 18 51 52
Hot pressed MgF_2

Coherent Auburn Group
2303 Lindbergh Street
Auburn CA 95602-9595
530-8988-5107
infrared and diffractive optics

Corning Glass Works
35 W. Market Street
Corning NY 14830
800-525-2524
glasses and optical fibers

Crystal Systems
27 Congress Street
Salem MA 01970-5597
978-745-0088
sapphire, Si

Crystar Research
721 Vanalman Avenue
Victoria, British Columbia
Canada V8Z 3B6
604-479-9922
sapphire, GaAs

De Beers Industrial Diamond Division
Charters, Sunninghill, Ascot
Berkshire SL5 9PX
United Kingdom
+44-1344-623456
CVD diamond

Deposition Sciences, Inc.
386 Tesconi Court
Santa Rosa CA 95401
707-579-2008
coatings

Diamonex
7150 Windsor Drive
Allentown PA 18106-9328
610-366-7100
CVD diamond, diamondlike carbon coatings

Dynasil Corp.
385 Copper Road
W. Berlin NJ 08091-9145
609-767-4600
fused silica

Eagle-Picher Technologies
P.O. Box 737
737 Highway 69A
Quapaw OK 74363
918-673-1650
Ge, Si

ELCAN Optical Technologies
450 Leitz Road
Midland, Ontario
Canada L4R5B8
705-526-5401
optical finishing and assemblies

Evaporated Metal Films Corp.
701 Spencer Road
Ithaca NY 14850
800-456-7070
coatings

Ferson Optics
2006 Government Street
Ocean Springs MS 39564
601-875-8146
dome and optics fabrication

Exotic Materials
36570 Briggs Road
Murietta CA 92563-2347
909-926-2994
coatings, electromagnetic shielding

GFI Advanced Technologies
379 Winthrop Road
Teaneck NJ 07666
201-833-8530
GaP, Ge, CaF_2, MgF_2, SrF_2, SiC

Harris Diamond Corp.
(no relation to the author!)
100 Stierli Court, Suite 106
Mount Arlington NJ 07856-1312
201-770-1420
diamond importer

Infrared Optical Products
P.O. Box 292
Farmingdale NY 11735
516-694-6035
Ge, Si, ZnSe, CaF_2, BaF_2, KRS-5, fused silica, MgF_2, LiF, GaAs, ZnS

Insaco
P.O. Box 9006
Quakertown PA 18951-9006
215-536-3500
sapphire, optical fabrication

Laser Power Optics
12777 High Bluff Drive
San Diego CA 92130
619-755-0700
ZnSe, GaAs, Ge, Si

Lattice Materials Corp.
516 E. Tamarack Street
Bozeman MT 59715
406-586-2122
Si, Ge

Meller Optics
120 Corliss Street
Providence RI 02904
401-331-3717
sapphire, optical fabrication

Mho/Acec Union Minière
A. Greinerstraat 14
B-2660 Hoboken, Belgium
32-3-8297929
Ge

Morton Advanced Materials
185 New Boston Street
Woburn MA 01950
781-933-9243
ZnS, ZnSe, Tuftran, SiC

OFC Corp.
2 Mercer Road
Natick MA 01760
508-655-1650
coatings

Ohara Corp.
23141 Arroyo Vista
Rancho Santa Margarita CA 92688
949-858-5700
optical glasses and glass-ceramics

Optical and Conductive Coatings
428 N. Buchanan Circle, No. 8
Pacheco CA 94553
415-798-6066
coatings

Optical Coating Laboratory
2789 Northpoint Parkway
Santa Rosa CA 95407-7397
707-525-7011
coatings

Optical Coating Technologies
193 Northampton Street
Easthampton MA 01027-1019
413-527-6307
coatings

Optical Corporation of America
170 Locke Drive
Marlboro MA 01752
508-481-9860
Si

Opticoat Associates
10 Kidder Road, Suite 6
Chelmsford MA 01824
508-250-8115
coatings

Optovac
24 East Brookfield Road
P.O. Box 248
North Brookfield MA 01535-0248
508-867-6444
CaF_2, SrF_2, BaF_2, MgF_2, LiF,
CeF_3, PbF_2, NaCl, KCl, KBr

Pacific Optical
2660 Columbia Street
Torrance CA 90503
213-328-5840
windows and coatings

Pennsylvania Optical
234 S. Eighth Street
P.O. Box 1217
Reading PA 19603-1217
215-376-5701
optical fabrication

Phase4 Infrared
3 Foundry Street
Lowell MA 01852
978-458-8328
ZnS, ZnSe, Ge

Pilkington Optronics
Barr & Stroud Ltd.
1 Linthouse Road
Glasgow, Scotland G51 4BZ
FAX: 0141-440-4001
ZnS

Rafael Optical Coatings
Dept. 27
P.O. Box 2250
Haifa Israel 31021
FAX: 972-4-8792890
ZnS, coatings, conductive coatings

Raytheon Advanced Materials Group
131 Spring Street
Lexington MA 02421
781-860-3061
ALON, ZnS, CVD diamond,
polycrystalline Si, yttria?, spinel?

Raytheon Optical Systems
100 Wooster Heights Road
Danbury CT 06810-7589
203-797-6141
coatings, electromagnetic shielding

Raytheon Systems Co.
Advanced Optical Materials, MS 55
P.O. Box 655012
Dallas TX 75265
972-344-8001
GaP, GaAs, chalcogenide glass,
polyethylene optics

Raytheon Systems Co.
Custom Optics Services, MS 8731
P.O. Box 660199
Dallas TX 75266
972-480-1806
infrared components, coatings, diamond
turning

Rotem Industries
Kerem Division
P.O. Box 9046
Beer-Sheva 84190
Israel
FAX: 972 7 656 8770
sapphire

Saphikon
33 Powers Street
Milford NH 03055
603-673-5831
sapphire

Schott Glass Technologies
400 York Avenue
Duryea PA 18642
717-457-7485
Schott Glaswerke
Postfach 2480
Mainz 55014, Germany
FAX: 49 61 31 66 20 03
calcium aluminate, germanate glass,
Zerodur

Silicon Casting
1741 Saratoga Avenue
San Jose CA 95129
408-252-0733
Si

Sumitomo Electric Industries
Itami Research Laboratories
1-1-1, Koya-kita, Itami
Hyogo, 664 Japan
FAX: 0727-70-6727
spinel, ZnS, MgF_2

Thermal American Fused Quartz Co.
Route 202
Montville NJ 07045
201-334-7770
fused silica

II-VI, Inc.
375 Saxonburg Boulevard
Saxonburg PA 16056
412-352-5207
ZnS, ZnSe

Zygo Corp.
Laurel Brook Road
Middlefield CT 06455
860-347-8506
optical fabrication

Sources of information on infrared materials

Books devoted to materials:

SPIE Symposium Proceedings, especially Volumes 50, 297, 505, 618, 683, 929, 968, 1112, 1326, 1535, 1760, 2286, 3060, 3705 and CR39.

P. Klocek, ed., *Handbook of Infrared Optical Materials*, Marcel Dekker, New York (1991).

J. A. Savage, *Infrared Optical Materials and Their Antireflection Coatings*, Adam Hilger, Bristol, England (1985).

S. Musikant, *Optical Materials: An Introduction to Selection and Application*, Marcel Dekker, New York (1985).

E. D. Palik, ed., *Handbook of Optical Constants of Solids*, Academic Press, Orlando, Florida, Volumes I, II and III (1985, 1991, 1998).

Books with some chapters on materials:

J. S. Accetta and D. L. Shumaker, eds, *The Infrared and Electro-Optical Systems Handbook* (8 volumes), Environmental Research Institute of Michigan, Ann Arbor, Michigan, and SPIE Press, Bellingham, Washington (1993).

M. E. Bass, E. W. van Stryland, D. R. Williams, and W. L. Wolfe, eds. *Handbook of Optics* (2nd edition, 2 volumes), McGraw-Hill, New York (1995).

W. L. Wolfe and G. J. Zissis, eds., *The Infrared Handbook*, Environmental Research Institute of Michigan, Ann Arbor, Michigan (1985).

Appendix C

OPTICAL PROPERTIES OF INFRARED MATERIALS

Table C.1. Refractive index (n) and absorption coefficient (α) near room temperature, and temperature dependence of n (dn/dT)

Material[a]	Refractive index[b] λ=0.5 µm	Refractive index[b] λ=4 µm	Refractive index[b] λ=10 µm	dn/dT (10^{-6} K^{-1})	Flat plate transmittance[c]	Absorption coefficient (cm^{-1}) λ=4 µm	Absorption coefficient (cm^{-1}) λ=10 µm
ALON[2,4,72]	1.797	1.702	------	2.8 (4 µm) 15 (0.63 µm)[d]	0.874	0.10[40] 0.16[87]	------
(Polycrystalline aluminum oxynitride has nominal composition $9Al_2O_3 \cdot 5AlN$. Actual composition varies from $9Al_2O_3 \cdot 2.80AlN$ to $9Al_2O_3 \cdot 4.41AlN$.[87] Optical properties depend on composition. For experimental data, see Ref. 87.)							
Aluminum nitride (AlN)[4,33,94]	2.168 (n_o) 2.216 (n_e)	2.056 2.092	------ ------	------ ------	0.787 ------	~0.09[93] ------	------ ------
	2.11 (chemical vapor deposited thin layer)[95]						
AMTIR-1[1,4,11] (Ge/As/Se glass)	------	2.513	2.497	72 (10 µm) 101 (1 µm)	0.687	0.002[38]	
Arsenic trisulfide glass[89] (As_2S_3)	2.5976 (0.644 µm)	2.4116	2.3822	9 (5 µm)[89] 1 (5.1 µm)[90]	------	0.014	0.85
Barium fluoride (BaF_2)[4,5,7,29,94]	1.478	1.457	1.401	-14.5 (10.6 µm) -15.9 (3.39 µm) -16.0 (0.63 µm)	0.933	<0.001[40]	0.083[40]
Beryllium oxide (BeO)[2,4,5,35]	1.723 (n_o) 1.634 (n_e)	1.739 1.639	------ ------	8.2 (n_o, 0.6 µm) 13.4 (n_e, 0.6 µm)	0.868 ------	<0.001[40]	------
Boron nitride (cubic BN)[94]	2.11	1.99	2.77	------	0.80	0.01	150
Boron phosphide (BP)[8]	3.251		------	------	0.562 (0.5 µm)	------	------
Cadmium sulfide (CdS)[2,56,36] (polycrystalline) (dn/dT is for single-crystal hexagonal CdS)	------	2.265	2.230	58.6(n_o, 10.6 µm)	0.739	------	------
Calcium aluminate glass							
(Barr & Stroud 37A)[37]	1.678	1.607	------	------	0.897	0.19	------
(Schott IRG 11)[36]	1.690	1.621	------	------	0.894	------	------

Material[a]	Refractive index[b] λ=0.5 µm	λ=4 µm	λ=10 µm	dn/dT (10^{-6} K^{-1})	Flat plate transmittance[c]	Absorption coefficient (cm^{-1}) λ=4 µm	λ=10 µm
Calcium fluoride (CaF_2)[2,12,39,99]	1.436	1.410	1.300	-8.1 (3.39 µm) -10.4 (0.66 µm)	0.944	<0.001[40]	36[39]
Calcium lanthanum sulfide[41,42] (stoichiometry $CaLa_{2.7}S_{5.05}$)		~2.4[43]	~2.4[43]		~0.71		0.12 (10.6 µm)
Cesium bromide (CsBr)[5,13,20]	1.709	1.668	1.662	-75.8 (30 µm) -84.7 (0.63 µm)	0.882	------	------
Diamond (Type IIa)[44,45,46,47]	2.432	2.377	2.376	15.6 (10 µm) 17.6 (2.3 µm)	0.715	3.1[40]	0.04 (10.6 µm)
Fused silica (SiO_2 glass)[3,15,24,99]	1.462	1.389	------	10[48]	0.948	1.2[40]	------
Gallium arsenide (GaAs)[3]	4.305	3.304	3.274	150	0.555	<0.01[50]	<0.01[50]
Gallium nitride (β-GaN, cubic)[73]	2.349	------	------	------	0.721(0.5 µm)	------	------
Gallium nitride (hexagonal)[4,74]	2.44 (n_o)	2.27	------	------	0.738	------	------
	------	2.26	------	------	------	------	------
Gallium phosphide (GaP)[3,4,17]	3.590	3.014	2.964	137[97]	0.598	<0.01[97,49]	0.12[97,49]
Germanate glass[51] (Corning 9754, Ge-Al-Ca-Ba-Zn-oxide)	1.68	1.605	------	------	0.898	------	------
Germanium (Ge)[3,53]	4.388	4.025	4.004	424 (4 µm)[58,98]	0.468	<0.01[52]	0.03[52]
	------	------	------	404 (10 µm)[58,98]	------	------	------
(Absorption coefficients apply to material with a resistivity of 34.6 ohm·cm)							
KRS-5 ($TlI_{0.543}Br_{0.457}$)[91]	2.624 (0.58 µm)	2.382	2.371	-237[92]	0.714	------	------
Lithium fluoride (LiF)[3,5,20]	1.394	1.349	------	-15	0.957	<0.0024[40]	69[40]
Magnesium fluoride (MgF_2)[2,5]	1.380 (n_o)	1.349 (n_o)	------	1	0.957	0.0006[40]	15[40]
	1.392 (n_e)	1.359 (n_e)	------	0.6	------	------	------
Hot pressed Irtran-1[9]	------	1.353	------	------	0.956	0.015[54]	------

Material[a]	Refractive index[b] λ=0.5 μm	Refractive index[b] λ=4 μm	Refractive index[b] λ=10 μm	dn/dT (10^{-6} K^{-1})	Flat plate transmittance[c]	Absorption coefficient (cm^{-1}) λ=4 μm	Absorption coefficient (cm^{-1}) λ=10 μm
Magnesium oxide (MgO)[2,22,9]	1.745	1.668	-------	19[5]	0.882	0.001-0.01[2]	68[40]
Potassium bromide (KBr)[2,20]	1.570	1.536	1.526	-41.1 (10.6 μm)[29] -41.9 (1.15 μm)[29] -39.3 (0.46 μm)[29]	0.914	-------	-------
Quartz (α-SiO_2)[3,23]	1.549 (n_o)	1.466	-------	-6.2 (0.55 μm)	0.931	<0.0001[40]	-------
	1.558 (n_e)	1.473	-------	-7.0 (0.55 μm)	-------	<0.0001[40]	-------
Sapphire (Al_2O_3)[40,55,72,85,94]	1.774 (n_o)	1.677 (n_o)	-------	+6[85,e]	0.880	0.047[40]	670[94]
	1.766 (n_e)	1.667 (n_e)	-------	12[85]	-------	0.03[86]	460[94]
Silicon (Si)[10,53,58]	-------	3.425	3.418	159 (5 μm) 157 (10.6 μm)	0.538	<0.005[3]	0.9[66]
Silicon carbide (cubic β-SiC)[2]	2.664	2.489[88]	-------	-------	0.691	-------	-------
Hexagonal 6H SiC[3]	2.686 (n_o)	2.516 (n_o)	-------	-------	0.686	20[3]	-------
	2.732 (n_e)	-------	-------	-------	-------	-------	-------
CVD SiC[88]	-------	2.489	-------	37 (2-4 μm)	0.692	1.9	-------
Silicon nitride (Si_3N_4)[3,34]	2.04	-------	-------	-------	~0.79 (0.5 μm)	-------	-------
Sodium chloride (NaCl)[20,5]	1.552	1.522	1.495	-36.3 (3.39 μm) -34.2 (0.46 μm)	0.918	<0.0001[40]	0.0006[40,59]
Spinel[2,72] (polycrystalline $MgAl_2O_4$)	1.723	1.639	-------	3[f]	0.889	0.018[40]	-------
Yttria (Y_2O_3)[2,74]	1.947	1.859	-------	8[2], 31[65,g]	0.834	0.04 (3.4 μm)[60]	-------
Yttria (La-doped)[60,72] (polycrystalline $0.09La_2O_3 \cdot 0.91Y_2O_3$)	1.937	1.847	-------	32(3.4 μm)[64,h] 50(0.36 μm)[64]	0.837	0.04 (3.4 μm)[60]	-------

Material[a]	Refractive index[b] λ=0.5 μm	λ=4 μm	λ=10 μm	dn/dT (10^{-6} K^{-1})	Flat plate transmittance[c]	Absorption coefficient (cm^{-1}) λ=4 μm	λ=10 μm
Zinc selenide (CVD ZnSe)[2,63]	2.7	2.433	2.406	60 (10 μm) 63 (4 μm) 143 (0.55 μm)	0.703	0.0016[61] (3.8 μm)	0.001[69]
Zinc sulfide (CVD ZnS)[63,3]	2.420	2.252	2.200	41 (10 μm) 43 (4 μm) 71 (0.5 μm)	0.742	0.02[61]	0.08[67] 0.23 (10.5 μm)[67]
Multispectral CVD ZnS	------ ------	------ ------	------ ------	------ ------	------ ------	~0.001[62]	0.013 (9.3 μm)[68] 0.19 (10.6 μm)[68]
Zirconia (ZrO_2)[31] (ZrO_2/12 mol % Y_2O_3)	2.179	2.051		6 (1.7 μm) 9 (0.5 μm)	0.788	------	------

[a]Materials are single crystal unless otherwise noted.

[b]n_o is the refractive index in the ordinary direction, which is the highest axis of symmetry, also called the optical axis. n_e is the refractive index in the extraordinary direction, which is a 2-fold symmetry axis perpendicular to the optical axis.

[c]At 4 μm wavelength calculated with the equation transmittance = $2n/(n^2+1)$.

[d]ALON at 0.633 μm wavelength:[72] $dn/dT = 1.7900 + 14.6(\pm 0.1) \times 10^{-6} T - 0.0030 \times 10^{-6} T^2$ (T = °C)

[e]Sapphire at 0.633 μm wavelength:[72] $dn_o/dT = 1.7659 + 11.7(\pm 0.2) \times 10^{-6} T - 0.0050 \times 10^{-6} T^2$ (T = °C)

[f]Spinel at 0.633 μm wavelength:[72] $dn/dT = 1.6837 + 13.0(\pm 0.4) \times 10^{-6} T - 0.0036 \times 10^{-6} T^2$ (T = °C)

[g]Yttria at 0.633 μm wavelength:[72] $dn/dT = 1.9253 + 8.28(\pm 0.3) \times 10^{-6} T - 0.0038 \times 10^{-6} T^2$ (T = °C)

[h]La-doped yttria at 0.633 μm wavelength:[72] $dn/dT = 1.8965 + 8.36(\pm 0.1) \times 10^{-6} T - 0.0040 \times 10^{-6} T^2$ (T = °C)

Table C.2. Dispersion equations for refractive index

Material	Dispersion equation ($\lambda = \mu m$)	Wavelength range (μm)	Temperature (°C)
ALON[2]	$n^2 - 1 = \frac{2.1375\lambda^2}{\lambda^2 - 0.10256^2} + \frac{4.582\lambda^2}{\lambda^2 - 18.868^2}$	0.4-2.3	room
	(The composition of single phase, transparent ALON varies from $9Al_2O_3{\cdot}2.80AlN$ to $9Al_2O_3{\cdot}4.41AlN$.[87] The refractive index and other optical properties vary as the chemical composition varies. For experimental data, see Ref. 87.		
Aluminum nitride (AlN)[4,33]	$n_o^2 = 3.1399 + \frac{1.3786\lambda^2}{\lambda^2 - 0.1715^2} + \frac{3.861\lambda^2}{\lambda^2 - 15.03^2}$	0.22-5.0	room
	$n_e^2 = 3.0729 + \frac{1.6173\lambda^2}{\lambda^2 - 0.1746^2} + \frac{4.139\lambda^2}{\lambda^2 - 15.03^2}$		
AMTIR-1[4,11]	$n^2 - 1 = \frac{5.298\lambda^2}{\lambda^2 - 0.29007^2} + \frac{0.6039\lambda^2}{\lambda^2 - 32.022^2}$	1-14	room
Arsenic trisulfide glass[90]	$n^2 - 1 = \frac{1.8983678\lambda^2}{\lambda^2 - 0.150^2} + \frac{1.9222979\lambda^2}{\lambda^2 - 0.250^2} + \frac{0.8765134\lambda^2}{\lambda^2 - 0.350^2}$	0.58-11.8	room
(As_2S_3)	$+ \frac{0.1188704\lambda^2}{\lambda^2 - 0.450^2} + \frac{0.9569903\lambda^2}{\lambda^2 - 27.3861^2}$		
Barium fluoride (BaF_2)[7]	$n^2 - 1 = \frac{0.643356\lambda^2}{\lambda^2 - 0.057789^2} + \frac{0.506762\lambda^2}{\lambda^2 - 0.10968^2} + \frac{3.8261\lambda^2}{\lambda^2 - 46.3864^2}$	0.26-10.4	25
Beryllium oxide (BeO)[2,4]	$n_o^2 - 1 = \frac{1.92274\lambda^2}{\lambda^2 - 0.07908^2} + \frac{1.24209\lambda^2}{\lambda^2 - 9.7131^2}$	0.44-7.0	room
	$n_e^2 - 1 = \frac{1.969392\lambda^2}{\lambda^2 - 0.08590^2} + \frac{1.67389\lambda^2}{\lambda^2 - 10.4797^2}$	0.44-7.0	room
Boron phosphide (BP)[8]	$n^2 - 1 = \frac{6.841\lambda^2}{\lambda^2 - 0.267^2}$	0.45-0.63	27

Table C.2 (continued)

Material	Dispersion equation ($\lambda = \mu m$)	Wavelength range (μm)	Temperature (°C)
Cadmium sulfide (CdS)[10]	$n_o^2 - 1 = \frac{3.96582820\lambda^2}{\lambda^2 - 0.23622804^2} + \frac{0.18113874\lambda^2}{\lambda^2 - 0.48285199^2}$	0.51-1.4	room
	$n_e^2 - 1 = \frac{3.97478769\lambda^2}{\lambda^2 - 0.22426984^2} + \frac{0.26680809\lambda^2}{\lambda^2 - 0.46693785^2} + \frac{0.00074077\lambda^2}{\lambda^2 - 0.50915139^2}$		
Cadmium sulfide (polycrystal)[57]	$n^2 - 1 = \frac{3.08470\lambda^2}{\lambda^2 - 0.23192^2} + \frac{1.05629\lambda^2}{\lambda^2 - 0.29050^2} + \frac{1.09792\lambda^2}{\lambda^2 - 27.57220^2} + \frac{0.75609\lambda^2}{\lambda^2 - 68.08738^2}$	2-14	room
Calcium aluminate (IRG 11 glass)[32]	$n^2 - 1 = \frac{0.67880\lambda^2}{\lambda^2 - 0.06584^2} + \frac{1.07826\lambda^2}{\lambda^2 - 0.13752^2} + \frac{2.00510\lambda^2}{\lambda^2 - 16.24454^2}$	0.48-4.6	room
Calcium fluoride (CaF_2)[10,12]	$n^2 - 1 = \frac{0.5675888\lambda^2}{\lambda^2 - 0.050263605^2} + \frac{0.4710914\lambda^2}{\lambda^2 - 0.1003909^2} + \frac{3.8484723\lambda^2}{\lambda^2 - 34.649040^2}$	0.23-9.7	24
Calcium fluoride (Irtran-3)[9]	$n = 1.4278071 + \frac{0.0022807}{\lambda^2 - 0.028} - \frac{0.00009194}{(\lambda^2 - 0.028)^2} - 0.00111658\lambda^2 - 0.00000159\lambda^4$	1.0-10.0	room
Cesium bromide (CsBr)[5,13]	$n^2 - 1 = \frac{0.9533786\lambda^2}{\lambda^2 - 0.0905643^2} + \frac{0.8303809\lambda^2}{\lambda^2 - 0.1671517^2} + \frac{2.847172\lambda^2}{\lambda^2 - 119.0155^2}$	0.36-39.0	27
Diamond[46,14]	$n = 2.37553 + \frac{0.0336440}{\lambda^2 - 0.028} - \frac{0.0887524}{(\lambda^2 - 0.028)^2} - 0.00000240455\lambda^2 - 0.00000000221390\lambda^4$	2.5-25	room
	$n^2 - 1 = \frac{4.3356\lambda^2}{\lambda^2 - 0.1060^2} + \frac{0.3306\lambda^2}{\lambda^2 - 0.1750^2}$	0.225-0.644	room
Fused silica (SiO_2)[10,15,24]	$n^2 - 1 = \frac{0.6961663\lambda^2}{\lambda^2 - 0.0684043^2} + \frac{0.4079426\lambda^2}{\lambda^2 - 0.1162414^2} + \frac{0.8974794\lambda^2}{\lambda^2 - 9.896161^2}$	0.21-3.71	20

Table C.2 (continued)

Material	Dispersion equation ($\lambda = \mu m$)	Wavelength range (μm)	Temperature (°C)
Gallium arsenide (GaAs)[4,16]	$n^2 = 3.5 + \frac{7.4969\lambda^2}{\lambda^2 - 0.4082^2} + \frac{1.9347\lambda^2}{\lambda^2 - 37.17^2}$	1.4-11	room
Gallium nitride (β-GaN, cubic)[73]	$n^2 = 3.99499 + \frac{0.912375\lambda^2}{\lambda^2 - 0.314551^2} + 0.055137\lambda^2$	0.41-1.5	room
Gallium nitride (hexagonal)[4,74]	$n_o^2 = 3.60 + \frac{1.75\lambda^2}{\lambda^2 - 0.256^2} + \frac{4.1\lambda^2}{\lambda^2 - 17.86^2}$	0.4-10	room
	$n_e^2 = 5.35 + \frac{5.08\lambda^2}{\lambda^2 - 18.76^2}$	?-10	room
Gallium phosphide (GaP)[4,17]	$n^2 - 1 = \frac{1.390\lambda^2}{\lambda^2 - 0.172^2} + \frac{4.131\lambda^2}{\lambda^2 - 0.234^2} + \frac{2.570\lambda^2}{\lambda^2 - 0.345^2} + \frac{2.056\lambda^2}{\lambda^2 - 27.52^2}$	0.8-10	room
Germanium (Ge)[18,19]	$n^2 = 9.28156 + \frac{6.72880\lambda^2}{\lambda^2 - 0.44105^2} + \frac{0.21307\lambda^2}{\lambda^2 - 3870.1^2}$	2-12	room
KRS-5 ($TlI_{0.543}Br_{0.457}$)[91]	$n^2 - 1 = \sum_{i=1}^{3} \frac{S_i\lambda^2}{\lambda^2 - \lambda_i^2}$ $S_1 = 3.902110 - 2.225 \times 10^{-3}\ T$ $S_2 = 0.789182 + 1.098 \times 10^{-3}\ T$ $S_3 = 13.704349 - 4.794 \times 10^{-2}\ T$ $\lambda_1 = 0.213778 - 4.693 \times 10^{-5}\ T$ $\lambda_2 = 0.383628 - 6.998 \times 10^{-5}\ T$ $\lambda_3 = 171.9424 - 0.26021\ T$ (T = 19-31 °C)	0.6-39	
Lithium fluoride (LiF)[20]	$n^2 - 1 = \frac{0.92549\lambda^2}{\lambda^2 - 0.07376^2} + \frac{6.96747\lambda^2}{\lambda^2 - 32.79^2}$	0.1-11	20
Magnesium fluoride (MgF_2)[21]	$n_o^2 - 1 = \frac{0.48755108\lambda^2}{\lambda^2 - 0.04338408^2} + \frac{0.39875031\lambda^2}{\lambda^2 - 0.09461442^2} + \frac{2.3120353\lambda^2}{\lambda^2 - 23.793604^2}$	0.20-7.04	19
	$n_e^2 - 1 = \frac{0.41344023\lambda^2}{\lambda^2 - 0.03684262^2} + \frac{0.50497499\lambda^2}{\lambda^2 - 0.09076162^2} + \frac{2.4904862\lambda^2}{\lambda^2 - 12.771995^2}$		

Table C.2 (continued)

Material	Dispersion equation ($\lambda = \mu m$)	Wavelength range (μm)	Temperature (°C)
Magnesium fluoride (Irtran-1)[9]	$n = 1.3776955 + \frac{0.00135155}{\lambda^2 - 0.028} + \frac{0.00021254}{(\lambda^2 - 0.028)^2} - 0.00150412\lambda^2 - 0.00000441\lambda^4$	1.0-9.0	room
Magnesium oxide (MgO)[5,22]	$n^2 - 1 = \frac{1.111033\lambda^2}{\lambda^2 - 0.0712465^2} + \frac{0.8460085\lambda^2}{\lambda^2 - 0.1375204^2} + \frac{7.808527\lambda^2}{\lambda^2 - 26.89302^2}$	0.36-4.4	20
Magnesium oxide (Irtran-5)[9]	$n = 1.7200516 + \frac{0.00561194}{\lambda^2 - 0.028} - \frac{0.00001099}{(\lambda^2 - 0.028)^2} - 0.00309946\lambda^2 - 0.00000961\lambda^4$	1.0-9.0	room
Potassium bromide (KBr)[20]	$n^2 = 1.39408 + \frac{0.79221\lambda^2}{\lambda^2 - 0.146^2} + \frac{0.01981\lambda^2}{\lambda^2 - 0.173^2} + \frac{0.15587\lambda^2}{\lambda^2 - 0.187^2} + \frac{0.17673\lambda^2}{\lambda^2 - 60.61^2} + \frac{2.06217\lambda^2}{\lambda^2 - 87.72^2}$	0.2-42	20
Quartz (α-SiO_2)[23]	$n_o^2 - 1 = \frac{0.663044\lambda^2}{\lambda^2 - 0.060^2} + \frac{0.517852\lambda^2}{\lambda^2 - 0.106^2} + \frac{0.175912\lambda^2}{\lambda^2 - 0.119^2} + \frac{0.565380\lambda^2}{\lambda^2 - 8.844^2} + \frac{1.675299\lambda^2}{\lambda^2 - 20.742^2}$	0.18-0.71	room
	$n_e^2 - 1 = \frac{0.665721\lambda^2}{\lambda^2 - 0.060^2} + \frac{0.503511\lambda^2}{\lambda^2 - 0.106^2} + \frac{0.214792\lambda^2}{\lambda^2 - 0.119^2} + \frac{0.539173\lambda^2}{\lambda^2 - 8.792^2} + \frac{1.807613\lambda^2}{\lambda^2 - 197.70^2}$		
Sapphire (Al_2O_3)[5,25]	$n_o^2 - 1 = \frac{1.4313493\lambda^2}{\lambda^2 - 0.0726631^2} + \frac{0.65054713\lambda^2}{\lambda^2 - 0.1193242^2} + \frac{5.3414021\lambda^2}{\lambda^2 - 18.028251^2}$	0.2-5.5	20
	$n_e^2 - 1 = \frac{1.5039759\lambda^2}{\lambda^2 - 0.0740288^2} + \frac{0.55069141\lambda^2}{\lambda^2 - 0.1216529^2} + \frac{6.5927379\lambda^2}{\lambda^2 - 20.072248^2}$		
Silicon (Si)[10]	$n^2 - 1 = \frac{10.6684293\lambda^2}{\lambda^2 - 0.301516485^2} + \frac{0.003043475\lambda^2}{\lambda^2 - 1.13475115^2} + \frac{1.54133408\lambda^2}{\lambda^2 - 1104.0^2}$	1.36-11	room

Table C.2 (continued)

Material	Dispersion equation ($\lambda = \mu m$)	Wavelength range (µm)	Temperature (°C)
Silicon carbide (cubic β-SiC)[26]	$n^2 - 1 = \frac{5.5705\lambda^2}{\lambda^2 - 0.1635^2}$	0.47-0.69	room
Silicon carbide (6H SiC)[27]	$n_o^2 - 1 = \frac{5.5515\lambda^2}{\lambda^2 - 0.16250^2}$	0.49-1.06	room
	$n_e^2 - 1 = \frac{5.7382\lambda^2}{\lambda^2 - 0.16897^2}$		
CVD SiC[88] (chemical vapor deposited)	$n^2 - 1 = [5.558 + 1.843 \times 10^{-4}(T\text{-}289)] \frac{6.1035^2\lambda^2}{6.1035^2\lambda^2 - 1} + 3 \frac{0.0830^2\lambda^2}{0.0830^2\lambda^2 - 1}$	2.5 - 4	T = kelvins
Silicon nitride (Si_3N_4)[34]	$n^2 = 2.12477 + \frac{1.83281\lambda^2}{\lambda^2 - 0.15733^2}$	0.21-1.24	room
Sodium chloride (NaCl)[20]	$n^2 = 1.00055 + \frac{0.19800\lambda^2}{\lambda^2 - 0.050^2} + \frac{0.48398\lambda^2}{\lambda^2 - 0.100^2} + \frac{0.38696\lambda^2}{\lambda^2 - 0.128^2} + \frac{0.25998\lambda^2}{\lambda^2 - 0.158^2} + \frac{0.08796\lambda^2}{\lambda^2 - 40.50^2} + \frac{3.17064\lambda^2}{\lambda^2 - 60.98^2} + \frac{0.30038\lambda^2}{\lambda^2 - 120.34^2}$	0.2-30	20
Spinel ($MgAl_2O_4$)[2]	$n^2 - 1 = \frac{1.8938\lambda^2}{\lambda^2 - 0.09942^2} + \frac{3.0755\lambda^2}{\lambda^2 - 15.826^2}$	0.35-5.5	—
Yttria (Y_2O_3)[4,28]	$n^2 - 1 = \frac{2.578\lambda^2}{\lambda^2 - 0.1387^2} + \frac{3.935\lambda^2}{\lambda^2 - 22.936^2}$	0.2-12	20

Table C.2 (continued)

Material	Dispersion equation ($\lambda = \mu m$)	Wavelength range (μm)	Temperature (°C)
Zinc selenide (ZnSe)[63,29,9] (CVD, chemical vapor deposited)	$n^2 = 9.01536 + \frac{0.24482}{\lambda^2 - 0.29934^2} + \frac{7229.93}{\lambda^2 - 48.38^2}$	0.55-18	room
Zinc sulfide[63,29,30] (CVD, chemical vapor deposited)	$n^2 = 8.34096 + \frac{0.14540}{\lambda^2 - 0.23979^2} + \frac{4321.39}{\lambda^2 - 36.525^2}$	0.5-14	room
Multispectral[70,71]	$n^2 = 8.393 + \frac{0.14383}{\lambda^2 - 0.2421^2} + \frac{4431.0}{\lambda^2 - 36.71^2}$	0.4-13	20
Zirconia (ZrO_2/12 mol % Y_2O_3)[31]	$n^2 - 1 = \frac{1.347091\lambda^2}{\lambda^2 - 0.062543^2} + \frac{2.117788\lambda^2}{\lambda^2 - 0.166739^2} + \frac{9.452943\lambda^2}{\lambda^2 - 24.320570^2}$	0.36-5.1	25

Table C.3. Absorption coefficients of selected materials calculated by *OPTIMATR®*[40]

ALON (polycrystalline aluminum oynitride)									
	Calculated Absorption Coefficient (cm^{-1}) at Indicated Temperature								
Wavelength (μm)	300K	400K	500K	600K	700K	800K	900K	1000K	1300K
3.0	0.002	0.003	0.003	0.007	0.007	0.012	0.032	0.092	1.208
3.5	0.019	0.023	0.029	0.048	0.048	0.063	0.094	0.16	1.25
4.0	0.10	0.12	0.15	0.24	0.24	0.30	0.37	0.50	1.72
4.5	0.47	0.54	0.64	0.93	0.93	1.12	1.34	1.61	3.31
5.0	1.51	1.74	2.06	2.90	2.90	3.41	3.98	4.62	7.51
5.5	3.67	4.22	4.95	6.84	6.84	7.96	9.18	10.50	15.50
6.0	9.20	10.27	11.70	15.28	15.28	17.34	19.52	21.84	29.86

Lanthana-Doped Yttria (polycrystalline)									
Wavelength (μm)	300K	400K	500K	600K	700K	800K	900K	1000K	1300K
4.0	0.000	0.000	0.000	0.000	0.001	0.001	0.001	0.002	0.005
4.5	0.001	0.001	0.002	0.003	0.004	0.006	0.009	0.013	0.030
5.0	0.004	0.006	0.010	0.015	0.022	0.031	0.043	0.058	0.12
5.5	0.019	0.029	0.043	0.062	0.089	0.12	0.16	0.21	0.42
6.0	0.070	0.10	0.15	0.21	0.29	0.39	0.51	0.65	1.19

Magnesium Oxide (single-crystal)									
Wavelength (μm)	300K	400K	500K	600K	700K	800K	900K	1000K	1300K
3.5	0.000	0.000	0.000	0.000	0.000	0.000	0.001	0.001	0.002
4.0	0.000	0.001	0.001	0.002	0.002	0.003	0.004	0.006	0.013
4.5	0.003	0.005	0.007	0.010	0.013	0.018	0.024	0.031	0.062
5.0	0.017	0.023	0.031	0.043	0.058	0.076	0.099	0.12	0.23
5.5	0.064	0.084	0.11	0.15	0.20	0.25	0.32	0.39	0.67
6.0	0.224	0.29	0.37	0.47	0.60	0.74	0.90	1.09	1.74

Sapphire (single-crystal, o-ray)									
	Calculated Absorption Coefficient (cm^{-1}) at Indicated Temperature								
Wavelength (μm)	300K	400K	500K	600K	700K	800K	900K	1000K	1300K
3.0	0.001	0.001	0.001	0.001	0.002	0.003	0.004	0.005	0.010
3.5	0.007	0.008	0.011	0.015	0.019	0.025	0.033	0.042	0.080
4.0	0.047	0.058	0.074	0.096	0.12	0.16	0.20	0.24	0.42
4.5	0.24	0.28	0.35	0.44	0.54	0.67	0.81	0.97	1.58
5.0	0.96	1.13	1.36	1.65	1.99	2.39	2.83	3.31	5.01
5.5	2.66	3.13	3.76	4.53	5.43	6.43	7.53	8.72	12.72
6.0	6.53	7.54	8.88	10.49	12.32	14.35	16.53	18.86	26.51

For an example of the good agreement between calculated and measured optical properties, see S. Kaplan and L. Hanssen, "Normal Infrared Spectral Emittance of Al_2O_3," *Proc. SPIE*, **3425**, 120-125 (1998).

Spinel (polycrystalline)									
Wavelength (μm)	300K	400K	500K	600K	700K	800K	900K	1000K	1300K
3.0	0.000	0.000	0.000	0.000	0.001	0.001	0.003	0.011	0.24
3.5	0.002	0.003	0.004	0.005	0.007	0.010	0.015	0.025	0.25
4.0	0.018	0.023	0.031	0.041	0.053	0.069	0.090	0.12	0.41
4.5	0.099	0.12	0.16	0.20	0.26	0.32	0.40	0.50	1.04
5.0	0.46	0.55	0.68	0.84	1.03	1.26	1.52	1.81	3.03
5.5	1.49	1.78	2.18	2.66	3.22	3.87	4.58	5.37	8.23
6.0	3.70	4.39	5.32	6.43	7.72	9.15	10.72	12.42	18.26

Yttria (polycrystalline)									
Wavelength (μm)	300K	400K	500K	600K	700K	800K	900K	1000K	1300K
3.5	0.000	0.000	0.000	0.000	0.000	0.000	0.000	0.000	0.001
4.0	0.000	0.000	0.000	0.000	0.001	0.001	0.002	0.003	0.007
4.5	0.001	0.001	0.002	0.004	0.006	0.008	0.012	0.017	0.038
5.0	0.005	0.008	0.013	0.019	0.028	0.040	0.055	0.074	0.16
5.5	0.026	0.038	0.056	0.081	0.11	0.16	0.21	0.27	0.52
6.0	0.091	0.13	0.19	0.27	0.37	0.49	0.64	0.81	1.47

Calcium Fluoride (single-crystal)									
	Calculated Absorption Coefficient (cm^{-1}) at Indicated Temperature								
Wavelength (μm)	300K	400K	500K	600K	700K	800K	900K	1000K	1300K
5.0	0.000	0.000	0.000	0.001	0.001	0.002	0.003	0.004	0.011
6.0	0.002	0.004	0.007	0.012	0.018	0.027	0.038	0.052	0.11
7.0	0.024	0.040	0.064	0.097	0.14	0.19	0.26	0.34	0.63
8.0	0.15	0.23	0.35	0.50	0.69	0.91	1.17	1.46	2.49
9.0	0.69	1.01	1.43	1.95	2.55	3.23	3.99	4.81	7.56
10.0	2.43	3.44	4.69	6.14	7.76	9.53	11.42	13.40	19.69

Lithium Fluoride (single-crystal)									
Wavelength (μm)	300K	400K	500K	600K	700K	800K	900K	1000K	
4.0	0.002	0.004	0.006	0.009	0.013	0.018	0.024	0.032	
5.0	0.046	0.066	0.095	0.13	0.18	0.24	0.31	0.39	
6.0	0.40	0.55	0.75	0.99	1.29	1.63	2.01	2.42	
7.0	2.32	2.98	3.83	4.82	5.95	7.17	8.48	9.85	
8.0	8.05	10.25	12.93	15.97	19.25	22.71	26.28	29.91	
9.0	23.11	28.02	33.79	40.08	46.69	53.47	60.31	67.11	
10.0	69.11	79.98	92.07	104.60	117.10	129.40	141.30	152.80	

Magnesium Fluoride (single-crystal, o-ray)									
Wavelength (μm)	300K	400K	500K	600K	700K	800K	900K	1000K	1300K
4.0	0.001	0.001	0.001	0.002	0.003	0.005	0.006	0.008	0.019
5.0	0.011	0.016	0.023	0.034	0.046	0.062	0.081	0.10	0.18
6.0	0.10	0.13	0.19	0.25	0.33	0.42	0.52	0.63	1.01
7.0	0.53	0.70	0.92	1.18	1.48	1.81	2.17	2.54	3.72
8.0	2.03	2.63	3.35	4.18	5.07	6.00	6.97	7.95	10.87
9.0	5.53	6.97	8.66	10.51	12.46	14.47	16.50	18.51	24.30
10.0	15.37	18.31	21.61	25.09	28.63	32.14	35.59	38.94	48.21

Table C.4. Change of refractive index (n) for isotropic compression under pressure P

Material	Wavelength (nm)	dn/dP (kbar^{-1})
Diamond (C)[96]	---	-0.86×10^{-4}
Fused silica (SiO_2)[78]	0.589	$+9.2 \times 10^{-4}$ (sign?)*
Gallium arsenide (GaAs)[79]	---	-23×10^{-4}
Gallium nitride (GaN)[82]	---	-7.3×10^{-4}
Germanium (Ge)[79]	---	-28 to -40×10^{-4}
Lithium fluoride (LiF)[81]	---	$+1.98 \times 10^{-4}$
Magnesium fluoride (MgF_2)[84]	---	(anisotropic crystal values in Ref. 84)
Magnesium oxide (MgO)[76]	0.589	-1.58×10^{-4}
	0.546	-1.65×10^{-4}
	0.436	-1.91×10^{-4}
	0.405	-2.06×10^{-4}
Potassium bromide (KBr)[81]	---	$+24.4 \times 10^{-4}$
α-Quartz (SiO_2)[77]	0.589	$+10.3 \times 10^{-4}$ (n_o)
	0.589	$+10.7_5 \times 10^{-4}$ (n_e)
Silicon (Si)[79]	---	-10×10^{-4}
Sodium chloride (NaCl)[81]	---	$+11.7 \times 10^{-4}$
Spinel ($MgAl_2O_4$)[75]	0.589	$-0.9_1 \times 10^{-4}$
CVD-Zinc selenide (ZnSe)[83]	0.633	-9.6×10^{-4}
Zinc sulfide (CVD ZnS?)[80]	---	$+9 \times 10^{-4}$ (sign?)*
Zinc sulfide (crystalline ZnS?)[79]	---	-2×10^{-4}
	10.6	-3.1×10^{-4}

*There may be confusion in the literature in the sign of dn/dP. A positive value of dn/dT (T = temperature) means that when the lattice expands, n increases. A positive value of dn/dT implies that the value of dn/dP should be negative, since the lattice will contract as P increases. Comparing dn/dP in Table C.3 to dn/dT in Table C.1 suggests that the signs of dn/dP for fused silica and CVD ZnS may be wrong.

References for Appendix C

1. W. L. .Wolfe and G. J. Zissis, eds., *The Infrared Handbook*, Environmental Research Institute of Michigan, Ann Arbor, Michigan (1985).
2. E. D. Palik, ed., *Handbook of Optical Constants of Solids II*, Academic Press, San Diego, California (1991).
3. E. D. Palik, ed., *Handbook of Optical Constants of Solids*, Academic Press, Orlando, Florida (1985).
4. W. J. Tropf, M. E. Thomas and T. J. Harris, "Properties of Crystals and Glasses," in *Handbook of Optics* 2nd ed. (E. vanStryland, D. Williams and W. L. Wolfe, eds.), Vol. II, McGraw-Hill, New York (1995).
5. M. J. Dodge, "Refractive Index," in *CRC Handbook of Laser Science and Technology* (M. J. Weber, ed.), Vol. IV, CRC Press, Boca Raton, Florida (1986).
6. P. Klocek, ed., *Handbook of Infrared Optical Materials*, Marcel Dekker, New York (1991).
7. I. H. Malitson, "Refractive Properties of Barium Fluoride," *J. Opt. Soc. Am.*, **54**, 628 (1964).
8. W. Wettling and J. Windscheif, "Elastic Constants and Refractive Index of Boron Phosphide," *Solid State Comm.*, **50**, 33-34 (1984).
9. *Kodak Irtran Infrared Optical Materials*, Kodak Publication U-72, Eastan Kodak Co., Rochester, New York (1971).
10. B. Tatian, "Fitting Refractive-Index Data with Sellmeier Dispersion Formula," *Appl. Opt.*, **23**, 4477-4485 (1984).
11. Amorphous Materials, Inc., Garland, Texas.
12. I. H. Malitson, "A Redetermination of Some Optical Properties of Calcium Fluoride," *Appl. Opt.*, **2**, 1103-1107 (1963).
13. W. S. Rodney and R. J. Spidler, "Refractive Index of Cesium Bromide for Ultraviolet, Visible and Infrared Wavelengths," *J. Res. Nat. Bur. Stand*, **51**, 123-126 (1953).
14. F. Peter, "Refractive Indices and Absorption Coefficients of Diamond Between 644 and 226 mμ," *Z. Phys.*, **15**, 358-368 (1923).
15. W. S. Rodney and R. J. Spidler, "Index of Refraction of Fused-Quartz Glass for Ultraviolet, Visible and Infrared Wavelengths," *J. Res. Nat. Bur. Stand*, **53**, 185-189 (1954).
16. A. H. Kachare, W. G. Spitzer and J. E. Fredrickson, "Refractive Index of Ion-Implanted GaAs," *J. Appl. Phys.*, **47**, 4209-4212 (1976). The refractive index given by the dispersion formula differs by about 0.02 from the tabulated data in Ref. 5.
17. D. F. Parsons and P. D. Coleman, "Far Infrared Optical Constants of Gallium Phosphide," *Appl. Opt.*, **10**, 1683-1685 (1971).
18. N. P. Barnes and M. S. Piltch, "Temperature-Dependent Sellmeier Coefficients and Nonlinear Optics Average Power Limit for Germanium," *J. Opt. Soc. Am.*, **69**, 178-180 (1979).
19. H. W. Icenogle, B. C. Platt and W. L. Wolfe, "Refractive Indexes and Temperature Coefficients of Germanium and Silicon," *Appl. Opt.*, **15**, 2348-2351 (1976).
20. H. H. Li, "Refractive Index of Alkali Halides and its Wavelength and Temperature Derivatives," *J. Phys. Chem. Ref. Data*, **5**, 329-528 (1976).
21. M. Dodge, "Refractive Properties of Magnesium Fluoride," *Appl. Opt.*, **23**, 1980-1985 (1984).
22. R. E. Stephens and I. H. Malitson, "Index of Refraction of Magnesium Oxide," *J. Nat. Bur. Stand.*, **49**, 249-252 (1952).
23. T. Radhakrishnan, "Further Studies on the Temperature Variation of the Refractive Index of Crystals," *Proc. Indian. Acad. Sci.*, **A33**, 22-34 (1951).

24. I. H. Malitson, "Interspecimen Comparison of the Refractive Index of Fused Silica," *J. Opt. Soc. Am.*, **55**, 1205-1209 (1965).
25. I. H. Malitson and M. Dodge, "Refractive Index and Birefringence of Synthetic Sapphire," *J. Opt. Soc. Am.*, **62**, 1405 (1972).
26. P. T. B. Schaffer, "Refractive Index, Dispersion, and Birefringence of Silicon Carbide Polytypes," *Appl. Opt.*, **10**, 1034-1036 (1971).
27. S. Singh, J. R. Potopowicz, L. G. Van Uitert and S. H. Wemple, "Nonlinear Optical Properties of Hexagonal Silicon Carbide," *Appl. Phys. Lett.*, **19**, 53-56 (1971).
28. Y. Nigara, "Measurement of the Optical Constants of Yttrium Oxide," *Jap. J. Appl. Phys.*, **7**, 404-408 (1968).
29. A. Feldman, D. Horowitz, R. M. Walker and M. J. Dodge, "Optical Materials Characterization Final Technical Report," *Nat. Bur. Standards Technical Note 993*, February 1979.
30. M. Herzberger and C. D. Salzberg, "Refractive Indices of Infrared Optical Materials and Color Correction of Infrared Lenses," *J. Opt. Soc. Am.*, **52**, 420-427 (1962).
31. D. L. Wood and K. Nassau, "Refractive Index of Cubic Zirconia Stabilized with Yttria," *Appl. Opt.*, **21**, 2978-2981 (1982).
32. Fit to data from Ref. 8, p. 436, for Schott Glass Technologies IRG 11 glass.
33. J. Pastrňák and L. Roskovcová, "Refractive Index Measurements on AlN Single Crystals," *Phys. Stat. Sol.*, **14**, K5-K8 (1966); D. A. Yas'kov and A. N. Pikhtin, "Refractive Index and Birefringence of Semiconductors with the Wurtzite Structure," *Sov. Phys. Semicond.*, **15**, 8-12 (1981).
34. Fit to data from Ref. 5, p. 774, which is derived from H. R. Philipp, "Optical Properties of Silicon Nitride," *J. Electrochem. Soc.*, **120**, 295-300 (1973). Data for amorphous Si_3N_4 thin film on silicon.
35. H. W. Newkirk, D. K. Smith and J. S. Kahn, "Synthetic Bromellite. III. Some Optical Properties," *Am. Mineralogist*, **51**, 141-151 (1966).
36. R. Weil and D. Neshmit, "Temperature Coefficient of the Indices of Refraction and the Birefringence in Calcium Sulfide," *J. Opt. Soc. Am.*, **67**, 190-195 (1977).
37. *Calcium Aluminate Type WB37A Data Sheet*, Sassoon Advanced Materials Ltd., Dumbarton, Scotland.
38. AMTIR-1 Data Sheet, Amorphous Materials, Inc., Garland, Texas.
39. W. Kaiser, W. G. Spitzer, R. H. Kaiser and L. E. Howarth, "Infrared Properties of CaF_2, SrF_2 and BaF_2," *Phys. Rev.* **127**, 1950-1954 (1962).
40. *OPTIMATR*®, ARSoftware, 8201 Corporate Drive, Suite 1110, Landover MD 20785 U.S.A.; Phone: 301-459-3773.
41. M. E. Hills, *Preparation, Properties, and Development of Calcium Lanthanum Sulfide as an 8- to 10-micrometer Transmitting Ceramic*, Naval Weapons Center TP 7037, China Lake CA, September 1989.
42. R. L. Gentilman, M. B. Dekosky, T. Y. Wong, R. W. Tustison and M. E. Hills, "Calcium Lanthanum Sulfide as a Long Wavelength IR Material," *Proc. SPIE*, **929**, 57-64 (1988).
43. Estimated from the transmission value of 71% in Figure 3 of J. A. Savage, K. L. Lewis, B. E. Kinsman, A. R. Wilson and R. Riddle, "Fabrication of Infrared Optical Ceramics in the $CaLa_2S_4$ - La_2S_3 Solid Solution System," *Proc. SPIE*, **683**, 79-84 (1986).
44. K. Harris, G. L. Herrit, C. J. Johnson, S. P. Rummel and D. Scatena, "Infrared Optical Characteristics of Type 2A Diamonds," *Appl Opt.*, **30**, 5015-5017 (1991); *Ibid.*, **32**, 4342 (1992).
45. D. C. Harris, "Properties of Diamond for Window and Dome Applications," *Proc. SPIE*, **2286**, 218-228 (1994).

46. D. F. Edwards and E. Ochoa, "Infrared Refractive Index of Diamond," *J. Opt. Soc. Am.*, **71**, 607-608 (1981).
47. M. E. Thomas, W. J. Tropf and A. Szpak, "Optical Properties of Diamond," *Diamond Films & Technol.*, **5**, 159-180 (1995).
48. *Synthetic Fused Silica Data Sheet*, Dynasil Corp., Berlin, NJ.
49. P. Klocek, J. T. Hoggins and M. Wilson, "Broadband IR Transparent Rain Erosion Protection Coating for IR Windows," *Proc. SPIE*, **1760**, 210-223 (1992).
50. P. Klocek, M. W. Boucher, J. M. Trombetta and P. A. Trotta, "High Resistivity and Conductive Gallium Arsenide for IR Optical Components," *Proc. SPIE*, **1760**, 74-85 (1992).
51. W. H. Dumbaugh, "Infrared Transmitting Germanate Glass," *Proc. SPIE*, **297**, 80-85 (1981).
52. J. C. Richter, C. R. Poznich and D. W. Thomas, "Minimization of IR Absorption by Germanium at Elevated Temperature," *Proc. SPIE,* **1326**, 106-119 (1990).
53. H. W. Icenogle, B. C. Platt and W. L. Wolfe, "Refractive Indexes and Temperature Coefficients of Germanium and Silicon," *Appl. Opt.*, **15**, 2348-2351 (1976).
54. Derived from an emittance of 0.005 from 0.30-cm-thick MgF_2 reported in Fig. 7 of M. E. Thomas, T. M. Cotter and K. T. Constantikes, "Infrared Properties of Polycrystalline Magnesium Fluoride," *Proc. 6th DoD Electromagnetic Windows Symp.*, pp. 464-476, Huntsville, AL, October 1995.
55. M. E. Thomas, R. M. Sova and R. I. Joseph, "Temperature Dependence of the Infrared Refractive Index of Sapphire," *Proc. SPIE,* **2286**, 522-530 (1994).
56. A. B. Francis and A. I. Carlson, "Cadmium Sulfide Infrared Optical Material," *J. Opt. Soc. Am.*, **50**, 118-121 (1960).
57. Fit to data in references 4 and 58.
58. H. H. Li, "Refractive Index of Silicon and Germanium and its Wavelength and Temperature Derivatives," *J. Phys. Chem. Ref. Data*, **9**, 561-658 (1980).
59. S. Allen and J. A. Harrington, "Optical Absorption in KCl and NaCl at Infrared Laser Wavelengths," *Appl. Opt.*, **17**, 1679-1680 (1978).
60. D. C. Harris, *Development of Yttria and Lanthana-Doped Yttria as Infrared-Transmitting Materials*, Naval Weapons Center TP 7140, China Lake CA, March 1991.
61. N. C. Fernelius, G. A. Graves and W. Knecht, "Characterization of Candidate Laser Window Materials," *Proc. SPIE*, **297**, 188-195 (1981).
62. C. A. Klein, *Compendium of Property Data for Raytran Zinc Selenide and Raytran Zinc Sulfide*, Report RAY/RD/T-1154, 31 Aug 1987, Raytheon, Co., Lexington, MA.
63. H. H. Li, "Refractive Index of ZnS, ZnSe, and ZnTe and its Wavelength and Temperature Derivatives," *J. Phys. Chem. Ref. Data*, **13**, 103-150 (1984).
64. G. C. Wei, C. Brecher, M. R. Pascucci, E. A. Trickett and W. H. Rhodes, "Characterization of Lanthana-Strengthened Yttria Infrared Transmitting Materials," *Proc. SPIE*, **929**, 50-56 (1988).
65. D. C. Harris and W. R. Compton, *Optical, Thermal and Mechanical Properties of Yttria and Lanthana-Doped Yttria*, Naval Weapons Center TP 7002, China Lake CA, September 1989.
66. C. R. Poznich and J. C. Richter, "Silicon for Use as a Transmissive Material in the Far IR?" *Proc. SPIE*, **1760**, 112-120 (1992).
67. C. A. Klein and R. N. Donadio, "Infrared-Active Phonons in Cubic Zinc Sulfide," *J. Appl. Phys.*, **51**, 797-800 (1980).
68. J. A. Savage, *Infrared Optical Materials and Their Antireflection Coatings*, Adam Hilger, Bristol, p. 125 (1985).

69. C. A. Klein, R. P. Miller and D. L. Stierwalt, "Surface and Bulk Absorption Characteristics of Chemically Vapor-Deposited Zinc Selenide in the Infrared," *Appl. Opt.*, **33**, 4304-4313 (1994). In addition to bulk absorption, ZnSe has significant surface absorption.
70. C. A. Klein, "Room-Temperature Dispersion Equations for Cubic Zinc Sulfide," *Appl. Opt.*, **25**, 1873-1875 (1986).
71. M. Debenham, "Refractive Indices of Zinc Sulfide in the 0.405-13-μm Wavelength Range," *Appl. Opt.*, **23**, 2238-2239 (1984).
72. C. H. Lange and D. D. Duncan, "Temperature Coefficient of Refractive Index for Candidate Optical Windows," *Proc. SPIE*, **1326**, 71-78 (1990).
73. Refractive index dispersion equation fit to data in M. A. Vidall, G. Ramirez-Flores, H. Navarro-Contreras, A. Lastras-Martínez, R. C. Powell and J. E. Greene, "Refractive Indices of Zincblende Structure β-GaN(001) in the Subband-Gap Region (0.7-3.3 eV)," *Appl. Phys Lett.*, **68**, 441-443 (1996).
74. A. S. Barker and M. Ilegems, "Infrared Lattice Vibrations and Free-Electron Dispersion in GaN," *Phys. Rev. B*, **7**, 743-750 (1973).
75. K. Vedam, J. L. Kirk and B. N. N. "Piezo- and Thermo-Optic Behavior of Spinel ($MgAl_2O_4$)," *J. Solid State Chem.*, **12**, 213-218 (1975).
76. K. Vedam and E. D. D. Schmidt, "Variation of Refractive Index of MgO with Pressure to 7 kbar," *Phys. Rev.*, **146**, 548-554 (1966).
77. K. Vedam and T. A. Davis, "Nonlinear Variation of the Refractive Indices of Quartz with Pressure," *J. Opt. Soc. Am.*, **57**, 1140-1145 (1967).
78. K. Vedam, E. D. D.Schmidt and R. Roy, "Nonlinear Variation of Refractive Index of Vitreous Silica with Pressure to 7 Kbars," *J. Am. Ceram. Soc.*, **49**, 531-535 (1966).
79. D. L. Camphausen, G. A. Neville Connel and W. Paul, "Calculation of Energy-Band Pressure Coefficients from the Dielectric Theory of the Chemical Bond," *Phys. Rev. Lett.*, **26**, 184-188 (1971).
80. Sassoon Advanced Materials FLIR Grade ZnS Data Sheet.
81. B. Bendow, P. D. Gianino, Y.-F. Tsay and S. S. Mitra, "Pressure and Stress Dependence of the Refractive Index of Transparent Crystals," *Appl. Opt.*, **13**, 2382-2396 (1982).
82. P. Perlin, Y. Gorczyca, N. E. Christensen, Y. Grzegory, H. Teisserye and T. Suski, "Pressure Studies of Gallium Nitride: Crystal Growth and Fundamental Electronic Properties," *Phys. Rev. B*, **45**, 13307-13313 (1992).
83. A. Feldman, D. Horowitz, R. Waxler and M. Dodge, "Optical Materials Characterization: Final Technical Report," National Bureau of Standards Technical Note 993, U. S. Government Printing Office, Washington DC (1979). The piezo-optic constants (q_{11} and q_{12}) from this reference were cited in C. A. Klein, "Compendium of Property Data for Raytran Zinc Selenide and Raytran Zinc Sulfide," Raytheon Co. Report RAY/RD/T-1154, 31 Aug 1987. The value of dn/dP was computed from the piezo-optic constants by the relation $dn/dP = (n^3/2)(q_{11} - 2q_{12})$, where n is the initial refractive index.
84. I. I. Afanas'ev, L. K. Andrianova, I. Ya. Mamontov and V. M. Reiterov, "Photoelastic Properties and Residual Stresses in Magnesium Fluoride Crystals," *Soviet Phys. Solid State*, **17**, 2006-2007 (1976).
85. M. E. Thomas, S. K. Andersson, R. M. Sova, and R. I. Joseph, "Frequency and Temperature Dependence of the Refractive Index of Sapphire," *Infrared Phys. & Technol.*, **39**, 235-249 (1998); M. E. Thomas, S. K. Andersson, R. M. Sova, and R. I. Joseph, "Frequency and Temperature Dependence of the Refractive Index of Sapphire," *Proc. SPIE*, **3060**, 258-269 (1997). For the o-ray of sapphire, there is a strong temperature dependence for dn/dT at 4 μm wavelength. At 300K, $dn/dT \approx 0$.

86. P. W. Wayland, M. E. Thomas, M. J. Linevsky, and R. I. Joseph, "Multiphonon Extraordinary Ray Absorption Coefficient for Sapphire," *Appl. Opt.*, submitted for publication (1999).
87. T. M. Hartnett, S. D. Bernstein, E. A. Maguire, and R. W. Tustison, "Optical Properties of ALON (Aluminum Oxynitride)," *Proc. SPIE*, **3060**, 284-295 (1997).
88. S. K. Andersson and M. E. Thomas, "Infrared Properties of CVD β-SiC," *Infrared Phys. & Technol.*, **39**, 223-234 (1998); S. K. Andersson and M. E. Thomas, "Infrared Properties of CVD β-SiC," *Proc. SPIE*, **3060**, 306-319 (1997).
89. A. R. Hilton, Sr., A. R. Hilton, Jr., J. McCord, and G. Whaley, "Production of Arsenic Trisulfide Glass," *Proc. SPIE*, **3060**, 335-343 (1997); A. R. Hilton and C. E. Jones, "The Thermal Change in the Nondispersive Refractive Index of Optical Materials," *Appl. Opt.*, **6**, 1513-1517 (1967).
90. W. S. Rodney, I. H. Malitson, and T. A. King, "Refractive Index of Arsenic Trisulfide," *J. Opt. Soc. Am.*, **48**, 633-636 (1958).
91. A. Z. Tropf, M. E. Thomas, and W. J. Tropf, "Optical Properties of KRS-5," *Proc. SPIE*, **3060**, 344-355 (1997).
92. W. S. Rodney and I. H. Malitson, "Refraction and Dispersion of Thallium Bromide Iodide," *J. Opt. Soc. Am.*, **46**, 956-961 (1958).
93. Estimated from transmission of 4.3 mm thick crystal. Unpublished data from L. Schowalter and I. Bhat (1998). The absorption coefficient at the peak absorption at 5.05 μm was 5.3 cm^{-1}.
94. E. D. Palik, ed., *Handbook of Optical Constants of Solids III*, Academic Press, San Diego, California (1998).
95. X. Tang, Y. Yuan, K. Wongchotigul, and M. G. Spencer, "Dispersion Properties of Aluminum Nitride as Measured by an Optical Waveguide Technique," *Appl. Phys. Lett.*, **70**, 3206-3208 (1997).
96. J. Fontanella, R. L. Johnston, J. H. Colwell and C. Andeen, "Temperature and Pressure Variation of the Refractive Index of Diamond," *Appl. Opt.*, **16**, 2949-2951 (1977).
97. J. Trombetta, Raytheon Systems Co., to be published.
98. A. R. Hilton, Sr., "Infrared Refractive Index Measurement Results for Single Crystal and Polycrystal Germanium," *Proc. SPIE*, **1498**, 128-137 (1991).
99. R. Gupta, J. H. Burnett, U. Griesmann and M. Walhout, "Absolute Refractive Indices and Thermal Coefficients of Fused Silica and Calcium Fluoride Near 193 nm," *Appl. Opt.*, **38**, 5964-5968 (1998).

Appendix D
DEFINITIONS FROM RADIOMETRY

Radiometry is the science of the measurement of electromagnetic radiation. Several quantities that appear in discussions of radiometry are defined below.

Radiant energy, Q, measured in joules (J), is the quantity of electromagnetic energy incident on or emerging from a surface.

Spectral radiant energy, Q_λ, measured in joules per nanometer (J/nm), is the radiant energy per unit wavelength interval at a specific wavelength. The total radiant energy in the wavelength interval from λ_1 to λ_2 is

$$Q = \int_{\lambda_1}^{\lambda_2} Q_\lambda \, d\lambda \, . \tag{D-1}$$

Radiant flux, Φ, also called *radiant power*, is the radiant energy per unit time passing through a surface. The units are watts (W), equal to joules per second. The radiant energy passing through the surface in the time interval from t_1 to t_2 is

$$Q = \int_{t_1}^{t_2} \Phi \, dt \, . \tag{D-2}$$

Spectral radiant flux, Φ_λ, also called *spectral radiant power*, is the radiant flux (power) per unit wavelength passing through a surface. The units are watts per nanometer (W/nm). The radiant flux in the wavelength interval from λ_1 to λ_2 passing through a surface is

$$\Phi = \int_{\lambda_1}^{\lambda_2} \Phi_\lambda \, d\lambda \tag{D-3}$$

and the spectral radiant energy passing through a surface in the time interval from t_1 to t_2 is

$$Q_\lambda = \int_{t_1}^{t_2} \Phi_\lambda \, dt \,. \tag{D-4}$$

Irradiance, E, is the radiant flux (power) per unit area (W/m^2) incident on or emerging from a surface. If an element of surface area is *da*, the flux incident on or emerging from the entire surface is the integral of irradiance over the entire area:

$$\Phi = \int E \, da \,. \tag{D-5}$$

The irradiance leaving the surface is also called the *exitance*, which was discussed under blackbody radiation in Section 0.2. Irradiance includes all of the radiation in a hemisphere around an infinitesimal patch of surface.

Spectral irradiance, E_λ, is the radiant flux (power) per unit area per unit wavelength (W/[m^2·nm]) at a specific wavelength incident on or emerging from a surface. The irradiance in the wavelength interval from λ_1 to λ_2 is

$$E = \int_{\lambda_1}^{\lambda_2} E_\lambda \, d\lambda \,. \tag{D-6}$$

The radiant flux incident on or emerging from the entire surface is the integral of spectral irradiance over wavelength and area:

$$\Phi = \iint E_\lambda \, d\lambda \, da \,. \tag{D-7}$$

Radiant intensity, I, is the radiant flux (power) per unit solid angle, Ω, passing through a point in a specific direction. The units are watts per steradian (W/sr). Radiant intensity usually refers to the radiant flux from a point source of electromagnetic radiation. The flux from the point source in a certain solid angle is

$$\Phi = \int I \, d\Omega \,. \tag{D-8}$$

Spectral radiant intensity, I_λ, is the radiant flux per unit solid angle per unit wavelength passing through a point in a specific direction. The units are watts per steradian per nanometer (W/[sr·nm]). The flux from the point source in a certain solid angle is the integral of spectral radiant intensity over solid angle and wavelength:

$$\Phi = \iint I_\lambda \, d\lambda \, d\Omega \,. \tag{D-9}$$

Radiance, L, is the radiant flux (power) per unit projected area and per unit solid angle incident on or emerging from a point on a surface. The units are watts per square meter per steradian (W/[m²·sr]). The flux emerging from the surface in Fig. D.1 within the solid angle Ω is

$$\Phi = \iint L \, da \, d\Omega = \iint L \, (da_0 \cos\theta) \, d\Omega \, . \tag{D-10}$$

Radiance is a function of position on the surface and the direction of the flux from the surface. If the radiance is independent of location on the surface, then radiance can be expressed as a function of just θ and ϕ in Fig. D.1. Irradiance integrates the radiance over a complete hemisphere (2π steradians) around a point on the surface:

$$E = \int_{2\pi} L(\theta,\phi) \cos\theta \, d\Omega \, . \tag{D-11}$$

The factor $\cos\theta$ comes from projecting the area da_0 into the area da in Fig. D.1.

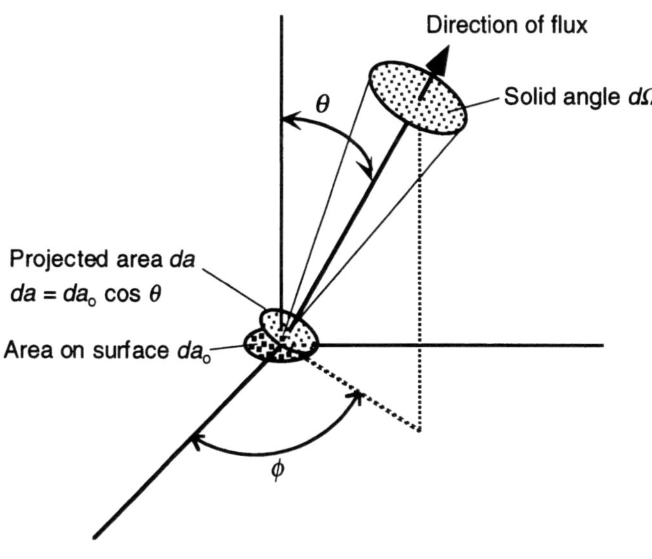

Fig. D.1. Radiance is the radiant flux per unit projected area per unit solid angle from the surface element *da*.

A *Lambertian radiator* is an ideal surface whose radiance is independent of direction. This turns out to be equivalent to saying that irradiance is proportional to $\cos\theta$ in Fig. D.1:

$$\text{Lambertian radiator:} \quad E(\theta) = E_\perp \cos\theta \tag{D-12}$$

where $E_\perp$ is the irradiance in the direction perpendicular to the surface. For a Lambertian radiator, the irradiance is equal to π times the radiance ($E = \pi L$) if L is a constant, independent of θ and ϕ.

Appendix E
ELASTIC CONSTANTS

Chapter 3 described the elastic behavior of isotropic materials — those whose properties are identical in every direction. Single-crystal materials and some polycrystalline materials with a preferential alignment of the crystallites in one direction are not isotropic. For example, the sapphire crystal in Fig. 1.6 has different physical properties, such as refractive index or elastic modulus, along the *c* and *a* axes. This Appendix tells how we describe the mechanical properties of anisotropic materials such as sapphire.

Figure E.1 shows the notation for stresses (σ) acting on a general volume element within a substance. The second subscript is *x* for all stresses acting on face *x* (normal to the *x* direction) at the left side of the cube. Similarly, stresses acting on face *y* end in the subscript *y* and stresses acting on face *z* end in the subscript *z*. The first subscript on each stress gives the direction of the force. Thus stress σ_{xz} acts in direction *z* on face *x*. The stresses σ_{xx}, σ_{yy}, and σ_{zz} are tensile or compressive. Shear stresses have two different subscripts, such as σ_{xy} and σ_{zx}. To prevent rotation of the volume element, the following relations must be true: $\sigma_{xy} = \sigma_{yx}$, $\sigma_{xz} = \sigma_{zx}$ and $\sigma_{yz} = \sigma_{zy}$. Therefore, there are only six distinct components of stress: three of the type σ_{ii} and three of the type σ_{ij}.

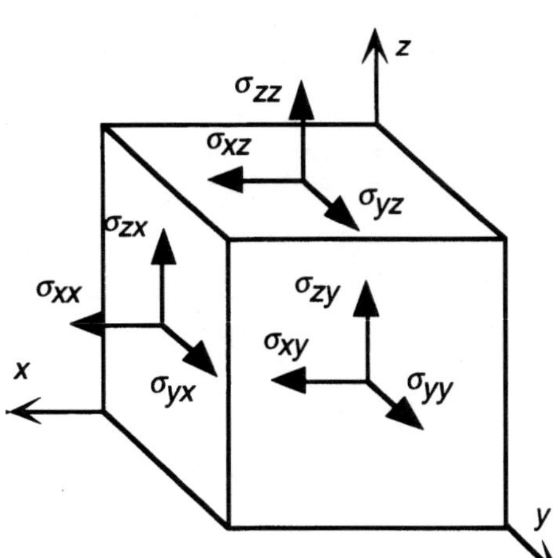

Fig. E.1. Notation for stresses, acting on a general volume element. σ_{ii} are tension or compression and σ_{ij} are shear stresses.

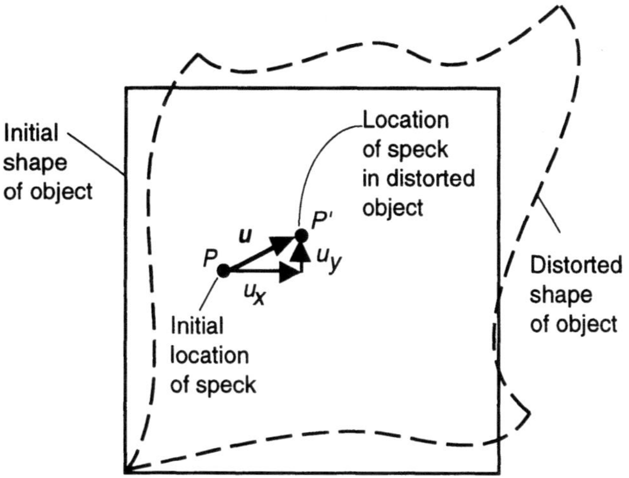

Fig. E.2. Distortion of square moves a speck of dust embedded in the square from point P to point P'.

Elastic constants relate the applied stresses to the resultant strains in a material. For a general definition of strain in a 3-dimensional object, first consider the 2-dimensional object in Fig. E.2. Suppose that the initially square object has a speck of dust embedded in it at point P. After applying stresses to distort the object, the speck of dust has moved to P'. The vector connecting P to P' is $\boldsymbol{u}$ with components u_x and u_y. If the object were 3-dimensional, there would also be a component u_z.

In general, each point P in the undistorted object can have a different displacement vector $\boldsymbol{u}$ when the object is distorted. The strain in the x direction, ε_{xx}, at the point P is defined as the change in u_x with respect to x, evaluated at the point P:

$$\text{Strain in } x \text{ direction} \equiv \varepsilon_{xx} = \frac{\partial u_x}{\partial x} . \tag{E-1}$$

Similarly, the strains in the y and z directions are $e_{yy} = \partial u_y/\partial y$ and $\varepsilon_{xx} = \partial u_z/\partial z$.

Shear strains are slightly more complicated. Recall in Fig. E.1 that shear is generated, for example, by a force in the x direction acting parallel to a plane that is normal to the y direction. There are three distinct shear strains defined by the equation

$$\varepsilon_{xy} = \varepsilon_{yx} = \frac{1}{2}\left(\frac{\partial u_x}{\partial y} + \frac{\partial u_y}{\partial x}\right) \qquad \varepsilon_{xz} = \varepsilon_{zx} = \frac{1}{2}\left(\frac{\partial u_x}{\partial z} + \frac{\partial u_z}{\partial x}\right)$$

$$\varepsilon_{yz} = \varepsilon_{zy} = \frac{1}{2}\left(\frac{\partial u_y}{\partial z} + \frac{\partial u_z}{\partial y}\right) . \tag{E-2}$$

So just as with stress, there are only 6 distinct components of strain: three of the form ε_{ii} and three of the form ε_{ij}.

When an object such as the dashed square in Fig. E.3 is sheared, the engineering shear strain, γ, is usually defined as the sum of the two shear angles: $\gamma = \phi_1 + \phi_2$. This definition makes γ twice as big as ε_{xy} defined in Eq. E-2.

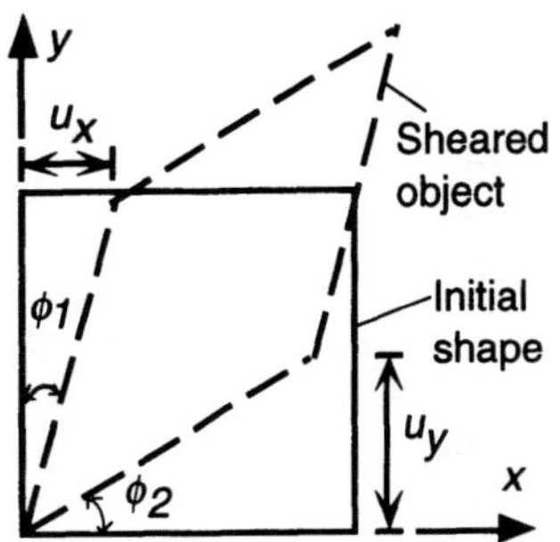

Fig. E.3. Engineering shear strain is defined as $\gamma = \phi_1 + \phi_2$.

For linear elastic behavior, which is an idealization for small strains in optical ceramics, Hooke's law tells us that stress is proportional to strain. For an isotropic material, we wrote Eq. 3-3: stress (σ) = Young's modulus × strain = $E\varepsilon$.

For an anisotropic material, we write an equation analogous to Eq. 3-3 to relate the six components of stress to the six components of strain:

$$\begin{pmatrix} \sigma_{xx} \\ \sigma_{yy} \\ \sigma_{zz} \\ \sigma_{xy} \\ \sigma_{xz} \\ \sigma_{yz} \end{pmatrix} = \begin{pmatrix} c_{11} & c_{12} & c_{13} & c_{14} & c_{15} & c_{16} \\ c_{12} & c_{22} & c_{23} & c_{24} & c_{25} & c_{26} \\ c_{13} & c_{23} & c_{33} & c_{34} & c_{35} & c_{36} \\ c_{14} & c_{24} & c_{34} & c_{44} & c_{45} & c_{46} \\ c_{15} & c_{25} & c_{35} & c_{45} & c_{55} & c_{56} \\ c_{16} & c_{26} & c_{36} & c_{46} & c_{56} & c_{66} \end{pmatrix} \begin{pmatrix} \varepsilon_{xx} \\ \varepsilon_{yy} \\ \varepsilon_{zz} \\ 2\varepsilon_{xy} \\ 2\varepsilon_{xz} \\ 2\varepsilon_{yz} \end{pmatrix} \quad \text{(E-3)}$$

The coefficients c_{ij} are called *stiffness* coefficients. It turns out that the stiffness matrix must be symmetric, so there are a maximum of 21 distinct stiffness coefficients, not 36. Equation (E-3) describes a fully anisotropic material — one that has no symmetry at all. The shear strains in Eq. (E-3) are written with a factor of two to make them consistent with the engineering strain.

We can also write that the strains are proportional to the stresses:

$$\begin{pmatrix} \varepsilon_{xx} \\ \varepsilon_{yy} \\ \varepsilon_{zz} \\ 2\varepsilon_{xy} \\ 2\varepsilon_{xz} \\ 2\varepsilon_{yz} \end{pmatrix} = \begin{pmatrix} s_{11} & s_{12} & s_{13} & s_{14} & s_{15} & s_{16} \\ s_{12} & s_{22} & s_{23} & s_{24} & s_{25} & s_{26} \\ s_{13} & s_{23} & s_{33} & s_{34} & s_{35} & s_{36} \\ s_{14} & s_{24} & s_{34} & s_{44} & s_{45} & s_{46} \\ s_{15} & s_{25} & s_{35} & s_{45} & s_{55} & s_{56} \\ s_{16} & s_{26} & s_{36} & s_{46} & s_{56} & s_{66} \end{pmatrix} \begin{pmatrix} \sigma_{xx} \\ \sigma_{yy} \\ \sigma_{zz} \\ \sigma_{xy} \\ \sigma_{xz} \\ \sigma_{yz} \end{pmatrix} \quad \text{(E-4)}$$

The coefficients s_{ij} are called *compliance* coefficients. The matrix of compliance coefficients is the inverse of the matrix of stiffness coefficients. We will show the form of the stiffness matrix for three cases of high symmetry: cubic, trigonal, and isotropic.

Cubic materials: For a cubic material, such as diamond or sodium chloride, symmetry dictates that most of the coefficients are zero and there are only 3 nonzero coefficients. The stiffness and compliance matrices have the following forms:

$$\begin{pmatrix} c_{11} & c_{12} & c_{12} & 0 & 0 & 0 \\ c_{12} & c_{11} & c_{12} & 0 & 0 & 0 \\ c_{12} & c_{12} & c_{11} & 0 & 0 & 0 \\ 0 & 0 & 0 & c_{44} & 0 & 0 \\ 0 & 0 & 0 & 0 & c_{44} & 0 \\ 0 & 0 & 0 & 0 & 0 & c_{44} \end{pmatrix} \qquad \begin{pmatrix} s_{11} & s_{12} & s_{12} & 0 & 0 & 0 \\ s_{12} & s_{11} & s_{12} & 0 & 0 & 0 \\ s_{12} & s_{12} & s_{11} & 0 & 0 & 0 \\ 0 & 0 & 0 & s_{44} & 0 & 0 \\ 0 & 0 & 0 & 0 & s_{44} & 0 \\ 0 & 0 & 0 & 0 & 0 & s_{44} \end{pmatrix}$$

Stiffness matrix for a cubic material such as diamond — Compliance matrix for a cubic material such as diamond

Stiffness and compliance constants for several cubic infrared optical materials are listed in Table E.1.

Let's use the compliance matrix for the cubic crystal, diamond, to compute the strains when a stress of σ_{xx} = 100 MPa is applied along the x axis only.

$$\begin{pmatrix} \varepsilon_{xx} \\ \varepsilon_{yy} \\ \varepsilon_{zz} \\ 2\varepsilon_{xy} \\ 2\varepsilon_{xz} \\ 2\varepsilon_{yz} \end{pmatrix} = \begin{pmatrix} s_{11} & s_{12} & s_{12} & 0 & 0 & 0 \\ s_{12} & s_{11} & s_{12} & 0 & 0 & 0 \\ s_{12} & s_{12} & s_{11} & 0 & 0 & 0 \\ 0 & 0 & 0 & s_{44} & 0 & 0 \\ 0 & 0 & 0 & 0 & s_{44} & 0 \\ 0 & 0 & 0 & 0 & 0 & s_{44} \end{pmatrix} \begin{pmatrix} 100 \times 10^6 \text{ Pa} \\ 0 \\ 0 \\ 0 \\ 0 \\ 0 \end{pmatrix} \tag{E-5}$$

where $s_{11} = 0.9524 \times 10^{-12}$ Pa^{-1}, $s_{12} = -0.0991 \times 10^{-12}$ Pa^{-1} and $s_{12} = -0.0991 \times 10^{-12}$ Pa^{-1} from Table E.1. The results are

$\varepsilon_{xx} = 9.524 \times 10^{-5}$ $\quad \varepsilon_{yy} = -0.991 \times 10^{-5}$ $\quad \varepsilon_{zz} = -0.991 \times 10^{-5}$
$\varepsilon_{xy} = 0$ $\quad \varepsilon_{xz} = 0$ $\quad \varepsilon_{yz} = 0$

The sensible results say that pulling along the x axis extends the crystal in the x direction and shrinks the crystal in the y and z directions. No shear strains are created.

Trigonal materials: A crystal such as sapphire, has a 3-fold axis of symmetry (Fig. 1.6). The stiffness and compliance matrices have 6 distinct elastic constants:

$$\begin{pmatrix} c_{11} & c_{12} & c_{13} & c_{14} & 0 & 0 \\ c_{12} & c_{11} & c_{13} & -c_{14} & 0 & 0 \\ c_{13} & c_{13} & c_{33} & 0 & 0 & 0 \\ c_{14} & -c_{14} & 0 & c_{44} & 0 & 0 \\ 0 & 0 & 0 & 0 & c_{44} & c_{14} \\ 0 & 0 & 0 & 0 & c_{14} & {}^{1}/{}_{2}(c_{11}-c_{12}) \end{pmatrix}$$

Stiffness matrix for a trigonal material such as sapphire

$$\begin{pmatrix} s_{11} & s_{12} & s_{13} & s_{14} & 0 & 0 \\ s_{12} & s_{11} & s_{13} & -s_{14} & 0 & 0 \\ s_{13} & s_{13} & s_{33} & 0 & 0 & 0 \\ s_{14} & -s_{14} & 0 & s_{44} & 0 & 0 \\ 0 & 0 & 0 & 0 & s_{44} & 2s_{14} \\ 0 & 0 & 0 & 0 & 2s_{14} & 2(s_{11}-s_{12}) \end{pmatrix}$$

Compliance matrix for a trigonal material such as sapphire

Elastic constants for several trigonal infrared optical materials are listed in Table E.2.

Table E.1. Elastic constants of some cubic infrared materials near 20°C*

	Stiffness (GPa)			Compliance (10^{-12} Pa^{-1})		
Crystal	c_{11}	c_{12}	c_{44}	s_{11}	s_{12}	s_{44}
ALON (aluminum oxynitride)[a]	~393	~108	~119	~2.89	~-0.62	~8.40
Barium fluoride (BaF_2)[b]	90.7	41.0	25.3	15.2	-4.7	39.6
Boron phosphide (BP)[c]	315	100	160	3.75	-0.90	6.25
Cadmium telluride (CdTe)[e]	53.8	37.4	20.18	43.24	-17.73	49.55
Calcium fluoride (CaF_2)[b]	165	46	33.9	6.94	-1.53	29.5
Calcium lanthanum sulfide[d]	~98	~47	~50	~15	~-5	~20
Cesium iodide (CsI)[b]	24.5	6.6	6.3	46.1	-9.7	158
Diamond (C)[f]	1076	125	577	0.9524	-0.0991	1.733
Gallium arsenide (GaAs)[b]	118	53.5	59.4	11.75	-3.66	16.8
Gallium phosphide (GaP)[b]	142	63	71.6	9.60	-2.93	14.0
Germanium (Ge)[b]	129	48	67.1	9.73	-2.64	14.9
KRS-5 ($TlBr_{0.5}I_{0.5}$)[b]	34.1	13.6	5.79	38.0	-10.8	173
Lithium fluoride (LiF)[b]	112	46	63.5	11.6	-3.35	15.8
Magnesium oxide (MgO)[g]	297.8	95.1	155.8	3.97	-0.96	6.42
Potassium bromide (KBr)[b]	34.5	5.5	5.10	30.3	-4.2	196
Silicon (Si)[b]	165	64	79.2	7.74	-2.16	12.6
Silicon carbide (β-SiC)[h]	350	142	256	3.18	-0.85	3.91
Sodium chloride (NaCl)[b]	49.1	12.8	12.8	22.9	-4.8	78.3
Spinel ($MgAl_2O_4$)[g]	282.9	155.4	154.8	5.79	-2.05	6.49
Yttria (Y_2O_3)[a]	~233	~101	~67	~5.82	~-1.76	~14.93
Zinc selenide (ZnSe)[i]	81.0	48.8	44.0	26.3	-9.89	22.7
Zinc sulfide (ZnS)[i]	104.6	65.3	46.1	18.4	-7.07	21.7

*From W. J. Tropf, M. E. Thomas and T. J. Harris, "Properties of Crystals and Glasses," in *Handbook of Optics* 2nd ed. (E. van Stryland, D. Williams and W. L. Wolfe, eds.), Vol. II, McGraw-Hill, New York (1995).

a. Properties *estimated* from engineering moduli by Tropf *et al.*

b. K.-H. Hellwege & A. M. Hellwege (eds.), *Landolt-Börnstein Numerical Data & Functional Relationships in Science & Technol. New Series; Group III: Crystal & Solid State Physics; Vol. 11: Elastic, Piezoelectric, Pyroelectric, Piezooptic, Electro-optic Constants & Nonlinear Susceptibilities of Crystals*, Springer-Verlag, Berlin (1979).

c. W. Wettling and J. Windscheif, "Elastic Constants and Refractive Index of Boron Phosphide," *Solid State Commun.*, **50**, 33-34 (1984).

d. *Estimates* based on data from M. E. Hills, *Preparation, Properties, and Development of Calcium Lanthanum Sulfide as an 8- to 12-Micrometer Transmitting Ceramic*, Naval Weapons Center Report TP 7073, China Lake, California, September 1989.

e. R. D. Greenough and S. B. Palmer "The Elastic Constants and Thermal Expansion of Single-Crystal CdTe," *J. Phys. D*, **6**, 587-592 (1973).

f. M. Grimsditch and A. Ramdas, "Brillouin Scattering in Diamond," *Phys. Rev. B*, **11**, 3139-3148 (1975).

g. A. Yoneda, "Pressure Derivatives of Elastic Constants of Single Crystal MgO and $MgAl_2O_4$," *J. Phys. Earth*, **38**, 19-55 (1990).

h. W. R. L. Lambrecht, B. Segall, M. Methfessel and M. van Schilfgaard, "Calculated Elastic Constants and Deformation Potentials of Cubic SiC, " *Phys. Rev. B*, **44**, 3685-3694 (1991).

i. D. Berlincourt, H. Jaffe and L. Shiozawa, "Electroelastic Properties of the Sulfides, Selenides, and Tellurides of Zinc and Cadmium," *Phys. Rev.*, **129**, 1009-1017 (1963).

Table E.2. Stiffness and compliance coefficients of sapphire at different temperatures*

ij:	*11*	*33*	*44*	*12*	*13*	*14*
296K:						
Stiffness, c_{ij} (GPa)	497.3	500.9	146.8	162.8	116.0	-21.90
Compliance, s_{ij} (10^{-12} Pa^{-1})	2.349	2.173	6.948	-0.7000	-0.3819	0.4549
600K:						
Stiffness, c_{ij} (GPa)	486.0	489.2	139.2	163.1	113.0	-23.26
Compliance, s_{ij} (10^{-12} Pa^{-1})	2.424	2.223	7.361	-0.7490	-0.3870	0.5303
900K:						
Stiffness, c_{ij} (GPa)	472.3	476.0	131.2	162.4	109.6	-23.92
Compliance, s_{ij} (10^{-12} Pa^{-1})	2.516	2.282	7.843	-0.8043	-0.3941	0.6053
1200K:						
Stiffness, c_{ij} (GPa)	457.3	461.1	123.2	160.7	105.4	-24.32
Compliance, s_{ij} (10^{-12} Pa^{-1})	2.620	2.352	8.388	-0.8647	-0.4012	0.6878

*From T. Goto, O. L. Anderson, I. Ohno, and S. Yamamoto, "Elastic Constants of Corundum up to 1825 K," *J. Geophys. Res.*, **94** [B6], 7588-7602 (1989). See also W. E. Tefft, "Elastic Constants of Synthetic Single Crystal Corundum," *J. Res. National Bureau of Standards*, **70A**, 277-280 (1966) and J. B. Wachtman, Jr., W. E. Tefft, D. G. Lam, Jr., and R. P. Stinchfield, "Elastic Constants of Synthetic Single Crystal Corundum at Room Temperature," *J. Res. National Bureau of Standards*, **64A**, 213-228 (1960).

Now let's compute the strains in sapphire when a stress of σ_{zz} = 100 MPa is applied along the z axis, which is the 3-fold axis of symmetry in Fig. 1.6.

$$\begin{pmatrix} \varepsilon_{xx} \\ \varepsilon_{yy} \\ \varepsilon_{zz} \\ 2\varepsilon_{xy} \\ 2\varepsilon_{xz} \\ 2\varepsilon_{yz} \end{pmatrix} = \begin{pmatrix} s_{11} & s_{12} & s_{13} & s_{14} & 0 & 0 \\ s_{12} & s_{11} & s_{13} & -s_{14} & 0 & 0 \\ s_{13} & s_{13} & s_{33} & 0 & 0 & 0 \\ s_{14} & -s_{14} & 0 & s_{44} & 0 & 0 \\ 0 & 0 & 0 & 0 & s_{44} & 2s_{14} \\ 0 & 0 & 0 & 0 & 2s_{14} & 2(s_{11}-s_{12}) \end{pmatrix} \begin{pmatrix} 0 \\ 0 \\ 100 \times 10^6 \text{ Pa} \\ 0 \\ 0 \\ 0 \end{pmatrix} \quad \text{(E-6)}$$

where the compliance constants at room temperature are given by the second row in Table E.2. The results are

$\varepsilon_{xx} = -3.819 \times 10^{-5}$ $\varepsilon_{yy} = -3.819 \times 10^{-5}$ $\varepsilon_{zz} = 21.73 \times 10^{-5}$

$\varepsilon_{xy} = 0$ $\varepsilon_{xz} = 0$ $\varepsilon_{yz} = 0$

Pulling along the z *axis* extends the crystal in the z direction and shrinks the crystal in the x and y directions by equal amounts because of symmetry in the xy plane. If the same 100 MPa of tension were applied along the x axis, the strains would be

$\varepsilon_{xx} = 23.49 \times 10^{-5}$ $\varepsilon_{yy} = -7.000 \times 10^{-5}$ $\varepsilon_{zz} = -3.819 \times 10^{-5}$

The strains in the y and z directions are not equal because these two axes are not equivalent in the crystal.

Table E.3. Elastic constants of isotropic infrared materials near 20°C*

Material	Young's modulus (GPa)	Poisson's ratio
AMTIR-1 glass[a]	21.9	0.266
BK7 glass[a]	81	0.208
BS39B glass[a]	104	0.29
Corning 9754[a]	84.1	0.290
Fused silica (SiO_2)[a]	72.6	0.164
IRG-11 glass[a]	107.5	0.284
ALON (aluminum oxynitride)[b]	323	0.24
Diamond (chemical vapor deposited)[c]	1143	0.069
Lanthana-doped yttria ($0.09La_2O_3 \cdot 0.91Y_2O_3$)[d]	170	0.30
Magnesium fluoride (hot pressed MgF_2)[e]	142	0.271
Spinel ($MgAl_2O_4$)[f,g]	275	0.26
Yttria (Y_2O_3)	173[d], 181.1[h]	0.299[h]
Zinc selenide (ZnSe)[i]	70	0.28
Zinc sulfide (ZnS)[i]	74	0.29
Zinc sulfide (multispectral ZnS)[i]	88	0.32

*See Table 1.1 for glass compositions.

a. W. J. Tropf, M. E. Thomas and T. J. Harris, "Properties of Crystals and Glasses," in *Handbook of Optics* 2nd ed. (E. vanStryland, D. Williams and W. L. Wolfe, eds.), Vol. II, McGraw-Hill, New York (1995).

b. E. A. Maguire, J. K. Rawson and R. W. Tustison, "Aluminum Oxynitride's Resistance to Impact and Erosion," *Proc. SPIE*, **2286**, 26-32 (1994).

c. C. A. Klein and G. F. Cardinale, "Young's Modulus and Poisson's Ratio of CVD Diamond," *Diamond and Related Mater.*, **2**, 918-923 (1993). Calculated for randomly oriented grains. Most chemical vapor deposited diamond has a preferential orientation, so the elastic properties will be anisotropic.

d. W. J. Tropf and D. C. Harris, "Mechanical, Thermal, and Optical Properties of Yttria and Lanthana-Doped Yttria," *Proc. SPIE*, **1112**, 9-19 (1989).

e. D. M. Bailey, F. W. Calderwood, J. D. Greiner, O. Hunter, Jr., J. F. Smith and R. J. Schiltz, "Reproducibilities of Some Physical Properties of MgF_2," *J. Am. Ceram. Soc.*, **58**, 489-492 (1975).

f. T. M. Hartnett and R. L. Gentilman, "Optical and Mechanical Properties of Highly Transparent Spinel and ALON Domes," *Proc. SPIE*, **505**, 15-22 (1984).

g. D. W. Roy and G. G. Martin, Jr., "Advances in Spinel Optical Quality, Size/Shape Capability and Applications," *Proc. SPIE*, **2286**, 213 (1992).

h. O. Yeheskel and O. Tevet, "Elastic Moduli of Transparent Yttria," *J. Am. Ceram. Soc.*, **82**, 136-144 (1999).

i. C. A. Klein and C. Willingham, "Elastic Properties of Chemically Vapor-Deposited ZnS and ZnSe," in *Basic Properties of Optical Materials*, pp. 137-140, National Bureau of Standards Special Publication 697, Washington D.C. (1985).

Isotropic materials: The physical properties of an isotropic material are the same in every direction. The compliance (and stiffness) matrix have the same form as for a cubic material, except that

$$s_{12} = s_{11} - \frac{1}{2} s_{44} . \tag{E-7}$$

That is, there are only 2 independent elastic constants, not 3. If you know s_{11} and s_{14}, you can compute s_{12}. The compliance matrix takes the following form:

$$\begin{pmatrix} 1/E & -\nu/E & -\nu/E & 0 & 0 & 0 \\ -\nu/E & 1/E & -\nu/E & 0 & 0 & 0 \\ -\nu/E & -\nu/E & 1/E & 0 & 0 & 0 \\ 0 & 0 & 0 & 1/\mu & 0 & 0 \\ 0 & 0 & 0 & 0 & 1/\mu & 0 \\ 0 & 0 & 0 & 0 & 0 & 1/\mu \end{pmatrix}$$

Compliance matrix for an isotropic material

where E is Young's modulus, ν is Poisson's ratio, and μ is the shear modulus in Eq. (3-6). Only two of the constants are independent. For example, we can express the shear modulus in terms of Young's modulus and Poisson's ratio:

$$\mu = \frac{E}{2 + 2\nu} . \tag{E-8}$$

For completeness, the bulk modulus, K, in Eq. (3-4) is related to Young's modulus and Poisson's ratio by the equation

$$K = \frac{E}{3 - 6\nu} . \tag{E-9}$$

Table E.3 gives elastic constants for some polycrystalline materials, which are nearly isotropic, and glasses which are isotropic.

Appendix F
THE WEIBULL DISTRIBUTION

The Weibull model for ceramic failure is based on the idea that a ceramic breaks when its weakest element fails. In addition to this weakest link assumption, Weibull had to assume a form for the probability of failure function. We now explore mathematical aspects of the Weibull distribution for the probability of ceramic failure.[1]

F.1 Weibull probability distribution

To describe the probability of failure, we can conceptually divide the ceramic into many small elements, as in a finite element analysis. If failure originates at the surface, then the elements would each include a portion of the surface. If the part fails within the bulk, then we divide the part into many tiny volume elements. For the sake of discussion, let's consider a ceramic object divided into n tiny volume elements of volume δV such that the total volume is $V = n\,\delta V$.

Let P_{fi} be the probability of failure of the ith volume element due to stress in that element. For simplicity, suppose that the distribution of flaws in the ceramic component is uniform, which means that all volume elements have equal probability of failure if the stress is uniform, as in the cylinder in Fig. 3.3. The probability of failure plus the probability of survival, P_{si}, of each element must be unity, so the probability of survival is $1 - P_{fi}$. The probability of survival of n elements is the product of the probability of survival of each element:

$$\text{Total probability of survival} = P_s = (P_{si})^n = (1 - P_{fi})^n . \tag{F-1}$$

That is, if the probability of survival of each element is 0.999 and if there are 1000 elements, then the probability of survival of the entire component is $(0.999)^{1000}$ = 0.368. The component has a 36.8% probability of survival and a 63.2% probability of failure.

Now we play some tricks. Let's multiply and divide the term P_{fi} by V and then rewrite V in the denominator as $V = n\,\delta V$:

$$P_s = (1 - P_{fi})^n = (1 - \frac{VP_{fi}}{V})^n = (1 - \frac{VP_{fi}}{n\,\delta V})^n . \tag{F-2}$$

The quotient $P_{fi}/\delta V$ is the probability of failure per unit volume. As the element size gets smaller, the term $P_{fi}/\delta V$ ought to approach a limit that we will designate ϕ:

$$\lim_{n \to \infty} P_s = \left(1 - \frac{V\phi}{n}\right)^n . \quad \text{(F-3)}$$

Now you might remember from your childhood that

$$\lim_{n \to \infty} \left(1 - \frac{x}{n}\right)^n = e^{-x} \quad \text{(F-4)}$$

where e is the base of the natural logarithm. So the probability of survival of the entire component in the limit as it is divided into smaller and smaller pieces is just

$$P_s = e^{-V\phi} . \quad \text{(F-5)}$$

Equation (F-5) is the result of assuming that the component fails when its weakest link fails.

At this point, Weibull needed to guess some form for the function ϕ, which is the probability of failure per unit volume. Certainly ϕ is a function of the stress, σ, in each element. For agreement with many types of experimental data, Weibull *assumed* that ϕ has the form

3-parameter Weibull function: $$\phi = \left(\frac{\sigma - \sigma_u}{\sigma_o}\right)^m \quad (\text{for } \sigma \geq \sigma_u) \quad \text{(F-6)}$$

where σ_o is called the *Weibull scaling factor*, σ_u is the *critical stress* below which the failure probability is zero, and m is called the *Weibull modulus*. For $\sigma < \sigma_u$, the probability of failure of the element is taken as zero. The values of σ_o, σ_u and m are chosen to fit a particular set of experimental data. Frequently we use a two-parameter probability function instead of the three-parameter function.

2-parameter Weibull function: $$\phi = \left(\frac{\sigma}{\sigma_o}\right)^m . \quad \text{(F-7)}$$

The 2-parameter function is more conservative than the 3-parameter function in that the 2-parameter function presumes that there is always a finite probability of failure — even with very low stress.

To summarize the important result so far, the probability of survival of the entire component is $P_s = e^{-V\phi}$, where ϕ could be given by the 3- or 2-parameter Weibull functions (F-6) or (F-7). For simplicity of writing equations, and because it is so common, we will only write the 2-parameter function in most of the following discussion.

Weibull probability of survival for uniform stress: $$P_s = e^{-V\phi} = e^{-V(\sigma/\sigma_o)^m} . \quad \text{(F-8)}$$

Now comes the subtle issue of dimensions. The exponent in Eq. (F-8) must be dimensionless. We could give σ_o dimensions that cancel the volume, but it is easier and more common to express the volume as V/V_o, where V_o is some chosen unit volume such as 1 cm^3 or 1 $inch^3$. In many Weibull analyses, the factor V in the exponent of Eq. (F-8) is ignored. Ignoring V is equivalent to taking the volume of the test specimen as the unit volume. When the time comes to extrapolate the probability of failure to specimens of other sizes, the volume of the test specimen must be explicitly considered.

So far we have assumed that the stress is uniform in the test specimen. If stress is not uniform, then the Weibull probability of survival should be expressed as an integral over volume. For the 2-parameter function (F-7), this integral has the form

Weibull probability of survival for nonuniform stress: $$P_s = e^{-\int (\sigma/\sigma_o)^m dV} . \qquad \text{(F-9)}$$

For the 3-parameter function, the integrand in Eq. (F-9) would be given by Eq. (F-6).

F.2 Effective volume or area

Equation (F-9) can be cast into a convenient form for scaling probabilities of failure for specimens of different size. First we multiply and divide the term σ/σ_o by $\sigma_{max}/\sigma_{max}$, where σ_{max} is the maximum stress anywhere in the specimen.

$$P_s = e^{-\int (\sigma_{max}\sigma/\sigma_{max}\sigma_o)^m dV} = e^{-(\sigma_{max}/\sigma_o)^m \int (\sigma/\sigma_{max})^m dV} . \qquad \text{(F-10)}$$

Defining the dimensionless number k as

$$k = \frac{1}{V} \int \left(\frac{\sigma}{\sigma_{max}} \right)^m dV \qquad \text{(F-11)}$$

we recast Eq. (F-10) as

$$P_s = e^{-kV(\sigma_{max}/\sigma_o)^m} . \qquad \text{(F-12)}$$

The product kV is called the *effective volume*. If we were testing a specimen that is under the uniform tension σ_{max}, then $k = 1$ and kV is the actual volume. The effective volume can be thought of as the volume of a specimen if it were under the uniform tension σ_{max}. If we were concerned with surface failure instead of volume failure, we would substitute area, A, for volume in Eqns. (F-11) and (F-12) and the product kA would be called the *effective area*.

F.3 Weibull equations for different kinds of test specimens[2]

Figure F.1 shows rectangular bars tested in pure tension, 4-point bending and 3-point bending. The height (d), width (b) and outer load span (L) are the same for all three bars. If all three bars were fabricated from identical material, the strengths measured in the three different tests would not be equal because the stress distribution is different in each test. We can use Eqns. (F-11) and (F-12) to compare the expected strength in each kind of test.

First consider the tensile test at the top left of Fig. F.1. The stress everywhere in the bar is equal. The fraction σ/σ_{max} in Eq. (F-11) is unity because $\sigma = \sigma_{max} = \sigma_{tension}$.

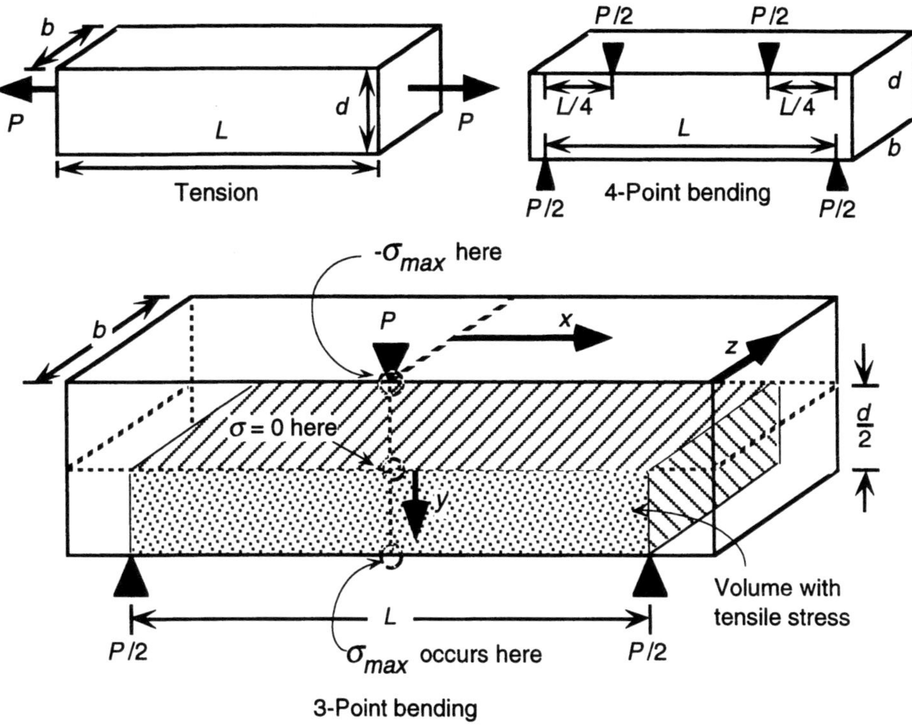

Fig. F.1. Geometries of rectangular bars for tension, 4-point bending and 3-point bending strength measurements. The lower bar is drawn at twice the scale as the upper bars to show the details of the 3-point bending test. *P* is the applied force.

Therefore $k = 1$ in Eqns. (F-11) and (F-12). The probability of survival, P_S, is given by the first entry in Table F.1. The mean strength is calculated by evaluating the integral

$$\bar{\sigma} \equiv \int_0^\infty \sigma \left(\frac{dP_S}{d\sigma}\right) d\sigma = \frac{\sigma_o}{(kV)^{1/m}}\, \Gamma(1 + \frac{1}{m}) \qquad \text{(F-13)}$$

whose value is also shown in Table F.1. The mean strength depends on the effective volume, kV, and a numerical factor, Γ (a function of the Weibull modulus), that is close to unity and tabulated in Table F.2.

For a series of identical test specimens of any geometry, the factor kV in Eq. (F-13) is unity, because all specimens have the same volume and stress state. Therefore, the mean strength is predicted to be $\bar{\sigma} = \sigma_o \Gamma$, where Γ is taken from Table F.2 for the measured Weibull modulus. For example, in Table 3.1, 13 zinc sulfide disks have a mean strength of 92.8 MPa and a Weibull scaling factor of 100.6 MPa. With a measured Weibull modulus of 5.43, the interpolated value of Γ in Table F.2 is 0.9225. The predicted mean strength is $\bar{\sigma} = \sigma_o \Gamma$ = (100.6 MPa)(0.9225) = 92.8 MPa, which happens in this fortuitous example to be equal to the observed mean strength.

Table F.1. Weibull equations for volume scaling of tensile and flexure tests*

Tensile test (Fig. F.1, upper left)

$P_s = e^{-kV(\sigma/\sigma_o)^m}$ $k = 1$ $\sigma = \dfrac{P}{db}$ $V \equiv Ldb$

Mean strength: $\bar{\sigma} = \dfrac{\sigma_o}{(kV)^{1/m}}\Gamma(1+\dfrac{1}{m})$

The expression for mean strength is identical for all tests in this table

4-Point flexure test (Fig. F.1, upper right, inner load span = L/2)

$P_s = e^{-kV(\sigma_{max}/\sigma_o)^m}$ $\sigma_{max} = \dfrac{3PL}{4d^2b}$ $V \equiv Ldb$

$k = \dfrac{(m+2)}{4(m+1)^2}$ if gauge section is entire region between outer load points

$k = \dfrac{1}{4(m+1)}$ if gauge section is region between inner load points only

4-Point flexure test (inner load span = L/4)

$P_s = e^{-kV(\sigma_{max}/\sigma_o)^m}$ $\sigma_{max} = \dfrac{9PL}{8d^2b}$ $V \equiv Ldb$

$k = \dfrac{(m+4)}{8(m+1)^2}$ if gauge section is entire region between outer load points

$k = \dfrac{1}{8(m+1)}$ if gauge section is region between inner load points only

3-Point flexure test (Fig. F.1, bottom)

$P_s = e^{-kV(\sigma_{max}/\sigma_o)^m}$ $\sigma_{max} = \dfrac{3PL}{2d^2b}$ $V \equiv Ldb$

$k = \dfrac{1}{2(m+1)^2}$ (gauge section is the region between outer load points)

*For the same material, the Weibull parameters m and σ_o should be the same for all three types of test specimens. The function $\Gamma(z)$ is equal to $(z-1)!$ (factorial) if z is an integer. This function is listed in Table F.2 for a range of values of Weibull modulus.

Table F.2. Gamma function: $\Gamma(z) = \int_0^\infty t^{z-1}e^{-t}dt$

m	$\Gamma(1+\frac{1}{m})$	m	$\Gamma(1+\frac{1}{m})$	m	$\Gamma(1+\frac{1}{m})$	m	$\Gamma(1+\frac{1}{m})$
3	0.8930	6.5	0.9318	10	0.9514	17	0.9693
3.5	0.8997	7	0.9354	11	0.9551	18	0.9708
4	0.9064	7.5	0.9387	12	0.9583	19	0.9722
4.5	0.9126	8	0.9417	13	0.9611	20	0.9735
5	0.9182	8.5	0.9445	14	0.9635	22	0.9757
5.5	0.9232	9	0.9470	15	0.9657	25	0.9784
6	0.9277	9.5	0.9493	16	0.9676	30	0.9818

Next, consider the 3-point flexure test at the bottom of Fig. F.1. The tensile stress is maximum at the lower center of the specimen. It decreases linearly to zero between the center line and the outer load lines. It also decreases linearly to zero from the lower surface of the specimen to the middle of the specimen. The stress in the upper half of the specimen is compressive, not tensile. We assume that only tensile stress contributes to failure and ignore the regions under compressive stress. The stress in the tensile volume of the bar as a function of the position (x,y) is therefore

$$\sigma = \sigma_{\max}\left(\frac{x}{L/2}\right)\left(\frac{y}{d/2}\right) . \quad \text{(F-14)}$$

To find the effective volume for 3-point flexure, we evaluate the integral in Eq. (F-11):

$$k = \frac{1}{V}\int\left(\frac{\sigma}{\sigma_{max}}\right)^m dV = \frac{1}{V}\int\left(\frac{x}{L/2}\right)^m\left(\frac{y}{d/2}\right)^m dV$$

$$= \frac{1}{Ldb}\int_0^{L/2}\left(\frac{x}{L/2}\right)^m dx \int_0^{d/2}\left(\frac{y}{d/2}\right)^m dy \int_0^b dz = \frac{1}{2(m+1)^2} . \quad \text{(F-15)}$$

The effective volume is $kV = V/[2(m+1)^2]$, where V is the volume of the specimen between the outer load points ($V = Ldb$). The mean strength turns out to be identical to the expression in Eq. (F-13).

Table F.1 also gives the effective volume for different 4-point flexure experiments. The two cases considered are with inner spans equal to 1/2 or 1/4 of the outer load span, L. In both cases, the volume is $V = Ldb$. The effective volume, kV, depends on whether we consider just the region between the central loads or the full region between the outer loads. If you test a set of specimens and reject those that do not fracture between the central loads, then you should use the expression for k between the inner loads only. If you do not know the fracture origin, it would be appropriate to use the expression for k between the outer loads. The region in which failure is considered "acceptable" for a test is called the *gauge section* in Table F.1.

F.4 Relative strengths of different kinds of test specimens

Table F.1 allows us to predict the relative strengths that should be observed when we change the type of mechanical test or the specimen size. Suppose that we have many coupons made from one lot of material. If we perform many tensile tests with the geometry at the upper left in Fig. F.1, we would observe a Weibull distribution with a modulus, m, a scaling factor, S_o, and a mean strength $\bar{S}$. We have replaced σ for stress in Table F.1 with S for strength measured in an experiment. According to the Weibull failure model, the mean strength should be given by the expression for the tensile test in Table F.1 with $k = 1$:

$$\bar{S} = \frac{S_o}{(kV)^{1/m}}\Gamma(1+\frac{1}{m}) = \frac{S_o}{(1 \cdot Ldb)^{1/m}}\Gamma(1+\frac{1}{m}) . \quad \text{(F-16)}$$

Eq. (F-16) predicts what will happen if the size of the tensile specimen is changed. The values of m and S_o depend on the material and not on its size. If the volume were changed from V_1 to V_2, the predicted change in the mean strength is

$$\frac{\bar{S}_2}{\bar{S}_1} = \frac{\dfrac{S_o}{(k_2V_2)^{1/m}}\,\Gamma(1+\dfrac{1}{m})}{\dfrac{S_o}{(k_1V_1)^{1/m}}\,\Gamma(1+\dfrac{1}{m})} = \left(\frac{k_1V_1}{k_2V_2}\right)^{1/m} = \left(\frac{V_1}{V_2}\right)^{1/m} . \qquad \text{(F-17)}$$

If the Weibull modulus were 5 and the dimensions of the tensile specimen were all doubled, the volume would increase by 2^3 and the mean strength is predicted to be reduced to 66% of the strength of the smaller samples:

$$\frac{\bar{S}_2}{\bar{S}_1} = \left(\frac{V_1}{8V_1}\right)^{1/5} = 0.66 . \qquad \text{(F-18)}$$

Eq. (F-18) contains a general result that you can rationalize from inspection of the expressions for mean strength of different kinds of specimens in Table F.1. The mean strength is related inversely to the effective volume raised to the $1/m$ power:

$$\text{Comparing strengths in different tests:} \quad \frac{\bar{S}_2}{\bar{S}_1} = \left(\frac{k_1V_1}{k_2V_2}\right)^{1/m} . \qquad \text{(F-19)}$$

Example: Comparing strengths in tensile and flexure tests. Suppose that we measure the tensile strengths of many rectangular prisms with dimensions of 3 × 4 × 40 mm and observe $\bar{S}$ = 100 MPa and m = 5. Let's use Eq. (F-19) to predict the flexure strength of a set of 3 × 4 × 45 mm 4-point flexure bars tested with an inner load span of 20 mm and an outer load span of 40 mm in Fig. F.1. The numbers were chosen so that the volume of the tensile bars is the same as the volume of the flexure bar between the outer load points. This volume is $V = 3 \times 4 \times 40 = 480$ mm^3. If we only consider specimens that break within the inner load span, the value of k from the second section of Table F.1 is $k = 1/[4(m+1)] = 1/24 = 0.04167$.

$$\frac{\bar{S}_{4\text{-}pt}}{\bar{S}_{tension}} = \left(\frac{k_{tension}\cancel{V}}{k_{4\text{-}pt}\cancel{V}}\right)^{1/m} = \left(\frac{1}{0.04167}\right)^{1/5} = 1.89$$

Since $\bar{S}_{tension}$ = 100 MPa, we predict $\bar{S}_{4\text{-}pt}$ = 189 MPa. The flexure specimen is stronger because less of its volume is exposed to the high stress seen throughout the entire volume of the tensile specimen.

Example: Comparing strengths in 4-point and 3-point flexure tests. What is the expected mean strength of a material in 4-point flexure if the mean strength in 3-point flexure is 100 MPa, the Weibull modulus is 5, and the volumes between the outer load points are equal? If the 4-point flexure test is done with an inner load span of $L/2$ in Table F.1 and if failure only between the inner loads is considered, then $k_{4\text{-}pt} = 1/[4(m+1)] = 1/24$. For 3-point flexure, $k_{3\text{-}pt} = 1/[2(m+1)^2] = 1/72$. The predicted ratio of strengths is

$$\frac{\bar{S}_{4\text{-}pt}}{\bar{S}_{3\text{-}pt}} = \left(\frac{k_{3\text{-}pt}\cancel{V}}{k_{4\text{-}pt}\cancel{V}}\right)^{1/m} = \left(\frac{1/72}{1/24}\right)^{1/5} = 0.80 \quad \Rightarrow \bar{S}_{4\text{-}pt} = 80 \text{ MPa} .$$

The 4-point flexure specimens are expected to have 80% of the strength of the 3-point specimens because there is less stress throughout the volume in the 3-point flexure test.

If we choose to do a 4-point flexure test with an inner load span of $L/4$ instead of $L/2$, then only half as much of the specimen is stressed and Table F.1 tells us that $k_{4\text{-}pt} = 1/[8(m+1)] = 1/48$. In this case, the 4-point flexure specimens are predicted to have 92% of the strength of the 3-point flexure specimens.

F.5 Weibull scaling by area instead of volume

Optical ceramics most frequently fail from the surface, not the volume. In this case, we substitute area for volume in the Weibull expression for the probability of survival:

$$P_s = e^{-kA(\sigma_{max}/\sigma_o)^m} . \tag{F-20}$$

The *effective area* is kA, where k is given by an equation analogous to (F-11) with area substituted for volume.

Example: Effective area of a 3-point flexure specimen. Let's find the effective area of a 3-point flexure specimen that fails only on the tensile surface at the bottom of the specimen in Fig. F.2. The stress decreases linearly from the central load to the outer loads: $\sigma = \sigma_{max}[x/(L/2)]$. The total area of the gauge surface is $A = Lb$ and the integral analogous to Eq. (F-15) is

$$k = \frac{1}{A}\int\left(\frac{\sigma}{\sigma_{max}}\right)^m dA = \frac{1}{Lb}\int_0^{L/2}\left(\frac{x}{L/2}\right)^m dx \int_0^b dz = \frac{1}{m+1} .$$

The value $k = 1/(m+1)$ computed in the preceding example is found in the bottom section of Table F.3, which also gives expressions for 4-point flexure tests. The equations for flexure tests in Table F.3 assume that failure originates on the lower tensile surface only, as in Fig. F.2. The equations would have additional terms if failure on the side surfaces is also considered.

References

1. J. B. Wachtman, *Mechanical Properties of Ceramics*, Chap. 7, Wiley, New York (1996).
2. N. A. Weil and I. M. Daniel, "Analysis of Fracture Probabilities in Nonuniformly Stressed Brittle Materials," *J. Am. Ceram. Soc.*, **47**, 268-274 (1964).

Table F.3. Weibull equations for area scaling of tensile and flexure tests*

Tensile test (Fig. F.1, upper left)

$P_s = e^{-kA(\sigma/\sigma_o)^m}$ $k = 1$ $\sigma = \frac{P}{db}$ Area = $2Ld + 2Lb$

Mean strength: $\bar{\sigma} = \frac{\sigma_o}{(kA)^{1/m}} \Gamma(1 + \frac{1}{m})$

The expression for mean strength is identical for all tests in this table

4-Point flexure test (Fig. F.1, upper right, inner load span = $L/2$)

$P_s = e^{-kA(\sigma_{max}/\sigma_o)^m}$ $\sigma_{max} = \frac{3PL}{4d^2b}$ $A \equiv Lb$

$k = \frac{m + 2}{2(m + 1)}$ if gauge section is between outer load points

$k = 1/2$ if gauge section is between inner load points only

4-Point flexure test (inner load span = $L/4$)

$P_s = e^{-kA(\sigma_{max}/\sigma_o)^m}$ $\sigma_{max} = \frac{9PL}{8d^2b}$ $A \equiv Lb$

$k = \frac{m + 4}{4(m + 1)}$ if gauge section is between outer load points

$k = 1/4$ if gauge section is between inner load points only

3-Point flexure test (Fig. F.2, bottom)

$P_s = e^{-kA(\sigma_{max}/\sigma_o)^m}$ $\sigma_{max} = \frac{3PL}{2d^2b}$ $A \equiv Lb$

$k = \frac{1}{m + 1}$ (gauge section is between outer load points)

*Equations for k for flexure tests presume that failure occurs only on the tensile surface in Fig. F.2. k would be greater if failure can also occur on the side surfaces of the bar. Values of the function $\Gamma(z)$ are given in Table F.2.

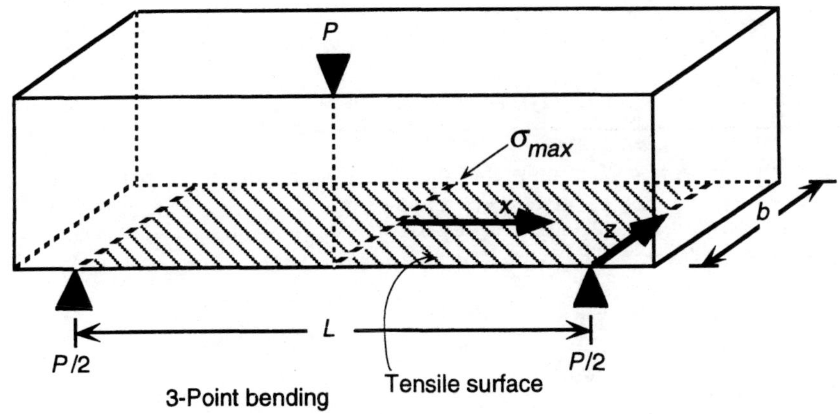

Fig. F.2. Geometry of rectangular bars for 3-point bending strength measurement. The tensile surface is the shaded area at the bottom.

Appendix G

THERMAL PROPERTIES OF SELECTED MATERIALS

This Appendix provides thermal expansion coefficients (α), thermal conductivity (k), and heat capacity (C_p, also called specific heat) as a function of temperature for selected window materials. These properties were described in Sections 4.1 and 4.2. The heat capacity, C_p, is measured at constant applied pressure. For solid materials, the heat capacity, C_V, measured at constant volume, is nearly the same as C_p.

With respect to thermal expansion, sometimes the change in length (ΔL in Fig. 4.1) of a material is expressed as a multiple of its initial length (L_o) at some reference temperature (such as 298 K) by a polynomial in temperature (T):

$$\Delta L/L_o = a + bT + cT^2 \quad \Rightarrow \quad \Delta L = L_o(a + bT + cT^2). \tag{G-1}$$

From the relation

$$L = L_o + \Delta L = L_o + L_o(a + bT + cT^2) = L_o(1 + a + bT + cT^2) \tag{G-2}$$

we can write

$$dL/dT = L_o(b + 2cT)\,. \tag{G-3}$$

The thermal expansion coefficient, α, is

$$\alpha \equiv \frac{1}{L}\frac{dL}{dT} = \frac{b + 2cT}{1 + a + bT + cT^2}\,. \tag{G-4}$$

Equation (G-4) allows us to calculate the expansion coefficient starting from an expression for $\Delta L/L_o$.

If we begin with an expression for the expansion coefficient such as

$$\alpha = A + BT + CT^2 \tag{G-5}$$

we can derive a formula for length as a function of temperature by integrating the definition of α in Eq. (G-4):

$$\alpha = \frac{1}{L}\frac{dL}{dT} \Rightarrow \frac{dL}{L} = \alpha\, dT \Rightarrow \int_{L_o}^{L}\frac{dL}{L} = \int_{T_0}^{T}\alpha\, dT \Rightarrow \ln\frac{L}{L_o} = \int_{T_0}^{T}\alpha\, dT . \qquad \text{(G-6)}$$

If $\alpha = A + BT + CT^2$, then the integral on the right side of Eq. (G-6) is

$$\ln\frac{L}{L_o} = A\,(T - T_o) + \frac{1}{2}B\,(T^2 - T_o^2) + \frac{1}{3}C\,(T^3 - T_o^3) . \qquad \text{(G-7)}$$

Alphabetical data section

ALON, Aluminum oxynitride ($Al_{23-x/3}O_{27+x}N_{5-x}$) (Melting point ≈ 2420 K) (0.429 < x < 2) (x is typically 1)[1]

Thermal expansion:[2]

Average expansion coefficient from 300 to 473 K = 5.8×10^{-6} K^{-1}
Average expansion coefficient from 300 to 1173 K = 7.8×10^{-6} K^{-1}

Thermal conductivity:[2]

12.6 W/(m·K) at 300 K

Specific heat (J K^{-1} g^{-1}):[3]

T (K)	300
Specific heat	0.77

1. For a general review of ALON and its properties, see N. D. Corbin, "Aluminum Oxynitride Spinel: A Review," *J. Eur. Ceram. Soc.*, **5**, 143-154 (1989).
2. T. M. Hartnett and R. L. Gentilman, "Optical and Mechanical Properties of Highly Transparent Spinel and ALON Domes," *Proc. SPIE*, **505**, 15-22 (1984).
3. P. Klocek, ed., *Handbook of Infrared Optical Materials*, Marcel Dekker, New York (1991).

Calcium fluoride (CaF_2) (Melting point = 1775 K; phase change at 1424 K)

Thermal expansion (T = 293 to 900 K, L_o at 293 K):[1]

$\Delta L/L_o = -0.00564 + 1.991 \times 10^{-5}\,T - 5.582 \times 10^{-9}\,T^2 + 1.109 \times 10^{-11}\,T^3$

Thermal conductivity (W/(m·K)):[2]

Crystal 1:

T (K)	319	383	421
Conductivity	9.1	7.1	5.9

Crystal 2:

T (K)	331	379	463	569	651	748	848
Conductivity	5.4	4.5	3.6	3.2	3.0	2.9	2.8

Specific heat (J K^{-1} g^{-1}):[3]

T (K)	300	400	500	600	700	800	900
Specific heat	0.90	0.93	0.96	1.00	1.04	1.07	1.11
T (K)	1000	1100	1200	1300	1400		
Specific heat	1.15	1.19	1.23	1.27	1.30		

1. Y. S. Touloukian, R. K. Kirby, R. E. Taylor and T. Y. R. Lee, *Thermophysical Properties of Matter*, Vol. 13, Thermal Expansion, IFI/Plenum Press, New York (1977).
2. Y. S. Touloukian, R. W. Powell, C. Y. Ho and P. G. Klemens, *Thermophysical Properties of Matter*, Vol. 2, Thermal Conductivity, IFI/Plenum Press, New York (1970).
3. Y. S. Touloukian and E. H. Buyco, *Thermophysical Properties of Matter*, Vol. 5, Specific Heat, IFI/Plenum Press, New York (1970).

Diamond (Type IIa) (C)

Thermal conductivity (W/m·K) "representative values":[1]

T(K)	3	4	5	6	7	8	9	10	11	12
Conductivity	11.1	26.1	49.4	82.0	124	177	241	317	400	500
T(K)	13	14	15	16	18	20	25	30	35	40
Conductivity	610	732	865	1000	1320	1680	2710	3890	5180	6590
T(K)	45	50	60	70	80	90	100	123.2	150	173.2
Conductivity	7930	9210	11200	11900	11700	10900	10000	7920	6020	4930
T(K)	200	223.2	250	273.2	298.2	300				
Conductivity	4030	3470	2970	2620	2320	2300				

For more thermal conductivity and other thermal properties, see Sections 9.2.3 and 9.2.4.

1. C. Y. Ho, R. W. Powell and P. E. Liley, "Thermal Conductivity of the Elements," *J. Phys. Chem. Ref. Data*, **1**, 279-421 (1972).

Gallium arsenide (GaAs) (Melting point = 1511 K)

Thermal expansion:[1,2]

$$\alpha\ (K^{-1}) = 4.24 \times 10^{-6} + 5.82 \times 10^{-9} T - 2.82 \times 10^{-12} T^2 \qquad (T = 200\text{-}1000\ K)$$

$$\Delta L/L_o = -0.00147 + 4.239 \times 10^{-6} T + 2.916 \times 10^{-9} T^2 - 9.360 \times 10^{-13} T^3$$

(T = 200 to 1600 K, L_o at 293 K)

Thermal conductivity:[1]

conductivity (W/m·K) = $54400T^{-1.2}$ ($T \approx$ 75-1000 K)
(for *n*-type GaAs with $<5 \times 10^{16}$ carriers/cm^3)

Specific heat (J K^{-1} g^{-1}):[3]

T (K)	50	100	150	200	273	400	500
Specific heat	0.0904	0.199	0.264	0.296	0.317	0.339	0.342

T (K)	600	700	800	900	1000	1100	1200
Specific heat	0.345	0.348	0.351	0.354	0.357	0.360	0.363

1. Texas Instruments (now Raytheon Systems Co.) data sheet.
2. Y. S. Touloukian, R. K. Kirby, R. E. Taylor and T. Y. R. Lee, *Thermophysical Properties of Matter*, Vol. 13, Thermal Expansion, IFI/Plenum Press, New York (1977).
3. Y. S. Touloukian and E. H. Buyco, *Thermophysical Properties of Matter*, Vol. 5, Specific Heat, IFI/Plenum Press, New York (1970).

Gallium phosphide (GaP) (Melting point = 1740 K)

Thermal expansion:[1]

T (K)	298	323	373	423	473	523
α (10^{-6}/K)	3.5	3.7	4.1	4.6	5.0	5.4

T (K)	573	623	673	723	773
α (10^{-6}/K)	5.7	5.9	6.1	6.2	6.2

Thermal conductivity (W/m·K):[1]

T (K)	298	323	373	473	573	673
Conductivity	$86._4$	$76._3$	$66._5$	$46._5$	$37._3$	$30._2$

T (K)	773	823
Conductivity	$25._8$	$24._3$

Specific heat (J K^{-1} g^{-1}):[2]

T (K)	298	373	473	573	673	773
Specific heat[1]	0.44	0.46	0.48	0.49	0.50	0.52

1. J. Trombetta, Raytheon Systems Co., to be published.
2. Data from Ref. 1. These data are approximately half of the corresponding values listed in Y. S. Touloukian and E. H. Buyco, *Thermophysical Properties of Matter*, Vol. 5, Specific Heat, IFI/Plenum Press, New York (1970). The new data are thought to be correct.

Germanium (Ge) (Melting point = 1211 K)

Thermal expansion:[1,2]

$$\alpha\ (10^{-6}/\text{K}) = -3.386 + 0.08232\,T - 3.032 \times 10^{-4}\,T^2 + 6.066 \times 10^{-7}\,T^3 - 6.688 \times 10^{-10}\,T^4 + 3.831 \times 10^{-13}\,T^5 - 8.911 \times 10^{-17}\,T^6 \quad (T = 200 \text{ to } 1100 \text{ K})$$

$$\Delta L/L_o = 5.790 \times 10^{-6}(T - 293) + 1.768 \times 10^{-9}(T - 293)^2 - 4.562 \times 10^{-13}(T - 293)^3 \quad (T = 293 \text{ to } 1200 \text{ K},\ L_o \text{ at } 293 \text{ K})$$

Thermal conductivity (W/m·K) recommended values:[3]

(Values below 290 K are typical, but actual values are highly sensitive to small chemical and physical variations among specimens. Values above 290 K are considered accurate to ±10%.)

T (K)	100	123.3	150	173.2	200	223.2	250
Conductivity	232	168	132	113	96.8	85.9	74.9
T (K)	273.2	298.2	300	323.2	350	373.2	400
Conductivity	66.7	60.2	59.9	54.8	49.5	46.5	43.2
T (K)	473.2	500	573.2	600	673.2	700	773.2
Conductivity	35.9	33.8	28.8	27.3	23.7	22.7	20.4
T (K)	800	873.2	900	973.2	1000	1073.2	1100
Conductivity	19.8	18.5	18.2	17.6	17.4	17.1	17.0
T (K)	1173.2	1200					
Conductivity	17.2	17.4					

Specific heat:[1]

$$\text{Heat capacity } (C_p,\ \text{J K}^{-1}\,\text{g}^{-1}) = 0.156858 + 9.82450 \times 10^{-4}\,T - 1.93455 \times 10^{-6}\,T^2 + 1.69594 \times 10^{-9}\,T^3 - 5.23981 \times 10^{-13}\,T^4 \quad (T \approx 200\text{-}1200 \text{ K})$$

1. C. C. Gibson, D. L. Taylor and R. H. Bogaard, *Databook on Properties of Selected Infrared Window and Dome Materials*, High Temperature Materials Information Analysis Center Report HTMIAC 27, September 1996. (Available from Defense Technical Information Center.)
2. Y. S. Touloukian, R. K. Kirby, R. E. Taylor and P. D. Desai, *Thermophysical Properties of Matter*, Vol. 12, Thermal Expansion, IFI/Plenum Press, New York (1975).
3. C. Y. Ho, R. W. Powell and P. E. Liley, "Thermal Conductivity of the Elements," *J. Phys. Chem. Ref. Data*, **1**, 279-421 (1972).

Lanthana-Doped Yttria ($0.09La_2O_3 \cdot 0.91Y_2O_3$) (Melting point ≈ 2670 K)[1]

Thermal expansion:[1]

$$\Delta L/L_o = -1.71622 \times 10^{-3} + 5.92378 \times 10^{-6}\,T + 1.31095 \times 10^{-9}\,T^2$$

$(T \approx 300\text{-}2300 \text{ K},\ L_o \text{ at } 273 \text{ K})$

The average expansion coefficient from 273 K to the indicated temperature is the same as the values for undoped yttria to 2 decimal places:

T (K)	300	400	500	600	800	1000	1500	2000
α (10^{-6}/K)	6.5	6.7	6.9	7.0	7.3	7.5	8.2	8.9

Thermal conductivity:

Gibson *et al.*[2] recommend the following equation for lanthana-doped yttria:

$1/\text{conductivity (W/m·K)} = 0.128 + 2.07 \times 10^{-4}\, T$ $(T \approx 300\text{-}800 \text{ K})$

Tropf and Harris[1] recommend

$\text{conductivity (W/m·K)} = 100.3\, T^{-0.5557} + 1.10$ $(T \approx 270\text{-}1400 \text{ K})$

Specific heat (same as undoped yttria):[1]

Heat capacity (C_p, J K^{-1} g^{-1}) $= 0.441 + 1.284 \times 10^{-4}\, T$ $(T \approx 310\text{-}2000 \text{ K})$

1. W. J. Tropf and D. C. Harris, "Mechanical, Thermal and Optical Properties of Yttria and Lanthana-Doped Yttria," *Proc. SPIE*, **1112**, 9-19 (1989).
2. C. C. Gibson, D. L. Taylor and R. H. Bogaard, *Databook on Properties of Selected Infrared Window and Dome Materials*, High Temperature Materials Information Analysis Center Report HTMIAC 27, September 1996. (Available from Defense Technical Information Center.)

Lithium fluoride (LiF_2) (Melting point = 1115 K)

Thermal expansion (T = 293 to 1100 K, L_o at 293 K):[1]

$$\Delta L/L_o = -0.01035 + 3.424 \times 10^{-5}\, T + 1.733 \times 10^{-9}\, T^2 + 6.745 \times 10^{-12}\, T^3$$

Thermal conductivity (W/(m·K)):[2]

T (K)	250	300	500
Conductivity	19	14	7.5

Specific heat (J K^{-1} g^{-1}):[3]

T (K)	300	350	400	450	500	550	600
Specific heat	1.61	1.71	1.78	1.84	1.89	1.94	1.98

T (K)	700	800	900	1000	1100
Specific heat	2.05	2.11	2.19	2.28	2.39

1. Y. S. Touloukian, R. K. Kirby, R. E. Taylor and T. Y. R. Lee, *Thermophysical Properties of Matter*, Vol. 13, Thermal Expansion, IFI/Plenum Press, New York (1977).
2. W. J. Tropf, M. E. Thomas and T. J. Harris, "Properties of Crystals and Glasses," in *Handbook of Optics* (M. Bass, E. W. van Stryland, D. R. Williams and W. L. Wolfe, eds.), Vol. II, Chap. 33, Table 18, McGraw-Hill, New York (1995).
3. Y. S. Touloukian and E. H. Buyco, *Thermophysical Properties of Matter*, Vol. 5, Specific Heat, IFI/Plenum Press, New York (1970).

Magnesium fluoride, polycrystalline (MgF_2) (Melting point = 1536 K)

Thermal expansion:[1]

Polycrystalline material (T = 293 to 1300 K, L_o at 293 K):

$$\Delta L/L_o = -0.00275 + 7.864 \times 10^{-6} T + 5.235 \times 10^{-9} T^2 + 3.107 \times 10^{-13} T^3$$

Single crystal (T = 293 to 900 K, L_o at 293 K):

$$\Delta L/L_o \parallel c = -0.00394 + 1.171 \times 10^{-5} T + 6.383 \times 10^{-9} T^2 - 1.433 \times 10^{-12} T^3$$

$$\Delta L/L_o \parallel a = -0.00243 + 7.439 \times 10^{-6} T + 2.042 \times 10^{-9} T^2 + 2.653 \times 10^{-12} T^3$$

Thermal conductivity:[2]

conductivity (W/m·K) = $15.45 - 0.02868\, T + 2.728 \times 10^{-5} T^2 - 8.818 \times 10^{-9} T^3$
(T in °C from 40-840°C)

Specific heat ($J\ K^{-1}\ g^{-1}$):[3]

T (K)	200	300	400	500	600	700	800
Specific heat	0.77	1.02	1.11	1.16	1.20	1.23	1.25

T (K)	900	1000	1100	1200	1300	1400	1500
Specific heat	1.27	1.29	1.31	1.33	1.35	1.37	1.38

1. Y. S. Touloukian, R. K. Kirby, R. E. Taylor and T. Y. R. Lee, *Thermophysical Properties of Matter*, Vol. 13, Thermal Expansion, IFI/Plenum Press, New York (1977).
2. D. M. Bailey, F. W. Calderwood, J. D. Greiner, O. Hunter, Jr., J. F. Smith and R. J. Schlitz, Jr., "Reproducibilities of Some Physical Properties of MgF_2," *J. Am. Ceram. Soc.*, **58**, 489-492 (1975).
3. Y. S. Touloukian and E. H. Buyco, *Thermophysical Properties of Matter*, Vol. 5, Specific Heat, IFI/Plenum Press, New York (1970).

Sapphire (Al_2O_3) (Melting point = 2313 K)

Thermal expansion:[1]

L = length of specimen; L_o = length at 273 K; $L = L_o[1 + \alpha(T-273)]$

$\alpha = A + BT - Ce^{-D(T-273)}$ T in K from 273 to 2073 K

	c-axis	a-axis
A	8.026×10^{-6}	7.419×10^{-6}
B	8.17×10^{-10}	6.43×10^{-10}
C	3.279×10^{-6}	3.211×10^{-6}
D	2.91×10^{-3}	2.59×10^{-3}

Alternative equations for the expansion coefficient (α, 10^{-6}/K) from T = 200 to 1000 K are:[2]

$$\alpha \parallel c = -3.312 + 0.05064\ T - 8.692 \times 10^{-5}\ T^2 + 7.368 \times 10^{-8}\ T^3 - 2.440 \times 10^{-11}\ T^4$$

$$\alpha \perp c = -3.549 + 0.04682\ T - 7.586 \times 10^{-5}\ T^2 + 5.923 \times 10^{-8}\ T^3 - 1.781 \times 10^{-11}\ T^4$$

Thermal conductivity (W/m·K) parallel to c-axis:[3]

T (K)	298	473	573	673	873	1073	1273
Conductivity	36.8	20.6	16.4	14.1	10.8	8.9	7.9

Bogaard[4] collected thermal conductivity data from several sources and plotted them in a graph of thermal resistivity (= 1/conductivity) versus temperature. Bogaard's data plus the China Lake data[3] listed above are plotted below. The China Lake data (open circles) diverge from Ref. 5 (open squares) at 1073 and 1273 K. The least-squares fits to the total data for each axis are shown in the graph below.

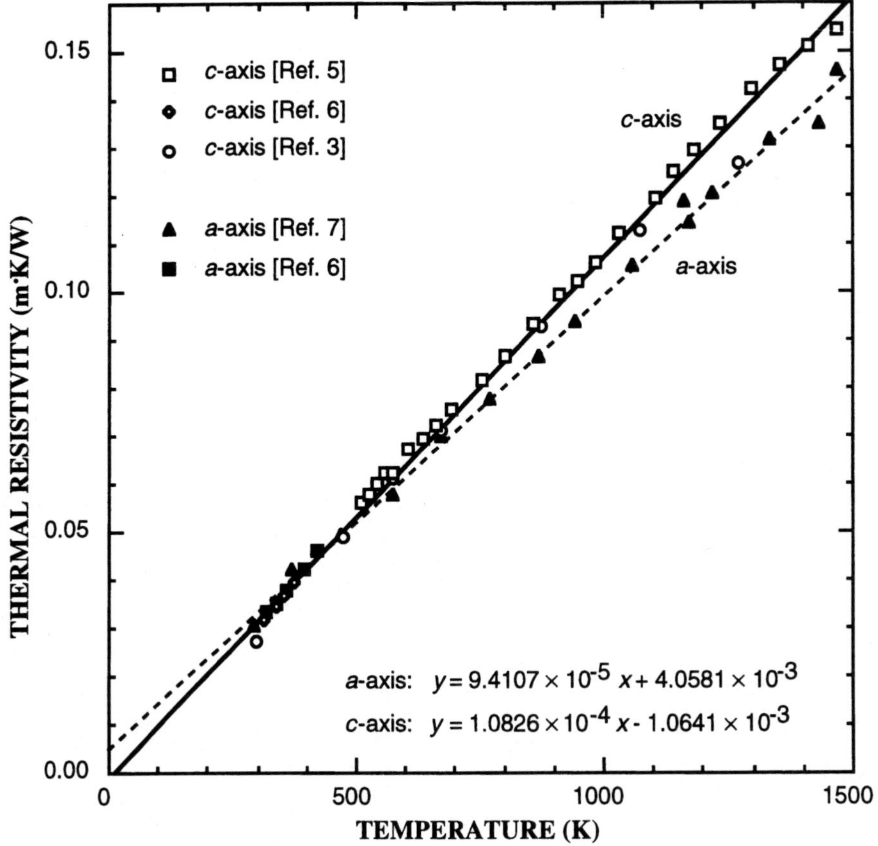

Specific heat ($J\ K^{-1}\ g^{-1}$):[8]

T (K)	50	100	150	200	250	300	350
Specific heat	0.0148	0.1259	0.3133	0.5014	0.6577	0.7792	0.8721

T (K)	400	500	600	700	800	900	1000
Specific heat	0.9429	1.041	1.104	1.147	1.178	1.203	1.223

T (K)	1100	1200	1300	1400	1500
Specific heat	1.240	1.255	1.269	1.282	1.294

1. R. G. Munro, "Evaluated Material Properties for a Sintered α-Alumina," *J. Am. Ceram. Soc.*, **80**, 1919-1928 (1997).
2. C. C. Gibson, D. L. Taylor and R. H. Bogaard, *Databook on Properties of Selected Infrared Window and Dome Materials*, High Temperature Materials Information Analysis Center Report HTMIAC 27, September 1996. (Available from Defense Technical Information Center.)
3. China Lake data obtained by laser flash method at Holometrix (Bedford, MA) using sapphire disks from Crystal Systems (Salem, MA).
4. R. H. Bogaard, "Toward a Thermophysical Property Database: Consolidation and Updating of Material Property Data Files," *Proc. 24th International Thermal Conductivity Conf.*, P. Goal, ed., Technomic Publishers, Lancaster, PA (1998).
5. S. P. Howlett, R. Taylor and R. Morrell, "Heat Pulse Thermal Diffusivity Measurements on Transparent Materials," in *Thermal Conductivity 17, Proc. 17th International Thermal Conductivity Conf.* (J. G. Hust, ed), pp. 447-457, Plenum Press, New York (1983).
6. J. H. Koenig, *Rutgers University Ceramic Research Station Progress Report 1* (1953).
7. B. Schulz, "High Temperature Thermal Conductivity of Irradiated and Non-Irradiated Alpha-Aluminum Oxide," *J. Nucl. Mater.*, **155-157**, 348-351 (1988).
8. D. G. Archer, "Thermodynamic Properties of Synthetic Sapphire Standard Reference Material 720," *J. Phys. Chem. Reference Data*, **22**, 1441-1453 (1993).

Silicon (Si) (Melting point = 1687 K)

Thermal expansion (T = 293 to 1600 K, L_o at 293 K):[1]

$$\Delta L/L_o = -0.00071 + 1.887 \times 10^{-6} T + 1.934 \times 10^{-9} T^2 - 4.544 \times 10^{-13} T^3$$

Thermal conductivity (W/m·K) recommended values:[2]

(Values below 300 K are typical, but actual values are highly sensitive to small chemical and physical variations among specimens. Values from 300-1000 K are considered accurate to ±5% and values from 1000-1685 K are accurate to ±10%.)

T (K)	100	123.3	150	173.2	200	223.2	250
Conductivity	884	599	409	330	264	225	191
T (K)	273.2	298.2	300	323.2	350	373.2	400
Conductivity	168	149	148	133	119	108	98.9
T (K)	473.2	500	573.2	600	673.2	700	773.2
Conductivity	81.4	76.2	65.1	61.9	53.6	50.8	44.2
T (K)	800	873.2	900	973.2	1000	1073.2	1100
Conductivity	42.2	37.4	35.9	32.3	31.2	28.6	27.9
T (K)	1173.2	1200	1273.2	1300	1373.2	1400	1473.2
Conductivity	26.2	25.7	24.7	24.4	23.7	23.5	22.9
T (K)	1500	1573.2	1600	1673.2	1685		
Conductivity	22.7	22.3	22.1	22.0	22.0		

Specific heat ($J\ K^{-1}\ g^{-1}$):[3]

T (K)	50	100	150	200	250	273	373
Specific heat	0.0787	0.257	0.427	0.556	0.654	0.690	0.770

T (K)	473	573	673	773	873	973	1073
Specific heat	0.824	0.847	0.864	0.881	0.897	0.912	0.927

T (K)	1173	1273	1373
Specific heat	0.941	0.958	0.981

1. Y. S. Touloukian, R. K. Kirby, R. E. Taylor and T. Y. R. Lee, *Thermophysical Properties of Matter*, Vol. 13, Thermal Expansion, IFI/Plenum Press, New York (1977).
2. C. Y. Ho, R. W. Powell and P. E. Liley, "Thermal Conductivity of the Elements," *J. Phys. Chem. Ref. Data*, **1**, 279-421 (1972).
3. Y. S. Touloukian and E. H. Buyco, *Thermophysical Properties of Matter*, Vol. 4, Specific Heat, IFI/Plenum Press, New York (1970).

Spinel ($MgAl_2O_4$) (Melting point = 2408 K)

Thermal expansion (10^{-6}/K) (T = 100 to 1250 K):[1]

$$\alpha = -0.7549 + 0.01558\,T + 6.094 \times 10^{-5}\,T^2 - 2.471 \times 10^{-7}\,T^3 + 3.664 \times 10^{-10}\,T^4 - 2.562 \times 10^{-13}\,T^5 + 6.989 \times 10^{-17}\,T^6$$

Thermal conductivity:[2]

Based on data from Refs. 2 and 3, Gibson *et al.*[1] recommend the following equation:

$$1/\text{conductivity (W/m·K)} = 0.03790 + 1.037 \times 10^{-4}\,T \qquad (T \approx 300\text{-}1500\ \text{K})$$

Specific heat ($J\ K^{-1}\ g^{-1}$):

T (K)	295
Specific heat[3]	0.879

T (K)	294	811	1089	1366	1644	1922
Specific heat[4]	0.92	1.15	1.27	1.39	1.47	1.55

1. C. C. Gibson, D. L. Taylor and R. H. Bogaard, *Databook on Properties of Selected Infrared Window and Dome Materials*, High Temperature Materials Information Analysis Center Report HTMIAC 27, September 1996. (Available from Defense Technical Information Center.)
2. R. L. Gentilman, "Current and Emerging Materials for 3-5 Micron IR Transmission," *Proc. SPIE*, **783**, 2-11 (1986).
3. D. W. Roy and G. C. Martin, Jr., "Advances in Spinel Optical Quality, Size and Shape Capability and Applications," *Proc. SPIE*, **1760**, 2-13 (1992).
4. J. R. Koenig, *Thermostructural Evaluation of Four Infrared Seeker Dome Materials*, Naval Weapons Center Report TP 6539, Part 2, China Lake, CA, April 1985.

Yttria (Y_2O_3) (Melting point = 2710 K)[1]

Thermal expansion:[1]

$$\Delta L/L_o = -1.69277 \times 10^{-3} + 5.83245 \times 10^{-6} T + 1.33104 \times 10^{-9} T^2$$
$(T \approx 300\text{-}2300 \text{ K}, \; L_o = \text{length at 273 K})$

The average expansion coefficient from 273 K to the indicated temperature is

T (K)	300	400	500	600	800	1000	1500	2000
α (10^{-6}/K)	6.5	6.7	6.9	7.0	7.3	7.5	8.2	8.9

Thermal conductivity:

Gibson *et al.*[2] recommend

1/conductivity (W/m·K) = $0.0041 + 2.43 \times 10^{-4} T$ $(T \approx 300\text{-}800 \text{ K})$

Tropf and Harris[1] recommend

conductivity (W/m·K) = $85660\, T^{-1.577} + 2.86$ $(T \approx 270\text{-}1800 \text{ K})$

Specific heat:[1]

Heat capacity (C_p, J K^{-1} g^{-1}) = $0.441 + 1.284 \times 10^{-4} T$ $(T \approx 310\text{-}2000 \text{ K})$

1. W. J. Tropf and D. C. Harris, "Mechanical, Thermal and Optical Properties of Yttria and Lanthana-Doped Yttria," *Proc. SPIE*, **1112**, 9-19 (1989).
2. C. C. Gibson, D. L. Taylor and R. H. Bogaard, *Databook on Properties of Selected Infrared Window and Dome Materials*, High Temperature Materials Information Analysis Center Report HTMIAC 27, September 1996. (Available from Defense Technical Information Center.)

Zinc selenide (ZnSe) (Melting point ≈ 1790 K)

Thermal expansion (α, 10^{-6}/K) (T = 200 to 800 K):[1]

$$\alpha = 1.007 + 0.04022\, T - 9.192 \times 10^{-5} T^2 + 1.126 \times 10^{-8} T^3 - 4.956 \times 10^{-11} T^4$$

Thermal conductivity:[2]

Based on data from Refs. 2-5, Gibson *et al.*[1] recommend the following equation:

1/conductivity (W/m·K) = $0.0051 + 2.07 \times 10^{-4} T$ $(T \approx 150\text{-}450 \text{ K})$

Specific heat (J K^{-1} g^{-1}):[6]

T (K)	294	533	811	1089	1255
Specific heat	0.33	0.38	0.44	0.51	0.54

1. C. C. Gibson, D. L. Taylor and R. H. Bogaard, *Databook on Properties of Selected Infrared Window and Dome Materials*, High Temperature Materials Information Analysis Center Report HTMIAC 27, September 1996. (Available from Defense Technical Information Center.)
2. G. A. Slack, "Thermal Conductivity of II-VI Compounds and Phonon Scattering by Iron(2+) Impurities," *Phys. Rev. B*, **6**, 3791-3800 (1972).
3. C. J. Johnson and F. J. Kramer, *Optics Catalog*, II-VI, Inc., Saxonburg, PA (1994).
4. C. A. Klein, *Compendium of Property Data for Raytran Zinc Selenide and Raytran Zinc Sulfide*, Raytheon Report RAY/RD/T-1154, August 1987.
5. J. C. Wurst and T. P. Graham, *Thermal, Electrical, and Physical Measurements of Laser Window Materials*, U.S. Air Force Report AFML-TR-75-28 (1975).
6. J. R. Koenig, *Thermostructural Evaluation of Four Infrared Seeker Dome Materials*, Naval Weapons Center Report TP 6539, Part 2, China Lake, CA, April 1985.

Zinc sulfide (ZnS) (Melting point ≈ 1973 K; cubic → hexagonal phase change at 1293 K)

Thermal expansion (α, 10^{-6}/K) (T = 300 to 1200 K):[1]

$$\alpha = 3.469 + 0.01737\,T - 2.553 \times 10^{-5}\,T^2 + 2.015 \times 10^{-8}\,T^3 - 5.816 \times 10^{-12}\,T^4$$

Thermal conductivity:[2]

Based on data from Refs. 2-4, Gibson *et al.*[1] recommend the following equation for standard grade zinc sulfide:

1/conductivity (W/m·K) = $0.0156 + 1.44 \times 10^{-4}\,T$ ($T \approx$ 150-1200 K)

Specific heat (J K^{-1} g^{-1}):[5]

T (K)	294	533	811	1089	1255
Specific heat	0.481	0.502	0.523	0.544	0.565

1. C. C. Gibson, D. L. Taylor and R. H. Bogaard, *Databook on Properties of Selected Infrared Window and Dome Materials*, High Temperature Materials Information Analysis Center Report HTMIAC 27, September 1996. (Available from Defense Technical Information Center.)
2. G. G. Gadzhiev and G. N. Dronova, "Thermal Conductivity of Polycrystalline Zinc Sulfide," *Izv. Akad. Nauk SSSR, Neorg. Mater.*, **19**, 1087-1089 (1983).
3. G. A. Slack, "Thermal Conductivity of II-VI Compounds and Phonon Scattering by Iron(2+) Impurities," *Phys. Rev. B*, **6**, 3791-3800 (1972).
4. C. J. Johnson and F. J. Kramer, *Optics Catalog*, II-VI, Inc., Saxonburg, PA (1994).
5. J. R. Koenig, *Thermostructural Evaluation of Four Infrared Seeker Dome Materials*, Naval Weapons Center Report TP 6539, Part 2, China Lake, CA, April 1985.

INDEX

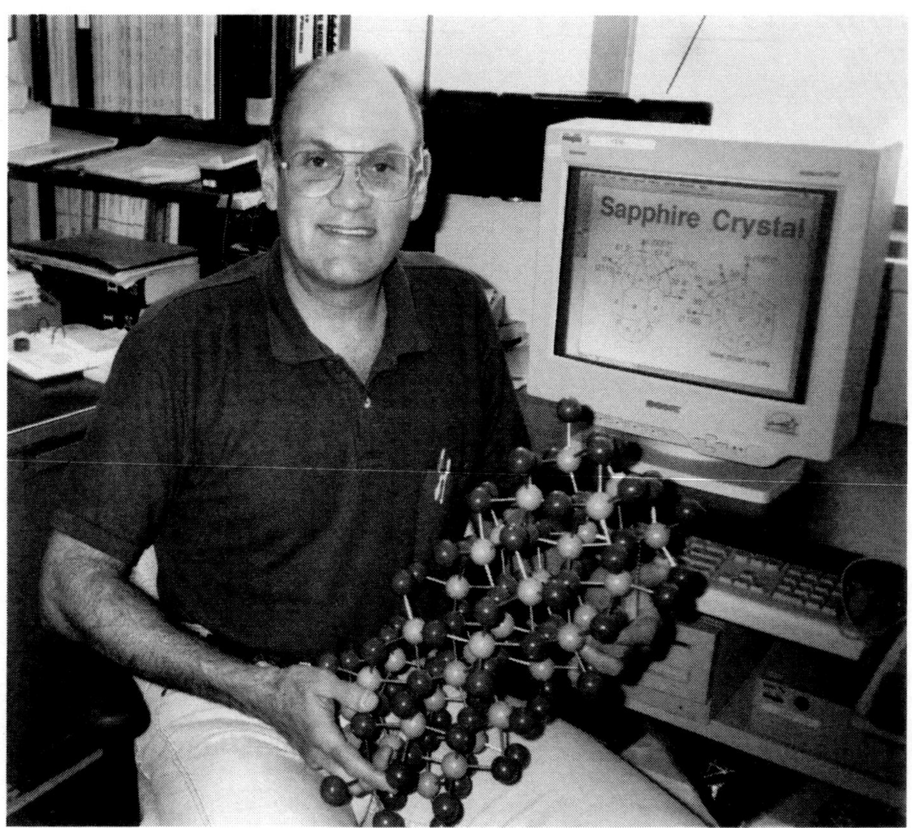

Daniel C. Harris is a Senior Scientist in the Chemistry and Materials Division of the Research Department at the Naval Air Warfare Center at China Lake, California, where his specialty is infrared window and dome materials. He holds a Bachelor's degree in chemistry from Massachusetts Institute of Technology and a Ph.D. in chemistry from California Institute of Technology. Prior to coming to the Naval Air Warfare Center in 1983, he taught at the University of California at Davis and Franklin and Marshall College in Lancaster, Pennsylvania. He is the author of the widely used undergraduate analytical chemistry text, *Quantitative Chemical Analysis* (5th edition, 1998), and has also written the textbook *Exploring Chemical Analysis* and co-authored *Symmetry and Spectroscopy*. On those very rare occasions when he is not at his desk or in an airport, he can be found hiking under the blue California sky.